1500621249

AF616344

# THE ORIGIN AND EVOLUTION OF NEUTRON STARS

INTERNATIONAL ASTRONOMICAL UNION

UNION ASTRONOMIQUE INTERNATIONALE

# THE ORIGIN AND EVOLUTION OF NEUTRON STARS

PROCEEDINGS OF THE 125TH SYMPOSIUM OF THE INTERNATIONAL ASTRONOMICAL UNION HELD IN NANJING, CHINA, MAY 26–30, 1986

EDITED BY

D. J. HELFAND

*Department of Astronomy and Physics, Columbia University, New York, U.S.A.*

and

J.-H. HUANG

*Astrophysics Institute, Nanjing University, Nanjing, People's Republic of China*

D. REIDEL PUBLISHING COMPANY

A MEMBER OF THE KLUWER  ACADEMIC PUBLISHERS GROUP

DORDRECHT / BOSTON / LANCASTER / TOKYO

**Library of Congress Cataloging in Publication Data**

International Astronomical Union. Symposium
(125th: 1986: Nanjing, China)
The origin and evolution of neutron stars.

Includes index.
1. Neutron stars—Congresses. 2. Pulsars—Congresses. 3. Supernovae—Congresses. I. Helfand, D. J. (David J.), 1950– . II. Huang, J. -H. (Jie-Hao), 1939– . III. Title.
QB843.N4I57 1986 523.8 87–12919
ISBN 90–277–2537–3
ISBN 90–277–2538–1 (pbk.)

---

*Published on behalf of*
*the International Astronomical Union*
*by*
*D. Reidel Publishing Company, P.O. Box 17, 3300 AA Dordrecht, Holland*



*Sold and distributed in the U.S.A. and Canada*
*by Kluwer Academic Publishers,*
*101 Philip Drive, Assinippi Park, Norwell, MA 02061, U.S.A.*

*In all other countries, sold and distributed*
*by Kluwer Academic Publishers Group,*
*P.O. Box 322, 3300 AH Dordrecht, Holland*



*Printed in The Netherlands*

***TABLE OF CONTENTS***


PREFACE .......... xi

INTRODUCTORY REMARKS .......... xv

LIST OF PARTICIPANTS .......... xvii

PHOTOGRAPHS .......... xxii


## *I. ROTATION-POWERED PULSARS*

*The Population*

CHAIR: V. Radhakrishnan


R. N. MANCHESTER: Pulsar Surveys .......... 3
D. C. BACKER: Millisecond Pulsar Surveys .......... 13
A. G. LYNE: The Kinematics of Pulsars .......... 23
J. M. CORDES: Interstellar Scintillations and Neutron Star Kinematics .......... 35
T. R. Clifton *et al.*: A Deep Search for Young Pulsars in the Galactic Plane at 1400 MHz (abstract) .......... 47
A. Blaauw: Progenitors of the Local Pulsars: Lower Mass Limit and Beaming Factor (abstract) .......... 48
S. Pineault: The Relation Between Radio Luminosity and Magnetic Field in Rotation-Powered Pulsars (abstract) .......... 49
T. Lu *et al.*: Does the Radio Luminosity of Pulsar Grow up in its Later Stage? (abstract) .......... 50
Z.-G. Deng *et al.*: A Few Remarks on the Two Types of Pulsars (abstract) .......... 51
J.-H. Huang *et al.*: The Kinematic Properties of Two Pulsar Types (abstract) .......... 52
J .M. Rankin: Toward an Empirical Theory of Pulsar Emission (abstract) .......... 53
J.-H. Huang *et al.*: Radio Emission Mechanisms for Two Types of Pulsars (abstract) .......... 54
K. Y. Chen and J. Shaham: Pulse Asymmetry of Millisecond Pulsars (abstract) .......... 55
J. J. Barnard: Pulsar Polarization Limiting Radii and the Evolution of Pulsar Beams (abstract) .......... 56
W. Sieber: Polarization Anglè Swings Rediscussed (abstract) .......... 57
J. A. Gil: Triplicity of Pulsar Profiles and Orthogonal Polarization Modes (abstract) .......... 58
X.-Y. Xia *et al.*: The Effects of Inverse Compton Scattering on the Pulsars' Radiation (abstract) .......... 59

X. J. Wu *et al.*: The Geometric Study of Drifting Subpulses (abstract)........ 60
T. Y. Shi: Pair Production in Intense Electromagnetic Fields of Pulsars (abstract)........ 61
L. F. Chen and S. R. Liang: A Modified Pulsar Model, Green Function, Period Distribution (abstract)........ 62
A. G. Lyne: A Massive Glitch in PSR 0355+54 (abstract)........ 63
C. S. Flanagan: A Large Timing Discontinuity in the Vela Pulsar, July 1985 (abstract)........ 64

*Pulsars and Supernova Remnants*
CHAIR: L. Woltjer

R. NARAYAN: The Galactic Pulsar Population and Neutron Star Birth.... 67
M. SALVATI and F. PACINI: Neutron Star Coupling to its Environment... 79
R. H. BECKER: Crab-Like Supernova Remnants........ 91
F. D. SEWARD: Neutron Stars in Twelve Supernova Remnants........ 99
G. SRINIVASAN and D. BHATTACHARYA: The Progenitors of Pulsars........ 109
H. Sun and Z.W. Li: Distribution of Two Types of Pulsars in Comparison with that of SNRs (abstract)........ 121
R. A. Chevalier and R. T. Emmering: The Structure of Pulsar Nebulae (abstract)........ 122
S. P. Reynolds and H. D. Aller: High-Resolution Radio Observations of the Crablike Supernova Remnant 3C 58 (abstract)........ 123
S. Krishnamohan *et al.*: Search for Plerions in the Direction of Two Young Pulsars (abstract)........ 124
M. J. Kesteven *et al.*: Supernova Remnants with Radio Jets (abstract)........ 125
J. R. Dickel: The Evolution of Young Supernova Remnants (abstract)... 126
R. Bandiera: On the Origin of Kepler's Supernova Remnant (abstract)........ 127
K. Koyama: The X-ray Spectra of Tycho and SN1006 (abstract)........ 128
J. L. Caswell *et al.*: The Galactic Sources G5.4-1.2 and G5.27-0.90 (abstract)........ 129
A. J. Turtle: Supernova Remnants in the Magellanic Clouds Observed at the Molonglo Observatory (abstract)........ 130
Z. W. Li *et al.*: A Statistical Study of the Correlation Between Galactic SNRs and Spiral Arms (abstract)........ 131

## *II. ACCRETION POWERED PULSARS*

*The Population*
CHAIR: D. Helfand

N. E. WHITE: Accreting Neutron Stars........ 135
G. TRINCHIERI: Population of Accreting Neutron Stars in External Galaxies........ 149
Y. TANAKA: Observations of X-ray Burst Sources........ 161
J. E. GRINDLAY: On the Origin of Neutron Stars in Globular Clusters........ 173

F. VERBUNT and P. HUT: The Globular Cluster Population of X-ray Binaries.......... 187
J. Shaham and M. Tavani: Hardness Ratio in Evolving Low-Mass X-ray Binary Systems (abstract).......... 199
F. Nagase: Results of the Timing Analyses of X-ray Pulsars Observed by Hakucho and Tenma (abstract).......... 200
P. Durouchoux: Herculis X1: Results and Interpretation (abstract).......... 201
Y. Q. Ma *et al.*: A Hard X-ray Observation of Cyg X-1 in 1985 (abstract).......... 202
A. N. Parmar *et al.*: The X-ray Transient EXO 2030+375 (abstract).......... 203
F. Z. Cheng and J. E. Grindlay: Analysis of X-ray Spectrum for Globular Cluster M15 X-ray Source (abstract).......... 204

*Theoretical Considerations*
CHAIR: K. L. Huang

J. ARONS: Accretion onto Magnetized Neutron Stars: Polar Cap Flow and Centrifugally Driven Winds.......... 207
D. Y. WANG: Line Radiation from Accreting Magnetized Neutron Stars... 227
H. INOUE: X-ray Bursting Neutron Stars.......... 233
J. Arons and R. I. Klein: Polar Cap Accretion onto Magnetized Neutron Stars: An Analytic Solution (abstract).......... 245
R. I. Klein and J. Arons: Radiation Gas Dynamics of Polar Cap Accretion onto Magnetized Neutron Stars (abstract).......... 246
R. Hoshi and H. Inoue: X-ray Irradiated Accretion Disk and Bimodal States (abstract).......... 247
G. J. Qiao *et al.*: An Inverse Compton Scattering Model for the Spectra of X-ray Pulsars (abstract).......... 248
T. Yi *et al.*: The X-ray Radiation Mechanism of the Compact (Neutron) Binary Stars (abstract).......... 249
T. Ebisuzaki: X-ray Spectra and Atmospheric Structures of Bursting Neutron Stars (abstract).......... 250
R. A. London *et al.*: Model Atmospheres for X-ray Bursting Neutron Stars (abstract).......... 251

***III. NEUTRON STAR FORMATION IN THEORETICAL SUPERNOVAE***
CHAIR: R. Chevalier

S. E. WOOSLEY: The Birth of Neutron Stars.......... 255
W. D. ARNETT: Numerical Experiments and Neutron Star Formation....... 273
K. NOMOTO: Neutron Star Formation in Theoretical Supernovae - Low Mass Stars and White Dwarfs -.......... 281
Z.-R. WANG: Ancient Guest Stars as Harbingers of Neutron Star Formation.......... 305

## IV. NEUTRON STELLAR EVOLUTION

### *Quasi Periodic Oscillations as a Clue to Field and Spin Evolution*

CHAIR: E. P. J. van den Heuvel

M. VAN DER KLIS: Quasi-Periodic Oscillations in Low-Mass X-ray Binaries.......... 321
G. HASINGER: Observations of Quasi-Periodic Oscillations in Cyg X-2.......... 333
J. SHAHAM: The Beat Frequency Model for QPOs.......... 347
W. H. G. LEWIN: Quasi-Periodic Oscillations.......... 363
D. Eichler and Z. Wang: On the Long Term Stability of Neutron Star Magnetic Fields (abstract).......... 375
L. S. Li: On the Evolution of Magnetic Inclination with Age (abstract).......... 376
S. Krishnamohan: Timescale for the Decay of Magnetic Fields of Pulsars (abstract).......... 377
C. N. Zhang *et al.*: Tortion Influence on the Magnetic Field of Rotating Neutron Stars (abstract).......... 379

### *Evolution in Binaries*

CHAIR: A. Blaauw

J. H. TAYLOR: Binary Pulsars: Observations and Implications.......... 383
E. P. J. VAN DEN HEUVEL: Millisecond Pulsar Formation and Evolution... 393
S. R. Kulkarni: Secondary Components of Binary Pulsars and Magnetic Field Decay in Neutron Stars (abstract).......... 407
R. J. Dewey and J. M. Cordes: Monte Carlo Simulations of Radio Pulsars and Their Progenitors (abstract).......... 408
E. H. P. Pylyser: Constraints to Possible Progenitor Systems of PSR 1831-00 (abstract).......... 409
M. Bailes: Geodetic Precession in Binary Pulsars (abstract).......... 410

## V. NEUTRON STAR PHYSICS

CHAIR: T. Weaver

C. ALCOCK: Neutron Star Interiors.......... 413
J.H. HUANG: Two Types of Pulsars.......... 425
N. ITOH: Neutron Star Cooling: Criticical Test of Dense Matter Physics.......... 439
Y. C. Li *et al.*: Neutron Stars Formed from Supernova Explosion and Quark Matter (abstract).......... 447
W. H. Huang and S. H. Gao: The Problem of Solidification in Neutron Stars (abstract).......... 448
J.-H. Huang *et al.*: Interior Structures for Two Types of Pulsars (abstract).......... 449
S. Shibata: Modes of Energy Loss from Isolated Magnetized Neutron Star (abstract).......... 450

Q. M. Wang and S. H. Gao: Effects of Magnetization on Structure Parameters of Neutron Stars (abstract)............ 451
S.-Y. An: Synchro-Curvature Radiation and Magnetic Pair Production of Relativistic Electrons in Strong Curved Magnetic Fields (abstract)............................................. 452
H. Chen *et al.*: Inverse Compton Scattering in Strong Magnetic Fields: Applied to the Radiation Mechanism of PSR 0531+21 (abstract)............................................. 453
C. R. Gwinn: The Nutations of Neutron Stars and Core-Crust Coupling (abstract)................................. 454
Y. Zhang: The Spin-Torsion Coupling Precession of Spin and its Effects on Single Pulses of Pulsars (abstract).............. 455
K. Nomoto and S. Tsuruta: Neutron Star Cooling and the Vela Pulsar (abstract)............................... 456
F. R. Harnden: Einstein Observatory Limits on Neutron Star Surface Temperatures (abstract)........................... 457
W. Brinkmann and H. Ögelman: Thermal Radiation from a Radio Pulsar: PSR1055-52 (abstract)......................... 458
R. W. Romani *et al.*: Model Atmospheres for Cooling Neutron Stars (abstract)..................................... 459
Y.-J. Wang and Q.-H. Peng: The Neutron Stars with Magnetic Charges (abtract)...................................... 460
M. Xinjie and T. Yi: The Nonlinear Dispersion Relation and the Relationship of the Forming Soliton Area to the Evolution of Pulsars (abstract)........................... 461

## *VI: OTHER MANIFESTATIONS OF NEUTRON STARS AND THEIR PLACE IN NEUTRON STAR EVOLUTION*

CHAIR: S. Colgate

G. F. BIGNAMI: Neutron Stars and Gamma-Rays........................ 465
W. D. EVANS and J.G. LAROS: A Review of Gamma-Ray Burst Observations................................................ 477
K. HURLEY: Some Constraints on Neutron Star Properties from Gamma Ray Burster Observations................................ 489
G. S. BISNOVATYI-KOGAN: Theory of Gamma Ray Bursters................ 501
J. J. BARNARD: Cygnus X-3 and Other Ultra-High-Energy Gamma-Ray Sources.................................................. 521
K. KOYAMA: A New Aspect of Galactic Ridge X-ray Emission - SNRs in a Tenuous Medium?.................................. 535
P. A. Caraveo and G. F. Bignami: EXOSAT News on Geminga (abstract)... 545
S. A. Colgate: The Origin of the Enhanced Dissipation "alpha" in Accretion Discs and its Relation to Gamma Bursts (abstract)............................................. 546
W. M. Howard and E. P. Liang: Inverse Compton Model of Gamma Ray Burst Spectra (abstract)............................. 547
C. Alcock *et al.*: A Model for the 1979 March 5 Gamma-Ray Transient (abstract)........................................ 548
E. P. Liang: New Distance Limit for the 3-5-79 Source (abstract)..... 549
J. M. Bonnet-Bidaud: What Type of Binary System is Cygnus X-3? (abstract)................................. 550

W. Q. Luo: On the Phases of Period Variation of SS433 (abstract)..... 551
A. K. Harding *et al.*: Production and Interaction of High Energy Neutrinos in Close X-ray Binaries (abstract).......... 552
Y. L. Jang *et al.*: An Experiment for Observing VHE Gamma Ray Sources (abstract).................................. 553
S. Miyaji: High Energy Cosmic Rays from Young Neutron Stars (abstract).............................................. 554
K. O. Thielheim: Cosmic Ray Particle Acceleration in Pulsar Magnetospheres (abstract)................................. 555
J. F. Dolan: Observations of Neutron Stars Planned by the High Speed Photometer Team Using Space Telescope (abstract)....... 556

## *EPILOGUE*

L. WOLTJER: Where Neutron Stars Come From, How Neutron Stars Evolve, and Where Neutron Stars Go.......................... 559

SUBJECT INDEX.......................................................... 563

## PREFACE

The publication of this volume coincides with the 55th anniversary of the discovery of the neutron and Landau's suggestion at the time that one could make stars out of the new particles. This year also marks the twenty-fifth anniversary of the detection of Sco X-1, the first known X-ray binary system, and follows by just twenty years Jocelyn Bell Burnell's discovery of that "little bit of scruff" on her chart record that led to the recognition of radio pulsars. As Q.Y. Qu, President of Nanjing University noted in his welcoming address, however, Chinese astronomers have been observing the consequences of neutron star formation for several millenia. It was appropriate, then, that this Symposium, the first International Astronomical Union meeting ever to be held in the Peoples Republic of China, be devoted to the topic of neutron stars.

IAU Symposium Number 125, "The Origin and Evolution of Neutron Stars", was convened on the morning of May 26, 1986 at Nanjing University, Nanjing, Peoples Republic of China. One hundred and thirty-nine participants from fifteen countries, including over eighty-five scientists who were visiting China for the first time, met each day for the following week to discuss where neutron stars come from, how they evolve, and where they go. The meeting was judged, by unanimous acclaim of the participants, to be a scientific, cultural, and culinary success. The thoughtful review articles, exciting abstracts, and vigorous discussion recorded in these pages attests to the scientific vitality of the endeavor, although the invaluable contacts and enduring ties between Chinese hosts and foreign guests will be remembered and cherished long after this volume is assigned to dead storage. As evidence of the culinary triumphs, the reader is referred to the partial menu of the Symposium Banquet reproduced below:

Sun-Flower-Shaped Cold Dishes
Peacock-Shaped Shelled Shrimps
Rice Crust with Three Delicacies
Specially-Cooked Chicken
Two-Different-Taste Prawn
Eight-Treasure Duck
Squirrel-Shaped Mandarin Fish
Seasonal Vegetables
Hot Candied Apple
Dessert

The Scientific Program for the meeting was designed by the Scientific Organizing Committee to reflect the aspects of neutron star observations and theory which are most germane to the origin and evolution of these remarkable objects. Thirty-seven invited speakers reviewed topics ranging from radio pulsar statistics to the theory of quasiperiodic oscillations, and from gamma ray burster observations to the possibility that neutron stars may, in fact, be made of quarks. More than seventy-five participants presented contributed poster papers (abstracts of which are included herein); each poster session was reviewed for the entire Symposium by an appointed interlocuter adding the latest in observational and theoretical developments to the general discussion. And throughout, the Local Organizing Committee produced meeting rooms for impromptu workshops, a profusion of tea, and a program of activities which fostered an ideal environment for the unfettered interchange of which this written version is but a partial reflection.

The field of neutron star astrophysics is in a state of heightened activity occasioned by a series of discoveries over the past few years which include millisecond pulsars, quasiperiodic oscillations, and ultrahigh energy emission from some accreting binary systems. With regard to the question of pulsar birthrates, the level of sophistication of the discussion has greatly improved recently; unfortunately, however, the increase in the error bars on birthrates resulting from this increasing sophistication has just compensated for the decrease in the uncertainties accompanying new and better data. It is my admittedly biased belief that the birthrate of neutron stars in the Galaxy is still uncertain by a fact of at least three. The formerly isolated fields of radio and X-ray pulsars and globular cluster X-ray sources have been drawn closer together by proposed scenarios for QPOs and millisecond pulsar evolution, although the genetic relationships are still far from determined. Gamma ray burster statistics are beginning to provide important constraints on such matters as magnetic field decay, binary evolution and neutron star space densities, while the discovery of a growing number of Crab-like and composite supernova remnants sets new limits on the properties of newly-born neutron stars. This increasing cross-fertilization amongst subfields bodes well for future progress in our understanding of the origin and evolution of neutron stars.

It is to be hoped that IAU Symposium Number 125, in some small way, helped foster such progress. At this writing, we are blind to the clues neutron stars are sending us in the infrared, ultraviolet, X-ray, and gamma-ray bands. By the time of publication, we will be back on the air, at least in X-rays, with the Japanese satellite ASTRO-C. In the coming decade, it remains our fond hope that a uniquely powerful series of satellite observatories will be in place to provide us with critical observations in these important wavelength regimes. Coupled with the increasing sensitivity of ground-based telescopes and new theoretical advances, both predicated on ever-cheaper computational power, future insight for neutron star astrophysicists is assured.

Ultimately, the success of this meeting will be judged by the degree to which it will hasten the obsolescence of these Proceedings.

David J. Helfand
Columbia University
New York, USA

INTRODUCTORY REMARKS

T. Lu
Department of Astronomy
Nanjing University
Nanjing
People's Republic of China

Ladies and Gentlemen:

It is a great pleasure for me to have this honor to give some introductory remarks related with this symposium.

First of all, I should like to join Prof. Qu in welcoming all of you to attend this meeting. We are very happy that so many distinguished astronomers and physicists have been able to come here and will give their interesting talks and/or posters on various aspects related with origin and evolution of neutron stars. This is a rather grand meeting in the field.

More than 1600 years ago, Wang Xizhi (321-379), a great calligrapher and writer of Eastern Jin Dynasty, used the Chinese words

群賢畢至，少長咸集

'lots of worthies are all coming and both younger and elder people are gathering here'

to describe a wonderful meeting in his famous essay "Preface of Anthology of Orchid Pavilion". I think, the very same words could also appropriately be used to describe our present symposium.

This symposium has just been opened by Prof. David J. Helfand. Perhaps, I may be allowed to say some words about the way this meeting has come about. Almost exactly three years ago, Prof. David J. Helfand first suggested to hold such a symposium in his letter to me of 10 June 1983. Prof. S. G. Wang, Prof. Q. Y. Qu, my colleagues and myself are all interested in holding such a meeting. Within a year or so, things were progressively going on. Heads of Nanjing University, Astronomy Department, Astrophysics Institute and Columbia Astrophysics Laboratory are willing to support it. Presidents of IAU Commissions No.40 and No.48 are also willing to sponsor it. By the summer of 1984, Scientific Organizing Committee and Local Organizing Committee were already really formed. As early as in June 1984, this meeting was formally approved by the Chinese Ministry of Education (now the Educational Committee of PRC). This symposium was originally suggested to be held in the autumn of 1985 just preceeding the IAU General Assembly in New Delhi. However, the Executive Committee of the IAU decided to sponsor it and name it officially as the IAU Symposium No.125 in February 1985, but postpone it to be held in 1986. This is the present symposium coming about.

Since the IAU Symposium No.95 on Pulsars, Bonn, FRG, August 1980, it has been nearly six years. This is really a rather long period for such a rapidly progressive field. People have observed the neutron stars through more and more "windows" and got much more information from X- and gamma-ray observations. Especially, the investigations on X- and gamma-ray bursts, both observational and theoretical, have provided important background for understanding the physics of neutron stars. The recently discovered phenomena of Quasi-Periodic Oscillations (QPO) may throw new light on their origin and evolution. Even in the field of pulsars themselves, the discoveries of milli-second pulsars have excited a lot of deep investigations on the ways that the pulsars come and go. We are very happy to have so many experts to talk about all these things to us.

**Thank you.**

## LIST OF PARTICIPANTS

| | |
|---|---|
| Ables, J. | ANRAO, AUSTRALIA |
| Alcock, C. | Massachusetts Institute of Technology, USA |
| An, C. | Hainan University, PRC |
| An, S. | Beijing Teacher's College, PRC |
| Arnett, D. | Enrico Fermi Institute, USA |
| Arons, J. | University of California, Berkeley, USA |
| Backer, D. | University of California, Berkeley, USA |
| Bailes, M. | Mt. Stromlo Observatory, AUSTRALIA |
| Bandiera, R. | Osservatorio Astrotisico di Arcetri, ITALY |
| Barnard, J. | NASA Goddard Space Flight Center, USA |
| Bazzano, A. | Istituto di Astrofisica Spaziale, ITALY |
| Becker, R. | University of California, Davis, USA |
| Belli, B. | Istituto di Astrofisica Spaziale, ITALY |
| Bignami, G. | Istituto di Astrofisica Spaziale, ITALY |
| Bisnovatyi-Kogan, G. | Space Research Institute, USSR |
| Blaauw, A. | Kapteyn Laboratorium, THE NETHERLANDS |
| Bonnet-Bidaud, J. | Service d'Astrophysique CEN Saclay, FRANCE |
| Brinkmann, W. | Max-Plank Institut, FRG |
| Burrows, A. | SUNY-Stony Brook, USA |
| Caraveo, P. | Istituto di Astrofisica Spaziale, ITALY |
| Caswell, J. | CSIRO, AUSTRALIA |
| Chechetkin, V. | Institute of Applied Mathematics, USSR |
| Chen, H. | Beijing University, PRC |
| Chen, L. | Beijing Normal University, PRC |
| Cheng, F. | University of Science and Technology, PRC |
| Chevalier, R. | University of Virginia, USA |
| Colgate, S. | Los Alamos National Laboratory, USA |
| Cordes, J. | Cornell University, USA |
| Deng, Z. | Chinese Academy of Sciences, PRC |
| Dewey, R. | Cornell University, USA |
| Dickel, J. | Los Alamos National Laboratory, USA |
| Dolan, J. | NASA Goddard Space Flight Center, USA |
| Durouchoux, P. | Service d'Astrophysique CEN Saclay, FRANCE |
| Ebisuzaki, T. | University of Tokyo, JAPAN |
| Eichler, D. | University of Maryland, USA |
| Emanuele, A. | Istituto Astrofisica Spaziale, ITALY |
| Evans, D. | Los Alamos National Laboratory, USA |
| Feng, L. | Nanjing University, PRC |
| Flanagan, C. | Hartebeesthoek RAO, SOUTH AFRICA |
| Frontera, F. | Istituto Tesre, C.N.R., ITALY |
| Gong, S. | Purple Mountain Observatory, PRC |

| | |
|---|---|
| Grindlay, J. | Harvard-Smithsonian Center for Astrophysics, USA |
| Gwinn, C. | Harvard-Smithsonian Center for Astrophysics, USA |
| Han, J. | Beijing Normal University, PRC |
| Hang, H. | Purple Mountain Observatory, PRC |
| Harding, A. | NASA Goddard Space Flight Center, USA |
| Harnden, F. | Harvard-Smithsonian Center for Astrophysics, USA |
| Hasinger, G. | Max-Planck Institut, FRG |
| Haynes, R. | CSIRO, AUSTRALIA |
| He, X. | Beijing Normal University, PRC |
| He, Z. | Institute of High Energy Physics, PRC |
| Helfand, D. | Columbia University, USA |
| Hoshi, R. | Rikkyo University, JAPAN |
| Howard, W. | Lawrence Livermore National Laboratory, USA |
| Huang, J. | Nanjing University, PRC |
| Huang, K. | Nanjing University, PRC |
| Huang, W. | Beijing Normal University, PRC |
| Hurley, K. | CESR, FRANCE |
| Inoue, H. | Institute of Space and Astronautical Science, JAPAN |
| Itoh, N. | Sophia University, JAPAN |
| Kesteven, M. | CSIRO, AUSTRALIA |
| Klein, R. | Lawrence Livermore National Laboratory, USA |
| Kong, X. | Hebei Teacher's University, PRC |
| Koyama, K. | Institute of Space and Astronautical Science, JAPAN |
| Krishnamohan, S. | Tata Institute of Fundamental Research Centre, INDIA |
| Kulkarni, S. | California Institute of Technology, USA |
| Lewin, W. | Massachusetts Institute of Technology, USA |
| Li, H. | Beijing Normal University, PRC |
| Li, L. | Northeast Normal University, PRC |
| Li, Y. | Hebei Teacher's University, PRC |
| Li, Z. | Beijing Normal University, PRC |
| Liang, E. | Lawrence Livermore National Laboratory, USA |
| Liang, S. | Beijing Normal University, PRC |
| Liu, J. | Shanghai Observatory, PRC |
| Liu, R. | Purple Mountain Observatory, PRC |
| Lo, W. | Sichuan Teacher's University, PRC |
| Lu, T. | Nanjing University, PRC |
| Lu, Z. | Institute of High Energy Physics, PRC |
| Lyne, A. | Nuffield Radio Astronomy Laboratories, UK |
| Ma, Y. | Institute of High Energy Physics, PRC |
| Manchester, R. | CSIRO, AUSTRALIA |
| Mao, X. | Beijing Normal University, PRC |
| Milne, D. | CSIRO, AUSTRALIA |
| Miyaji, S. | NASA Marshall Space Flight Center, USA |
| Nagase, F. | Nagoya University, JAPAN |
| Narayan, R. | University of Arizona, USA |

| | |
|---|---|
| Nomoto, K. | Brookhaven National Laboratory, USA |
| Ostgaard, E. | Fysisk Institutt AVH University, NORWAY |
| Pan, R. | Shanghai Observatory, PRC |
| Parmar, A. | European Space Operations Centre, FRG |
| Pineault, S. | University of Laval, CANADA |
| Qiao, G. | Beijing University, PRC |
| Qu, Q. | Nanjing University, PRC |
| Radhakrishnan, V. | Raman Research Institute, INDIA |
| Rankin, J. | University of Vermont, USA |
| Reynolds, S. | North Carolina State University, USA |
| Salvati, M. | Istituto di Astrofisica Spaziale, ITALY |
| Seward, F. | Harvard-Smithsonian Center for Astrophysics, USA |
| Shaham, J. | Columbia University, USA |
| Shen, C. | Institute of High Energy Physics, PRC |
| Shi, J. | Dalian Technology Institute, PRC |
| Shi, T. | Beijing Normal University, PRC |
| Shibata, S. | Tohoku University, JAPAN |
| Sieber, W. | Max-Planck-Institut, FRG |
| Srinivasan, G. | Raman Research Institute, INDIA |
| Sun, H. | Beijing Normal University, PRC |
| Tanaka, Y. | Institute of Space and Astronautical Science, JAPAN |
| Taylor, J. | Princeton University, USA |
| Thielheim, K. | Institut für Reine und Angewandte Kernphysik, FRG |
| Tong, Y. | Beijing Normal University, PRC |
| Trimble, V. | University of California, Irvine, USA |
| Trinchieri, G. | Osservatorio Astrofisico di Arcetri, ITALY |
| Turtle, A. | University of Sydney, AUSTRALIA |
| Ubertini, P. | Istituto di Astrofisica Spaziale, ITALY |
| van den Heuvel, E. | University of Amsterdam, THE NETHERLANDS |
| van der Klis, M. | European Space Agency/ESTEC, THE NETHERLANDS |
| Verbunt, F. | Max-Planck-Institut, FRG |
| Wang, D. | Purple Mountain Observatory, PRC |
| Wang, G. | Beijing Normal University, PRC |
| Wang, S. | Beijing Observatory, PRC |
| Wang, Y.J. | Changsha Railway Institute, PRC |
| Wang, Y.S. | Nanjing University, PRC |
| Wang, Z.R. | Nanjing University, PRC |
| Wang, Z.Z. | University of Maryland, USA |
| Weaver, T. | Lawrence Livermore National Laboratory, USA |
| White, N. | EXOSAT Observatory/ESOC, FRG |
| Woltjer, L. | European Southern Observatory, FRG |
| Woosley, S. | Lick Observatory, USA |
| Wu, X.J. | Beijing University, PRC |
| Wu, X.Y. | Columbia University, USA |
| Xia, X. | Tianjin Normal University, PRC |
| Xu, M. | Institute of Physics, PRC |
| Yang, J. | Purple Mountain Observatory, PRC |

| | |
|---|---|
| Yee, Z. | University of Science and Technology of China, PRC |
| You, J. | University of Science and Technology of China, PRC |
| Zhang, C. | Dalian Technology Institute, PRC |
| Zhang, S. | Institute of High Energy Physics, PRC |
| Zhao, Y. | Nanjing University, PRC |
| Zhu, C. | Nanjing University, PRC |

IAU SYMPOSIUM NO. 125:

THE ORIGIN AND EVOLUTION OF NEUTRON STARS

SCIENTIFIC ORGANIZING COMMITTEE:

Co-Chairpersons:
D. J. Helfand and S. G. Wang

R.H. Becker
R. Blandford
A. Lyne
T. Lu
R. Manchester
F. Pacini
Q.Y. Qu
V. Radhakrishnan
Y. Tanaka
E.P.J. van den Heuvel
S. Woosley

LOCAL ORGANIZING COMMITTEE:

Co-Chairpersons:
Q.Y. Qu and T. Lu

L.Z. Fang
K.L. Huang
Q.B. Li
Z.W. Li
Z.R. Wang
X.J. Wu
H.Q. Zhang

1. J. Grindlay
2. F. Verbunt
3. K. Thielheim
4. A. Harding
5. N. White
6. F.R. Harnden, Jr.
7. R. Becker
8. Z.R. Wang
9. J. Barnard
10. J. Dickel
11. J. Ables
12. D. Milne
13. R. Klein
41. Z. Lu
42. M. Howard
43. K. Hurley
44. S. Colgate
45. D. Eichler
46. B.C. Chiu
47. T. Ebisuzaki
48. M. Bailes
49. R. Dewey
50. M. Kesteven
51. S. Pineault
52. R. Narayan
53. A. Turtle
54. A. Parmar
87. F. Nagase
88. H. Sun
89. E. van den Heuvel
90. P. Durouchoux
91. V. Trimble
107. J.S. Huang
108. L.L. Feng
109. T. Shi
110. C. Wan
111. J. Dolan
112. X. Mao
113. Q. Wang
114. H. Li
115. S. Liang
116. V. Radhakrishnan
117. L. Woltjer
118. G. Bignami
119. G. Srinivasan

14. J. Cordes
15. A. Lyne
16. R. Manchester
17. J. Taylor
18. D. Backer
19. D. Arnett
20. R. Chevalier
21. S. Reynolds
22. A. Burrows
23. V. Chechetkin
24. R. Bandiera
25. G. Bisnovatyi-Kogan
26. Q.Y. Qu
55. P. Caraveo
56. F. Frontera
57. M. Belli
58. E. Østgaard
59. K. Koyama
60. R. Hoshi
61. L.S. Li
62. G. Hasinger
63. M. van der Klis
64. E. Basinska-Lewin
65. C.S. Zhu
66. J.H. You
67. D.G. Deng
68. J.H. Huang
69. D. Helfand
92. W. Brinkmann
93. H. Chen
94. X. Wu
95. G.J. Qiao
96. Y.S. Wang
120. J. Rankin
121. W. Sieber
122. J.-G. Shi
123. C.-M. Zhang
124. F.-G. Chang
125. F.-Z. Cheng
126. H. Inoue
127. X. Wu
128. J. Bonnet-Bidaud
129. J.-L. Han
130. Y. Tong
131. K. Nomoto
132. S. Woosley
133. Z.W. Li

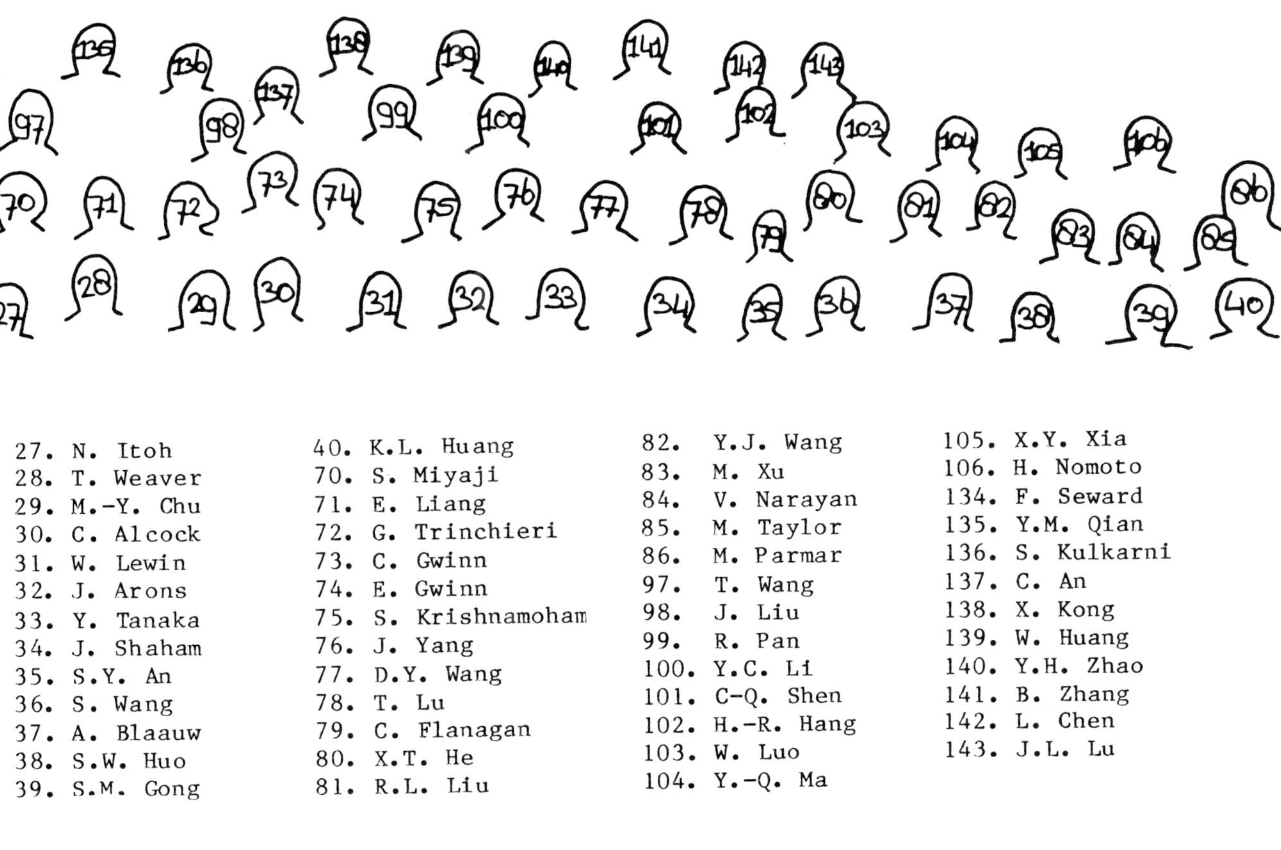


27. N. Itoh
28. T. Weaver
29. M.-Y. Chu
30. C. Alcock
31. W. Lewin
32. J. Arons
33. Y. Tanaka
34. J. Shaham
35. S.Y. An
36. S. Wang
37. A. Blaauw
38. S.W. Huo
39. S.M. Gong
40. K.L. Huang
70. S. Miyaji
71. E. Liang
72. G. Trinchieri
73. C. Gwinn
74. E. Gwinn
75. S. Krishnamoham
76. J. Yang
77. D.Y. Wang
78. T. Lu
79. C. Flanagan
80. X.T. He
81. R.L. Liu
82. Y.J. Wang
83. M. Xu
84. V. Narayan
85. M. Taylor
86. M. Parmar
97. T. Wang
98. J. Liu
99. R. Pan
100. Y.C. Li
101. C-Q. Shen
102. H.-R. Hang
103. W. Luo
104. Y.-Q. Ma
105. X.Y. Xia
106. H. Nomoto
134. F. Seward
135. Y.M. Qian
136. S. Kulkarni
137. C. An
138. X. Kong
139. W. Huang
140. Y.H. Zhao
141. B. Zhang
142. L. Chen
143. J.L. Lu

# *Ia. ROTATION-POWERED PULSARS*

## *The Population*

*CHAIR: V. Radhakrishnan*

# PULSAR SURVEYS

R.N. Manchester
Division of Radiophysics, CSIRO
PO Box 76, Epping, NSW 2121
Australia

ABSTRACT. The current situation regarding pulsar surveys is briefly reviewed. Most of the known pulsars have been found by major radio surveys that were unbiased in the sense of more-or-less uniformly covering a given area of sky. The results from two recent such surveys, the Green Bank 390 MHz survey of Stokes et al. (1985) and the Jodrell Bank 1400 MHz survey of Clifton and Lyne (1986) are compared. Conclusions are drawn regarding the effect of observing frequency on the results of pulsar surveys and the galactic distribution of pulsars and interstellar electrons.

## 1. PULSAR STATISTICS

At last count there were 437 pulsars known. (In this paper the term "pulsar" refers to rotationally powered pulsars and not to accretion-powered systems, which are generally detected at X-ray frequencies.) All but two of the known pulsars were discovered at radio frequencies. These two, PSR 1509-58 and PSR 0540-69, were discovered using the Einstein X-ray observatory (Seward and Harnden, 1982; Seward et al., 1984). Only four pulsars have been detected at frequencies above the radio band. These are the two pulsars mentioned above and the Crab and Vela pulsars, PSR 0531+21 and PSR 0833-45 respectively. All four of these pulsars are young and all are associated with known supernova remnants. They are in fact the only pulsars for which the association with a supernova remnant is reasonably definite. Only the Crab and Vela pulsars have been detected at gamma-ray frequencies and PSR 0540-69 has been detected in the optical band. Despite several intensive searches, this pulsar has not yet been detected in the radio band; the limiting mean flux density at a frequency of 1 GHz is <1 mJy. The situation regarding detection of pulsars in different frequency bands is summarized in Table I.

All but two of the known pulsars are located within our galaxy. Both of the exceptions are within the Large Magellanic Cloud; they are PSR 0540-69, discussed above, and PSR 0529-66, discovered in a radio search at Parkes (McCulloch et al., 1983). The distribution of pulsars in galactic coordinates is illustrated in Figure 1. The well-known concentrations toward the galactic equator and toward the galactic centre can be seen in this figure. These clearly show that pulsars are a galactic disk population and that, at least near the Sun, their spatial density increases with decreasing galactocentric radius. Lyne et al. (1985) have discussed



*D. J. Helfand and J.-H. Huang (eds.), The Origin and Evolution of Neutron Stars, 3–12.*

TABLE I - Detection of pulsars in different frequency bands*

| PSR | Radio | Optical | X-ray | γ-ray |
|---|---|---|---|---|
| Crab | D | x | x | x |
| Vela | D | x | | x |
| 1509-58 | x | | D | |
| 0540-69 | | x | D | |

*D indicates the discovery frequency band.

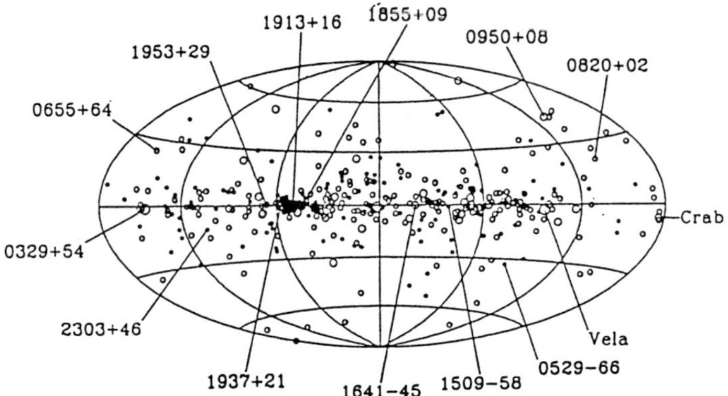


Figure 1. Distribution of 398 pulsars in galactic coordinates. The galactic centre is at the centre of the diagram and the size of the circles indicates the 400 MHz mean flux density. Locations of selected pulsars are marked. (From Taylor and Stinebring, 1986.)

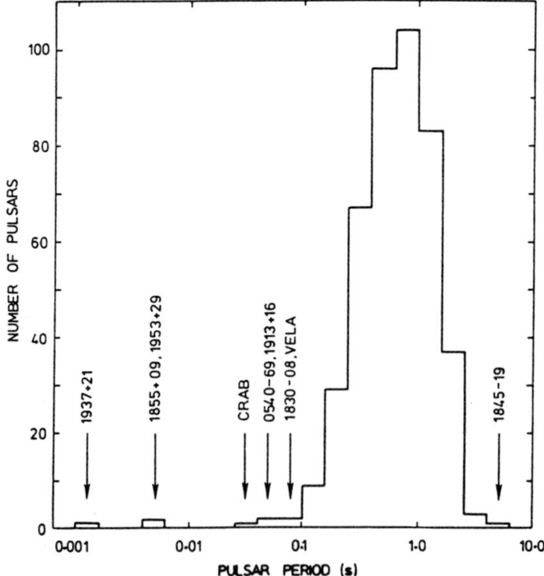


Figure 2. The observed period distribution of 437 pulsars.

the galactic distributions of pulsars that are implied by these results.

Figure 2 illustrates the period distribution of the observed sample of pulsars. The extreme range of pulsar periods is 0.0015 to 4.30 s but the vast majority of pulsars have periods in the range 0.2 to 2.0 s. The observed range was greatly extended by Backer et al.'s (1982) discovery of the millisecond pulsar PSR 1937+21. Despite a number of intensive searches since this initial discovery, only two more millisecond pulsars have been discovered, PSR 1953+29 (Boriakoff et al., 1983) and PSR 1855+09 (Segelstein et al., 1986). These have periods of 6.1 and 5.4 ms respectively.

For binary pulsars the picture is a little more encouraging. Since the discovery of PSR 1913+16 by Hulse and Taylor (1974) a further six binary pulsars have been found. The main parameters of the binary pulsars are summarized in Table II. They appear to fall into two distinct groups, those with high eccentricity and relatively large mass functions and those with almost circular orbits and low mass functions. For systems in the first group the companion is almost certainly a neutron star, whereas for those in the second group it is probably a white dwarf (van den Heuvel, 1984; Segelstein et al., 1986; Kulkarni, 1986). It is worth noting that two of the three known millisecond pulsars are members of binary systems.

TABLE II - Parameters of binary pulsars

| PSR | P (ms) | Log $\dot{P}$ | Log B (G) | $P_b$ (days) | e | Probable $m_2$ ($M_\odot$) |
|---|---|---|---|---|---|---|
| 1913+16 | 59.0 | -17.1 | 10.3 | 0.32 | 0.6171 | 1.4 |
| 0655+64 | 195.6 | -18.2 | 10.0 | 1.03 | <0.00005 | 0.7-1.3 |
| 1831-00 | 520.9 | <-17.0 | <10.9 | 1.81 | <0.005 | 0.06-0.13 |
| 1855+09 | 5.4 | -18.8 | 9.0 | 12.33 | 0.00002 | 0.2-0.4 |
| 2303+46 | 1066.4 | -15.4 | 11.8 | 12.34 | 0.6584 | 1.2-2.5 |
| 1953+29 | 6.1 | -19.5 | 8.6 | 117.35 | 0.0003 | 0.2-0.4 |
| 0820+02 | 864.9 | -16.0 | 11.5 | 1232.40 | 0.0119 | 0.2-0.4 |

## 2. MAJOR PULSAR SURVEYS

Of the 430 or so known pulsars over 390 were discovered in one or more of 11 major pulsar surveys. These surveys, the main parameters of which are summarized in Table III, all covered a given area of sky in a relatively unbiased way. Most of the pulsars were discovered in the 1970s and early 1980s in large-scale computer-based searches. These searches covered the whole of the celestial sphere to a reasonably uniform limiting mean flux density; the Arecibo searches are exceptional in that they were about an order of magnitude more sensitive owing to the large collecting area of this instrument. Most of these surveys had sampling intervals of 16 or 20 ms and so were insensitive to pulsars with periods of $\lesssim$100 ms. (In Table III the quoted minimum period is twice the data sampling interval and the limiting mean flux density quoted refers to pulsar periods greater than about five times this value.)

Following the discovery of the millisecond pulsar, searches have used faster sampling and hence have been sensitive to pulsars of shorter period. The Princeton-Green Bank (Stokes et al., 1985) and Jodrell Bank (Clifton and Lyne, 1986) surveys are discussed extensively below. The

TABLE III - Major pulsar surveys

| | Group | Antenna | Freq. (MHz) | Sky cover (sr) | $P_{min}$ (ms) | $S_{min}$ (mJy) | No. of PSRs discovered |
|---|---|---|---|---|---|---|---|
| 1 | Cambridge | Sq. array | 81.5 | ~6 | ~100 | ~100 | 6 |
| 2 | Sydney | Mills arm | 408 | 7 | >5 | >20 | 31 |
| 3 | Jodrell Bank | MkIA | 408 | 1.0 | 160 | 15 | 39 |
| 4 | U. Mass. | Arecibo | 430 | 0.05 | 33 | 1.5 | 40 |
| 5 | CSIRO/Syd. | Mills arm | 408 | 8.4 | 40 | 15 | 155 |
| 6 | U. Mass. | GB 92-m | 400 | 4 | 33 | 15 | 23 |
| 7 | Princeton | GB 92-m | 390 | 1.8 | 33 | 2 | 34 |
| 8 | Princeton | GB 92-m | 390 | 1.1 | 4.0 | 2 | 20 |
| 9 | Jodrell Bank | MkIA | 1400 | 0.06 | 4.0 | ~1 | 40 |
| 10 | Princeton | Arecibo | 430 | 0.10 | 0.6 | 3 | 5 |
| 11 | CSIRO/Pal./Pri./Syd. | MOST | 843 | 0.06 | 1.0 | 8 | ≥1 |

References: 1, Hewish et al. (1968); 2, Large and Vaughan (1971); 3, Davies et al. (1972, 1973); 4, Hulse and Taylor (1974, 1975); 5, Manchester et al. (1978); 6, Damashek et al. (1978, 1982); 7, Dewey et al. (1985); 8, Stokes et al. (1985); 9, Clifton and Lyne (1986); 10, Stokes et al. (1986); 11, D'Amico et al. (1986).

Princeton-Arecibo survey, described by Stokes et al. (1986), surveyed a small region near the galactic plane at a frequency of 430 MHz with good sensitivity for short-period pulsars and found a total of five pulsars, including the binary millisecond pulsar PSR 1855+09. A survey of a 2°-wide strip along the southern galactic plane has been carried out at Molonglo by D'Amico et al. (1986). The Molonglo instrument works at a frequency of 843 MHz and the survey had reasonable sensitivity down to periods of a few milliseconds. Analysis is not yet complete and so far one pulsar has been confirmed, PSR 0906-49, which has a period of 106 ms, a characteristic age of $\sim 1.2 \times 10^5$ years and a relatively strong interpulse.

Excepting the original Cambridge survey, most surveys have been at frequencies near 400 MHz. This frequency is a good compromise between interstellar effects, which tend to degrade the sensitivity at low frequencies, and the reduced sensitivity at high frequencies due to the steep intrinsic spectrum of pulsar radio emission. In addition, at these frequencies the antenna beam is relatively large and hence it is practicable to survey a large area of sky. However, the discovery at a higher radio frequency of several high-dispersion pulsars which are undetectable at frequencies around 400 MHz because of the effects of interstellar scattering (Manchester et al., 1985) showed that low-frequency surveys were missing a substantial portion of the galactic population of pulsars. These effects are most severe in directions along the galactic plane and especially towards the galactic centre and for short-period pulsars. To illustrate this it is instructive to compare two recent surveys for pulsars, one carried out at a frequency of 390 MHz and the other at 1400 MHz.

## 3. COMPARISON OF THE STOKES ET AL. (1985) AND CLIFTON AND LYNE (1986) PULSAR SURVEYS

Comparison of the Stokes et al. (S85) and Clifton and Lyne (CL86) surveys is of interest since these two surveys have a number of parameters in common but differ in observing frequency. The S85 survey at Green Bank was at a frequency of 390 MHz, whereas the Jodrell Bank survey of CL86 was made at 1400 MHz. Both surveys covered an area of sky along the galactic plane north of the galactic centre and both used a sampling interval of 2 ms. The basic parameters of the two surveys are summarized in Table IV. The limiting flux density is for directions away from the

TABLE IV - Basic parameters for the Stokes et al. (1985) and Clifton and Lyne (1986) surveys

| | Stokes et al. | Clifton and Lyne |
|---|---|---|
| Antenna | Green Bank 92-m | Jodrell Bank 76-m |
| Frequency | 390 MHz | 1400 MHz |
| Sampling interval | 2 ms | 2 ms |
| Filter bandwidths | 2x32x0.25 MHz | 2x8x5 MHz |
| Limiting flux density | ≳3 mJy | ≳1 mJy |
| Area surveyed | 3725 sq. deg | 200 sq. deg |
| Pulsars detected | 87 | 62 |
| Pulsars discovered | 20 | 40 |

galactic plane and for pulsar periods greater than ~10 times the sampling interval. Although at the observing frequency the CL86 survey has a lower limiting flux density than the S85 survey, when one compensates for the average spectral index of pulsars the S85 survey actually has about twice the sensitivity of the CL86 survey. It detected more pulsars than the CL86 survey, although to do this it had to cover nearly 20 times as much area of sky. Another notable difference between the two surveys is that for the S85 survey less than one-quarter of the pulsars detected were new discoveries, whereas for the CL86 survey nearly two-thirds of the pulsars detected were new.

Figure 3 illustrates the period distributions of the pulsars

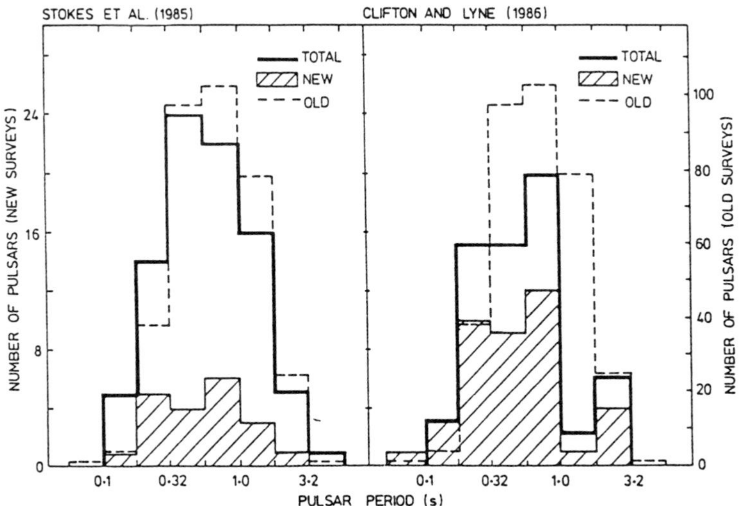


Figure 3. Period distributions for the pulsars detected (solid line), and pulsars discovered (hatched region) for the Stokes et al. (1985) and Clifton and Lyne (1986) surveys and for pulsars detected in previous major surveys (dashed line).

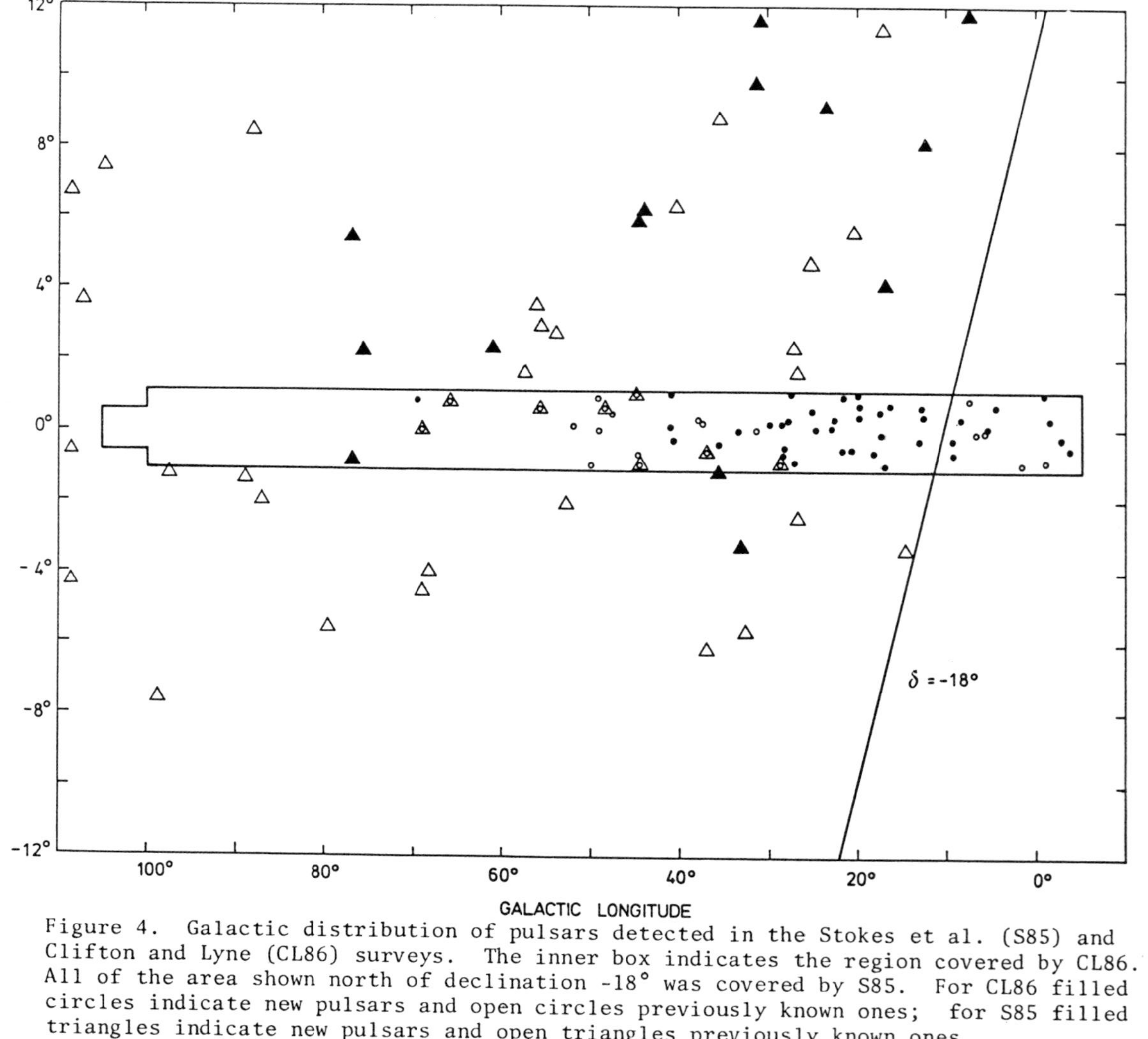


Figure 4. Galactic distribution of pulsars detected in the Stokes et al. (S85) and Clifton and Lyne (CL86) surveys. The inner box indicates the region covered by CL86. All of the area shown north of declination -18° was covered by S85. For CL86 filled circles indicate new pulsars and open circles previously known ones; for S85 filled triangles indicate new pulsars and open triangles previously known ones.

detected and discovered in the two surveys and for reference also gives the period distribution of pulsars detected in previous major surveys. The first point is that neither survey detected any millisecond pulsars despite having some, albeit reduced, sensitivity down to periods of 4 ms. In fact, only one pulsar with period $<100$ ms was detected, namely PSR 1830-08, which has a period of 85 ms, detected by CL86. Both surveys detected a higher proportion of short-period pulsars compared to earlier surveys, but the bias towards short periods is significantly greater for the CL86 survey. The median periods for the S85 pulsars, the CL86 pulsars and the previously known pulsars are respectively 0.60, 0.51 and 0.67 s.

As indicated above, the area density of detected pulsars is much higher for the CL86 survey compared to the S85 survey. In Figure 4 the distribution in galactic coordinates of pulsars detected in the two surveys is compared. A total of eight of the previously known pulsars were detected by both surveys, but none of the new pulsars are in common. For the S85 pulsars the distribution is essentially independent of longitude over the range shown in the figure whereas for the CL86 pulsars the distribution is quite different. No pulsars were detected at longitudes $>70°$ and all but five are at longitudes $<50°$. There is a marginal drop-off in density near the galactic centre, where the survey sensitivity is most affected by high galactic background temperatures and interstellar scattering. For the limited latitude range covered by the CL86 survey there is no significant latitude dependence of the observed distribution.

Associated with the very different sky distribution for the two surveys there is a very different distribution of dispersion measures (DM), as illustrated in Figure 5. The S85 pulsars have a distribution

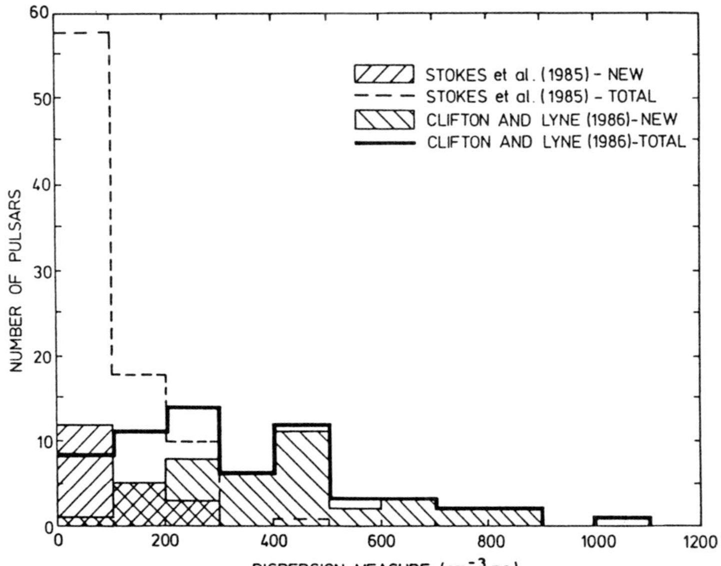


Figure 5. Distribution of dispersion measures for the new and previously discovered pulsars in the Stokes et al. (1985) survey and the Clifton and Lyne (1986) survey.

peaking at low dispersions very similar to that of previously known pulsars. The highest DM detected (for PSR 1859+03) was 403 $cm^{-3}$ pc and the highest value for a new pulsar was ~250 $cm^{-3}$ pc. In contrast, the CL86 pulsars have dispersions extending to over 1000 $cm^{-3}$ pc (for PSR 1758-23), with a broad peak in the distribution at around 400 $cm^{-3}$ pc.

The different sky and DM distributions of pulsars from the CL86 survey show that, on average, these pulsars are much more distant than both those from the S85 survey and those previously known. (It is however true that the distances derived for these pulsars from most models of the interstellar electron density distribution - for example, that derived by Lyne et al. (1985) - must in most cases be too large.) The CL86 survey in fact detected a different population of pulsars from that detected by previous surveys. Most of them are located in the inner part of the Galaxy, although it is not clear yet whether the true distribution peaks at the galactic centre or is annular in form. Despite the relatively high observing frequency, most of the CL86 pulsars with DM greater than about 500 $cm^{-3}$ pc are significantly affected by interstellar scattering. These pulsars would be quite undetectable at frequencies around 400 MHz. Even at 1400 MHz it is likely that some short-period pulsars have not been detected because of the combined effects of interstellar scintillation and high galactic background temperature. Consequently, even higher frequency searches may be required to determine the true galactic and period distributions of pulsars.

To summarize the conclusions derived from the results of these two surveys:

(i) There are few detectable pulsars in the Galaxy with periods between 10 and 100 ms. If pulsars evolve in period with constant luminosity (at least at these short periods), then only one or two pulsars would be expected in this period range from these surveys, so the statistical significance of this result may not be high.

(ii) High background temperatures and strong interstellar scattering, at least in the inner part of the Galaxy, are limiting factors for pulsar surveys at frequencies $<1$ GHz.

(iii) Determination of the true galactic and period distributions of pulsars will require extensive surveys at frequencies $\gtrsim 1$ GHz.

(iv) At galactocentric radii $\lesssim 5$ kpc, the interstellar electron density has a mean value at least a factor of three higher than the local value and fluctuations that cause strong scattering of radio signals.

I thank A.G. Lyne and J.H. Taylor for providing me with data on previously known pulsars detected in their surveys.

REFERENCES

Backer, D.C., Kulkarni, S.R., Heiles, C., Davis, M.M., and Goss, W.M.: 1982, Nature 300, 615.

Boriakoff, V., Buccheri, R. and Fauci, F.: 1983, Nature 304, 417.

Clifton, T.R., and Lyne, A.G.: 1986, Nature 320, 43.

Damashek, M., Taylor, J.H., and Hulse, R.A.: 1978, Astrophys. J. 225, L31.

Damashek, M., Taylor, J.H., Backus, P.R., and Burkhardt, R.: 1982, Astrophys. J. 253, L57.
D'Amico, N., Manchester, R.N., Taylor, J.H., Stokes, G.H., Durdin, J.M., and Brissenden, R.J.V.: 1986 (in preparation).
Davies, J.G., Lyne, A.G., and Seiradakis, J.H.: 1972, Nature 240, 229.
Davies, J.G., Lyne, A.G., and Seiradakis, J.H.: 1973, Nature Phys. Sci. 244, 84.
Dewey, R.J., Taylor, J.H., Weisberg, J.M., and Stokes, G.H.: 1985, Astrophys. J. 294, L25.
Hewish, A., Bell, S.J., Pilkington, D.H., Scott, P.F., and Collins, R.A.: 1968, Nature 217, 709.
Hulse, R., and Taylor, J.H.: 1974, Astrophys. J. 191, L59.
Hulse, R., and Taylor, J.H.: 1975, Astrophys. J. 195, L51.
Kulkarni, S.R.: 1986, Astrophys. J. (in press).
Large, M.I., and Vaughan, A.E.: 1971, Mon. Not. R. Astron. Soc. 151, 277.
Lyne, A.G., Manchester, R.N., and Taylor, J.H.: 1985, Mon. Not. R. Astron Soc. 213, 613.
McCulloch, P.M., Hamilton, P.A., Ables, J.G., and Hunt, A.J.: 1983, Nature 303, 307.
Manchester, R.N., Lyne, A.G., Taylor, J.H., Durdin, J.M., Large, M.I., and Little, A.G.: 1978, Mon. Not. R. Astron. Soc. 185, 409.
Manchester, R.N., D'Amico, N., and Tuohy, I.R.: 1985, Mon. Not. R. Astron. Soc. 212, 975.
Segelstein, D.J., Rawley, L.A., Stinebring, D.R., Fruchter, A.S., and Taylor, J.H.: 1986, Nature (submitted).
Seward, F.D., and Harnden, F.R.: 1982, Astrophys. J. 256, L45.
Seward, F.D., Harnden, F.R., and Helfand, D.J.: 1984, Astrophys. J. 287, L19.
Stokes, G.H., Taylor, J.H., Weisberg, J.M., and Dewey, R.J.: 1985, Nature 317, 787.
Stokes, G.H. Segelstein, D.J. Taylor, J.H., and Dewey, R.J.: 1986, Astrophys. J. (submitted).
Taylor, J.H., and Stinebring, D.R.: 1986, Annu. Rev. Astron. Astrophys. (in press).
van den Heuvel, E.P.J.: 1984, J. Astrophys. Astron. 5, 209.

## DISCUSSION

**S. Krishnamohan:** How can one explain the nondetection of the Jodrell pulsars discovered by Clifton and Lyne in the Green Bank Survey by Stokes *et al.*? Is interstellar scattering a sufficient reason or does one have to invoke a flatter spectrum for these pulsars?

**R. Manchester:** I believe that the combination of high background temperature and interstellar scattering is sufficient to account for the non-detection of Clifton and Lyne pulsars by Stokes *et al.* The background temperature is more important for low DM pulsars and the scattering for high DM pulsars.

**A. Lyne:** I think that without exception all the new pulsars having dispersion measures in excess of 600 pc $cm^{-3}$ suffer from the effects of multipath scattering. Thus, even this survey has reduced sensitivity to such objects and may be missing many. We may not therefore be seeing even as far as the galactic center and the electron density enhancement in the inner regions of the Galaxy may be even greater than you suggest. A still higher frequency is needed to determine this.

**R. Manchester:** I agree.

**A. Blaauw:** In comparing the period distributions of the two surveys, is there evidence for a larger percentage of longer periods in the Stokes et al. survey (which includes higher latitudes)?

**R. Manchester:** No in fact, the Clifton and Lyne survey found four pulsars with periods between 1.8 and 3.2 seconds whereas the Stokes et al. survey found only one, but the difference is not statistically significant.

**G. Bignami:** What fraction of the high-latitude sky has been searched in reasonably sensitive recent pulsar surveys?

**R. Manchester:** With the exception of some observations toward selected objects, surveys sensitive to pulsars with periods less than about 100 ms have been almost totally confined to within 15 degrees of the galactic plane. The 390 MHz survey of Dewey et al. (1985) covered 1.8 steradian of sky, mostly at high latitudes, but was only sensitive to pulsars with period greater than about 100 ms.

**J. Dolan:** Given the volume of phase space which has been investigated so far ($S_{min}$, $\nu$, Period, $\Omega$), and the number of pulsars discovered so far (390), can you estimate the total number of pulsars observable from the Earth with currently available instrumentation?

**R. Manchester:** On the assumption of a planar pulsar distribution, the total number of pulsars detected in a flux density-limited survey is proportional to the inverse of the limiting flux density. For relatively low-frequency surveys (~ 400 MHz say) the effective limiting flux density is dominated by high background temperature and interstellar scattering near the galactic plane. On the other hand, these surveys can cover a large area of sky more easily. The number of pulsars to be detected in the future will be largely limited by telescope time assignment committees and the stamina of the observers. I would guess that the number of pulsars discoverable in the future with currently available instrumentation will be about equal to the number currently known.

Millisecond Pulsar Surveys

D. C. Backer
Astronomy Department and Radio Astronomy Laboratory
University of California
Berkeley, CA 94720

ABSTRACT. In 1982 a new class of pulsars was defined by the discovery of a star with a millisecond rotation period, 1.6 ms. In the past 3.5 years two additional pulsars with millisecond periods have been discovered. The rapid spin of these pulsars is attributed to mass transfer in a low-mass binary progenitor system. This hypothesis is supported by the presence of companions in two of the three millisecond pulsars. These recent discoveries have led both to a deeper understanding of the final stages of stellar evolution in binary systems, and to closer ties between the observational study of neutron stars by radio, optical and X-ray techniques. In addition the millisecond pulsars provide precise astrophysical clocks that can be used to improve the solar-system ephemeredes and to search for a background of gravitational waves that may have been produced in the early stages of the visible universe. Old and ongoing searches for new millisecond pulsars are described in this paper.

## 1. INTRODUCTION

Three millisecond pulsars are presently known. Their periods range from 1.6 to 6.1 ms. Although the periods of all three objects are short, their ages probably exceed one million years since there is no sign of debris in their vicinity, and, in the case of the nearest one, the companion star is cool. In sharp contrast the next fastest pulsar, the Crab pulsar, is located in the midst of a brilliant nebula and is known from Chinese court records to have an age of 932 years (Wang 1987). Two of three millisecond pulsars have low-mass companions with orbital periods of 12 and 120 days. Consideration of the of low-mass binary evolution (van den Heuvel 1987 and references therein) provides a solid framework for understanding the existence of old pulsars with rapid rotation periods. Both observational evidence and theoretical discussions lend support to the identification of a subclass of pulsars – those with periods less than about 10 milliseconds.



*D. J. Helfand and J.-H. Huang (eds.), The Origin and Evolution of Neutron Stars, 13–21.*

Further progress in the understanding of the origin and evolution of fast pulsars will require deeper and more complete surveys than are presently available. Upon completion these surveys must be carefully analyzed to define the galactic population of fast pulsars following the early discussion by Gunn and Ostriker (1970) and the recent analysis by Lyne, Manchester and Taylor (1985). In the case of binary systems important information on the evolution of the system can be obtained from optical detection (or limits) of the companion as shown recently by Kulkarni (1986).

New millisecond pulsars will be important for the application of pulsar observations to other branches of science: timekeeping, solar system dynamics, celestial astrometry, interstellar medium, general relativity, and cosmology. Intense studies of the present millisecond pulsars have already contributed to these other subjects and new objects will certainly lead to further developments. In some areas new objects anywhere in space would do as long as they are bright; in other areas, such as timekeeping and cosmology, a distribution on the sky is needed. In all cases a survey needs only to be successful in new discoveries, since completeness is less important.

One area of inquiry was neglected in the preceding discussion: neutron star structure. This deserves special mention in this Symposium. Many of the interesting applications of millisecond pulsars are dependent on the stability of the rotation of the microwave beams attached to these neutron stars. While models of pulsar emission are still schematic, it is clear that stability of the beam rotation implies stability of the rotation of the neutron star to which it is attached; the reverse conclusion cannot be made with certainty. Despite the rapid rotation period of the 1.6-ms pulsar its observed rotation is extremely stable (Davis *et al.* 1985). This is remarkable since the neutron star has a considerable equatorial bulge which must decrease continually as the star spins down. If a 'glitch' were to appear in one of the millisecond pulsars, we would lose its use as a precise clock, but gain valuable insight into the structure of a neutron star in rapid rotation.

The three known millisecond pulsars were discovered by three different paths. The first, 1937+21, can be traced back to a peculiar complex of radio sources in the 4C catalog, 4C21.53 (Backer *et al.* 1982; Backer 1984). The second, 1953+29, was discovered by Boriakoff *et al.* (1983) by seaching one square degree fields centered on gamma ray source positions. The association of this pulsar with the target gamma ray source is questionable. If 1953+29 is not the gamma ray source, then the Boriakoff *et al.* search reduces to a deep general search of about four square degrees near the galactic plane. The most recent millisecond pulsar, 1855+09, was the result of a rapid survey of 800 square degrees along portion of the galactic plane that is visible with the sensitive Arecibo telescope (Seigelstein *et al.* 1986). The properties of these three pulsars are summarized in Table 1.

Table 1
Millisecond Pulsars

| PSR | Rotation Period (ms) | Period Derivative ($*10^{18}$) | Orbital Period (d) | Companion Mass ($M_o$) | Dispersion Measure (pc $cm^{-3}$) |
|---|---|---|---|---|---|
| 1855+09 | 5.436 | 0.16 | 12 | 0.3-0.4 | 13 |
| 1937+21 | 1.558 | 0.11 | – | – | 71 |
| 1953+29 | 6.133 | 0.03 | 120 | 0.2-0.4 | 105 |

## 2. SEARCH STRATEGIES

Pulsars have small radio luminosities, $10^{30}$erg $s^{-1}$, with logarithmic frequency indices steeper than -1, and are distributed throughout the galaxy with a scale height of 400 pc. As a result surveys for new pulsars would be most sensitive if conducted at low frequencies with wide bandwidths. The integration time would be as long as possible consistent with the need to cover the independent beam areas on the sky in a given amount of telescope time.

However the bandwidth is restricted by the dispersion of the signal resulting from interstellar plasma. In most surveys sensitivity is enhanced by sampling many channels whose bandwidth is proportional to the cube of the observing frequency and the maximum ratio of dispersion measure to period expected. The penalty one pays for multi-channel sampling is that the pulse search for each beam position requires a two-dimensional algorithm for which the computations increase with the square of the minimum period if the total bandwidth and maximum dispersion measure are fixed.

If the computational power of a search is fixed, then searches are nearly frequency independent: the loss of sensitivity with increasing frequency from the steep source spectrum is compensated by the increased allowed bandwidth per channel that results from reduced dispersion. Another important factor that contributes to the frequency insensitivity is the decrease in background noise from galactic synchrotron radiation as frequency increases. The penalty here is that as the frequency increases the beam solid angle decreases, and more observing time is required to cover the same search solid angle.

The discovery of millisecond pulsars has brought attention to yet another factor that limits pulsar surveys – the broadening of pulses that results from multipath propagation in the interstellar medium. An impulse of radiation from the star is received at the Earth as an exponential-shaped pulse with decay time of $\tau_s$ which varies with the fourth power of the radio wavelength and is a steep function of the dispersion measure. Figure 1 presents measured values of $\tau_s$ scaled to an observing frequency of 1 GHz as a function of the dispersion measure. The median broadening time between the plotted lines can be expressed as a function of the dispersion measure as follows:

$$\begin{aligned}\tau_s &= 10^{7.25}\text{ms}\ \nu^{-4} DM^2,\ 0 < DM < 125\ \text{pc cm}^{-3} \\ &= 10^{-10}\text{ms}\ \nu^{-4} DM^4,\ 125 < DM < 1000\ \text{pc cm}^{-3}\end{aligned}$$

As shown in Figure 1 individual objects can differ from the median by an order of magnitude in both directions. This figure clearly shows that surveys for millisecond pulsars out to modest dispersion measures must choose an observing frequency near, or above, 1 GHz. Search algorithms look for as many as 16 harmonics of the pulsar frequency, and it is the highest harmonic, or smallest fraction of the period, desired that must not be smeared by interstellar scattering. The survey by Clifton and Lyne (1986) at Jodrell Bank using 1.4 GHz provides an example of the importance of higher observing frequency. The higher frequency does not limit the detection capability for comparable observing time, as discussed above. However the beam solid angle is reduced and as a result the time required to complete a survey is increased.

The detectability of a millisecond pulsar in a fast binary system can be limited by the Doppler acceleration during an integration time. If we consider a circular orbit for a binary of two 1.4 $M_o$ neutron stars, then there is a limiting product of rotational period and orbital period below which the signal will be Doppler spread in a Fourier analysis. Figure 2 shows the known binary pulsars in a rotation period-binary period diagram. Three limiting lines are displayed for integration times of 100, 300 and 1000 s. These lines were drawn under the assumption that the fourth harmonic of the pulsar rotation frequency is smeared by the inverse of the integration time. The most interesting binary 1913+16 is within the region where existing surveys are compromised by Doppler acceleration during the integration. Similar limits apply to the low-mass binary companion case. The three lines in Figure 2 correspond to integration times of approximately 30, 100 and 300 s, respectively, for the low–mass case. In principle an algorithm could be devised to search in acceleration along with period and dispersion. To date no survey has explored this domain.

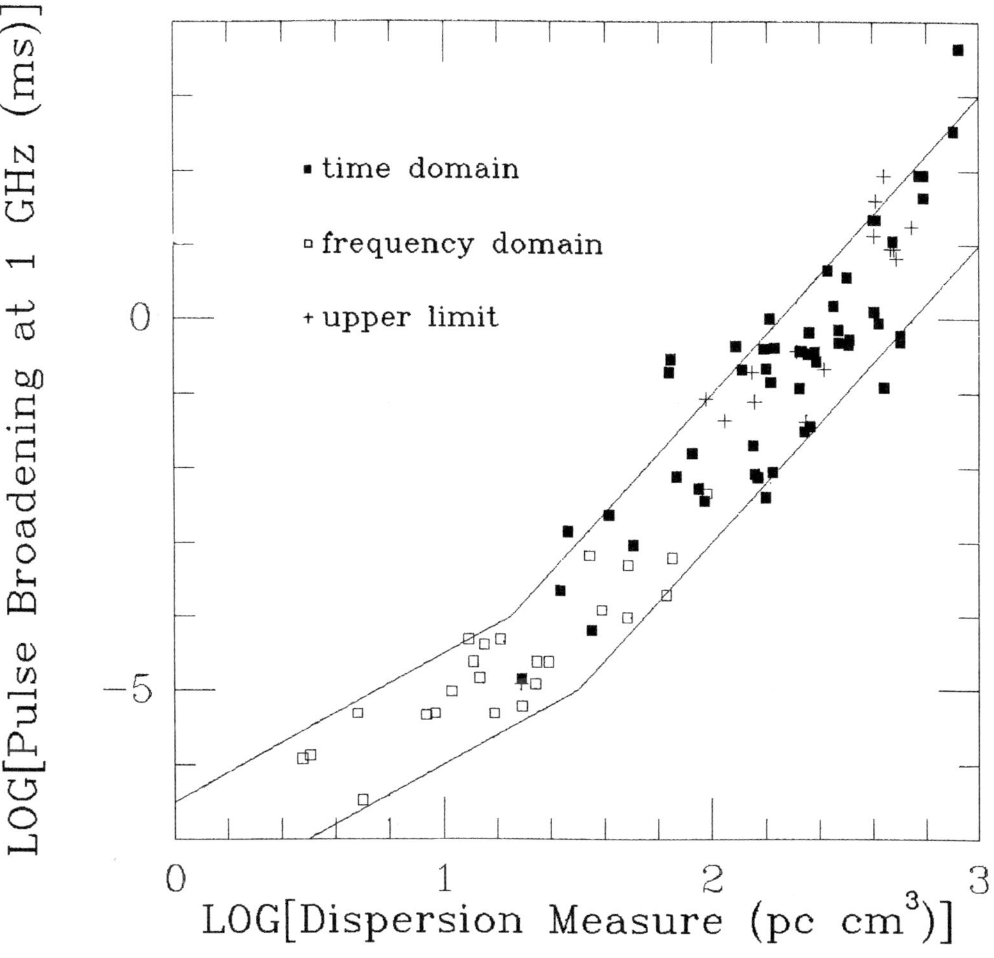


Figure 1 Pulse broadening by interstellar scattering as a function of the dispersion measure at 1 GHz observing frequency. Open squares are from measures of the decorrelation bandwidth and filled squares are from direct deconvolution of average pulse profiles. Plus symbols are upper limits to scattering time.

The difficulties in surveying the sky for millisecond pulsars discussed above led a number of investigators to try continuum techniques for identifying millisecond pulsar candidates. Several continuum techniques were suggested by aspects of the discovery story of 1937+21. One peculiarity of 4C21.53 was its small size indicated by interplanetary scintillation in spite of its low galactic latitude. Purvis *et al.* (1984) pursued other low-latitude 'scintars' without success. The steep meter-wavelength spectrum of 4C21.53 was also peculiar, and other compact sources with steep spectra have been identified by Erickson. While one or two of these remain interesting, no pulsars have been identified. Interstellar scintillation was observed from 1937+21 before it was known to be a millisecond pulsar. As a result a rapid survey of 3400 square degrees was conducted by Stevens *et al.* (1984) to identify scintillating sources that might be fast pulsars. Again they found none. The strong linear polarization of 1937+21 was early evidence in favor of its pulsar identification. Several surveys have now been conducted trying to identify at long wavelengths compact and highly polarized sources for subsequent pulse searches. Stevens (1986) has used the Westerbork array at 327 MHz to identify pulsar candidates, and then the Arecibo telescope to search for pulses. Lyne (1986) has used Jodrell Bank instruments at 408 MHz for a similar search. Neither search has led to new pulsars. A search for compact sources at the centers of globular clusters was conducted by Hamilton, Helfand and Becker (1985) based on the proposed connection between millisecond pulsars and the low-mass Xray binaries. One source in M28 remains as a potential hidden pulsar that is too weak or too scattered for present pulse detection techniques.

## 3. NEW MILLISECOND PULSAR SURVEYS

The difficult task of sifting through the multidimensional haystack of possibilities for millisecond pulsars – two sky coordinates, dispersion measure (distance) and period – has just begun. The first surveys that have reached the sensitivity capable of detecting a significant galactic population of millisecond pulsars are summarized in Table 2 below. The next year or so will lead to much firmer evidence on the number and possibly the distribution of these exotic stars.

Table 2
Millisecond Pulsar Surveys

| Investigators | Telescope (MHz) | Frequency Time (ms) | Sampling Angle* | Solid Time (s) | Integration |
|---|---|---|---|---|---|
| Seigelstein & Taylor | Arecibo | 430 | 0.3 | 10x30 | 39 |
| Manchester & D'Amico & Taylor | Molonglo | 843 | 0.5 | 2x105 | 130 |
| Lyne | Jodrell Bank | 610 | 0.3 | 20x110 | 79 |
| | | 1400 | 0.3 | 2x110 | 79 |
| Backer & Kulkarni & Clifton | Green Bank | 825 | 0.25 | 20x150 | 32 |
| | Deep Space Net | 2300 | 0.5 | 2x100 | 64 |

* sky coverage listed in degrees of latitude by degrees of longitude.

REFERENCES

Backer, D. C., Kulkarni, S. R., Heiles, C., Davis, M. M., and Goss, W. M. 1982, *Nature*, **300**, 615.

Backer, D. C. 1984, *J. Astron. Astrophys.*, **5**, 187.

Hamilton, T. T., Helfand, D. J., and Becker, R. H. 1985, *Astron. J.*, **90**, 606.

Boriakoff, V., Buccheri, R., and Fauci, F. 1983, *Nature*, **304**, 417.

Clifton, T. R., and Lyne, A. G. 1986, *Nature*, **320**, 43.

Davis, M. M., Taylor, J. H., Weisberg, J. M., and Backer, D. C. 1985, *Nature*, **315**, 547.

Gunn, J. E., and Ostriker, J. P. 1970, *Ap. J.*, **160**, 979.

Kulkarni, S. 1986, preprint.

Lyne, A. G. 1986, personal communication.

Lyne, A., Manchester, R. N., and Taylor, J. H. 1985, *M.N.R.A.S.*, **213**, 613.

Purvis, A., *et al.* 1984, NRAO Workshop on 'Millisecond Pulsars', edited by S. P. Reynolds and D. R. Stinebring, p. 252.

Seigelstein, D.J., et al., 1986, *Nature*, in press.

Stevens, M. *et al.* 1984, NRAO Workshop on 'Millisecond Pulsars', edited by S. P. Reynolds and D. R. Stinebring, **p. 242**.

Stevens 1986, Ph. D. thesis, University of California, Berkeley.

Wang, Z. R. 1987, IAU Symposium 125, this volume.

van den Heuvel, E. P. J. 1985, IAU Symposium, this volume.

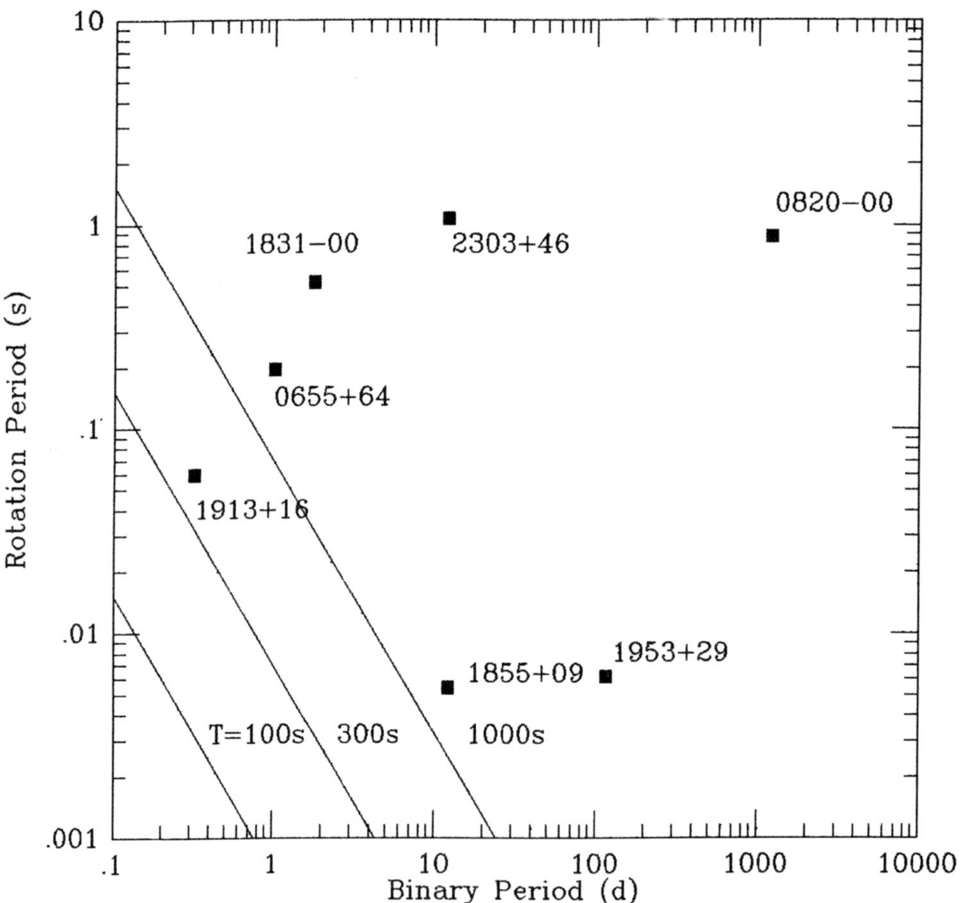


Figure 2 Binary pulsars in a rotation period - orbital period diagram. Slanting lines in lower left corner represent upper limits to parameters of pulsars detectable with integration times marked under the assumptions discussed in the text.

## DISCUSSION

**S. Kulkarni:** Regardng your point about loss of sensitivity towards fast, tight binaries I would like to just say that in principle one can circumvent the loss in sensitivity by including a variety of acceleration either before the Fourier transform or after it. Probably the increase in computing is no more than a factor of 10.

**D. Backer:** I agree that there are direct methods for adding a search in acceleration to the search in period, dispersion and celestial position.

# THE KINEMATICS OF PULSARS

A.G. Lyne
University of Manchester
Nuffield Radio Astronomy Laboratories
Jodrell Bank
Macclesfield
Cheshire SK11 9DL, United Kingdom

ABSTRACT. Pulsars have a galactic radial distribution similar to that of many galactic populations such as HII regions, massive stars and supernova remnants. However they are generally much further from the plane of the Galaxy than these objects. Proper motion measurements sho that this is because they are typically moving with high velocities. The measurements also indicate that most pulsars were formed a few million years ago close to the plane, within the normal Population I regions. Some pulsars will escape from the Galaxy, although the majority will end up in a halo population. The origin of the high velocities is not clear at present but may be due either to some asymme try in the formation event or to the disruption of a close binary system.

## 1. INTRODUCTION

The galactocentric radial distribution of pulsars (Lyne, Manchester and Taylor 1985) shows a falling surface density beyond the Sun's position and an increasing density towards the inner regions of the Galaxy. This is qualitatively similar to the distribution of many galactic Population I species of objects, notably massive stars, HII regions where massive stars are likely to be formed and supernova remnants, the likely birthplaces of pulsars.

These objects all have quite narrow galactic z-distributions, typically falling to half density at 50-100 pc from the plane. Pulsars on the other hand (Figure 1) have a half-density height of about 400 pc Gunn and Ostriker (1970) suggested that this might be due to a general migration from the galactic plane because of high velocities imparted to pulsars at birth. It remained for direct observation of the velocities to confirm this explanation.

This paper describes how we measure the motions of pulsars and what we can infer about their evolution, about their progenitor population and the origin of the high velocities.

## 2. METHODS OF MEASUREMENT

The three main methods of determining the velocities of pulsars involve pulse timing measurements, direct radio interferometry and



*D. J. Helfand and J.-H. Huang (eds.), The Origin and Evolution of Neutron Stars, 23–33.*

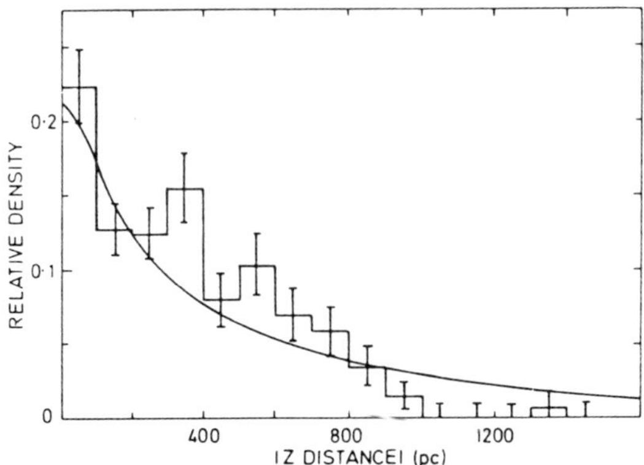


Figure 1. The | z |—distribution of pulsars.

measurements of interstellar scintillation. The first two methods determine proper motions which, combined with distance estimates, yield pulsar velocity vectors and it is these two which this paper will address. In the third method the motion of the scintillation pattern across the Earth is inferred and hence an estimate of the pulsar transverse speed obtained. This method is reviewed by Cordes (this volume).

## 2.1 Pulse Timing Measurements

The residuals of the observed times of arrival of pulses from a simple model of pulsar spin-down are very sensitive to the assumed position of the pulsar because of the annual barycentric correction term of about 500 seconds. The sensitivity is such that a 1 arcsec position error gives rise to an annual sinusoid in the residuals of up to 2.5 milliseconds. By fitting such sinusoids to the data, positions accurate to a small fraction of an arcsec can be obtained. From observations over several years, proper motions can be derived.

The first measurement of proper motion was carried out in this manner by Manchester, Taylor and Van (1974) and this illustrates the technique very nicely. Figure 2 shows an annual sinusoid increasing linearly with time as the pulsar moves away from the assumed position.

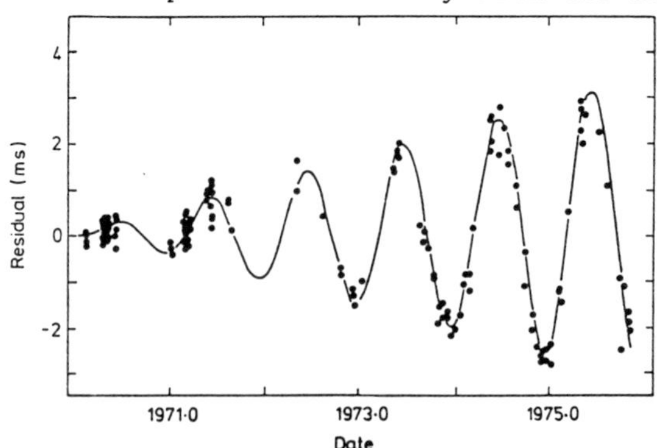


Figure 2. The timing residuals for PSR 1133+16 for an assumed constant pulsar position (Manchester, Taylor and Van 1974).

Such measurements have been carried out on a total of about 16 pulsars by Gullahorn and Rankin (1978) and by Helfand, Taylor, Backus and Cordes (1980), providing measurements with typical errors of 50 milliarcsec/year.

## 2.2 Radio interferometry

This method is akin to the conventional optical determination of proper motion which involves the measurement of motion relative to stationary background objects. In fact optical techniques have already been applied to the determination of the proper motion of the Crab pulsar (Wyckoff and Murray 1977).

In the radio case the measurement is relative to one or more nearby small diameter radio sources, usually lying within the primary beam of the instrument. At radio frequencies around 400 MHz, the ionosphere causes the image of the pulsar to move by up to several arc-seconds on timescales of a few minutes. However the reference source moves in a very similar way and relative positions can be measured with great accuracy. Figure 3 illustrates the motions of two pulsars

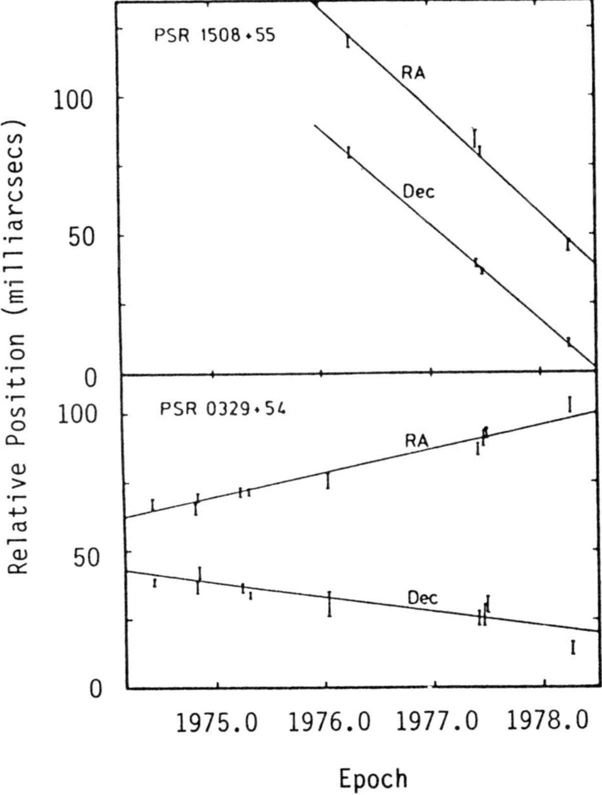


Figure 3. The positions of PSR 0329+54 and PSR 1508+55 relative to their reference sources as a function of epoch.

relative to nearby reference sources over periods of several years and shows that errors in position of only a few milliarcsec can be achieved (Lyne, Anderson and Salter 1982).

Two main experiments of this kind have been carried out to date: one at Jodrell Bank at 408 MHz (Lyne et al. 1982) on a total of 26 sources and one at Green Bank at 2700 MHz (Sramek and Backer 1981) on 5 sources, all of which were included in the Jodrell Bank list.

### 2.3 Comparison of measurements

The interferometric results from the Jodrell Bank and Green Bank experiments were in satisfactory agreement for all 5 sources in common. However, while some of the timing results also agreed with these results, for others the agreement was very poor indeed. In view of the consistency of the interferometric measurements we have to conclude that the timing measurements are prone to an unrecognized source of error. It seems that this error comes from the timing noise which arises in the irregularities in the rotation rates of pulsars. This timing noise may have significant spectral components with period of a year and these can masquerade as an apparent shift in the position of the pulsar. Although there are doubts about the general reliability of timing determinations of proper motion, for very stable pulsars such as the millisecond pulsar 1937+21, it seems that very precise measurements can be made (Davis et al. 1985).

The observations discussed here are therefore confined to the 26 pulsars contained in the list of Lyne et al. (1982) which encompass all the pulsars in the other experiments and also had rather smaller errors.

## 3. THE VELOCITIES

As suggested by Gunn and Ostriker (1970), the velocities are indeed high. The root mean square transverse velocity for the sample of 26 pulsars is 170 km/s implying that the mean space velocity is somewhat over 200 km/s. This velocity is an order of magnitude greater than the random velocities of most normal stellar populations. I will discuss the possible origin of these high velocities later.

## 4. THE MIGRATION FROM THE GALACTIC PLANE

The directions of the high velocities as seen on the sky are not random. Figure 4 shows how the directions of the velocity vectors are disposed relative to the galactic plane. For most of those objects which lie more than 50 or 100 pc away from the plane the motion is directed away from the plane confirming the picture of Gunn and Ostriker. The mean migration velocity from the plane is about 124 km/sec. The two pulsars which appear to be moving toward the galactic plane are both situated at high latitude so that their apparent motion on the sky is dominated by the component of velocity parallel to the plane. They may indeed have a velocity component directed away from the plane but this is mostly hidden in the unknown radial component of velocity. Most of the sources lie at low galactic latitudes where this effect is not important. Those pulsars which we see close to the galactic plane are mostly either young ones or those which have velocities parallel to the plane.

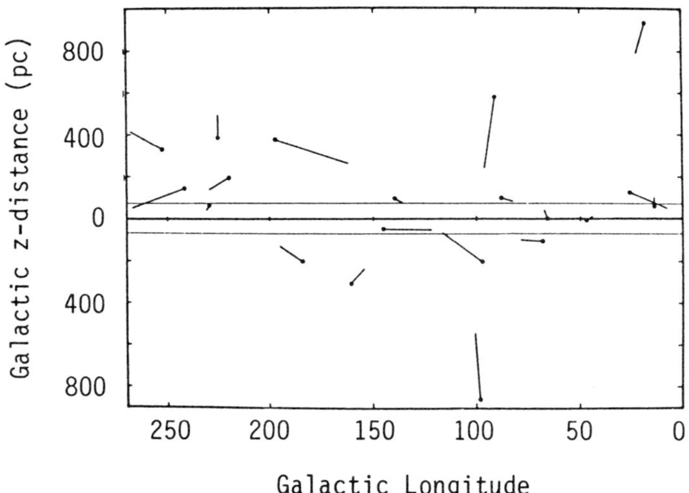


Figure 4. The positions and velocities of 20 pulsars relative to the galactic plane. The filled circles represent the observed positions of the pulsars as a function of galactic z-distance and longitude and the tails represent the distances travelled in the last million years.

## 5. KINETIC AGES

Accepting the model of pulsars being born on the galactic plane within a narrow Population I type of distribution and then given a high velocity at birth, extrapolation of their motions back to the time when they would have left the plane provides an estimate of the age of the pulsar, known as the kinetic age of the pulsar. There are some uncertainties in this estimate because the precise position of birth within the progenitor layer is not known and also because the radial component of velocity is not known. The latter means that the precise form of the trajectory may be poorly determined, although this effect is not important for the majority of pulsars in the sample as they lie at low latitudes.

Figure 5 shows the comparison of these kinetic ages with the characteristic ages $T=P/2\dot{P}$ which are determined from the spin-down rates of the pulsars. The characteristic age equals the true age if the pulsar is born with a very short period and the effective magnetic dipole moment of the neutron star remains constant. For pulsars with ages of less than a few million years the agreement is reasonably satisfactory. However for older objects, it is clear that the characteristic ages are substantial over-estimates of the 'true' ages as revealed by the kinetic ages. This discrepancy is most satisfactorily explained in terms of a decay in the effective braking torque of the pulsar, which may be due to decay of the magnetic field or alignment of the magnetic and rotation axes. The lines in Figure 5 indicate how a pulsar will evolve for various time-scales of exponential decay of braking torque. The data in this diagram suggests that the decay occurs on a time-scale of a few million years. This is in reasonable agreement with the

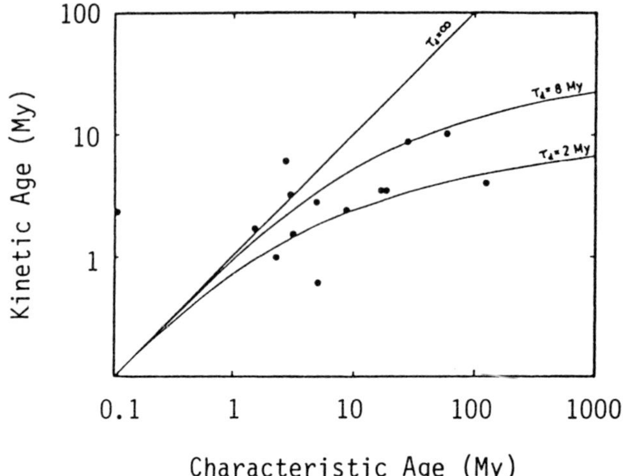


Figure 5. The kinetic and characteristic ages for 14 pulsars. The three lines show the paths expected for pulsars whose effective magnetic fields decay on the time-scales indicated.

time-scale of 9 million years obtained by Lyne, Manchester and Taylor (1985) using independent arguments.

## 6. THE PULSAR z-DISTRIBUTION AND LUMINOSITY DECAY

Lyne, Manchester and Taylor (1985) also provide evidence that the decay of effective magnetic field is accompanied by an exponential fall in luminosity on a timescale of about 4.5 million years. This assertion can now be checked by considering a model in which pulsars are born on the galactic plane with the observed velocity distribution and then suffer luminosity decay. This model provides a prediction of the steady-state z-distribution of pulsars exceeding a given radio luminosity. This distribution is compared with the experimentally determined one in Figure 1 which shows a satisfactory agreement between the two, indicating that the luminosity does indeed decay on that sort of timescale.

## 7. THE GALACTIC DISTRIBUTION OF DEAD PULSARS

The escape velocity from the Galaxy is about 300 km/s at the position of the Sun so that only about 10-20% of pulsars will leave the gravitational well of the Galaxy completely. The majority will continue their motion away from the plane long after they cease being radio pulsars and will reach maximum heights of several kpc above the plane after 30-60 million years. They will thus form a halo population containing between about $3 \times 10^8$ and $3 \times 10^9$ 'dead' neutron stars, the number depending upon rate of formation of neutron stars in the earlier life of the Galaxy.

## 8. THE ORIGIN OF THE HIGH VELOCITIES

Although a number of mechanisms have been proposed for the acceleration of pulsars to the high velocities we observe, it is not clear which one or ones are responsible, since no single explanation by itself is entirely satisfactory (see for example Anderson and Lyne 1983). In this review I consider the three most likely possibilities.

The first mechanism is that first proposed by Blaauw (1961) to explain the high velocities of the runaway stars, involving the disruption of a binary system by a symmetrical supernova explosion of the more massive component. If more than half of the mass of the system is lost in a time interval short compared with the orbital period, then the system will be disrupted. Any stellar remnant of the supernova will have a velocity whose magnitude is the same as its pre-supernova orbital velocity (Radhakrishnan and Shukre 1985). It seems that velocities comparable with those observed for pulsars could be produced by the disruption of appropriately close binary systems. Although many stars are formed in binaries, a substantial fraction are born as solitary stars. If these undergo symmetrical supernova explosions in the formation of pulsars, then there should be a class of low-velocity pulsars still lying within their progenitor population, presumably at small z-distances. There is no evidence (Figure 1) for such a population, any which does exist containing no more than about 20% of the known pulsars, despite the fact that a greater proportion of massive stars than this are probably solitary. Similarly there is no evidence of a bimodal distribution of transverse velocities as one might expect, although here the statistics are rather poor.

I think we must accept nevertheless that at least some high pulsar velocities are likely to be derived from this mechanism. For instance, the Crab pulsar has a transverse velocity of 100 km/s and is moving away from the galactic plane. However it is only 1000 years old and hence lies very close to its birthplace in the Crab Nebula at a z-distance of about 200 pc, well outside the proposed progenitor Population I region. It could well have resulted from the supernova of a runaway star which was originally in a binary system which was disrupted by the supernova of a more massive companion (Gott, Gunn and Ostriker 1970).

There is one rather puzzling piece of evidence first noted by Helfand and Tademaru (1977) which cannot be easily explained by a binary formation scenario, namely the apparent correlation of the measured transverse velocities with the pulsar magnetic field shown in Figure 6 (Anderson and Lyne 1983). One possibility (Helfand and Tademaru 1977 and Radhakrishnan 1984) is that there is no causal relationship here, but that there are two classes of pulsars, a high magnetic field, high velocity one and a low magnetic field, low velocity one. Perhaps there is some systematic difference in the magnetic moments of solitary and binary stars, so that the binary stars have much larger magnetic fields, leading to the observed correlation.

Another possible mechanism, proposed by Harrison and Tademaru (1975), involves asymmetric dipole radiation due to an off-axis magnetic moment. The resulting acceleration occurs during the first few months of the existence of the pulsar while the rotation rate is high, initial periods of a few milliseconds being required. It turns out that one would not expect any dependence of the final velocity upon the magnitude of the pulsar dipole moment, contrary to what is observed. A prediction of this model for the acceleration is that the proper motion will be

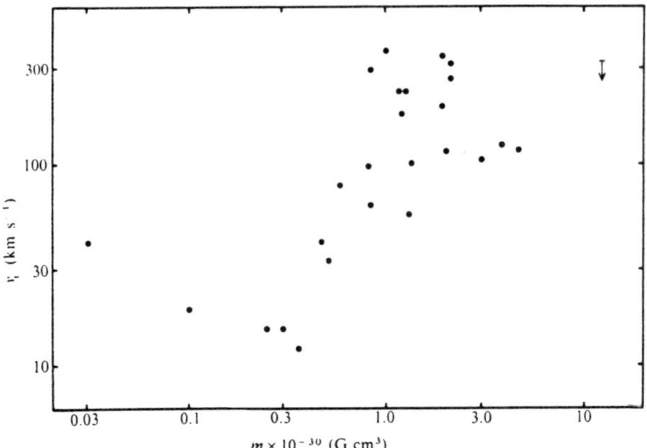


Figure 6. The transverse velocities of 26 pulsars as a function of the effective magnetic dipole moment (Anderson and Lyne 1983).

parallel to the rotation axis, whose direction should be revealed by the position angle of the linear polarization near to the centre of the radio pulse (Morris, Radhakrishnan and Shukre 1976). No clear relationship is observed. A further argument against this mechanism is that, although it may be able to explain the velocities of pulsars born in single stars, by itself it would be unable to disrupt most close binary systems, the main result being that the whole system would probably be dragged along. The small number of binary systems containing radio pulsars compared with the large number of stars in such systems suggests that the acceleration mechanism must also disrupt most systems.

The third mechanism is that the supernova explosion itself is asymmetric, giving a velocity impulse of the order of 200 km/s to the neutron star remnant (Shklovskii 1970). As far as I am aware there is no real theoretical basis for believing that this might or might not happen. Such an impulse would certainly disrupt most binary systems and explain why only about one percent of known pulsars are in binaries. Possible sources of asymmetry include the magnetic field configuration and gravitational distortion of the pre-supernova star by a binary companion. I think we are in need of more theoretical work on the possibilities for such asymmetrical events.

## 9. FUTURE OBSERVATIONS

Observations of the proper motions of pulsars can give much information on their physics and history. However, the presently available data are rather sparse and some are on local high latitude objects for which the kinetic age errors are likely to be large due to the unknown radial velocity component. A number of new programs are under way and these should provide 60 or 70 new proper motion determinations within the next three years or so. The main instruments being used for this work are MERLIN, the VLA and the Parkes-Tidbinbilla interferometer in Australia. These results should provide a more precise study of the magnetic field decay in neutron stars and also allow investigation of the nature of the magnetic field, velocity relationship.

REFERENCES

Anderson, B. and Lyne, A.G., 1983. Nature, 303, 597.
Blaauw, A., 1961. Bull. Astron. Inst. Netherlands, 15, 265.
Davis, M.M., Taylor, J.H., Weisberg, J.M. and Backer, D.C., 1984. Proceedings of Green Bank Workshop on Millisecond Pulsars.
Gott, J.R., Gunn, J.E. and Ostriker, J.P., 1970. Astrophys. J., 160, L91.
Gullahorn, G.E. and Rankin, J.M., 1978. Astrophys. J., 225, 963.
Gunn, J.E. and Ostriker, J.P., 1970. Astrophys. J., 160, 9.
Harrison, E.R. and Tademaru, E.P., 1975. Astrophys. J., 201, 447
Helfand, D.J. and Tademaru, E.P., 1977, Astrophys. J., 216, 842.
Helfand, D.J. Taylor, J.H., Backus, P.R. and Cordes, J.M., 1980. Astrophys. J., 237, 206.
Lyne, A.G., Anderson, B. and Salter, M.J., 1982. Mon. Not. R. astr. Soc., 201, 503.
Lyne, A.G., Manchester, R.N. and Taylor, J.H., 1985. Mon. Not. R. astr. Soc., 213, 613.
Manchester, R.N. and Taylor, J.H., 1977. Pulsars. (Freeman, San Francisco).
Manchester, R.N., Taylor, J.H. and Van, Y.Y., 1974. Astrophys. J., 189, L119.
Morris, D., Radhakrishnan, V. and Shukre, C.S., 1976. Nature, 260, 124.
Radhakrishnan, V., 1984. Proceedings of Green Bank Workshop on Millisecond Pulsars.
Radhakrishnan, V. and Shukre, C.S., 1985. Proc. of Workshop on Supernovae, their Progenitors and Remnants (Indian Acad. Sci., Bangalore, India), 155.
Shklovskii, I.S., 1970. Astr. Zh. 46, 715.
Sramek, R.A. and Backer, D.C., 1981. IAU Symp. No. 95, Pulsars, Reidel, Dordrecht, Holland, p.205.
Wyckoff, S. and Murray, C.A., 1977. Mon. Not. R. astr. Soc., 180, 717.

## DISCUSSION

**G. Bisnovatyi-Kogan:** How many old pulsars do exist within 100 parsec of the Sun? I mean the low-velocity pulsars accumulated in the galactic plane.

**A. Lyne:** Assuming that the formation of massive stars and hence pulsars has occurred at a constant rate for the life of the Galaxy, then about 5000 pulsars will have been formed within about 100 pc of the Sun. However, these are now spread over a scale height of a few kpc, so that I estimate that there are only about 200 dead pulsars within 100 pc of the Sun.

**S. Kulkarni:** In addition to the three methods for determining proper motion which you mentioned (interestellar scintillation, timing and radio interferometry) I would like to add another technique which works in the fortunate case where the pulsar system is identified at optical wavelengths. From optical data we hope to obtain proper motion of 1855+09, 0655+64, 0820+02 and Vela by next year. (In the first 3 cases, the optical candidate is the secondary white dwarf and in the last case the optical emission comes from the pulsar itself).

**A. Lyne:** Indeed. However, I think you are optimistic about the timescale!

**R. Manchester:** Could you comment on the importance of selection effects for the main conclusions you have drawn from the proper motion measurements?

**A. Lyne:** On the whole, selection effects do not seriously affect the main conclusions I have described. The most obvious one is that which reduces the observed number of low magnetic moment and hence low luminosity pulsar which have high velocities in the $(m, \upsilon_t)$ diagram (Anderson and Lyne 1983).

**J. Arons:** First a comment: A discrepancy between the $P/\dot{P}$ age and the kinematic age reflects a decay of the torque; such torque decay might or might not involve decay of the light cylinder field. I know of no model of surface field decay (or growth) which I find acceptable. It would be nice if the language more accurately reflected the observations, and one referred to torque decay, rather than biasing one's thoughts by referrring to field decay. Question: Does the luminosity decay smoothly, or is there a sharp cutoff?

**A. Lyne:** There is no evidence for anything other than a smooth decay of luminosity except in the small minority of pulsars which display nulling. In this group there may be a more sudden decrease in luminosity towards the end of their radio lives.

**R. Becker:** You should be careful in assuming that pulsars are born within 70 pc of the galactic plane. Half of the Crab-like SNR, which are indicators of where pulsars are born, are more than 75 pc from the plane.

**A. Lyne:** The proper motion measurements provide evidence that the acceleration to high velocity occurs within about 100 pc of the Galactic plane. This is the main justification in the subsequent discussion for assuming a Population I progenitor distribution.

**R. Becker:** Well, then, perhaps pulsars which form Crab-like SNR are a different population from most pulsars.

**A. Lyne:** I would not argue with that comment, although maybe the acceleration to high velocity still occurred at low z-distance in the disruption of a binary system by a supernova event. The pulsar was then formed some time after this from the supernova of the runaway star.

**A. Blaauw:** In the selection of pulsars for future proper motion programs, may I suggest that due attention be given to the nearby pulsars, say within 1 kpc, for these reasons:

- The direction of the (larger) proper motion is better determined thus narrowing down the choice of possible domains of the progenitor population.
- Only within these distances may we hope to obtain a reasonably complete inventory of the masses and evolutionary stages of the (B star) progenitor population.

# INTERSTELLAR SCINTILLATIONS AND NEUTRON STAR KINEMATICS

J.M. Cordes
Astronomy Department
Cornell University
Ithaca, NY 14853
U.S.A.

ABSTRACT The interstellar scintillation technique for measuring neutron star speeds is described and results are given for 71 radio pulsars. The mean transverse neutron star speed is 100 km $s^{-1}$ and the distribution extends to 300 km $s^{-1}$. The transverse speed correlates with the z velocity derived independently using distance from the galactic plane, consistent with most neutron stars having been born near the galactic plane. A correlation of transverse speed with the quantity $P\dot{P} \propto$ (magnetic moment)$^2$ is a general property of the neutron star population. Monte Carlo simulations of the progenitors of neutron stars show that the velocity distribution is inconsistent with the disruption of binary systems solely by *symmetric* supernova explosions. Either explosions are asymmetric or there are additional accelerations of neutron stars after their formation.

## 1. INTRODUCTION

As with any stellar population, the space velocities of neutron stars are informative about the processes by which the objects are formed. Radio pulsars move (translationally) roughly 10 times faster than their progenitor stars. The mechanisms which may explain the large velocities include the disruption of compact binaries by supernova explosions (Zwicky 1957; Blaauw 1961; Radhakrishnan 1984); asymmetry of supernova explosions (Anderson and Lyne 1983); and asymmetric magnetic dipole radiation (Harrison and Tademaru 1975). At present there is no concensus as to whether one or more of these processes dominates the situation. To make progress, a large sample of velocities is necessary to establish the generality of any derived distribution and to establish correlations with other pulsar parameters. This review discusses such a sample of measurements and its implications.

There are three methods for measuring transverse velocities of radio pulsars. These are (1) direct proper motion measurements using interferometry; (2) pulse timing measurements; and (3) interstellar scintillations, measurement of which yield the motion of a diffraction pattern which, in turn, is related to the neutron star speed. Lyne (this volume) has outlined the first two methods. In this paper I review the scintillation method, describe results obtained on a large sample of radio pulsars, and draw conclusions about mechanisms contributing to the space velocities.



*D. J. Helfand and J.-H. Huang (eds.), The Origin and Evolution of Neutron Stars, 35–46.*

## 2. INTERSTELLAR SCINTILLATIONS

Figure 1 shows the relevant geometry for scintillations. Pulsars are the most compact sources known in radio astronomy and are emitters of spatially coherent (but temporally incoherent) radiation. Propagation through turbulent plasma in the interstellar medium (ISM) distorts the wavefronts according to refractive index fluctuations described by the electron density and the cold plasma dispersion relation. Birefringence is negligible (Simonetti, Cordes, and Spangler 1984). Under the condition that the phase difference between wavefront components arriving from different angles is more than one radian, a deeply modulated diffraction pattern results, through which the Earth moves. The diffraction pattern is random but has well determined characteristic spatial and frequency scales even though the electron density itself appears to have no well defined spatial scale. In fact, the electron density is consistent with a Kolmogorov (power law) wavenumber spectrum (Armstrong, Cordes, and Rickett 1981; Cordes, Weisberg, and Boriakoff 1985). Diffraction scales are well determined because these are determined by the spatial scale in the medium that produces 1 radian of phase variation.

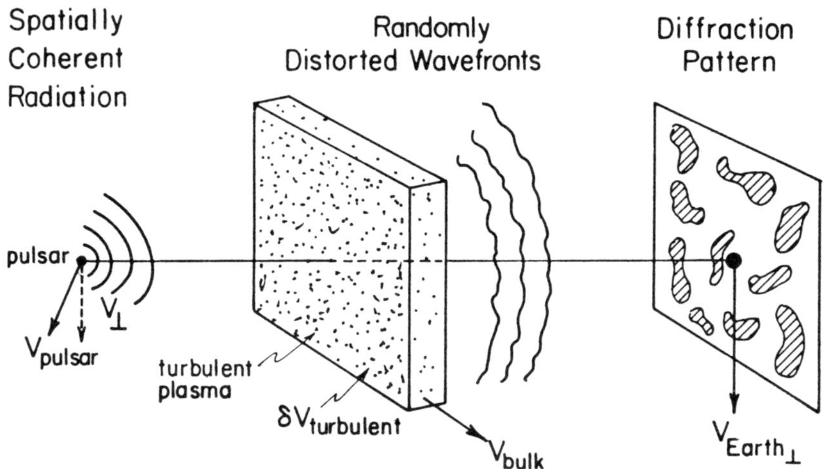


**Figure 1** Geometry for scattering of pulsar radiation.

Interstellar scattering is a topic of considerable interest because effects are manifest in many kinds of measurements. These include temporal broadening and time-of-arrival variations of pulsar pulses ($10^{-6}$ to 1 sec); angular broadening and angle-of-arrival variations of images of pulsars and active galactic nuclei ($10^{-6}$ to 1 arc sec); refractive intensity variations on time scales of days to years (Rickett, Coles, and Bourgois 1984); and the diffractive intensity variations that are of interest here and which have time scales of seconds to hours.

### 2.1 Observables

Measurements of intensity as a function of time and frequency ('dynamic spectra') show $\sim$ 100% modulations having characteristic time and frequency scales that are related to the spatial scale of the diffraction pattern. Dynamic spectra are usually obtained with an autocorrelation spectrometer (or sometimes an analog filter bank) that is gated synchronously with the periodic pulsar signal (for signal to noise enhancement), using a total bandwidth of 0.1 to 10 MHz, integration times of 10 sec, and total observation times between 20 min and 2 hr. Typical radio frequencies are from 0.3 to 1.4 GHz.

Dynamic spectra show fluctuations that are dominated by interstellar scintillations but also include intrinsic pulsar variations; interplanetary and ionospheric scintillations; and any gain variations of the telescope/receiver. All of these additional effects are broadband and therefore produce only additional *temporal* modulations of the intensity over the spectrometer bandwidth. By using different calibration schemes, we have shown that these additional effects have little influence on the velocities discussed below.

Estimation of the characteristic time and frequency scales is done by calculating the two dimensional intensity autocovariance function (ACV). Letting $\delta I(\nu, t)$ be the deviation of the intensity from the mean, the intensity ACV is $\langle \delta I(\nu, t) \delta I(\nu + \delta\nu, t + \delta t) \rangle$, where angular brackets denote averaging over the receiver bandwith and over time. Figure 2 shows a representative ACV. The widths (HWHM) of the slices along the time and frequency lag axes are the characteristic scintillation scales. These are denoted $\Delta t_d$ and $\Delta \nu_d$ where the subscript d implies the quantities are 'diffractive' in origin.

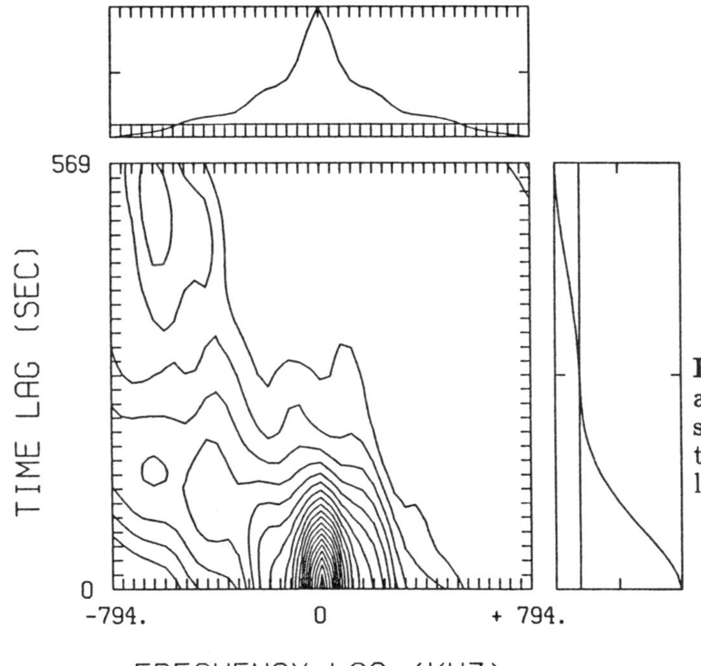


**Figure 2** Intensity autocovariance of dynamic spectra with slices along the time lag and frequency lag axes.

## 2.2 Theoretical Relations

Let $\theta_d$ be the characteristic diffraction angle and D be the distance to a pulsar. The diffraction pattern has a spatial scale $\ell_d \sim \lambda/(2\pi\theta_d)$ which is (for radio frequencies of interest) much less than the Fresnel scale $\sqrt{\lambda D} \sim 10^{11}$ cm. The diffractive frequency scale is $\Delta\nu_d \sim 2c/(D\theta_d^2) \ll \nu$. Temporal variations are caused by two effects: 1) convection of the diffraction pattern through the line of sight and 2) true time variations that would be seen if all bulk velocities are zero and would be due to turbulent motions of the medium. The velocities that must be considered include:

1. Pulsar velocities — $V_\perp = 0 \rightarrow 300$ km s$^{-1}$
2. Interstellar medium
   a. bulk (galactic rotation) — $V_{bulk} = 0 \rightarrow 150$ km s$^{-1}$
   b. random (turbulent) — $\delta V_{turbulent} \leq 10$ km s$^{-1}$
3. Earth orbital motion — $V_{Earth_\perp} \leq 30$ km s$^{-1}$

Differential galactic rotation is important for interferometric or timing determinations of proper motion (for distant pulsars), but is negligible for scintillation measurements (Cordes 1986; discussed below). Turbulent motions potentially affect all three kinds of measurements but in practice are probably negligible. Evidently, refractive image wandering or temporal wandering is too small to strongly influence interferometric and timing measurements.

For scintillations, we argue that convection is the dominant cause of time variation and that turbulent speeds in the ISM must be small compared to the average neutron star speed. If turbulent motions are negligible, the diffractive time scale is simply $\Delta t_d \sim \ell_d/V_\perp$, where $V_\perp$ is an effective velocity. Thus an estimator for this velocity is

$$V_\perp \equiv \frac{\ell_d}{\Delta t_d} \sim \frac{\sqrt{cD\Delta\nu_d}}{2\pi\nu\Delta t_d} \sim 10^{4.1} \text{ km s}^{-1}(\Delta\nu_\mathrm{d}\mathrm{D}_\mathrm{kpc})^{1/2}(\nu_\mathrm{GHz}\Delta\mathrm{t_d})^{-1}, \qquad (1)$$

where the last approximate equality holds for $\Delta\nu_d$ in MHz, $\Delta t_d$ in sec, D in kpc, and $\nu$ in GHz. In eqn (1), the spatial scale $\ell_d$ is estimated using the diffraction bandwidth and the distance. The velocity estimator has the same form independent of the electron density wavenumber spectrum, although the numerical coefficient in front depends on the spectrum and on how scattering material fills the line of sight. Typical numbers are $\theta_d \sim 10^{-3}$ arc sec, $\ell_d \sim 10^4$ km, $\Delta\nu_d \sim$ kHz $\rightarrow$ MHz, and $\Delta t_d \sim$ sec $\rightarrow$ min for $V_\perp \sim 100$ km s$^{-1}$.

The utility of the velocity estimator in equation (1) is that it is much more efficient to measure the frequency scale using a single antenna than to measure the spatial scale directly using intensity interferometry. Weak pulsars can also be measured using large antennas such as the 305 m Arecibo telescope.

How good is the velocity estimator? One test involves plotting the quantity $\sqrt{\Delta\nu_d}/\nu$ against $\Delta t_d$, as shown in Figure 3 for two pulsars. The plotted quantities are linearly related and demonstrate that the spatial scale may indeed be estimated from the diffraction bandwidth. This would not be true if turbulent motions were large. The linear relations therefore imply $\delta V_{turbulent} \ll$ bulk velocities. An additional test involves plotting $V_\perp$ against orbital phase for the binary pulsar PSR 0655+64 (Figure 4). The variation with orbital phase demonstrates that the neutron star speed dominates the motion of the diffraction pattern across the line of sight.

## 2.3 Comparison of Scintillations and Interferometry

Table 1 compares the scintillation and interferometric techniques. In contrast to interferometry, for which the spatial resolution at the source ($\propto D/\nu$) degrades

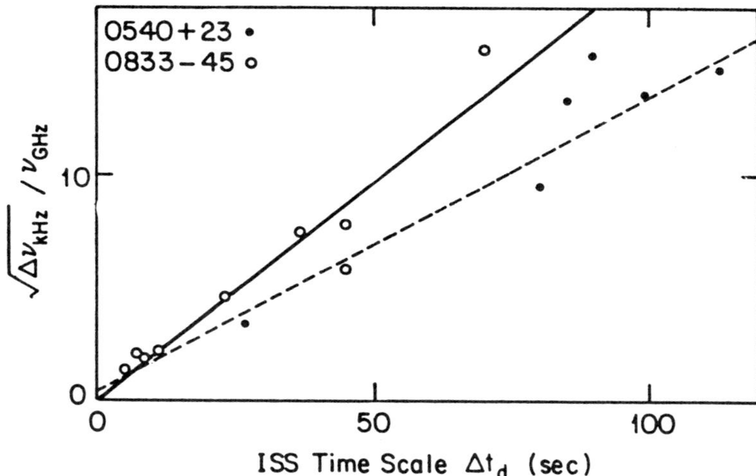


**Figure 3** Plot of $\sqrt{\Delta\nu_d}/\nu$ vs. $\Delta t_d$

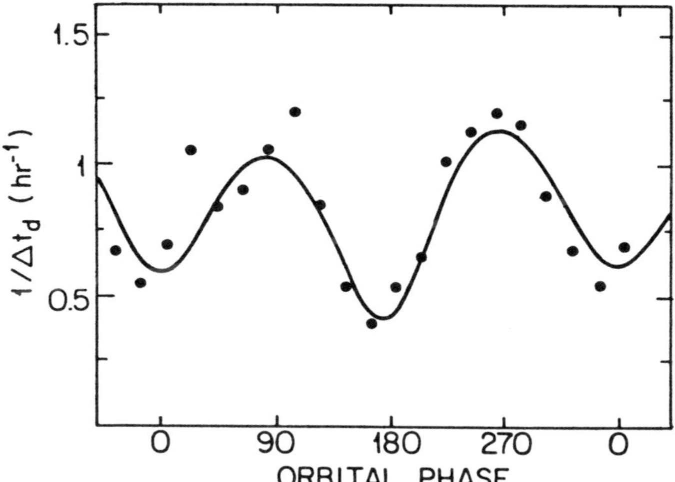


**Figure 4** Reciprocal diffraction time scale plotted against orbital phase for the binary pulsar 0655+64 (after Lyne 1984).

for distant pulsars, the scintillation resolution gets better for distant pulsars. For this reason, scintillations are applicable to distant, low-velocity pulsars whereas interferometry is not. Another advantage is that the scintillation velocity depends less strongly on the distance estimate. Most pulsar distances are estimated from the pulsar dispersion measure combined with a model for the interstellar electron density. Such distances, calibrated with a few independently known distances (from HI absorption, parallax, and optical measurements), are thought to have factor of 2 accuracy (e.g. Lyne, Manchester, and Taylor 1985).

Differential galactic rotation contributes to interometry derived proper motions because the interferometer phase is calibrated using extragalactic objects.

TABLE 1: COMPARISON OF TECHNIQUES

| PROPERTY | SCINTILLATIONS | INTERFEROMETRY |
|---|---|---|
| spatial resolution: (at source) | $10^4$km $\nu^{6/5} D^{-3/5}$ ($\nu$ in GHz, D in kpc) | $10^8$km $(b\nu)^{-1} D$ (b = 1000 km) |
| velocity estimates: | $V \propto D^{1/2}$ | $V \propto D$ |
| differential galactic rotation: | insensitive | includes |
| Selection against: | large V, large D | small V, large D |
| Advantages and Disadvantages: | 1 hour on large single dish antennas | $\geq 1$ year |
| | scalar velocity (vector possible) | vector velocity |
| | bias due to scattering leverage effects | |

With scintillations, however, differential galactic rotation is nearly cancelled out because the scintillation time scale is determined by relative motion of the line of sight and scattering material. Scintillations provide only a scalar velocity and may be biased by leverage effects if enhanced scattering occurs near the pulsar or near the Earth. It is known that distant, low galactic latitude objects are prone to enhanced scattering (Cordes, Weisberg, and Boriakoff 1985). For individual objects, enhanced scattering can be diagnosed by a smaller than expected diffraction bandwidth. It may be feasible to obtain vector velocities with scintillations (Cordes 1984) using multiepoch measurements, depending on the amplitude of refractive effects which also modulate the scintillation parameters on month to year time scales.

## 3. NEUTRON STAR VELOCITIES

Figure 5 shows the distribution of transverse speeds derived from scintillations for 71 objects and the distribution for the 26 objects studied with interferometry (Lyne, Anderson, and Salter 1982; hereafter LAS). The distributions are consistent with each other and show a large mean speed $\sim$ 100 km s$^{-1}$ and a tail extended to $\sim$ 400 km s$^{-1}$.

How do selection effects affect the shape of the derived distribution? Probably negligibly. The pulsars in the scintillation and the LAS samples were chosen primarily on the basis of their flux density and declination. Selection effects in the original pulsar searches are generally strong with respect to luminosity and galactic coordinates but these do not produce any strong velocity selection. Distance errors

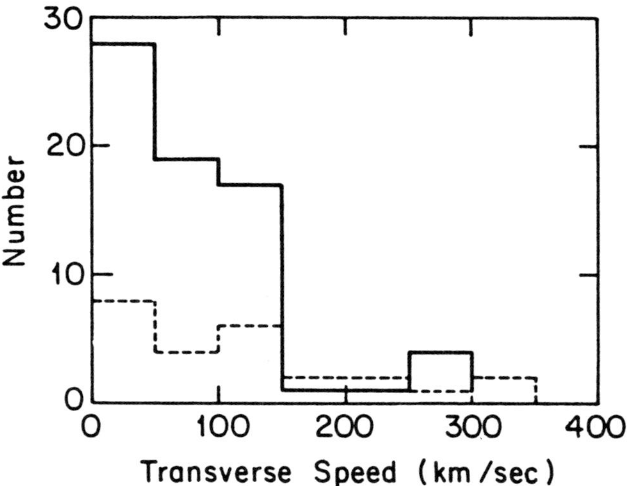


**Figure 5** Histogram of transverse neutron star speeds. (Solid line) Scintillation speeds for 71 objects (Dashed line) Interferometric speeds for 26 objects (Lyne, Anderson, and Salter 1982).

cause the velocity distribution to be biased toward *low* velocities because more objects are attributed distances too small than too large. This bias is stronger for the LAS sample but does not represent a major distortion of the distribution.

It is possible that there are subclasses of radio pulsars that have been severely biased against in pulsar searches, such as long lived millisecond pulsars (J.H. Taylor, this volume), which may rise to large distances from the galactic plane. Such a subclass is probably no more than about 10% of all pulsars, so any modification to the velocity distribution is minor (especially since this subclass is likely to be a low velocity component). Finally, although there are biases against precision measurement of high or low velocity objects inherent to the interferometry and scintillation techniques (as discussed above), these are at such extreme velocities that the derived distribution is essentially unaffected.

The velocity distribution is an important boundary condition for the stellar evolution of neutron star progenitors. Accounting for it with published models of evolution of single stars and of stars binary systems goes hand in hand with an accounting for the fraction of binary radio pulsars ($\sim$ 1.6%). This is discussed further below and in Dewey and Cordes (this volume).

## 4. CORRELATIONS OF $V_\perp$ WITH OTHER PULSAR PARAMETERS

Table 2 shows correlation coefficients (along with 99% confidence intervals) of transverse speed with other pulsar parameters, including $P\dot{P}$; the 'initial' value $P\dot{P}_i$ calculated assuming decay of the magnetic field on a time scale $\tau_B = 9Myr$; the spindown time $\tau_S \equiv P/2\dot{P}$; the chronological age $t \equiv 0.5\tau_B \ell n(1+2\tau_S/\tau_B)$; distance D; distance from the galactic plane $|z| = D \sin|b|$; the z velocity estimate $V_z \equiv |z|/t$; and radio 'luminosity' $L \equiv S_{400}D^2$, where $S_{400}$ is the 400 MHz flux density.

Correlation coefficients were calculated between *logarithms* of the various

TABLE 2 CORRELATONS OF $V_\perp$ WITH OTHER PARAMETERS

| Parameter | LAS Sample (26 objects) | Scintillation Sample (59 objects) |
|---|---|---|
| $P\dot{P}$ | 65 (+22, -42)% | 53 (+20, -29)% |
| $(P\dot{P})_i$ | 61 (+24, -44) | 33 (+27, -33) |
| $\tau_S$ | -39 (+51, -35) | -41 (+32, -24) |
| t | -29 (+52, -39) | -36 (+33, -26) |
| $\lvert z \rvert$ | 59 (+25, -45) | 14 (+31, -34) |
| D | 54 (+27, -47) | 27 (+28, -34) |
| $V_z$ | 61 (+24, -44) | 43 (+24, -31) |
| L | 41 (+34, -51) | 23 (+29, -33) |

quantities and are given for both the LAS sample of 26 interferometry derived velocities and for 59 of the scintillation velocities. The 59 object sample results from the removal of 'outlier' points which are objects with poorly measured velocities (probably due to leverage effects described above). Spearman rank correlations were also calculated and agree with those shown in Table 2, indicating that the coefficients are representative of all objects in the samples and are not dominated by just a few objects. Details of the correlation analysis may be found in Cordes (1986).

The velocity is correlated at significant levels with many of these parameters. Correlations with quantities involving distance are larger for the LAS sample than for the scintillation sample. This may be due to the fact that distances are in error and interferometer derived velocities are more sensitive to the distance. The correlation of velocity with $V_z$, however, is consistent with most pulsars being born in the galactic plane. A perfect correlation between transverse and z velocities is not expected because of geometrical decorrelation and from the fact that there exist runaway progenitors of neutron stars. The level of correlation is too large to be explained solely by errors in the distance scale. This may be checked by the correlation of $V_\perp/D$ with $\sin|b|$, which is at the level of 42% for the two samples. Since $\sin|b|$ is independent of distance, this level of correlation cannot be explained by errors in the distance scale.

For quantities in Table 2 involving $\dot{P}$, the velocity is most correlated with $P\dot{P}$, which is conventionally associated with the torque, and for magnetic dipole models of the magnetosphere, $P\dot{P} \propto$ (magnetic moment)$^2$. This correlation was first pointed out by Helfand and Tademaru (1977) and verified by Anderson and Lyne (1983). Figure 6 shows the correlation coefficient between $\log V_\perp$ and $\log P^\alpha \dot{P}$ plotted against $\alpha$. Although $\alpha$ is not well constrained, a value of 1 is certainly consistent with both velocity samples. The surprising result is that the velocity has *anything* to do with $P$ and $\dot{P}$.

Since the correlation with $P\dot{P}$ is potentially very interesting, I have investigated it for various subsamples of pulsars (Cordes 1986). It appears significant for all such subsamples and does not appear to arise from any peculiarity in those

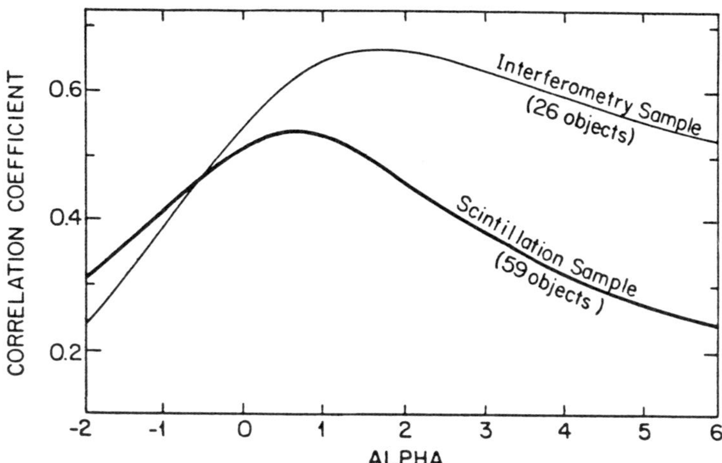


**Figure 6** Correlation coefficient between logarithms of $V_\perp$ and $P^\alpha \dot{P}$ as a function of $\alpha$.

objects that are near the Earth or to be due to any correlation of radio luminosity with $P\dot{P}$ combined with selection effects in pulsar searches. Cheng (1985) has attempted to account for the correlation with spindown models involving (velocity dependent) accretion of grains into the pulsar magnetosphere. Other aspects of the pulsar population are inconsistent with this model, however, such as a predicted relation of $\dot{P}$ with $|z|$. Thus, I conclude that the space velocity of a radio pulsar is generally correlated with $P\dot{P}$, possibly signifying a correlation of space velocity with magnetic moment of the neutron star itself.

## 5. DISCUSSION

The distribution of transverse space velocities is quite broad and it is clear that pulsars as a class are high velocity objects. Moreover, there are a number of significant correlations between transverse velocity and other parameters, especially the quantity $P\dot{P}$.

It has long been suspected that pulsar velocities are related to their genesis in (primarily) binary systems, most of which are disrupted by mass loss associated with supernova explosions (Gunn and Ostriker 1970). Only until now, however, has it become possible to test scenarios for neutron star formation against observational constraints, such as the velocity distribution. Given the large variety of evolutionary paths that may lead to the formation of radio pulsars, (e.g. van den Heuvel, these proceedings), it is likely that the velocity distribution is a mixture of distributions for different paths.

Dewey and Cordes (1986) have undertaken a Monte Carlo simulation of neutron star progenitors that uses the known statistics of massive stars (distributions in masses, binary membership, orbital period) and conventional ideas about mass exchange and mass loss to predict the velocity distribution and binary membership of radio pulsars. Selection effects have also been studied using simulations and the results show that the velocity distribution is largely unaffected.

Preliminary results show that if supernova explosions are symmetric, then (1) the resultant velocity distribution contains far too many *low* velocity objects and (2) there should be a much larger fraction of radio pulsars in binary systems than is currently known. To 'fix' the velocity distribution (while keeping the assumption of symmetric explosions), there would have to be many more compact binaries ($P_{orb} < 1$ day) than are implied in the work of Abt (1983). A simpler route involves relaxation of the assumption of perfectly symmetric explosions. It is found that an extra kick velocity $\sim 90$ km s$^{-1}$ is sufficient to account for both the velocity distribution and the number of binary radio pulsars (Dewey and Cordes, this volume).

I thank R. Dewey for many useful discussions. I also thank D. Backer, A. Blaauw, A. Lyne, R. Narayan, E. Tademaru, and A. Wolszczan for informative discussions. This research was supported by NSF grant AST-8311844 to Cornell University, by the Alfred P. Sloan Foundation, and by the National Astronomy and Ionosphere Center, which operates Arecibo Observatory under contract with the NSF.

## REFERENCES

Abt, H.A. 1983, *Ann. Rev. Ast. Ap.*, **21**, 343.
Anderson, B. and Lyne, A.G. 1983, *Nature*, 303, 597.
Armstrong, J.W., Cordes, J.M., and Rickett, B.J. 1981, *Nature*, **291**,561.
Blaauw, A. 1961, *B.A.N.*, **15**, 265.
Cheng, A. 1985, *Ap.J.*, **299**, 917.
Cordes, J.M., Weisberg, J.M., and Boriakoff, V. 1985, *Ap.J.*, **288**,221.
Cordes, J.M. 1984 in IAU Symposium 110 *VLBI and Compact Radio Sources*, eds. R.Fanti, K.Kellerman, and G. Setti (Dordrecht: Reidel), p. 303.
Cordes, J.M., 1986, *Ap.J.*, in press (1 Dec).
Gunn, J.E. and Ostriker, J.P. 1970, *Ap.J.*, **160**, 979.
Harrison, E.R. and Tademaru, E. 1975, *Ap.J.*, **201**, 447.
Helfand, D.J. and Tademaru, E. 1977, *Ap.J.*, **216**, 842.
Lyne, A.G. 1984 *Nature*, **310**, 300.
Lyne, A.G., Anderson, B., and Salter, M.J. 1982, *M.N.R.A.S.*, **201**, 503.
Lyne, A.G., Manchester, R.N., and Taylor, J.H. 1985, *M.N.R.A.S.*, **213**, 613.
Radhakrishnan, V. 1984 in *Millisecond Pulsars*, eds. S.P. Reynolds and D.R. Stinebring (NRAO: Green Bank), 130.
Rickett, B.J., Coles, W.A., and Bourgois, G. 1984, *Ast. Ap.*, **134**, 390.
Simonetti, J., Cordes, J.M., and Spangler, S.R. 1984, *Ap.J.*, **284**, 126.
Zwicky, F. 1957, *Morphological Astronomy* (Berlin: Springer-Verlag), 258.

## DISCUSSION

**E. Basinska-Lewin:** Ionospheric scintillations can cause significant variations in phase and intensity of radio signals at VHF frequencies. Is it possible that motion of ionospheric plasma irregularities could bias your results towards low velocities and contribute to the relatively larger number of pulsars with perpendicular velocities in your sample?

**J. Cordes:** Unlike interstellar scintillations, ionospheric scintillations are broadband and their influence on the observed fluctuations in time and frequency can be calibrated out. Moreover, even if there were residual leakage of ionospheric scintillations into the net results, they would tend to shorten the derived fluctuation time scale. This would imply a bias to large velocities. In any case, this bias is negligible.

**R. Becker:** Does the correlation between $\upsilon$ and $\upsilon_z$ indicate anything other than the fact that both depend on the assumed distance?

**J. Cordes:** Yes. The correlation is too large to be due to the distance dependence.

**D. Backer:** If $P\dot{P}$ is perfectly correlated with the total pulsar space velocity, what correlation coefficient is expected between $P\dot{P}$ and the perpendicular velocity? If this is so close to the observed correlation, can we then infer radial velocities from $P\dot{P}$ and $V_\perp$?

**J. Cordes:** As it turns out, the correlation of perpendicular velocity with $P\dot{P}$ would be identical with the correlation between total velocity and $P\dot{P}$. This is because, although geometrical effects lessen the average cross product of $V_\perp$ and $P\dot{P}$, they also lessen the standard deviation of $V_\perp$ with respect to total velocity. Thus the normalized correlation coefficient is unchanged.

**E. van den Heuvel:** There is something which I do not understand. From Lyne's talk I understand that $(P\dot{P})$ decreased in the course of time. If the correlation between $(P\dot{P})$ and V would be causal, this would mean that the velocity decreases in time. This seems rather odd to me.

**J. Cordes:** Your argument is amusing but not physical. Decay of $P\dot{P}$ decreases only the correlation coefficient between $V_\perp$ and $P\dot{P}$ (assuming that $V_\perp$ is correlated with $P\dot{P}$ at birth). Different pulsars of course have different ages and perhaps different decay times. Thus one should expect the correlation between $V_\perp$ and $P\dot{P}$ to be imperfect, especially since there are multiple physical contributions to $V_\perp$. I should add that the correlation coefficient $V_\perp$ and initial $P\dot{P}$ (assuming a universal decay time constant) is somewhat smaller than between $V_\perp$ and current $P\dot{P}$. I would argue that we do not understand torque decay well enough to extrapolate backwards in time.

**E. van den Heuvel:** One of my students, G. Stollman (1986), did a Monte Carlo simulation of the expected observable pulsar population, taking selection effects into account and found this leads to an expected observational correlation between the magnetic field strength B and space velocity. This is a purely spurious correlation, produced by selection effects, and of the same order of magnitude as the observed correlation.

**J. Cordes:** This is contrary to the results of simulations reported by Dewey and myself (this volume). As I recall from the paper (Stollman and van den Heuvel 1986) the level of correlation obtained with your treatment of selection effects is much smaller than the observed level of correlation. I think that saying the correlation is spurious is misrepresentative of your own simulation and certainly misrepresentative of the observational results.

**S. Kulkarni:** Does your model reproduce the observed fraction of binary pulsars? Does it reproduce the velocity distribution of runaway OB stars?

**J. Cordes:** The simulation involving symmetric supernova explosions produces far too many binary radio pulsars and B star velocities that are too small. Perhaps binary radio pulsars should be searched for, but other evidence (the pulsar velocity distribution) strongly suggests that other effects (e.g., asymmetric explosions) in addition to disrupted orbital motion are involved and that such binaries do not exist.

**D. Eichler:** The electromagnetic rocket mechanism (Harrison and Tademaru) would explain the $P\dot{P} - V_{\perp}$ correlation in a straightforward manner, wouldn't it?

**J. Cordes:** Perhaps. But remember the net translational velocity produced by this mechanism is independent of the magnetic moment (large moments produce a bigger acceleration, but for a shorter time). Thus one would need something like a relationship between magnetic moment and its offset from the neutron star axis to explain the correlation.

**R. Manchester:** The rocket mechanism for accelerating pulsars is unlikely to be effective as the force depends on the fifth power of the pulsar rotation frequency and there is now good evidence that pulsars are generally born with periods which are not very short.

**J. Cordes:** The question is what you mean by generally. Are you certain that no strong magnetic field pulsars are born with millisecond periods? I think the period distribution is still uncertain for small periods because I have not seen a complete treatment of selection effects. The evidence that pulsars are born with larger periods really is evidence that pulsars only spend small amounts of time at short periods. I contend that the jury is still out on the rocket mechanism.

**A. Burrows:** It might be pointed out that a velocity asymmetry is suggested in a scenario recently proposed by Stan Woosley and myself for PSR1913+16. It is encouraging to see indications of asymmetry from such independent work, though we, by no means, would conclude that all explosions are asymmetric in interesting ways.

**J. Cordes:** Yes. I recall from your preprint that your conclusion for 1913+16 depends critically on the mapping of progenitor mass to remnant mass. Let me point out that Ira Wasserman and I conclude that if the explosion was asymmetric in the 1913+16 system so as to give a kick out of the pre-explosion orbital plane, then geodetic precession of the pulsar spin axis must be occuring. We are presently looking for this effect in polarization measurements from the Arecibo Observatory.

# A DEEP SEARCH FOR YOUNG PULSARS IN THE GALACTIC PLANE AT 1400 MHz

T.R. Clifton, A.W. Jones and A.G. Lyne
University of Manchester
Nuffield Radio Astronomy Laboratories
Jodrell Bank
Macclesfield
Cheshire SK11 9DL, United Kingdom

ABSTRACT. A new survey of the Galactic plane for pulsars has now been completed at Jodrell Bank at the relatively high frequency of 1400 MHz (Clifton, T.R. and Lyne, A.G. 1986, Nature, 320, 43). A further 8 new objects have been detected. The properties of these are summarized below:

| PSR | P(ms) | err | DM | Long | Lat | err | Flux(mJy) |
|---|---|---|---|---|---|---|---|
| 1735-32 | 768.50 | 1 | 60 | 356.5 | -0.52 | 0.10 | 1 |
| 1806-21 | 702.413 | 5 | 380 | 9.4 | -0.69 | 0.10 | 1 |
| 1809-175 | 538.35 | 2 | 720 | 13.0 | 0.35 | 0.10 | 3 |
| 1828-10 | 405.025 | 2 | 180 | 20.9 | -0.52 | 0.10 | 1 |
| 1841-04 | 991.02 | 1 | 100 | 28.1 | -0.52 | 0.10 | 1 |
| 1842-04 | 162.251 | 1 | 280 | 28.2 | -0.69 | 0.10 | 3 |
| 1850+00 | 2180.3 | 1 | 840 | 33.6 | 0.00 | 0.10 | 5 |
| 1859+07 | 644.01 | 1 | 240 | 40.6 | 1.04 | 0.10 | 3 |

Although only 200 square degrees of the sky were searched, 62 pulsars in all were detected of which 40 were new discoveries. The new pulsars mostly have dispersion measures greater than 200 pc $cm^{-3}$ and, although longitudes between $l=-5$ and $l=100$ were surveyed, 39 of the 40 new objects lie at $l<41$. This shows that pulsars are far more common in the central regions of the Galaxy than hitherto appreciated. Most of the new pulsars will have escaped detection in previous surveys which were at around 400 MHz because of the extreme effects of interstellar scattering which broadens the pulses and because of the high galactic background emission at the low frequency. Many of the new pulsars still show the effects of interstellar scattering, even at this high frequency at which the broadening is reduced by a factor of about 150 compared with 400 MHz. This suggests that a still higher frequency is required in order to penetrate the innermost regions of the Galaxy.

Although the survey had good sensitivity for periods down to about 50 ms and reducing sensitivity down to periods of 4 ms, no pulsars were discovered having periods of less than 85 ms. However two of the new pulsars with periods of about 100 ms have characteristic ages of only about 20,000 years. Only 3 other radio pulsars have smaller ages and they all have supernova remnants associated with them. Since no strong radio emission is obvious in any continuum surveys around the two new pulsars, any associated supernova remnants probably have quite low surface brightness.



*D. J. Helfand and J.-H. Huang (eds.), The Origin and Evolution of Neutron Stars, 47.*

# PROGENITORS OF THE LOCAL PULSARS; LOWER MASS LIMIT AND BEAMING FACTOR

A Blaauw
Kapteyn Astronomical Institute
Groningen, The Netherlands

Following up on an earlier paper on the local pulsar population (Blaauw 1985; Paper I), in which it was shown that the majority of their progenitors must be sought among the local field stars between 6 and 10 solar masses (types B2, B3) - the local OB associations accounting for a minority only of the pulsar population - we consider implications of two further restrictions: raising the lower limiting mass for neutron star formation, $M_n$, above 6 solar masses, and raising the beaming factor, f, above 0.3.

For the estimated local production of <u>observed</u> pulsars we use the number of paper I, 33 pulsars per $kpc^2$ per 4.6 Myrs. Hence for values of f of 1.0, 0.5, and 0.3 the true number should be 33, 66 and 110, respectively.

Of the 17 (subgroups) of OB associations within 1 kpc, 14 are younger than 16 Myrs (see paper I), hence should still contain stars of 12 solar masses and below; their contribution per $kpc^2$ per 4.6 Myrs is estimated to be about 9 neutron stars. The three older subgroups currently convert stars around 8 solar masses at the rate of not more than about 6 per $kpc^2$ per 4.6 Myrs. Hence, for f = 1.0, at least some 18 pulsars per $kpc^2$ per 4.6 Myrs must originate from the local field stars, and some 50 and 95 for f = 0.5 and 0.3, respectively.

Assuming the field star population, estimated at 350 stars exceeding 6 solar masses within 500 pc (paper I), to have been formed at uniform rate between 50 and 20 million years ago, we estimate by means of the Evaporation Function E(t) (paper I) the number of neutron stars formed over the last 4.6 Myrs per $kpc^2$ as a function of $M_n$. For $M_n$ = 8.0 solar masses, the total yield, 24 neutron stars, is about the same as the required production for f = 0.8. For $M_n$ = 7.0 solar masses it is about 39, allowing f about 0.6, whereas for $M_n$ = 6.0 the yield, 73, allows f as low as 0.37.

More stringent specifications for $M_n$ and f than the above rather crude estimates should be possible by means of a more precise inventory of the evolutionary status of the local B star population.

REFERENCE: Blaauw, A. 1985, Birth and Evolution of Massive Stars and Stellar Groups, ed. E. Boland and H. van Woerden, p. 211-224 (Paper I).



*D. J. Helfand and J.-H. Huang (eds.), The Origin and Evolution of Neutron Stars, 48.*

# THE RELATION BETWEEN RADIO LUMINOSITY AND MAGNETIC FIELD IN ROTATION-POWERED PULSARS

Serge Pineault
Université Laval
Physics Department
Ste-Foy, Quebec
Canada G1K 7P4

ABSTRACT. The relationship between radio luminosity L and magnetic field B is re-investigated under the assumption that the new spindown mechanism proposed by Huang *et al*. (1982) becomes dominant at large periods ($\dot{P} \propto P^2$). Magnetic dipole braking is still assumed to provide the main torque at short periods ($\dot{P} \propto P^{-1}$). Both spindown torques depend on $B^2$ (assuming proportionality between internal and external fields) so that the resultant magnetic field in this hybrid model is given by (note the difference with the standard expression):

$$B^2 = 1.0 \times 10^{39}\ \dot{P}\ P\ /\ (1 + \gamma\ P^3)\ \text{Gauss.}$$

A value of 3.6 is used for $\gamma$ (Pineault 1986). For the radio luminosity we use the values tabulated by Manchester and Taylor (1981).

The results of the correlation analysis give log L = 0.80 log B + constant, with a correlation coefficient of 0.29, a highly significant result for the sample of 291 objects used (nominally $<$ 0.0001%). For comparison, in the standard model, the correlation coefficient is only 0.04 (this differs from the results of Lyne *et al*. (1985) because of a different luminosity definition). The implications of this analysis on the distribution of pulsar properties in models with luminosity evolution (Gunn and Ostriker 1970, Lyne *et al*. 1985) and on the apparent correlation between transverse velocity and magnetic field (Cordes 1987) are currently under study.

## REFERENCES

Cordes, J. M. 1987, these Proceedings
Gunn, J. E. and Ostriker, J. P. 1970, *Astrophys. J.*, **160**, 979
Huang, J. H., Lingenfelter, R. E., Peng, Q. H. and Huang, K. L. 1982, *Astron. Astrophys.*, **119**, 9
Lyne, A. G., Manchester, R. N. and Taylor, J. H. 1985, *Mon. Not. R. astr. Soc.*, **213**, 613
Manchester, R. N. and Taylor, J. H. 1981, *Astron. J.*, **86**, 953
Pineault, S. 1986, *Astrophys. J.*, **301**, 145



*D. J. Helfand and J.-H. Huang (eds.), The Origin and Evolution of Neutron Stars, 49.*

# DOES THE RADIO LUMINOSITY OF PULSAR GROW UP IN ITS LATER STAGE?

T. Lu
Department of Astronomy
Nanjing University
Nanjing
People's Republic of China

P. C. Zhu, J. S. Kui
Department of Physics
Nanjing University
Nanjing
People's Republic of China

In usual statistical analyses, because of diversities of proper parameters of pulsars, some interesting features might be smeared. In order to remove these diversities, we use the mean values for all quantities of pulsars, instead of values of individual pulsar, to do statistical analyses. $\log\dot{P}/P^3 - \log\tau$ and $\log L - \log\tau$ have been plotted, here $\tau = P/2\dot{P}$ and L denote the characteristic time scale and the radio luminosity of pulsars respectively. The most striking feature is that after its initial dropping to a dip at about $\tau \sim 10^6$ yrs, the radio luminosity of pulsar appears to grow up evidently and then redrop again. This feature is difficult to be understood in usual models. However, two tentative interpretations have been given in this paper.

First, if the radiation comes from the light cylinder of the pulsar, the radiating region ($\sim R_1{}^3 \sim P^3$) will be getting larger as pulsar spins down. This may increase the luminosity. However, the luminosity would redrop as P would approach a constant for very large $\tau$.

Second, the apparently growing up of radio luminosities may also be interpreted as due to a lot of new kind of pulsars with relative long initial periods and strong magnetic fields mixed in the procedures of doing average for large $\tau$ region rather than due to pulsars' luminosities themselves being growing up.



*D. J. Helfand and J.-H. Huang (eds.), The Origin and Evolution of Neutron Stars, 50.*

# A FEW REMARKS ON THE TWO TYPES OF PULSARS

Z.-G. Deng
Phys. Dept., Graduate School, Academia Sinica, Beijing, China
J.-H. Huang
Astrophys. Inst., Nanjing Univ., Nanjing, China
X.-Y. Xia
Phys. Dept., Tianjin Normal Univ., Tianjin, China

Huang et al have classified pulsars into two types according to the role played by the superfluid neutron vortexes in the interior of neutron stars. They have also presented the formule of $\dot{P}$ for these two types of pulsars. In this short remark, we shall compare the conclusions about the evolutions of these two types of pulsars with the observational distributions. The consequences obtained are summarized as follws:

a. The formula of Type II pulsars is a typical one with magnetic decay. The observational distribution of Type II pulsars shows this decay obviously, but the time scales of the field decay are dependent on the initial strengths of the surface field. From the observational distribution we can find out the upper and lower limits for the initial magnetic field strengths and the corresponding time scales. They are $B_{o1} = 6.6 \times 10^{12}$ Gauss, $\tau_{D1} = 1.6 \times 10^{6}$ yr, and $B_{o2} = 9.4 \times 10^{11}$ Gauss, $\tau_{D2} = 2.3 \times 10^{7}$ yr, respectively.

b. The relations between the ages and periods of these two types of pulsars are quite different from each other. Type II pulsars have a limit on their periods caused by the field decay, but the Type I pulsars have a limit on their life. When the period of a type I pulsar approaches to 2-3 seconds, the period of it increases faster and faster and the pulsar is going to stop rotating soon. Thus, there should be a clear boundary round 2-3 seconds in the period distribution of Type I pulsars. This consequence is conincident with the observational distribution of Type I pulsars.

c. The Type I pulsars have their special evolutionary tracks on the $\lg(\dot{P}/P)$ $-\lg(P)$ diagram. All these tracks reach some minimal values of $\dot{P}/P$ at about $P = 1$ second. $\dot{P}/P$ is the relative energy loss rate of pulsars. So, this minimum may imply that the pulsars radiate the least energy at about $P = 1$ second and will be fainter and harder to be observed. This feature may cause the appearance of the gap in the observational distribution on Type I pulsars round $P = 1$ second. Besides, this character may also suggest that the luminosities of Type I pulsars will increase with their periods if $P > 1$ second.



*D. J. Helfand and J.-H. Huang (eds.), The Origin and Evolution of Neutron Stars, 51.*

# THE KINEMATIC PROPERTIES OF TWO PULSAR TYPES

J.-H. Huang
Astrophys. Inst., Nanjing Univ., Nanjing, China
Z.-G. Deng
Phys. Dept., Graduate School, Academia Sinica, Beijing, China
X.-Y. Xia
Phys. Dept., Tianjin Normal Univ., Tianjin, China

The two types of pulsars are different from each other in many aspects. It may imply that they may have different progenitors. The kinematic analysis can gives us some clues to the properties of these progenitors. We have analysed the distributions of two types of pulsars with their distances $|z|$ from the galactic plane vs. their true ages

Type I:

$$T_I = \frac{(1+5P^3)}{3\ 5^{2/3} p\dot{p}}\left[\frac{1}{2}\ln\left(\frac{5^{2/3}p^2 - 5^{1/3}p+1}{(5^{1/3}p+1)^2}\right) + 3^{1/2}\left(tg^{-1}\frac{2\ 5^{1/3}p+1}{3^{1/2}} + \frac{\pi}{6}\right)\right]$$

Type II:

$$T_{II} = \frac{\tau_D}{2}\ \ln\left(\frac{p}{\tau_D \dot{p}} + 1\right).$$

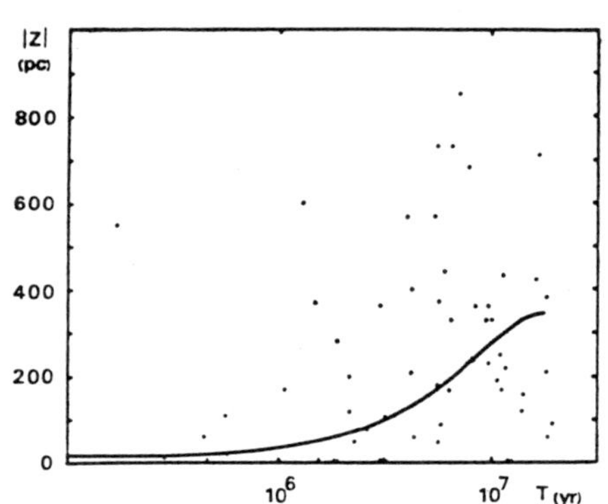


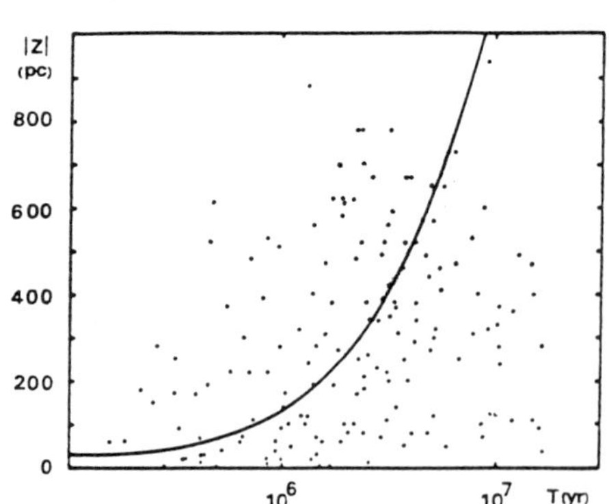


In fig.a) and b), we present plots of $|z|$ vs. T for type I and type II pulsars with ages less than $2 \times 10^7$ yr and the mean kinematic evolutionary tracks for each type, respectively.

The results show that the mean dynamic evolutions corresponding to 40 km/s for type I but 160 km/s for type II pulsars.

Fig. The plots of $|z|$ vs. T for a) type I, and b) for type II pulsars. The solid lines are their mean dynamic evolutionary tracks.



*D. J. Helfand and J.-H. Huang (eds.), The Origin and Evolution of Neutron Stars, 52.*

# TOWARD AN EMPIRICAL THEORY OF PULSAR EMISSION

Joanna M. Rankin
Department of Physics and Astronomy
University of Vermont
Burlington, Vermont 05405 USA

ABSTRACT. Polarized, pulsar profiles consist of two distinct types of emission units, core and conal components. Various configurations of these two basic units then underlie the 5 principal species of profile: conal and core single, conal double, triple, and multiple (5 comp.) profiles. Conal single profiles typically evolve through bifurcation into conal double forms at low frequency; whereas core single profiles evolve into triple forms at high frequency by adding pairs of conal "outriders". Conal components and outrider pairs exhibit polarization properties compatible with a hollow-conical emission zone which spreads weakly at low frequency. The central core components in each profile species, by contrast, often exhibit circular polarization of symmetrically alternating sense. The triple profile is then most generally prototypical of pulsar emission, and the physical distinction between its outriding conal components and its central core emission feature are primary--implying two distinct corresponding radiation mechanism. (Ap. J. 274, 333/359).

The isolated core components of stars with core single profiles exhibit little ordered modulation, nor do they null or evince of any sort of mode changing; many have featureless fluctuation spectra, while others exhibit low-frequency ($P_3$=15-50 periods/cycle) features with which no orderly "drift" is apparently associated. The core components in triple, double ("saddle" region), and multiple profiles also exhibit these stationary, low-frequency fluctuations. "Drifting" subpulses are then an exclusively conal phenomenon. Systematic subpulse modulation ($P_3$=2-15 periods/cycle) is associated with the conal components of stars with double, triple, and multiple profiles, but progressive, orderly drifting is observed only in conal single stars. The ensemble $P_3$ values associated with the conal components of the various profile species seem indistinguishable, again suggesting that these species are geometry-specific manifestations of a single physical configuration. Mode changing and pulse nulling are exclusively associated with neither core nor conal emission, but rather connect the two. Mode changing is typically manifest as a reorganization of the conal constituents of emission about the profile's core component. Finally, pulse nulling is not so closely associated with spindown age as has been thought. Pulsars which null are members of older profile species to be sure, but within a given species, those stars which null are no older than those that do not. (1986 Ap. J. 301, 901)



*D. J. Helfand and J.-H. Huang (eds.), The Origin and Evolution of Neutron Stars, 53.*

# RADIO EMISSION MECHANISMS FOR TWO TYPES OF PULSARS

J.-H Huang
Astrophys. Inst., Nanjing Univ., Nanjing, China
Z.-G. Deng
Phys. Dept., Graduate School, Academia Sinica, Beijing, China
X.-Y. Xia
Phys. Dept., Tianjin Normal Univ., Tianjin, China

We are trying to explore the radio emission mechanisms for our two types of pulsars from fitting the observational ( true ) age distribution for pulsars within 5 kpc of the sun with the theoretical one:

$$\frac{dN}{dt} \propto \frac{L(t)}{\left|\frac{dL}{dt}\right|} R(L)$$

here L is the radio luminosity of pulsars, $\frac{dL}{dt}$ the luminosity derivative and R the birth rate.

Assuming that the radio luminosity of pulsars is related to P and $\dot{P}$ with the following fomula

$$L = \alpha P^a \dot{P}^b (1+5 P^3)^c$$

here $\alpha$ is a coefficient, a, b and c are constants to be determined from the fitting.

The results are as follows: for Type I pulsars we should consider both the light cylinder model and the polar cap model, but for Type II pulsars, the only model we should consider is the polar cap model.



*D. J. Helfand and J.-H. Huang (eds.), The Origin and Evolution of Neutron Stars, 54.*

# PULSE ASYMMETRY OF MILLISECOND PULSARS

Kai You Chen and Jacob Shaham
Columbia Astrophysics Laboratory and Department of Physics
Columbia University
New York, New York 10027

By considering a simple model of slowly rotating neutron stars (for which the Schwartzschild metric could be used properly) whose emission is confined to circular antipolar caps, Pechenick, Ftaclas and Cohen (1983) studied the gravitational effects on the light curves. They found that for typical neutron star parameters, the light curves flattened out due to the bending of light in the neutron stellar gravity.

The results cannot, however, be directly applied to millisecond pulsars, particularly to 1937+214, because the effects of rotation are definitely not negligible.

By using the Kerr metric, we studied the effects of gravity and dragging of inertial frames on the light curves observed at infinity of fastly rotating neutron stars. When a = M (a being the stellar specific angular momentum in gravitational units) the corresponding period is 2.8 ms; when a = 2M it is 1.6 ms. (Note, that a > M poses no problem here because the Kerr metric only exists outwards of the stellar surface.) We adopted again the simple model that the emission of the neutron star is confined to circular polar caps and obtained numerical results by the Monte Carlo simulation of photon orbits.

For typical neutron star parameters, the results show that the flattening of light curves of millisecond pulsars is not as large as it is for non-rotating ones. Also, the peak is shifted to one side in an asymmetric way. It is no longer the center of the pulse or the polarization turning point. This is due to the Lorentz contraction of the polar region, while the dragging of the inertial frames produces the long tail following the peak. Therefore, if the neutron stars in galactic bulge sources are spinning fast, one cannot "smear out" their pulsation by the gravity. Furthermore, a comparison of our simulations with detailed pulse observations in 1937+214 may provide clues to the emission pattern.



*D. J. Helfand and J.-H. Huang (eds.), The Origin and Evolution of Neutron Stars, 55.*

PULSAR POLARIZATION LIMITING RADII AND THE EVOLUTION OF PULSAR BEAMS

John J. Barnard
Code 665
NASA-Goddard Space Flight Center
Greenbelt, MD 20771

ABSTRACT. We estimate the polarization limiting radius, $r_{pl}$, as a function of rotation period P, magnetic field strength B, and radio frequency ν, in radio pulsars assuming plasma parameters that are typical of polar-cap pair-creation models of pulsars. We find that $r_{pl} \simeq 9 \times 10^8 (P/1s)^{0.4}$ cm, for a surface magnetic field strength of $10^{12}$ G, and radio frequency of $10^9$ Hz. For short rotation periods, $r_{pl}$ approaches the light cylinder radius, $r_{lc}$. Here the magnetic field becomes more azimuthal, and the excursion in position angle over a pulse is less, on average, than when $r_{pl} \ll r_{lc}$. With the assumption of a vacuum magnetic field we calculate the polarization position angle as a function of pulse longitude, and the angles i (the angle between the magnetic moment and the rotation axis) and α (the angle between the line of sight and the rotation axis). We calculate the average change in polarization angle as a function of pulsar period, assuming a circular beam, and find consistency with the polarization data summarized by Narayan and Vivekand (1983). We conclude that the evidence is consistent with beams that are roughly constant in shape, providing an alternative to the evolving elliptical beam model of Narayan and Vivekand (1983). This interpretation is further supported by the frequency dependence of the polarization angle in the Crab Pulsar, the frequency of pulsars with double and multiple pulse components, the frequency of pulsars with interpulses, and the absence of pulsars in plerions. See Barnard (1986) for further details.

This work was supported by a National Research Council Resident Research Associateship.

REFERENCES

Barnard, J.J.: 1986, Astrophys. J., 303, p. 280.
Narayan, R. and Vivekanand, M.: 1983, Astron. Astrophys., 122, p. 45.



*D. J. Helfand and J.-H. Huang (eds.), The Origin and Evolution of Neutron Stars, 56.*

# POLARIZATION ANGLE SWINGS REDISCUSSED

W. Sieber
Max-Planck-Institut für Radioastronomie
Auf dem Hügel 69
D-5300 Bonn 1
Federal Republic of Germany

The polarization angle swing observed in some pulsars, e.g. PSR 0833−45, may be understood as the pattern which results when our line of sight cuts through the magnetic dipole field of a pulsar. In such a geometry the polarization angle $\chi$ is a function of $\mu$, the angle between rotation axis and magnetic field axis and $\Delta$, the angle between magnetic field and line of sight to the observer.

For real data, 90° polarization angle flips very often completely obscure smooth variations as expected from geometry. The determination of the contributions from competitive 90° modes is only possible by use of data with high signal-to-noise ratio; few such mesurements exist. Our analysis is based (see also Narayan and Vivekanand, 1983) on data sets published by Backer and Rankin (1980) and Bartel et al. (1982), which we used to fit model curves as given above ($\mu$ was kept constant: $\mu$ = 45°).

Remarkably good results were obtained for nearly all pulsars, especially convincing for pulsars with double peaked profiles. The polarization angle swing for PSR 0823+26 indicates that this pulsar may also be considered as a double peaked pulsar, the second component being very weak. A similar geometry might tentatively apply for PSR 1541+09, the main component being here on the trailing side of the profile. PSR 1237+25 and 0329+54 show nearly central cuts through the beam ($\Delta$ = 0°.2 and 0°.71 respectively). It becomes clear from PSR 0329+54 that mode-changes do obviously not influence the magnetic field geometry. Pulsars with high $\Delta$-values show very often considerable azimuthal structure in their profiles, perhaps due to absorption.

From the steepest gradient of $\chi$ it is possible to estimate an upper limit of $\Delta$, independent of $\mu$. The distribution of values gives direct evidence of detecting pulsars with a given separation between line of sight and magnetic field axis. Separations greater than 20° are extremely rare. We estimate that less than 10% of all pulsars can be detected due to beaming (one beam emission). It seems that the latitudinal-longitudinal beam asymmetry is about a factor of 2, not higher.

REFERENCES

Backer, D.C., Rankin, J.M.: 1980, Astrophys. J. Suppl. **42**, 143
Bartel, N., Morris, D., Sieber, W. et al.: 1982, Astrophys. J. **258**, 776
Narayan, R., Vivekanand, M.: 1983. Astron. Astrophys. **122**, 45



*D. J. Helfand and J.-H. Huang (eds.), The Origin and Evolution of Neutron Stars, 57.*

# TRIPLICITY OF PULSAR PROFILES AND ORTHOGONAL POLARIZATION MODES

J. A. Gil
Department of Physics and Astronomy
University of Kentucky
Lexington, Kentucky
USA

A mechanism for generation of two concentric pulsar beams corresponding to the core and conal pulsar emission is proposed. The inner beam originates close to the star, where the radius of curvature of the dipolar magnetic field lines is suitable for coherent curvature emission at pulsar radio frequencies. Further from the star where the outer beam originates, the radius of curvature of dipolar lines is probably too large, but the actual curvature can be dominated by the toroidal component of field lines twisted back due to the pulsar rotation. One can show that for even moderate twisting the actual radius of curvature leads again to pulsar radio frequencies. Thus, the pulsar emission is a superposition of two beams originating at widely radially separated locations. When the observer's line-of--sight cuts both beams, a three-component profile should be observed. Because of the retardational time delay, the inner (core) component should appear late with respect to the profile midpoint, that is , closer to the traling component. Such an asymmetry is indeed observed in complex profile pulsars. In pulsar magnetosphere, the radius of curvature of dipolar field lines depends on the radius of the emission region in the opposite way than that of the toroidal lines. This explains why core and conal components dominate the mean profile at low and high frequencies, respectively.

The proposed version of a dual-beam pulsar model explains different polarization modes in a natural way. The curvature emission of the inner (core) beam is polarized linearly in the planes of dipolar field lines while in the outer (conal) beam the polarization planes are determined by the toroidal component. The polarization vectors of the two pulsar beams should be mutually quasi-perpendicular in most cases, although nonorthogonal, phase dependent transitions are also expected. Thus, whithin this model the different polarization modes can be identified with the core and conal pulsar emission. The modes are superposed at any instant, which produces significant depolarization. The actual position angle is determined by the stronger beam. The evidence for the correspondence between the core and conal pulsar emission and different polarization modes are discussed.



*D. J. Helfand and J.-H. Huang (eds.), The Origin and Evolution of Neutron Stars, 58.*

# THE EFFECTS OF INVERSE COMPTON SCATTERING ON THE PULSARS' RADIATION

X.-Y. Xia
Phys. Dept., Tianjin Normal Univ., Tianjin, China
Z.-G. Deng
Phys. Dept., Graduate School, Academia Sinica, Beijing, China
G.-J. Qiao, X.-J. Wu and H. Chen
Geophys. Dept., Beijing Univ., Beijing, China

Our calculations show that the cross section of the inverse Compton scattering in strong magnetic fields may be larger than that of Thompson scattering by sevaral orders of magnitude in the case of polar cap surface of pulsars. We can also see that when the energy of $e^{\pm}$ exceeds a certain value, their energy loss caused by the inverse Compton scattering may be larger than the energy gain from electric field in the inner gap, which implies that the $e^{\pm}$ could not be accelerated to $\gamma = 10^6$ . Meanwhile, the electrostatic forces acting on the electrons will be balanced by the radiative pressure if temperature $T > 10^8$ K.

It is beleived that the surface temperarure for most of pulsars is less than $10^6$ K, in that case the ions of iron can not be emitted from the surface of pulsars. However, the temperarure at the polar cap can be increased to $3x10^6$ through the bombardment of electrons to the polar cap according to R-S model. This quasi-equilibrium state by self-regulating must make the coherent radio emission unstable on the contrary.

If we take the inverse Compton scattering effect into account, however, this problem can be solved.

Taking $B = 10^{12}$ Gauss and $T = 3x10^6$ K, the highest electron energy gained in the electric field is about $\gamma = 10^3 - 10^4$ . Therefore, the polar cap can not be heated by those electrons directly. Because the energy of scattered photons is about $4x\ \gamma^2$ times that of photons before scattering, those scattered photons with high energy will bombard the polar cap surface. It leads to the following results:

a. These photons will heat the polar cap.

b. They can directly kick out the ions of iron which in turn may produce $\gamma$-photons with energy of larger than 1 Mev by the inverse Compton scattering.These $\gamma$-photons then will give $e^{\pm}$ thru pair production, and the produced $e^{\pm}$ can produced high energy $\gamma$-photons by the inverse Compton scattering with thermal photons again and so on.

According to this process, we may have stationary ions of iron kicked out and the $e^{\pm}$ shower. then the radio emission can be kept stationary and avoid the difficulties in R-S model.



*D. J. Helfand and J.-H. Huang (eds.), The Origin and Evolution of Neutron Stars, 59.*

THE GEOMETRIC STUDY OF DRIFTING SUBPULSES

X.J. Wu, G.X. Deng, H. Chen
Geophysics Department, Peking University, Peking, China
X.Y. Xia
Physics Department, Tianjin Normal University, Tianjin, China

We use the polarization parameters to improve the geometric investigation of drifting subpulses. According to the geometry of the hollow cone emission, there exist certain relations between some parameters ($R, \Gamma, \alpha, G, \Phi, \vartheta, \beta$), here $R$ and $\Gamma/2$ are the half angular width of the emission cone of the integrated pulse and drifting subpulse, respectively, $\alpha$ and $G$ are the apparent beam width of the integrated pulse and drifting subpulse, respectively, $\Phi$ is the maximum of the gradient of the position angle of the linear polarization, $\vartheta$ is the angle between the sight line and rotation axis, $\beta$ is the magnetic inclination. Some interesting results we get are as follows:

a) The values of $\Gamma$ and $R$ change with $\beta$, but the parameter $Q_r(\Gamma/2R)$ is almost independent of $\beta$. Roughly, the $Q_r$ can be determined from the observational data.

b) Estimation of curvature radius: Comparing the $\vartheta_{min}$ and $\vartheta_{max}$ given by the R-S model with parameters R and $\Gamma/2$, we can get $\vartheta_{min} \leqq R - \ /2$ and $\vartheta_{max} \geqq R + \ /2$

$$Q_{cr} = \vartheta_{min}/\vartheta_{max} = P^{53/42} B_{12}^{10/7} \rho_6^{-47/21} \omega_{10}^{2/3}$$

$$(Q_r)_{ob} = (\Gamma/2R)_{ob} \leqq (1 - Q_{cr})/(Q_{cr} + 1) = (Q_r)_{RS}$$

where $P$ and $\omega_{10}$ are observed data, $B_{12}$ can be obtained from $P$ and $\dot{P}$. We can get the lower limit values of curvature radius for ten pulsars. There is a tight correlation between the curvature radius( $\rho_6$ ) and period ( P ).

c) Observational check of pulsars' classification: We can obtain the values of $Q_{cr}$ and $Q$ from observational data. $Q$ is a position parameter of sight line ( $Q = (\vartheta - \beta)/R$ ). If $Q > (Q_{cr})_{ob}$, the integrated pulse profile belongs to a simple class ( S ). If $Q < (Q_{cr})_{ob}$, it belongs to a double peak class ( C ). The results show that there are 9 pulsars conforming to this criterion except PSR 1919+21, but the integrated pulse profile of this pulsar is much similar to class S rather than class C.



*D. J. Helfand and J.-H. Huang (eds.), The Origin and Evolution of Neutron Stars, 60.*

# PAIR PRODUCTION IN INTENSE ELECTROMAGNETIC FIELDS OF PULSARS

T.Y. Shi
Department of Physics
Beijing Normal University, Beijing
China

The possible existence of strong electromagnetic fields in pulsars has motivated extensive interest in investigation of various quantum electrodynamics processes. In particular, the process of converting high energy photons into electron-positron pairs is of great significance in pulsar theory.

In order to obtain the absorption coefficient in external field, it is necessary to calculate the exact electron wave function through solving the Dirac equation in this field. However, this is too complicated to solve in the presence of electric and magnetic field. Eluding this difficulty Tsai uses the Schwinger's proper-time method directly to calculate the vacuum polarization in strong field (Tsai and Erber, 1974) by two ways — pertubative expansion and Green function — and then extracts the absorption coefficient of photons from the optical theorem.

The purpose of this paper is to calculate the vacuum polarization in electromagnetic field using the proper-time method with a way of pertubative expansion, then to evaluate the absorption coefficient and obtain some results in the weak-field approximation. The absorption coefficient for unpolarization is

$$\bar{\varkappa} = \frac{3\sqrt{3}}{16\sqrt{2}} \frac{\alpha e}{m} \sqrt{B^2 + E^2} \sin\vartheta \exp\left(-\frac{8}{3} \frac{m^3}{(\omega \sin\vartheta)\, e\, (B^2 + E^2)^{\frac{1}{2}}}\right)$$

It indicates that in the presence of both electric and magnetic field, a way to get absorption coefficient is to change (B+E) for B in the one obtained in the existence of B alone. By virtue of R-S model for pulsars, the electric field strength parallel to magnetic field is about $10^{10}$ V/cm in the gap. It is much lass than B, implying that we can ignore it at all because of its insignificant role indeed. Our results coincide with those obtained by Urrutia ( 1978 ) from the electron Green's function.

References

Urrutia, L.F.: 1978, Phys.Rev.D, **17**, 1977
Tsai, W., and Erber, T.: 1974, Phys.Rev.D, **10**, 492



*D. J. Helfand and J.-H. Huang (eds.), The Origin and Evolution of Neutron Stars, 61.*

# A MODIFIED PULSAR MODEL Green Function period distribution

L.F. Chen, S.R. Liang
Department of Physics
Beijing Normal University
Beijing, China

The downward accelerated $e^-$ in the " unfavorable " zone, like that in the " favorable " one, will emit $\gamma$-photons, which in turn convert into $e^\pm$ pairs in some places near the surface of stars. But what happens, which is different from that in the " favorable " zone, is that some $\gamma$-photons will travel through a long distance before their conversion. This makes it possible that some $\gamma$-photons arrive at the " diode " district in the " favorable " zone. The magnetic conversion of pairs is much easier to happen than that occured in the " favorable " zone, where the $\gamma$-photons are created by the primary $e^-$ beam. The existence of dense $e^\pm$ plasma near the surface of stars makes the $E_\parallel$ vanish at places where such plasma is present.

In this paper, we have introduced $R'_*$, which is the radius of the surface of dense $e^\pm$ plasma due to photon pair-production where $E_\parallel$ is shorted out. $R'_*$ will get large drastically when the pulsar period drops below a certain value.

Some new results have been obtained.

When $R'_*$ becomes large the electron acceleration mechanism looses its efficiency and there would be a substantial reduction in radiation emitted from pulsars. This mechanism provides an explanation for the observed distribution of pulsar periods.

When the angle between the magnetic dipole axis and the rotation axis approaches 90°, potential $|\Phi|$ increases and the outflowing particles in the " unfavorable " zone can produce $\gamma$-ray which can create dense $e^\pm$ plasma like the behavior of electrons in the " favorable " zone. The pulse we receive will come from both poles of the pulsars.

The energy of the outflowing positrons from the " unfavorable " zone can reach 1000 Gev ( if radiation reaction is neglected ). Such high energy positrons have been observed in interstellar cosmic rays.



*D. J. Helfand and J.-H. Huang (eds.), The Origin and Evolution of Neutron Stars, 62.*

# A MASSIVE GLITCH IN PSR 0355+54

A.G. Lyne
University of Manchester
Nuffield Radio Astronomy Laboratories
Jodrell Bank
Macclesfield
Cheshire SK11 9DL, United Kingdom

ABSTRACT. Pulse timing observations at Jodrell Bank have shown that the relatively old pulsar PSR 0355+54 has suffered two glitches in the last 18 months, the second glitch being larger than any previously observed in any pulsar.

PSR 0355+54 has a characteristic age of 0.6 My and until recently has showed a very steady slow-down with a low level of timing noise (e.g. Cordes and Downs 1985). Since 1980 the pulsar has been included in a series of timing observations at Jodrell Bank and these have revealed two glitches.

The first glitch was estimated as occurring within 7 days of 1985 Jan 14. The fractional change in period amounted to $\Delta P/P$=-5.5E-9. Jumps of this magnitude or greater have been observed in about 5 other pulsars.

The second glitch was 1000 times greater than this with $\Delta P/P$=-4.4E-6 and occurred some time between observations on 1986 Jan 03 and 1986 Mar 15. Both glitches were accompanied by increases in the period derivative, having $\Delta\dot{P}/\dot{P}$ values of 0.2% and about 10% respectively.

The size of the second glitch is greater even than any of the large glitches observed in the Vela pulsar, both in the value of $\Delta P/P$ and of $\Delta\dot{P}/\dot{P}$. The most striking implication of this event is that about 10% of the moment of inertia of the star became decoupled from the solid crust at the time of the glitch. This implies that at least this fraction of the mass of the star is superfluid (Pines and Alpar 1985). The first post-glitch observations also suggest that the value of $\dot{P}$ is recovering rapidly towards its pre-glitch value on a time-scale of about 40 days. However, so far only 0.5 percent of the period decrease has been recovered. There may be further recovery on much longer time-scales than is revealed in the two months' post-glitch data presently available.

REFERENCES

Cordes, J.M. and Downs, G.S., 1985, Astrophys. J. Suppl., 59, 343.
Pines, D. and Alpar, M.A., 1985. Nature, 316, 27.



*D. J. Helfand and J.-H. Huang (eds.), The Origin and Evolution of Neutron Stars, 63.*

# A LARGE TIMING DISCONTINUITY IN THE VELA PULSAR, JULY 1985

C S Flanagan
National Institute for Telecommunications Research, CSIR.
P O Box 3718, Johannesburg, South Africa.

A pulsar-monitoring programme has been running at Hartebeesthoek Radio Astronomy Observatory over the last three years, at 2.32 and (during the last year) at 1.67 GHz. Twenty pulsars are observed once or twice a fortnight, and PSRs 1641-45 and 0833-45 (the Vela pulsar) daily.

One result from this program is observations of a large timing discontinuity in PSR 0833-45, in July 1985. Preliminary analysis yields the following parameters for this event:

$$\Delta P / P = -1.593 \pm .001 \times 10^{-6}$$

$$\Delta \dot{P} / \dot{P} = 5.0 \pm .1 \times 10^{-3}$$



*D. J. Helfand and J.-H. Huang (eds.), The Origin and Evolution of Neutron Stars, 64.*

## *Ib. ROTATION-POWERED PULSARS*

## *Pulsars and Supernova Remnants*

*CHAIR: L. Woltjer*

# THE GALACTIC PULSAR POPULATION AND NEUTRON STAR BIRTH

Ramesh Narayan
Steward Observatory
University of Arizona
Tucson, AZ 85721, USA

ABSTRACT. The radio pulsars in the Galaxy are found predominantly in the disk, with a scale height of several hundred parsecs. After allowing for pulsar velocities, the data are consistent with the hypothesis that single pulsars form from massive stellar progenitors. The number of active single pulsars in the Galaxy is $\sim 1.5 \times 10^5$, and their birthrate is 1 per $\sim$ 60 yrs. There is some evidence that many single pulsars, particularly those with high magnetic fields, are born spinning slowly, with initial periods $\sim 0.5 - 1s$. This could imply an origin through binary "recycling" followed by orbit disruption, or might suggest that the pre-supernova stellar core efficiently loses angular momentum to the envelope through magnetic coupling. The birthrate of binary radio pulsars, particularly of the millisecond variety, seems to be much larger than previous estimates, and might suggest that these systems do not originate in low mass X-ray binary systems.

## 1. INTRODUCTION

More than 400 radio pulsars, including 7 binaries, have by now been discovered in the Galaxy, most by standardized surveys. The selection effects of these surveys are understood quite well. Therefore, a meaningful statistical analysis of the pulsar data should be possible, and may provide valuable information on basic questions concerning the origin and birth of neutron stars. Some recent results are discussed here. A more comprehensive review can be found in Taylor and Stinebring (1986).

Pulsar surveys measure the following basic parameters for each pulsar they detect: the galactic coordinates $\ell, b$, the pulse period $P$, the time derivative of the period $\dot{P}$ (this requires follow-up timing observations and is not available for very recently discovered pulsars), the dispersion measure $DM$, and the radio flux $S$. Using a suitable model of the free electron density in the Galaxy (e.g. Lyne, Manchester and Taylor 1985, henceforth LMT), the observed $DM$ of a pulsar can be used to infer its distance $d$, and thus its luminosity $L$, defined by

$$L = Sd^2 \ (\mathrm{mJy} \cdot \mathrm{kpc}^2) \tag{1}$$



*D. J. Helfand and J.-H. Huang (eds.), The Origin and Evolution of Neutron Stars, 67–78.*

If the slow-down of the pulsar is primarily due to magnetic dipole braking, then the surface magnetic field $B$ can be estimated by

$$B \sim (10^{15} P\dot{P})^{1/2} 10^{12}\ \mathrm{G} \tag{2}$$

assuming standard values for the radius and moment of inertia of the neutron star. Another useful number is the "characteristic age" of a pulsar,

$$\tau = P/2\dot{P} \tag{3}$$

This represents the true age, if the pulsar is born with with an initial period $P_i \ll P$, if it slows down primarily by magnetic dipole braking, and if its magnetic field has remained essentially constant during its life.

## 2. GALACTIC DISTRIBUTION OF SINGLE PULSARS

Let us for simplicity assume that the distribution function of single pulsars can be factorized in the form

$$\rho(R, z, L) dR dz d\ln L = \rho_R(R) dR \rho_z(z) dz \Phi(L) d\ln L \tag{4}$$

where $R, z$ are galactocentric cylindrical coordinates. This particular decomposition was first introduced by Large (1971) to analyze the early data from the First Molonglo survey. LMT have redone the analysis and fitted more recent data using an iterative self-consistent scheme that allows for selection effects. Their estimates for $\rho_R(R)$ and $\rho_z(z)$ are shown by the histograms in Figs. 1 and 2, where the Sun has been assumed to be located at $R = 10$ kpc. Note that the radial gradient of the pulsar number density is strongly negative in the solar vicinity, implying that most of the active pulsars in the Galaxy lie within the solar circle. The apparent deficit of pulsars at $R \lesssim$ 5kpc may not be real since the selection effects (particularly scatter-broadening) of the surveys on which LMT based their analysis are such that pulsars in this part of the Galaxy would be extremely hard to detect. A recent high frequency (1400 MHz) survey by Clifton and Lyne (1986) detected many new high DM pulsars towards the galactic center region, suggesting that the true number density there could be greater than Fig. 1 might suggest. A least squares fit of a Gaussian to LMT's results for $R \geq 5$ kpc gives

$$\rho_R(R) \propto \exp[-(R/8)^2] \tag{5}$$

The distribution as a function of $z$ (Fig. 2) shows that pulsars are definitely a disk population, but with a rather large scale height. A least squares fit gives

$$\rho_z(z) \propto \exp[-(z/0.61)^2] \tag{6}$$

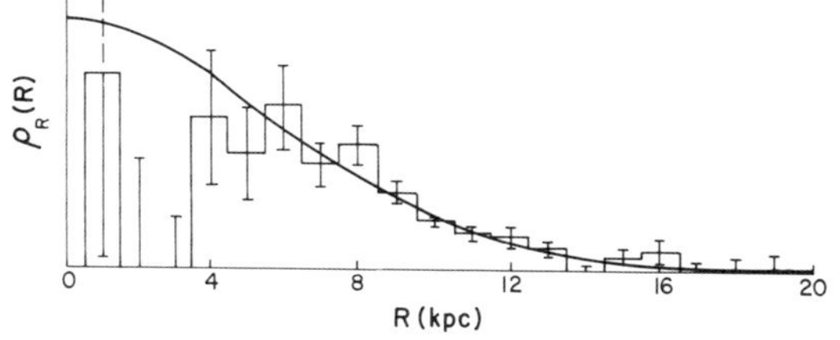


Fig. 1–Histogram of $\rho_R(R)$ estimated by LMT. The solid curve corresponds to eq. (5)

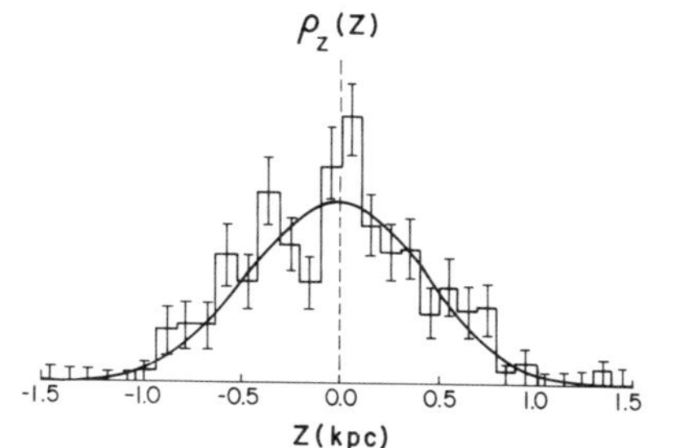


Fig. 2–Histogram of $\rho_z(z)$ estimated by LMT. The solid curve corresponds to eq. (6)

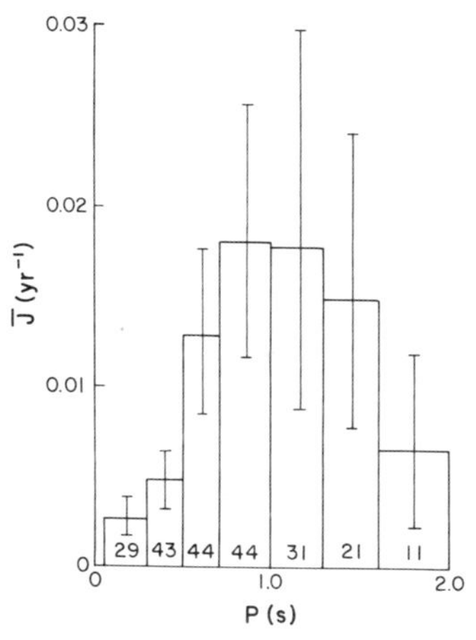


Fig. 3–Histogram of the mean pulsar current $\bar{J}(P_1, P_2)$, calculated using the scale factors $S(P, \dot{P})$.

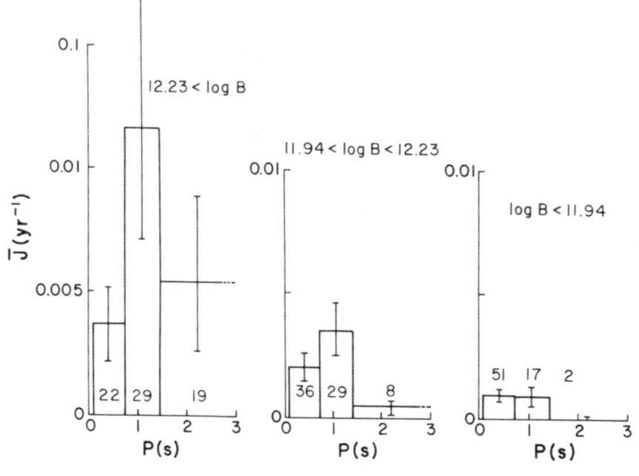


Fig. 4–Histograms of the mean pulsar current $\bar{J}(P_1, P_2)$ for pulsars corresponding to three ranges of magnetic field $B$, calculated using the scale factors $S(P, \dot{P})$.

which corresponds to an rms z height of 430 pc. The large scale height is believed to be almost entirely due to the large space velocities of pulsars (e.g., see the articles by Lyne and Cordes in this volume), and it is likely that pulsars originate with a scale height $< 100 - 200$ pc (LMT, Chevalier and Emmering 1986).

The above results are consistent with the hypothesis that single radio pulsars, and hence single neutron stars, are produced when young massive stars die. An additional piece of supporting evidence is due to del Romero and Gomez-Gonzalez (1981), who showed that radio pulsars are well correlated with HII regions, which are good tracers of young, hot stars in the Galaxy.

The solution for $\Phi(L)$ obtained by LMT shows that it varies as $L^{-1}$ for $L \gtrsim$ 10 mJy kpc$^2$. Unfortunately, since all surveys are flux-limited, very few pulsars with $L < 1$ mJy kpc$^2$ have been discovered, and so the luminosity function is poorly determined at the faint end. An important recent advance is the work of Dewey *et al.* (1985), who claim on the basis of more sensitive observations that the luminosity function definitely rolls over at $L < 1$ mJy kpc$^2$. Thus, we can now state with some confidence that there is unlikely to be a dominant class of single pulsars in the Galaxy that is completely unrepresented in the observed sample. Still, since the most numerous pulsars in the Galaxy are the faintest, and hence most poorly represented in the observed sample, therefore statistical studies of the galactic population are generally plagued by poor signal-to-noise. For instance, if $\Delta p$ is the error in some estimated parameter $p$, then $(p/\Delta p)^2$ will not be $\sim 400$ as one might naively expect for a sample of 400 pulsars, but may be $< 10$ in many cases. One frequently needs to introduce extra assumptions, usually in the form of a model of the time evolution of a pulsar, or a model of its radio luminosity, in order to improve the signal-to-noise ratio.

## 3. NUMBER OF ACTIVE SINGLE PULSARS IN THE GALAXY

A quantity of some interest is the number of active single pulsars $N_{psr}$ in the Galaxy. A straightforward way to obtain this is to normalize $\rho_R(R), \rho_z(z), \Phi(L)$ suitably so as to fit the observations, and then to integrate over these functions. (One would still need to allow for the beaming factor, discussed below.) However, we introduce here another approach (Narayan and Vivekanand 1981; Vivekanand and Narayan 1981, henceforth VN), which is equivalent as far as this discussion is concerned, but is of considerable value in the analysis of section 4.

Let us describe the present true population of pulsars in the Galaxy by the continuous function $\rho_t(P, \dot{P}, L) dP d\dot{P} dL$. The observed distribution $\rho_o(P, \dot{P}, L)$

differs from $\rho_t$ because of the selection effects of the surveys. Let us define the scale factor $S(P, L)$ as follows

$$S(P,L) = \frac{\int\int \rho_R(R)\rho_z(z)dRdz}{\int\int' \rho_R(R)\rho_z(z)dRdz} \tag{7}$$

where $\int\int'$ means that the integration is over only that part of the Galaxy where a pulsar with period $P$ and luminosity $L$ could have been detected by at least one of the surveys included in the calculation. A crucial requirement for the computation of $S(P, L)$ is a knowledge of all the selection effects of the surveys, including $P$-dependent effects arising from dispersion (Vivekanand *et al.* 1982), sampling (Dewey *et al.* 1984), and scatter-broadening (Manchester *et al.* 1985). The author has recently computed $S(P, L)$, including all known and suggested selection effects, for the following four surveys: Jodrell Bank Survey (Davies *et al.* 1972), U Mass-Arecibo Survey (Hulse and Taylor 1974), Second Molonglo Survey (Manchester *et al.* 1978), and U Mass-NRAO Survey (Damashek *et al.* 1978).

The meaning of $S(P, L)$ is that for every one pulsar of given $P$ and $L$ detected by the surveys considered, there are on average $S(P, L)$ similar ones in the whole Galaxy, which could be detected with sufficiently sensitive telescopes. The total number of potentially visible single pulsars $N_v$ in the Galaxy can thus be estimated to be

$$N_v = \sum_i S(P_i, L_i) \tag{8}$$

where the summation is over all single pulsars detected by the particular surveys with radio flux above the modeled minimum sensitivity limits of the surveys.

To obtain $N_{psr}$, we need to allow for one more factor *viz.*, the beaming fraction $f_i$, which represents the fraction of solid angle swept by the pulsar beam. If it is assumed that the conical beams that emerge from the two magnetic poles of a pulsar have circular cross-section, then one estimates $f \sim 0.2$ for all pulsars. However, there seems to be evidence that the beams are elongated in the meridional direction, with an elongation that is $P$-dependent (see Narayan, 1984, for a summary of earlier work). If we define $R$ to be the ratio of the NS to EW dimensions of the beam (where the directions are defined with respect to the rotation axis), polarization data suggest (Narayan and Vivekanand 1983) that

$$R \sim 1.8P^{-0.6} \tag{9}$$

Consequently, $f$ also has a $P$-dependence, with $f(P) \to 1$ at short periods $\lesssim 0.1s$, and decreasing to $f(P) \sim 0.2$ at long periods $\gtrsim 2s$. Including the beaming fraction, we thus estimate the total number of active single pulsars in the Galaxy to be

$$N_{psr} = \sum_i S(P_i, L_i)/f(P_i) = (1.5 \pm 0.5) \times 10^5 \tag{10}$$

The number of active single pulsars per kpc$^2$ in the solar vicinity is $150 \pm 50$. The errors quoted here and elsewhere in the article correspond to 95% confidence limits, and refer purely to the random statistical error. There could be additional systematic errors in the adopted distance scale and in the estimation of $f(P)$. The form of eq. (10) is obvious when one notes that the true and observed pulsar distributions are related by $\rho_t = \rho_o S(P, L)/f(P)$.

## 4. BIRTHRATE OF SINGLE RADIO PULSARS, AND "INJECTION"

If we take the pulsar population in the Galaxy to be in steady state, then we can follow VN and define the steady current of pulsars, $J(P)$, crossing from periods shorter than $P$ to larger then $P$ by

$$J(P) = \int\int \dot{P}\rho_t(P, \dot{P}, L)d\dot{P}dL \tag{11}$$

The mean current in the period-interval between $P_1$ and $P_2$ is then

$$\overline{J}(P_1, P_2) = \frac{1}{(P_2 - P_1)} \int_{P_1}^{P_2} J(P)dP \tag{12}$$

If all pulsars are born with periods $< P_{min}$ and die with periods $> P_{max}$, and if $P_{max} > P_{min}$, then the birthrate $BR$ of pulsars is given by $\overline{J}(P_{min}, P_{max})$ (see VN for a discussion). We have already seen that the factor $S(P, L)/f(P)$ relates the observed pulsar distribution $\rho_o$ to $\rho_t$. Thus, we can estimate $BR$ from the observed sample of pulsars as follows

$$BR \simeq \frac{1}{(P_{max} - P_{min})} \sum_i{}' \dot{P}_i S(P_i, L_i)/f(P_i) \tag{13}$$

where the summation $\sum'$ is over those pulsars detected by the various surveys with periods within the range $P_{min}$ to $P_{max}$. Arbitrarily selecting $P_{min} = 0.05s$, $P_{max} = 2s$, and using the current data, we find

$$BR \sim 1 \text{ psr in } 50^{+240}_{-30} \text{ yrs} \tag{14}$$

It should be noted that this method of estimating the birthrate is quite model-independent, in contrast to most other approaches, which assume some model regarding pulsar spindown (e.g. magnetic dipole braking), as well as a model for the variation of $L$ with $P$ and $\dot{P}$. The price we pay for the model-independence is that the statistical errors are extremely large.

To improve the signal-to-noise ratio, we now introduce the following model for pulsar radio luminosity

$$L \propto \dot{P}^{1/3} P^{-1} \tag{15}$$

Several previous studies have shown that $L$ increases with increasing $\dot{P}$ and decreases with increasing $P$, with exponents similar to those in (15) (Lyne *et al.* 1975; VN; Proszynski and Przybicien 1984). The particular form chosen here is somewhat natural since (i) $\dot{P}/P^3 \propto \Omega\dot{\Omega}$ is the total energy loss rate of a pulsar, and (ii) pulsars seem to switch off in the radio when they fall below a particular value of $\dot{P}/P^3$ (Taylor and Stinebring 1986).

The model (15) only gives the *mean* luminosity of a pulsar observed with period $P$ and period derivative $\dot{P}$. There is a great dispersion around this mean, which should also be included in the model. This can be done (the details will be reported elsewhere), and one can calculate a scale factor $S(P, \dot{P})$, which is defined in a manner similar to $S(P, L)$. Thus, the quantity $S(P, \dot{P})/f(P)$ represents, for every pulsar detected with a given $P, \dot{P}$, the expected mean number of similar pulsars in the whole Galaxy. Using these scale factors, we can again calculate the mean current $\overline{J}(P_1, P_2)$ between periods $P_1$ and $P_2$ as follows

$$\overline{J}(P_1, P_2) = \frac{1}{(P_2 - P_1)} \sum_i {}' \dot{P}_i S(P_i, \dot{P}_i)/f(P_i) \tag{16}$$

Fig. 3 shows the estimated $\overline{J}$ for the same four surveys considered earlier, binned suitably in $P$ in order to have a reasonable signal-to-noise ratio within each bin. The current appears to have a flat top between $P \sim 0.7s$ and $\sim 1.4s$. Using these values as $P_{min}$ and $P_{max}$, we then estimate

$$BR \sim 1 \text{ psr in } 60 \pm 20 \text{ yrs} \tag{17}$$

This is in reasonable accord with other recent determinations of pulsar birthrate (e.g., Vivekanand 1984; LMT), and is consistent with the rate of occurrence of supernovae, rate of formation of supernova remnants, and the rate of death of massive stars in the Galaxy (e.g. see LMT for a discussion). As recently as 5 years ago, estimates of $BR$ were embarrassingly high, typically 1 psr in ~7 yrs (e.g., Lyne 1981, Phinney and Blandford 1981). The key advances that have led to the present reduced estimates are (i) the realization by J. H. Taylor (see VN) that the variation of $L$ with $P$ and $\dot{P}$ is important and should be allowed for (this reduces $BR$ by a factor $\sim 2-3$), (ii) the introduction of non-circular beams (reduction factor $\sim 2$), and (iii) a modification of the distance scale (LMT). The pulsar birthrate per kpc$^2$ in the solar vicinity is estimated to be 1 per $\sim 7 \times 10^4$ yrs.

A surprising feature of Fig. 3 is that the pulsar current does not reach its maximum value until $P \sim 0.7s$. It is difficult to reconcile this result with the standard assumption that all new pulsars are, like the Crab pulsar, born spinning rapidly with $P \lesssim 20$ ms. VN noted this effect and termed the phenomenon as "injection", since new pulsars are being "injected" into the pulsar "stream" at intermediate points. It is unlikely that "injection" is due to a selection effect, since all suggested $P$-dependent effects have been included in the present calculation. Neither is it due to the assumed luminosity model (eq. 15), since a similar result is obtained even with a model-free calculation using $S(P, L)$, except with a much poorer signal-to-noise ratio. Also, the effect is present even if we eliminate the variable beaming fraction $f(P)$ and use a constant $f = 0.2$. We note that "injection" has been confirmed in an independent analysis by Chevalier and Emmering (1986). Also, Stokes *et al.* (1985) found fewer fast pulsars than expected in a new survey and concluded that many pulsars may be born spinning slowly. Srinivasan *et al.* (1984) studied the rate of formation of Crab Nebula-like plerions and again concluded that very few pulsars could be born spinning as rapidly as the Crab pulsar.

A further indication that "injection" is not the result of a statistical fluctuation, but possibly due to a real physical effect, is provided by Figure 4. Here the pulsars have been ordered according to the strength of their magnetic field (eq. 2), and divided into three equal groups. We see that "injection" is very strong for the high $B$ pulsars, not so prominent for the intermediate $B$ pulsars, and absent for the low $B$ pulsars. Using this as a clue, one could suggest three possible explanations for "injection". i) Many single pulsars may be "recycled" in massive binary systems before being released (e.g. articles by Srinivasan and van den Heuvel in this volume). High $B$ pulsars would be naturally "recycled" to longer periods. ii) Neutron stars with strong fields may be born slower because of magnetic braking during the red giant phase. Although the magnetic energy of a neutron star is negligibly small compared to the rotation energy, the two energies could be comparable *before* collapse, because they scale differently with radius, and so a magnetic coupling between the stellar core and its envelope is not unlikely. iii) Neutron stars may be born fast, but could turn on as radio pulsars only after slowing down (VN; Phinney and Blandford 1981), perhaps because magnetic fields build up after birth (Blandford, Applegate and Hernquist 1983). The results of Fig. 4 would require that high magnetic fields take longer to build up than low fields.

## 5. BIRTHRATE OF BINARY PULSARS

Of the 7 known binary radio pulsars, 6 were discovered in major surveys. It is therefore possible to roughly estimate the rate of formation of these systems. Van den Heuvel (e.g., this volume) divides the binary pulsars into two classes. The pulsars 0655+64, 1913+16, and 2303+46, have companions with masses $\gtrsim 1\ M_\odot$, and are believed to be produced from massive X-ray binaries (MXRB). The approximate birthrate of this class of binary radio pulsars in the Galaxy is 1 in $\sim 10^4 - 10^5$ yrs. The MXRB's in the Galaxy form at a much larger rate and can comfortably maintain this rate. The second group, consisting of 0820+02,

1831-00, 1855+09, and 1953+29, have companions with masses $\lesssim 0.4\ M_{\odot}$, and are believed to originate in low mass X-ray binaries (LMXB). Here there may be a problem with the birthrate (Kulkarni and Narayan, in preparation).

Consider the millisecond binary pulsar 1855+09, with $P = 5.362$ ms, $B \sim 10^9$ G, discovered by Segelstein *et al.* (1986) in their recent Princeton-Arecibo survey. Allowing for the volume of the Galaxy surveyed so far for such objects, we estimate $S(P, L)$ for this pulsar to be $\sim 10^5$. This somewhat astonishing result implies that there may be as many active binary millisecond pulsars in the Galaxy as ordinary single pulsars.

To estimate the birthrate of these objects, we need to know the age of 1855 +09. If the magnetic field is still decaying, then its age is $\tau \lesssim 10^8$ yrs, assuming a decay timescale of $10^7$ yrs. However, Kulkarni (1986) (see also Srinivasan, this volume) suggests that fields stop decaying after they reach a low value $\sim 10^8 - 10^{10}$ G. If this has happened to 1855+09, then it is slowing down by magnetic braking, and its age is $\sim 10^9$ yrs. Using this latter estimate, we find that the birthrate of 1855+09-like objects in the Galaxy is 1 in $\sim 10^4$ yrs. Although this is much smaller than the rate of formation of single pulsars, it is still much larger than the rate that can be maintained by LMXBs. The number of known LMXBs in the Galaxy is $\sim 30$, and they are presumed to live $\sim 3 \times 10^8$ yrs, which suggests that they can support a birthrate of only 1 in $\sim 10^7$ yrs. Unless we say that Segelstein *et al.* (1986) were extremely lucky to have found 1855+09 (which is unlikely since 1953+29, a similar system, was also discovered while searching a very small volume of the Galaxy, J. H. Taylor, private communication), we may be forced to conclude that either the galactic LMXBs are not the progenitors of 1855+09-like binary radio pulsars, or that the observed sample of LMXBs in the Galaxy is much more incomplete than we think.

## 6. REFERENCES

Blandford, R.D., Applegate, J.H., and Hernquist, L. (1983), *M.N.R.A.S.*, **204**, 1025.

Chevalier, R.A., and Emmering, R.T. (1986), *Ap.J.*, in press.

Clifton, T.R., and Lyne, A.G. (1986), *Nature*, **320**, 43.

Damashek, M., Taylor, J.H., and Hulse, R.A. (1978), *Ap.J.*, **225**, L31.

Davies, J.G., Lyne, A.G., and Seiradakis, J.H. (1972), *Nature*, **240**, 229.

del Romero, A., and Gomez-Gonzalez, J. (1981), *Astr. Ap.*, **104**, 83.

Dewey, R.J., Stokes, G.H., Segelstein, D.J., Taylor, J.H., and Weisberg, J.M. (1984), *Millisecond Pulsars*, Eds. S.P. Reynolds, D.R. Stinebring, NRAO, p. 234.

Dewey, R.J., Taylor, J.H., Weisberg, J.M., and Stokes, G.H. (1985), *Ap.J.*, **294**, L25.

Hulse, R.A., and Taylor, J.H. (1974), *Ap.J.*, **191**, L59.

Kulkarni, S.R. (1986), preprint.

Large, M.I. (1971), *The Crab Nebula, IAU Symp. No. 46*, Dordrecht: Reidel, p. 165.

Lyne, A.G. (1981), *Pulsars, IAU Symp. No. 95*, Eds. W. Sieber, R. Wielebinsky, Dordrecht: Reidel, p. 421.

Lyne, A.G., Ritchings, R.T., and Smith, F.G. (1975), *M.N.R.A.S.*, **171**, 579.

Lyne, A.G., Manchester, R.N., and Taylor, J.H. (1985), *M.N.R.A.S.*, **213**, 613 (LMT).

Manchester, R.N., and Taylor, J.H. (1977), *Pulsars*, San Francisco: Freeman.

Manchester, R.N., Lyne, A.G., Taylor, J.H., Durdin, J.M., Large, M.I., and Little, A.G. (1978), *M.N.R.A.S.*, **185**, 409.

Manchester, R.N., D'Amico, N., and Tuohy, I.R. (1985), *M.N.R.A.S.*, **212**, 975.

Narayan, R. (1984), *Millisecond Pulsars*, Eds. S.P. Reynolds, D.R. Stinebring, NRAO, p. 279.

Narayan, R., and Vivekanand, M. (1981), *Nature*, **290**, 571.

Narayan, R., and Vivekanand, M. (1983), *Astr. Ap.*, **122**, 45.

Phinney, E.S., and Blandford, R.D. (1981), *M.N.R.A.S.*, **194**, 137.

Proszynski, M., and Przybycien, D. (1984), *Millsecond Pulsars*, Eds. S.P. Reynolds, D.R. Stinebring, NRAO, p. 151.

Segelstein, D.J., Rawley, L.A., Stinebring, D. R., Fruchter, A.S., and Taylor, J.H. (1986), preprint.

Srinivasan, G., Bhattacharya, D., and Dwarakanath, K.S. (1984), *J. Ap. Astr.*, **5**, 403.

Stokes, G.H., Taylor, J.H., Weisberg, J.M., and Dewey, R.J. (1985), *Nature*, **317**, 787.

Taylor, J.H., and Stinebring, D.R. (1986), *Ann. Rev. Astr. Ap.*, in press.

Vivekanand, M. (1984), Ph.D. Thesis, Bangalore University.

Vivekanand, M., and Narayan, R. (1981), *J. Ap. Astr.*, **2**, 315 (VN).

Vivekanand, M., Narayan, R., and Radhakrishnan, V. (1982), *J. Ap. Astr.*, **3**, 237.

## DISCUSSION

**J. Taylor:** I like your suggestion that coupling to outer layers may explain the slow initial rotation rates of "injected" pulsars, but wouldn't coupling to an ejected supernova shell be as good as coupling to the outer layers of a red giant?

**R. Narayan:** Yes, it would, and the same coupling might also explain the orign of pulsar velocities and the correlation of velocity with magnetic field (Lyne, this volume). However, once the neutron star is formed, its rotation energy is so much larger than its magnetic energy that it is hard to visualize the field doing anything important. It is for this reason that I feel the extraction of angular momentum should take place before collapse, possibly over a long time in the red giant phase.

**S. Kulkarni:** I object to your extrapolation to small galactocentric radius (R) because pulsars are primarily young objects and hence their distribution should follow the molecular gas distribution. It is well known that, apart from some molecular gas at the Galactic center, there is a great deficiency of $H_2$ inwards of the "5-kpc ring".

**R. Narayan:** If we assume the galactic electron density model of LMT, then the results of the 1400 MHz survey of Clifton and Lyne (1986) suggest that there is little or no deficit of pulsars at small R. However, the electron density model was determined from previous data which did not probe the inner Galaxy. A self-consistent analysis of the new data to simultaneously solve for the electon density as well as the pulsar number density at small R might well show a concentration of pulsars in the 5-kpc ring. However, I believe that the functional form of $\rho_R(R)$ in eq. (5) is quite adequate for the limited purpose of estimating $N_{psr}$ and BR.

**F. Verbunt:** The X-ray sources in M31 are distributed in a central group (probably Pop II) and a ring (probably pulsars) around the center. The X-ray pulsars in our galaxy are also absent in the direction of the galactic center; in fact almost all X-ray pulsars are in the Carina arm. If there really are many radio pulsars in the galactic center area, doesn't this cause problems for the idea that radio and X-ray pulsars originate from the same O-star progenitors? What do we know about the O-star distribution near the galactic center?

**R. Narayan:** I believe the distribution of O stars in our Galaxy is traced by the giant HII regions, which are known to peak in the 5 kpc ring. Therefore, there would indeed be a problem if the pulsar distribution did not decrease in the galactic center region. The situation will become clearer after the data from the Clifton and Lyne (1986) survey are analyzed.

**J. Arons:** The argument that beams are elongated because of the lack of polarization sweep assumes the polarization-limiting radius is always deep in the dipole field region, independent of the rotation period and light cylinder distance. However, if the plasma density in the magnetosphere is sufficient to lead to collective emission, as in the case for the pair creation models, the polarization-limiting radius in the shorter period objects is well into the region where toroidal fields reduce the polarization sweep just as well as the elongated beam models. Therefore, the elongated beam model is not model independent, and is, in my opinion, less likely than the toroidal field model worked out by Barnard.

**R. Narayan:** I agree that Barnard's work nicely explains the polarization data without invoking an elongated beam. However, as reviewed in Narayan (1984), there are other arguments that support the elongated beam hypothesis. I would mention in particular the fact that interpulses occur very often in short period pulsars, and hardly at all in slower pulsars. This finds a natural explanation with period-dependent beam elongation, but would need additional hypotheses with a circular beam.

**W. Sieber:** The computation of pulsar luminosities assumes pulsar beam shapes which are box-like, i.e. there is either emission or no emission. Real pulsar beams are much smoother. Should one take this "smoothing" into account.

**R. Narayan:** You have raised an important problem for which I have so far found no simple solution. The hope is that the box-like beam can replace the true beam (which has a smooth fall-off of intensity) in some average sense, so that estimates of beam elongation, luminosity model, birthrate, etc. are not affected very much. It is important to make a more careful analysis of this effect.

**A. Blaauw:** Where would you suggest lies the dividing line in mass between the larger and the smaller mass groups in the binary pulsars?

**R. Narayan:** The division of the binary pulsars into two groups has been suggested by van den Huevel (this volume). In the smaller mass group, the companion white dwarf has a mass $\lesssim 0.4\ M_{\odot}$. These systems are believed to have evolved from binary stars with masses $\lesssim 8-10\ M_{\odot}$. In the larger mass group, the companion is either a neutron star with mass $\sim 1.4\ M_{\odot}$, or a heavy white dwarf with mass $\gtrsim 1\ M_{\odot}$. These systems are believed to originate from binary systems with masses $\gtrsim 10\ M_{\odot}$.

NEUTRON STAR COUPLING TO ITS ENVIRONMENT

Marco Salvati
Institute of Space Astrophysics Frascati, Italy

Franco Pacini
Arcetri Astrophysical Observatory
University of Florence, Italy

ABSTRACT. We review facts, myths and theories related to the formation of neutron stars and their coupling to the environment.

## 1. Introduction

In this paper we will consider only rotation-powered neutron stars, of the kind which sooner or later in their lifetime become "normal" radio pulsars. Of course, neutron stars in accreting binary systems are also very strongly coupled to their environment, but they belong to a different class of astrophysical objects (accretion-powered sources).

We will discuss mainly the outward-bound flow generated by the rotating neutron star, rather than a complete two-way interaction. The discussion will include the identification of the various components of the flow (particles, waves, and time steady electromagnetic fields), the estimate of their densities, and the assessment of their contributions to the global energetics. The one-sidedness of our approach is justified because during most of the active lifetime the outflow occurs in the form of a highly supersonic wind, which cannot carry upstream the reaction of the environment. One important exception is the birth of the neutron star in supernova explosion, when the ejecta are contiguous to the collapsing core and the interaction proceeds effectively in both directions.

Following the introduction, the paper is divided in three sections, arranged in chronological order. Section 2 deals with the very beginning of a neutron star lifetime, and encompasses a few year interval from its birth. Here, as we said, the coupling is very strong and determines the initial conditions, which in turn determine the subsequent evolution of the star through the pulsar phase. The observational information relevant to Section 2 is scarce and indirect. One can try and deduce something about the initial conditions from the present-day distribution of pulsars in parameter space. One can also hope to learn something about newly born pulsars



*D. J. Helfand and J.-H. Huang (eds.), The Origin and Evolution of Neutron Stars, 79–89.*

from the radio emission which characterizes Population I Supernovae, provided that such emission is indeed attributable to pulsar activity.

In Section 3, we discuss young objects, aged between $10^3$ and $10^4$ years. Only a few such objects are known, but they are very active, and are the site of a number of different phenomena. Among these, most relevant to our understanding of the global output are the high frequency optical and X-ray pulsed emission from the star, and the time steady spatially extended emission from the plerion, i.e., the synchrotron nebula which the pulsar inflates within the surrounding medium.

Section 4 is devoted to aged neutron stars, from $10^5$ years onward. This age bin contains most of the observed radio pulsars; the majority of them show up only through their radio pulsed emission. This radio emission is poorly understood and therefore it is not a good indicator of the large scale properties of the outflow. An exception is provided by the few cases where a small synchrotron nebula is observed; its origin and information content are analogous to the nebulae associated with the younger pulsars.

## 2. Pulsars At Birth

According to the commonly accepted scenario for neutron star formation, the evolution of the progenitor star is terminated by the collapse of the core and the explosive ejection of the envelope. In a conservative collapse the core angular velocity increases, and the magnetic field strength is amplified because of compression and differential rotation. It is thus expected that large magnetic stresses are built up between core and envelope, leading to a strong coupling. Although most Supernova models investigated until now do not incorporate the effects of rotation and magnetic fields in the collapse of the core and the expulsion of the outer layers, certainly those effects cannot be overlooked in real life.

Actually, from a purely energetic point of view, there is no difficulty in explaining through electromagnetic effects the expulsion of the outer layers and the entire SN phenomenon. This idea was put forward by Kardashev (1970), and elaborated upon by Ostriker and Gunn (1971), and Maceroni et al. (1974). Furthermore, in context of radio Supernovae, Bandiera et al. (1983) have pointed out that large stresses could induce the early fragmentation of the ejecta, and open low-opacity lines-of-sight to a central region.

The coupling between rotation and magnetic fields could also affect, in a major way, the final state of the collapsing core. We can give here two examples.

First, a simple minded expectation is that a strong magnetic field would result in a more effective removal of the core angular momentum, so that in a newly born neutron star field and rotation frequency could be anticorrelated. However, the anti-correlation between field and initial frequency can be argued against (Vivekanand and Narayan, 1981; Salvati, 1986) and may depend upon the ill-known details of the collapse.

Second, the speed of collapse could be substantially different depending on whether the magnetic field is able or not to rapidly remove the stellar angular momentum. If the field is strong and the angular momentum is removed quickly, rotation will not affect the stellar collapse which may take place very rapidly. The explosion of the outer layers would then be rapid and any binary system would be disrupted leading to the formation of fast moving pulsars. The fact that most pulsars are single stars has been interpreted as an evidence for the disruption of binary systems during the explosion of Supernovae. However, it has been pointed out that a relatively low magnetic field on the surface of the collapsing star would lead to much longer time-scales for the collapse and the explosion (Pacini 1983). Indeed the time-scale for removing angular momentum is proportional to $B^{-2}$ and weakly magnetized neutron stars ($10^8$ - $10^{11}$ gauss, instead of the "standard" $10^{12}$ gauss) would be formed in a time ~ months (or even longer). This could be longer than the orbital period in a close binary system and the system would not be disrupted. Pulsars formed in this way would have low velocities. This scenario is consistent with the available evidence that all binary pulsars have weak fields and small scale heights, and with the observed correlation between fields and velocities (Anderson and Lyne, 1973). The same scenario provides a framework in which to discuss the origin of weakly magnetized, millisecond pulsars different from the generally accepted notion that these objects have been "rejuvenated" and spun up after an accretion phase (see, e.g. Van den Heuvel 1984). In particular, we note that in this framework one expects that millisecond pulsars should, in a large percentage of cases (reflecting the original situation), be in binary systems and lie close to the galactic plane (this fact is regarded as a puzzle in the generally accepted theory, see Bhattacharya and Srinivasan 1986).

The uncertainties in the theory could be reduced by more direct observations of newly born pulsars. In this respect, the radio Supernova phenomenon could prove of crucial importance. A comprehensive summary of all the available data, and a discussion of the main competing models can be found in Weiler et al. (1986). Here it will suffice to say that non-thermal radio emission appears to be associated with all SNe of type II and type I peculiar; the luminosity rises with time scales of days to months, first at high, then at low frequencies; when the opacity effects are negligible, the evolution is by and large approximated by a negative power of the elapsed time. Modeling the phenomenon as a newly born plerion (Pacini and Salvati, 1981) has the advantage of economy , since a single assumption is invoked for the radio SNe and the evolved Crab-like SN Remnants; within this scheme one naturally expects a continuity between the two classes of objects.

Indeed, the radio SN light curves can be extrapolated smoothly into Crab-type luminosities.

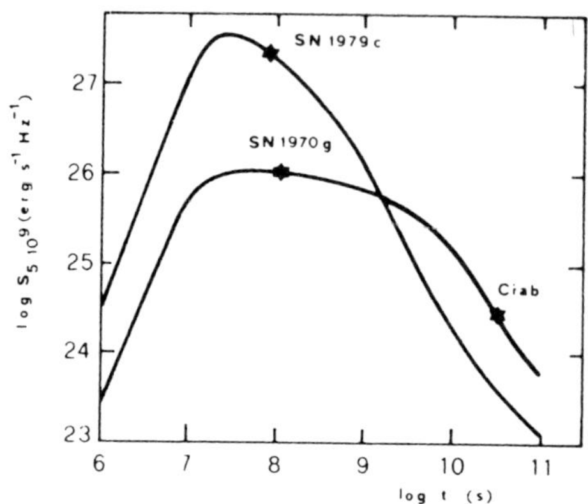


Fig. 1. Evolution of the radio flux at $5\times10^9$ Hz from two plerions. The upper curve postulates a central pulsar with $P_o \sim$ ms inside a remnant expanding with $v \sim 10^9$ cm $s^{-1}$. The lower curve postulates a central pulsar with $P_o \sim 16$ ms inside a remnant expanding with $v \sim 10^8$ cm $s^{-1}$ (Crab-type parameters). Note how the Crab Nebula had initial properties very similar to those of the Supernova 1970 g.

In the mini plerion model the total non-thermal luminosity and its time decline are clues to the initial period and field of the central pulsar. The critical point remains the difficulty of detecting the newly born pulsar directly; the reliability of the model thus rests solely on the combined statistics of plerions and radio SNe, and especially on the (still very poor) statistics of intermediate age objects. These are predicted to fill-in the gap, or to be completely absent, according to the scenario which is adopted (Salvati, 1987).

## 3. Young, Energetic Pulsars

By the time an isolated neutron star reaches the age of Crab or Vela, its coupling to the surroundings has taken the comparatively simple form of a supersonic outflow. Also, the debris of the explosion have been swept away from the central zone, and the star is clearly visible as a pulsar. The radio pulses are not suited to the study of the global output: their luminosity is only a minute fraction of the total pulsed luminosity, not to mention the total rotational $\dot{E}$, and their interpretation is especially cumbersome since the radiation mechanism is coherent. The incoherent, high frequency pulses, instead, carry most of the power, and provide useful constraints on the emitting region.

### 3.1 Inferences from Observations

Several arguments suggest that the incoherent optical emission from fast pulsars is due to the synchrotron process, involving particles streaming away from the neutron star. The emitting region is thought to be in proximity of the "speed of light distance", or at least at a distance from the central star which is a constant fraction of this characteristic distance (Shklovsky, 1970; Pacini, 1971). The time averaged spectrum of the Crab pulsar turns over gradually from the UV to the optical and to the near IR (e.g., Middleditch et al., 1983). A reasonable interpretation is that the onset of synchrotron self-absorption is being observed. Then at the turn-over frequency the flux can be expressed in two independent ways as a function of the particle density per unit energy interval, the orthogonal component of the magnetic field, and the emitting volume. One has to make assumptions about the latter: for instance, the geometry appropriate to a fan beam would be an azimuthal slice, with thickness proportional to $\delta$P ($\delta$ is the pulse duty cycle) and height and depth proportional to P. It follows that the particle number flux, $\phi$, and the typical Lorentz factor, $\gamma$, are

$$\phi \ \tilde{\geq} \ a \frac{L_\nu^3}{\nu^5 P^5 \delta^2}; \qquad \gamma \ \tilde{\geq} \ b \frac{L_\nu}{\nu^2 P \delta} \tag{1}$$

Here the constants a and b include universal constants, as well as our parameterization of the emitting volume; the inequality signs would apply if no turn-over were observed down the frequency $\nu$. For the Crab pulsar one obtains the canonical numbers $\phi \simeq 10^{40}$ s$^{-1}$, $\gamma \simeq 1.3\ 10^2$. Since Shklovsky's paper, though, two additional pulsars have been detected at optical frequencies, and the argument can, in principle, be extended to them. The Vela pulsar is very weak intrinsically, and no spectral information is available beside broad band colors (Manchester et al, 1978); eq. (1) reduces to a lower limit, so low as to be meaningless. The Magellanic Cloud pulsar PSR 0540-693 is quite similar to the Crab (Middleditch and Pennypacker, 1985); the lower critical frequency implied by the red color and the larger duty cycle tend to compensate in Eq. (1), so that Crab-like values for $\phi$ and $\gamma$ do not appear unreasonable. Note that the power associated with the particle flux, mc$^2$ $\gamma\phi$, is less than the total available $\dot{E}$.

The outflow produced by a young, active pulsar inflates a bubble within the surrounding medium; the bubble emits synchrotron radiation, and exhibits a characteristic filled-center morphology. These so-called plerions are catalogued and discussed elsewhere, for instance in Weiler (1983). Four of them contain an observed pulsar, and represent the only known associations between pulsars and SNRs of any kind. They are the Crab Nebula, LMC 0540-693, MSH 15-52, and the Vela SNR (associated with PSR 0531+21 0540-693, 1509-58, and 0833-45, respectively ).

The plerions work as integrators of the global pulsar output, the integration time being different for particles of different energies.

Detailed models have been built along these lines (Pacini and Salvati, 1973; Bandiera et al., 1984), and by parameter fitting one can discern the various components injected by the pulsars. For our purpose a simplified approach will be sufficient.

The spectrum of a typical plerion is flat in the radio domain, then it becomes steeper and steeper in the optical and in the X-rays. The bend is far from sharp, and does not correspond to the canonical change 0.5; nonetheless, we assume it to be due to synchrotron losses. We further assume that the age of the system is the slowing down time scale of the pulsar, and arrive at the following expressions

$$\gamma \simeq 1.6\ 10^5\ \nu_{13}^{2/3}\ t_{10.5}^{1/3}; \qquad \phi \simeq \frac{L}{mc^2\gamma}; \qquad B \simeq 7.1\ 10^{-7}\frac{\nu}{\gamma^2}\ \text{(Gauss)} \tag{2}$$

When applied to the Crab nebula Eq. (2) gives $\gamma \sim 1.6\ 10^5$, $\phi \sim 7.8\ 10^{38}\ s^{-1}$, and $B \sim 2.3\ 10^{-4}$ G; for LMC 0540-693 the same quantities are $4.1\ 10^4$, $8.0\ 10^{38}\ s^{-1}$, and $4.2\ 10^{-4}$ G, respectively; in both cases there is approximate equipartition between particle and field energy. On the other hand the plerion component of MSH 15-52 is not detected, and the Vela plerion is so weak as to be reminiscent of the older nebulae described in Section 4 (Harnden et al., 1985). Notwithstanding all the uncertainties, the data indicate that the particle number flux at the nebula is roughly the same as at the pulsar. The associated power, instead, is much larger, and comparable with $\dot{E}$. We deduce that at the pulsar most of the energy flux must reside in carriers different from particles, either large amplitude waves or time steady e.m. fields.

We thus obtain a picture where the global outflow runs from the pulsar to the nebula with conserved total energy and total number flux. The individual particles are accelerated somewhere along the road, so that $\sigma$ (the ratio of field energy to particle energy) starts $\gg 1$ and arrives $\sim 1$. For the sake of the following discussion, we stress that the latter condition is "observed" to hold only in the main body of the nebula: without a certain amount of modeling, nothing can be said about the behavior of $\sigma$ in the intermediate zone.

### 3.2 Theories

We start from the pulsar side with a brief reminder of the basic electrodynamics (Gold, 1968; Pacini, 1968; Goldreich and Julian, 1969). The pulsar is modeled as a magnetized, rotating neutron star. Rotation and magnetization, plus a highly conductive constituent material, imply large potential differences, and most of the theory is about the ways of tapping this potential. The locus of the points where rigid corotation would take place at the speed of light is the light cylinder, and the field lines which do or do not cross the cylinder define the open and closed magnetosphere, respectively. Only the open magnetosphere is connected to the outside world, and the pulsar outflow originates from there; hence the relevant potential drop is the one occurring across the open magnetosphere, that is, in a dipole geometry,

$$\Delta V \sim 6.6\ 10^{12}\ \frac{B_{12}}{P^2}\ \text{(Volts)};\quad I \sim 5.9\ 10^{11}\frac{B_{12}}{P^2}\text{(Amps)};\quad \dot{E} \sim I\Delta V \qquad (3)$$

In Eq. (3) a characteristic current is also given; it corresponds to the sign reversal of the light cylinder field (again estimated in a dipole geometry) over a light cylinder radius. It has long been known that for the Crab pulsar $I/e \sim 1.3 10^{34}\ s^{-1}$, much less than the particle flux estimated from the optical pulses (the real currents must not be completely charge separated). At any rate, the power I $\Delta V$ is a good approximation to $\dot{E}$ in all variants of the model.

In an empty magnetosphere $\Delta V$ would appear both across and along the field lines; since along the lines there is negligible resistivity, a real magnetosphere becomes filled up with plasma, distributed so as to cancel out $\Delta V$ (parallel). The asymptotic condition $\Delta V$ (parallel) = 0 cannot be reached everywhere, and the regions where it is violated-the gaps-are suited to particle acceleration. With some subtle exceptions, $\Delta V$ (gap) is substantially smaller than $\Delta V$ (pulsar). Its growth is limited by electro-optical cascades, ignited by gap-energized primary particles. Then I (gap) and the energy production rate depend on how many field lines cross the accelerating region.

The various types of gaps, their location, their efficiency and relevance within the global machinery are the subject of a specialized literature (see, for instance, Ruderman and Sutherland, 1975; Ruderman, 1987; Arons and Scharlemann, 1979; Arons, 1983). A few important points can be extracted, on which a satisfactory consensus has been reached.

The gaps occurring close to the star surface produce very low powers; they might be relevant to the radio pulsed emission, and under this assumption one reproduces the death line as observed in the P-$\dot{P}$ diagram. The gaps occurring at higher altitudes are wider, and produce larger powers; this is still less than $\dot{E}$, but is sufficient to account for the high frequency pulsed radiation of young pulsars. Even if the overall energetics are settled, the details of the radiation mechanisms are not. The particle number flux is much larger than the minimum charge-separated flux estimated before. The multiplication is due to electro-optical cascades, so the main constituents are $e^+$ $e^-$ pairs. Without straining the model parameters too much, it is possible to reproduce the "observed" fluxes; note however that an agreement on integral quantities-energy and number-does not necessarily imply an agreement on distribution shape.

As for the theory of plerions, we have seen that kinematical models allow one to quantify the various components to be injected. In order to connect the plerion to the pulsar wind one needs dynamical models, which take into account spatial gradients (Rees and Gunn, 1974; Kennel and Coroniti, 1984). The incoming wind is highly relativistic, and must be slowed down to the speed of the plerion boundary; this is achieved through a shock, which can be strong enough, i.e. effective enough in decelerating the flow, only if the upstream energetics are dominated by the particles, $\sigma \ll 1$. After the shock $\sigma$ is also small, but the particle pressure does work on the field until in the main body of the nebula a rough equipartition is established.

We see that theory and observation are in order-of-magnitude agreement at the two ends of the flow; only, we do not know how to put the two pieces together. There is a region outside the light cylinder and inside the inner plerion boundary where, all other quantities conserved, $\sigma$ changes from $\gg 1$ to $\ll 1$, and we have no information, theoretical or observational, on how this change comes about.

It might be that large amplitude waves plays a role in this context; they cannot propagate in a density much larger than the charge-separated one (Asseo et al., 1981), and their energy must be transferred to the particles. The details are unclear, and at any rate at the light cylinder one expects approximate equipartition between wave component and steady e.m. components: even if the radiation reaction were negligible, this would imply $\sigma \sim 1$, only half-way to what is needed.

## 3.3 Acceleration of Heavy Ions

If spin axis and magnetic axis point to the same hemisphere, so that electrons are required above the polar caps, the presence or absence of ions in the pulsar outflow depend on the ill-known structure of the return currents. However, one pulsar out of two should be born with antipodal axes, and should inject a ion flux equal to the charge-separated minimum. No multiplication occurs, since the ions cannot ignite electro-optical cascades; for this one needs stray electrons, which guarantee a pair plasma and a gap structure analogous to the previous case (Arons, 1983). The ions go through the same potential differences as the pairs, so the energy flux associated with them is likewise $\ll \dot{E}$.

A distinguishing feature of the ions is that they can attain the maximum energy corresponding to the given $\Delta V$, whereas their lighter counterparts are limited by radiation reaction. Hence the $10^{15}$ eV radiation from the Crab pulsar could only arise from ions, and-if confirmed-would provide a strong indication that heavy particles are being accelerated.

## 4. X-ray Nebulae Around Aged Pulsars

The Einstein satellite has surveyed a sample of radio pulsars, old enough for the SN Remnant to be completely dissipated. The sample members were selected according to small distance and large $\dot{E}$, so as to maximize the probability of detection. At the position of three of them-namely, PSR 0355+54, 1055-52, and 1642-03-time steady, spatially extended emission was detected (Helfand, 1983). The author interprets his results as evidence of a threshold effect, whereby all and only the pulsars with $\dot{E} > 10^{34}$ erg $s^{-1}$ maintain around them synchrotron nebulae roughly 1% efficient.

The nebulae differ by orders of magnitude in surface brightness, but even the most diffuse ones could hardly be confined by the static pressure of the interstellar medium. The source around PSR 1055-52 is definitely out of balance; a dynamic confinement is possible, if the pulsar is moving at a few 100 km $s^{-1}$ in a density of a few $cm^{-3}$ (Cheng, 1983). If we take an equipartition field of the order of $10^{-4}$-$10^{-5}$ gauss, the production of X-rays via the synchrotron process requires accelerating potentials of $\simeq 10^{14}$ volts. It is interesting to note that, according to Eq. (3), the maximum available $\Delta V$ falls in this range precisely when $\dot{E}$ is around $10^{34}$ erg $s^{-1}$.

In our view the agreement provides comforting evidence that the pulsar outflow continues into the middle age; that the global properties of the magnetosphere are described by eq. (3); and, finally, that the models based on the gap idea are right in predicting less and less pair production in older and older pulsars: in the end, the gaps widen to include the entire magnetosphere and the full $\Delta V$ (pulsar) becomes available locally.

## 5. Conclusions

The basic principles of a pulsar's working are, we think, rather firmly established. The structure of the magnetosphere, the pair production, the fate of the large amplitude waves, the asymptotic behavior of the wind are at least qualitatively understood.

There remain a few key points which have not yet been worked out, albeit grossly. One such point is the problem of the initial conditions in which a neutron star is born; here the most plausible approach is the improvement of the statistics and the assessment of the selection effects both in the P-$\dot{P}$ diagram and in the radio SN evolution.

The presence of heavy ions in the flow can be proven in a clearcut way by the detection of pulsed ultra high energy radiation, although in practice the issue is not so straightforward.

Finally, the problems about the flow properties in the intermediate region are well exemplified by the actual observations of the Crab nebula. There the region from the pulsar to the wisps (which are believed to mark the position of the shock and the beginning of the plerion) is completely black. The same blackness is a good representation of the state of our understanding of the detailed physical process produced by the pulsars but taking place in the extended volume which surrounds the central star.

Acknowledgements

One of us (F.P.) gratefully acknowledges the hospitality of the Center for Astrophysics, Cambridge Massachusetts and the partial support of NASA Contract 8-30751 while this paper was being written.

References

Anderson, B., and Lyne, A.G.: 1983, Nature 303, 597
Arons, J.: 1983, Astrophys. J. 266, 215
Arons, J., and Scharlemann, E.T.: 1979, Astrophys.J. 231, 854
Asseo, E., Llobet, X., and Pellat, R.: 1981, in "Pulsars", Proceedings of IAU Symp. No. 95, eds. W. Sieber and Wielebinski, Reidel, Dordrecht, p. 53
Bandiera, R., Pacini, F., and Salvati, M.: 1983, Astron. Astrophys. J. 126, 7
Bandiera, R., Pacini, F., and Salvati, M.: 1984, Astrophys. J. 285, 134
Bhattacharya, D., and Srinivasan, G.: 1986, Curr. Sci. 55, 327
Cheng, A.F.: 1983, Astrophys. J. 275, 790
Gold, T.: 1968, Nature 218, 731
Goldreich, P., and Julian, W.H.: 1969, Astrophys. J. 157, 869
Harnden, F.R., Grant, P.D., Seward, F.D., and Kahn, S.M.: 1985, Astrophys. J. 299, 828
Helfand, D.J.: 1983, in "Supernova Remnants and their X-ray Emission", Proceedings of IAU Symp. no. 101, eds. I.J. Danziger and P. Gorenstein, Reidel, Dordrecht, p. 471
Kardashev, N.S.: 1970, Soviet Astron. AJ 14, 375
Kennel, C.F., and Coroniti, F.V.: 1984, Astrophys J. 283, 694
Maceroni, C., Salvati, M., and Pacini, F.: 1974, Astrophys. Space Sci. 28, 205
Manchester, R.N., et al.: 1978, MNRAS 184, 159
Middleditch, J., and Pennypacker, C., and Burns M.S.: 1983, Astrophys. J. 273, 261
Ostriker, J., and Gunn, J.: 1971, Astrophys. J. 164, L95
Pacini, F.: 1968, Nature 219, 145
Pacini, F.: 1983, Astron. Astrophys. 126, L11
Pacini, F.,and Salvati, M.: 1973, Astrophys. J. 186, 249
Pacini, F., and Salvati, M.: 1981, Astrophys. J. 245, L107
Rees, M.J., and Gunn, J.E.: 1974, MNRAS 167, 1
Ruderman, M.A.: 1987, in "High Energy Phenomena around Collapsed Stars", Proceedings of a NATO ASI held in Cargese in September 1985, ed. F. Pacini.
Ruderman, M.A., and Sutherland, P.: 1975, Astrophys. J. 199, 51
Salvati, J.: 1987, in "High Energy Phenomena around Collapsed Stars", Proceedings of a NATO ASI held in Cargese in September 1985, ed. F. Pacini.
Shklovsky, I.S.: 1970, Astrophys. J. 159, L77

van den Huevel, E.P.J.: 1984, J. Astrophys. Astron. 5, 209
Vivekanand, M., and Narayan, R.: 1981, J. Astrophys. Astron. 2, 315
Weiler, K.W.: 1983, in "Supernova Remnants and their X-ray Emission", Proceedings of IAU Symp. no. 101, eds. I.J. Danziger and P. Gorenstein, Reiden, Dordrecht, p. 299
Weiler, K.W., Sramek, R.A., Panagia, N., van der Hulst, J.M., and Salvati, M.: 1986, Astrophys. J. 301, 790

# CRAB-LIKE SUPERNOVA REMNANTS

R.H. Becker
Physics Department
University of California
Davis, CA 95616
and
Institute of Geophysics and Planetary Physics
Lawrence Livermore National Laboratory

ABSTRACT. Crab-like SNR are markers for recently formed pulsars. The current catalog of 15 objects allows a direct measure of the space distribution of pulsars, the beaming factor of pulsars, and the energetics at young pulsars. The 15 Crab-like SNR are equally divided between those with surroundings shells and those without. No other properties appear to correlate with the presence or lack of a shell. The ratio of 10 to 1 between all remnants and Crab-like remnants has important implications for the formation of pulsars.

## 1. INTRODUCTION

Although we know of hundreds of pulsars, only four of them are demonstrably younger than $\sim 10^4$ years: the Crab pulsar, the Vela pulsar, MSH15-52, and 0540-693. This is a small number in comparison with the ~150 known galactic supernova remnants (SNR). It even lags behind the number of galactic historical supernovae. The uncertainty in the selection effects which hinder detection of pulsars makes it difficult to extrapolate from the four observed young pulsars to the population at large.

We can extend our knowledge of young pulsars by studying the associated population of Crab-like SNR of which ~15 are currently known in the galaxy. In this paper, I will present the current status of the observational data for Crab-like SNR and indicate what the implications are for young pulsars.

## 2. A CATALOG

Studies of Crab-like SNR have developed rather slowly over the past 15 years with a fair amount of confusion and backtracking. The IAU Symp. 46 on The Crab Nebula held in 1970 only contains reference to 3 Crab-like objects. At that time, Minkowski (1971) pointed out



*D. J. Helfand and J.-H. Huang (eds.), The Origin and Evolution of Neutron Stars, 91–97.*

that "The frequency of pulsars .... is indeed in complete agreement with the observed frequency of supernovae. One might wonder why there is not a pulsar in every supernova remnant." He could well have asked the same question in regard to Crab-like SNR. The absence of pulsars has generally been explained in terms of beaming, high dispersion measures, and low sensitivity. The absence of Crab-like SNR remains unsolved.

By 1982, at the IAU Symposium #101 entitled SNR and Their x-ray Emission, the number of proposed Crab-like SNR had grown to ~20 and two subclasses had emerged (Weiler 1971, Becker 1971). In addition to sources similar to the Crab Nebula, some Crab-like objects had been discovered within more traditional SNR shells. For the rest of this paper I will refer to the two subclasses as "Crabs" and "Crab Shells". Of the ~20 aforementioned objects, 8 were Crabs and 12 were Crab Shells.

In retrospect, many of the identifications assert in 1982 were premature and today I would put the number of secure identifications at 15, 8 Crabs and 7 Crab Shells. (This count excludes extragalactic members of the class). In Table I, I list the 15 sources and also

TABLE I. Catalog of Crab-like SNR

| Crabs |
|---|
| G20.0+0.2 (Becker & Helfand 1985) |
| G21.5-0.9 (Davelaar et al. 1986) |
| G54.1+0.3 (Reich et al. 1985) |
| G74.9+1.2 (Wilson 1980) |
| G130.7+3.1 (Green 1986) |
| G184.6-5.8 (Velusamy 1985) |
| G291.0-0.1 (Wilson 198 ) |
| G328.4+0.2 (Caswell et al. 1980) |
| **Crab Shells** |
| G0.9+0.1 (Helfand & Becker 1986) |
| G24.7+0.6 (Reich et al. 1984) |
| G29.7-0.3 (Becker & Helfand 1984) |
| G263.9+2.8 (Harnden et al. 1985) |
| G320.4-1.2 (Seward et al. 1984) |
| G326.3-1.8 (Milne et al. 1985) |
| G351.2+0.1 (Becker & Helfand 1986) |
| **Other SNR with compact sources** |
| G27.4+0.0 (Kriss, G. et al. 1985) |
| G39.7-2.0 (Downes et al. 1986) |
| G68.9+2.8 (Strom & Blair 1985) |
| G109.1-1.0 (Hughes et al. 1984) |
| G332.4-0.4 (Tuohy et al. 1983) |

include 5 addition SNR which appear to contain neutron stars which are not pulsars.

Of the 15 Crab-like objects 3 contain pulsars and a fourth contains an unresolved x-ray source at its center. Does this imply a beaming factor of 4? More likely 4 should be taken as a lower limit in so far as most of these sources have not undergone exhaustive searches for pulsars. Such searches should be of the upmost importance for they would permit a direct measure of the beaming factor of young pulsars.

## 3. CRABS AND CRAB SHELLS

We do not know if Crabs and Crab Shells are truly two distinct phenomena with different types of progenitors or if they instead a continuous gradation of core vs. shell luminosity. In figure 1, I show a histogram of the number of Crab Shells as a function of the ratio of core to shell luminosities. In addition I have plotted the lower limit of the ratio for two Crabs, the Crab Nebula (Velusamy 1983) and 3C58 (Reynolds and Allen 1985). Among the Crab Shells, the ratio spans 3 orders of magnitude while the inclusion of Crabs brings the range to over 5 orders of magnitude. The lower end of the range may be an observational limit. It would be difficult to observe a Crab-like component which only contributes <.001 of the total luminosity. I see no compelling rationale for distinguishing between the two kinds of Crab-like SNR based on the presence or absence of a shell.

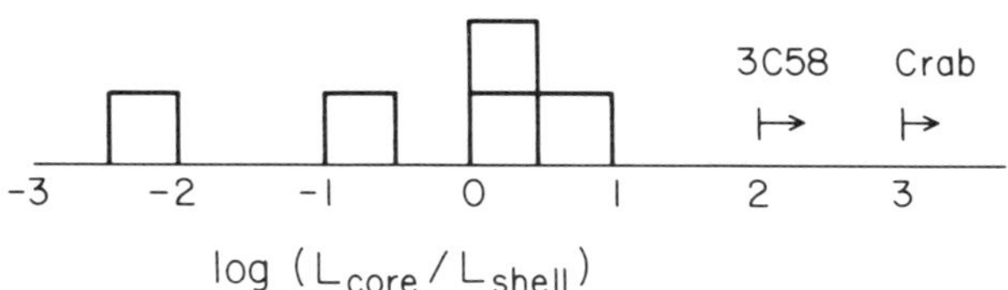


Figure 1. Ratio of core luminosity to shell luminosity.

Figure 2 displays the distribution in galactic latitude of Crabs and Crab Shells. In fact, they are surprising similar in so far as the vastly different selection effects involved in their discovery. The Crab Shells are surrounded by steep spectrum shells which are readily discernible at low frequencies while the Crabs have flat spectra and are easily confused with HII regions. In Figure 3, the z-distribution of Crabs and Crab Shells are plotted. Again we see little significant difference. The distribution of B stars has been plotted as well. There may be some evidence that Crab-like SNR are formed further from the plane than massive stars, the commonly accepted progenitors.

Recently, it has become popular to consider the ratio of x-ray to radio luminosity as a way to rank Crab-like objects. In Table II, the

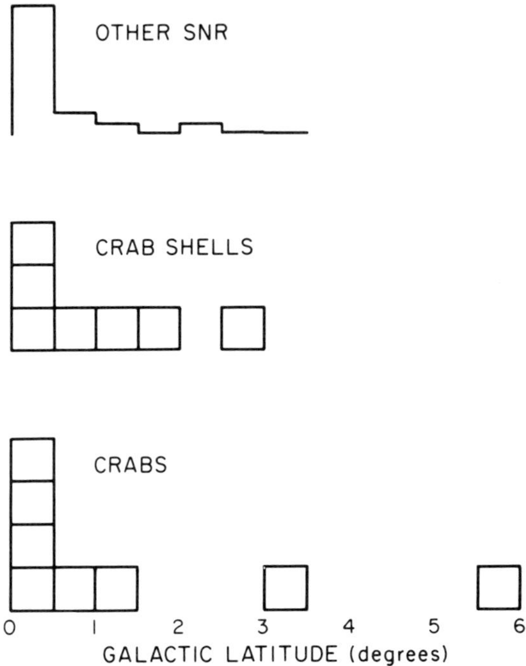


Figure 2. Latitude distribution of Crabs, Crab Shells, and other SNR.

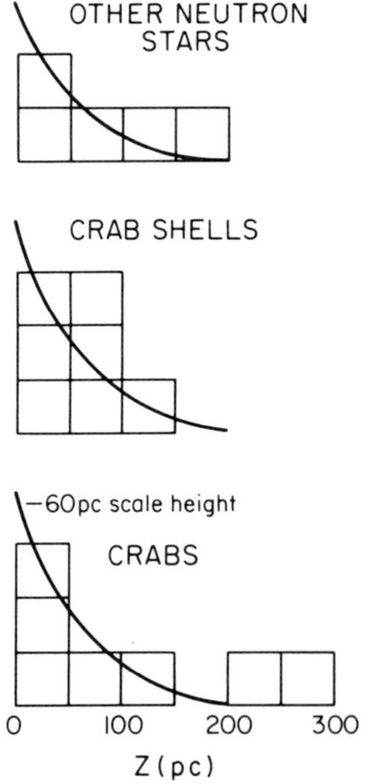


Figure 3. Z-distribution of Crabs, Crab shells, and other SNR containing neutron stars.

values of $L_x/L_R$ for Crabs and Crab Shells alike are listed in descending order. Again no clear difference between Crabs and Crab Shells emerge.

TABLE II. Ratio of X-ray to Radio Luminosity for Crab-like SNR

| Source | $L_x/L_r$ |
|---|---|
| G29.7-0.3 | 500 |
| Crab | 100 |
| G21.5-0.9 | 15 |
| MSH 15-52 | $\geq$10 |
| G74.9+1.2 | 1.3 |
| G291.0-0.1 | 1 |
| G0.9+0.1 | $\leq$.5 |
| 3c58 | .5 |
| Vela X | .1 |

We would like to use Crab-like SNR as a means of studying pulsar energetics. This is most reliably done by measuring the x-ray flux from Crab-like remnants. Since the x-ray emitting electrons rapidly radiate away their energy, the current x-ray luminosity should be closely correlated to the energy being put into the acceleration of particles by the pulsar. Ideally, we would then relate this number to the total energy less rate of the pulsar. Helfand (1984) has tabulated the x-ray luminosity of Crab-like SNR and several older pulsars and the energy loss rate of the associated pulsar if there is one. It seems that in most cases 1-5% of the pulsar energy losses are converted into x-ray emission over a wide range of pulsar ages. Therefore, we can reasonably estimate pulsar energetics even when the pulsar is not observable.

## 4. PULSAR BIRTH RATES

Previously, pulsar birthrates have been calculated based on pulsar surveys. These estimates depend on the value of the beaming factor used. Since the emission from Crab-like SNR are not beamed, they might provide a less model-dependent birthrate estimate. Presently we know of ~15 SNR which show evidence for a young pulsar. This represents ~10% of the total number of known galactic SNR implying a birthrate only ~20% of that inferred from pulsar statistics (assuming a beaming factor of 5). How can we reconcile this inconsistency? I will suggest four possible solutions.

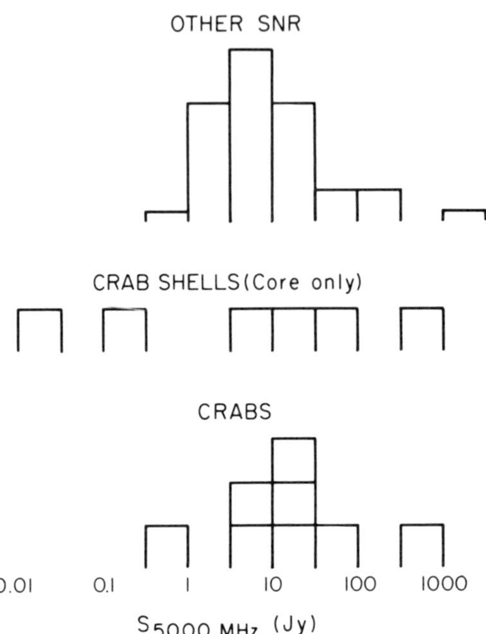


Figure 4. Radio flux at 5000 MHz of Crabs, Crab shells, and other SNR.

The most trivial solution is that there are many more Crab-like SNR yet to be discovered but remain hidden due to observational selection effects. This could be the case if most Crab-like remnants have 6 cm fluxes below several Janskies. The histograms in Fig. 4 show the distribution of 6 cm flux for the known Crabs and Crab Shells. Note particularly that the cores of two Crab Shells have fluxes below 0.5 Jy. It would be easier to accommodate a large increase in Crab Shells from the known population of shell remnants because this would not change the overall birthrate estimates. The discovery of ~100 weak Crabs would upset the current equality of SN with SNR birthrates. Alternatively, the small percentage of Crab-like SNR could result if Crab-like SNR have lifetimes short compared to shell remnants. This possibility is in some sense a variant of suggesting that most Crab-like remnants are too weak to observe.

Others have suggested that the majority of pulsars are born as slow rotators (period > 0.3 sec) and hence are never energetic enough to create synchrotron nebula. These conclusions are supported by studying the observed population of old pulsars and evolving them backwards to infer the properties of the initial population. This argument would gain greater support if a slow pulsar were found associated with a SNR. To date, no such occurrence has been found.

Lastly, we could accept the observational result that only ~10% of SNR contain pulsars. That is to say only 10% of galactic SN result in the formation of a pulsar. Reconciliation of this result with the observed space density of pulsars would require a beaming factor of order unity rather than the nominal value of 5.

## REFERENCES

Becker, R.H. 1983 in Supernova Remnants and Their X-ray Emission (ed. by Danziger and Gorenstein) D. Reidel, Dordrecht, p. 329.
Becker, R.H. and Helfand, J. 1984 Ap. J. 281, 650.
Becker, R.H. and Helfand, J. 1985 Ap. J. (Letters) 297, L25.
Becker, R.H. and Helfand, J. 1986 in preparation.
Caswell et al. 1980, MNRAS 190, 882.
Davelaar et al. 1986 Ap. J. (Letters) 300, L59.
Downes et al. 1986 MNRAS 218, 393.
Green 1986 MNRAS 218, 533.
Harnden et al. 1985 Ap. J. 299, 828.
Helfand, D.J. 1984, Adv. Space Res. 3, 29.
Helfand, D.J. and Becker, R.H. 1986 Ap. J., in press.
Hughes et al. 1984, Ap.J. 283, 147.
Kriss, G. et al. 1985 Ap. J. 288, 703.
Milne, D. et al. 1985 Proc. Astron. Soc. Aust. 6, 78.
Minkowski, R. 1971 in The Crab Nebula (ed. by Davies and Smith) D. Reidel, Dordrecht, p. 241.
Reich et al. 1985 Astron. Astrophys. 151, L10.
Reich et al. 1984 Astron. Astrophys. 133, L4.
Reynolds, S.P. and Aller, H. 1985 A.J. 90, 2312.
Seward, F. et al. 1984 Ap. J. 281 650.
Strom, R. and Blair, W. 1985 Astron. Astrophys. 149, 254.
Tuohy, et al. 1983 Ap. J. 268, 778.
Velusamy, T. 1985 in The Crab Nebula and Related Supernova Remnants (ed. by Kafatos and Henry) Cambridge Univ. Press, p. 115.
Weiler, K.W. 1983 in Supernova Remnants and Their X-Ray Emission (ed. by Danziger and Gorenstein) D. Reidel, Dordrecht, p. 299.
Wilson, A. 1980 Ap. J. (Letters) 241, L19.
Wilson, A. 1986 Ap.J.

## DISCUSSION

**S. Woosley:** Is there any case of a pulsar existing in an oxygen-rich SNR. Could pulsar-containing remnants all be compositionally like the Crab, i.e., helium-rich and little else? Is the rotation rate of the pulsar especially slow in such cases (if they exist)?

**R. Becker:** The LMC pulsar 0540-693 is very similar to the Crab in all its nonthermal properties but is surrounded by an oxygen-rich shell. Many of the other Crab-like nebulae are in remnants that have not been detected optically.

# NEUTRON STARS IN TWELVE SUPERNOVA REMNANTS

Frederick D. Seward
Harvard Smithsonian Center for Astrophysics
60 Garden Street, Cambridge MA 02138

ABSTRACT. X-ray observations of selected SNR are summarized. Five contain internal spinning neutron stars--four isolated and one in a binary system. Another seven contain central unresolved sources or bright nebulae. Observations of these nebulae, probably due to synchrotron emission, are used to estimate characteristics of the unseen pulsars.

## 1. Introduction

Neutron stars are believed to originate in the gravitational collapse of cores of massive stars, which is calculated to be very rapid, and deep within the interior of red giant stars. When the resulting shock reaches the outer layers, the visible light of a supernova is produced, and at much later times we observe energy from the interaction of the expanding stellar debris with surrounding material. The shells of supernova remnants are visible for thousands of years and most can be observed at radio, optical and X-ray wavelengths. The shells of remnants which contain pulsars (= neutron stars) can reveal something about the precursor star, perhaps the mass and composition of the outer layers at the time of collapse.

If the neutron star is isolated its period (P) and period derivative ($\dot{P}$) can be used to calculate its characteristic age (A), and the rate of rotational energy loss ($\dot{E}$). Rotational energy is transformed to a relativistic wind of charged particles and magnetic field. High energy electrons radiate synchrotron radiation in the magnetic field and young pulsars observed within supernova remnants are apparently always surrounded by diffuse synchrotron emission. Observations have been summarized in recent reviews listed by Seward (1985).



*D. J. Helfand and J.-H. Huang (eds.), The Origin and Evolution of Neutron Stars, 99–108.*

## 2. The Crab Nebula

The supernova of 1054 and the Crab Pulsar remain the only indisputable example of a supernova associated with the formation of a neutron star. The location of the remnant is the same as that of the SN, the characteristic age ($P/2\dot{P}$) of the pulsar is only 30% higher than the age of the remnant, and the calculated pulsar rotational energy loss is exactly that needed to power the bright nebular radiation observed at all wavelengths.

In the X-ray band the Crab is the brightest SNR in our galaxy. All-sky X-ray surveys have found no other SNR as luminous in the Milky Way. 96% of the soft X-ray luminosity comes from the synchrotron nebula, which has the highest surface brightness of any diffuse X-ray source observed by Einstein (Harnden & Seward, 1984). It is this bright synchrotron nebula that gives the Crab its distinct character. Why is this so strong?

Among SNR, the Crab has two anomalous features: the rapid pulsar, which deposits energy in the interior, and the low expansion velocity of the optical filaments. These filaments, formerly the outer layer of the star, confine the pulsar-deposited energy to a relatively small volume and the consequent high energy-density gives the Crab its unique properties. Optical and X-ray continuum radiation is confined to the volume bounded by the filaments. Radio emission is brightest close to bright filaments, indicating a compression of the nebular magnetic field flowing from the vicinity of the pulsar. The spectral measurements of Henry & McAlpine (1982) indicate that the filaments contain more than one solar mass of hydrogen and helium, and confirm the finding of others that helium is overabundant. Henry and McAlpine calculate that 60% of the mass is helium. This is the only evidence for a substantial enrichment over solar abundance in the debris from SN1054.

The X-ray synchrotron nebula occupies a small volume at the center. The brightest emission comes not from an area surrounding the pulsar, but is centered approximately 10" northwest of the pulsar, indicating the existence of processes which accelerate electrons within the nebula, rather than a simple flow of high-energy electrons from the pulsar which lose energy as they diffuse outward. The small size of the X-ray nebula shows the lifetime of the relativistic electrons producing X-rays is shorter than the time it takes these electrons to diffuse to the edge of the Crab.

After the discovery of the Crab pulsar in 1968, the early evolution of pulsars was thought to be understood: Pulsars form spinning rapidly and lose energy at a high rate. We have found, however, no other object in our galaxy with a synchrotron nebula as bright as that of the Crab. Pulsars and surrounding nebulae in other remnants are all less luminous than in the Crab. The neutron star within the Crab Nebula is the brightest object in its class rather than a typical young pulsar.

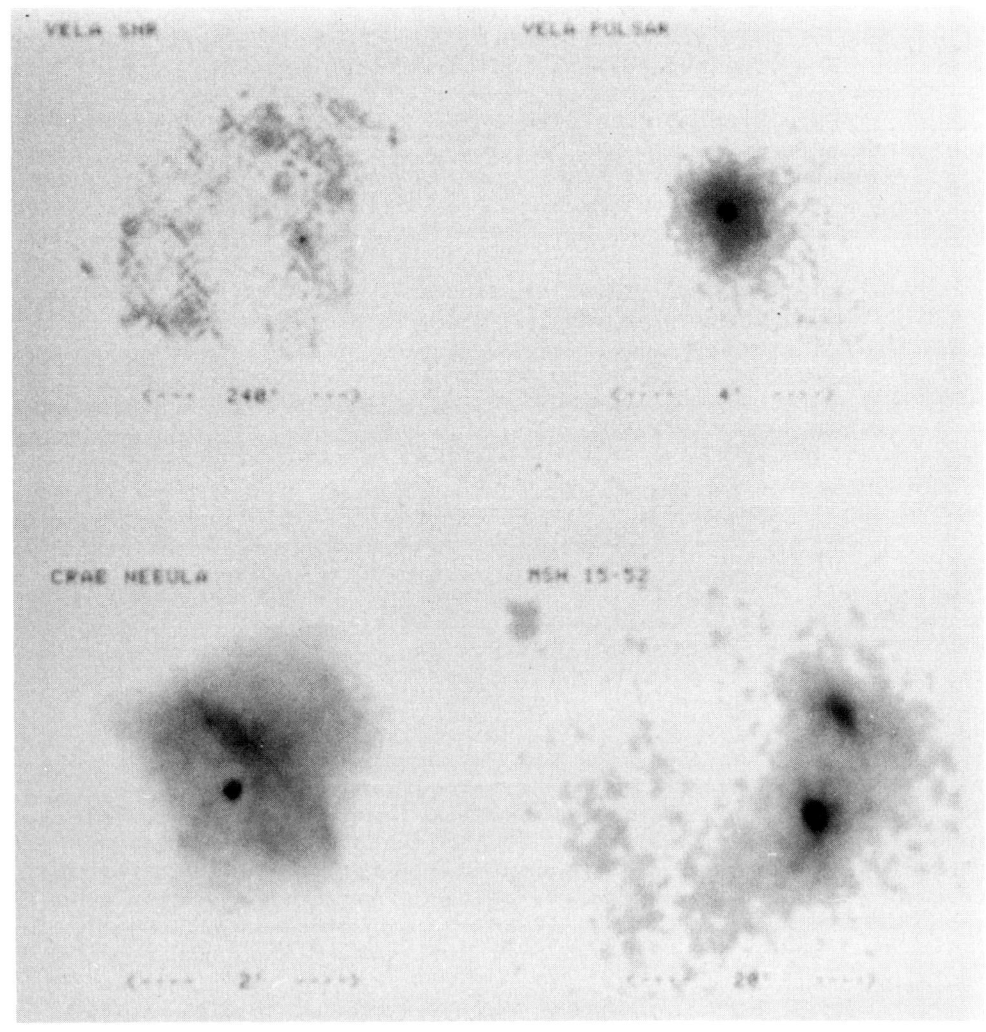


Fig. 1: Einstein pictures of SNR with internal isolated neutron stars.

## 3. The Vela Supernova Remnant

Soon after the pulsar in the Crab Nebula was discovered, a second was found within the Vela SNR. At a distance of 400 pc, the Vela remnant is over 5° in diameter, and intrinsically older and larger than the Crab. The Vela pulsar is the closest young, active pulsar to Earth and offers unique observing opportunities. Not only can low-luminosity features be detected, but the small column density of interstellar material allows transmission of soft X-rays that are unobservable from other remnants. Thus intense soft X-ray emission is detected from the shell of the Vela SNR and low surface brightness emission can be seen surrounding the pulsar itself (Harnden et al,

1985). The composite image in Figure 1 was made from 40 separate Einstein observations and shows the interior of the remnant filled with wisps of X-ray emitting plasma. The pulsar is clearly visible as a bright unresolved source about 1° from the remnant's center. The adjoining high-resolution image shows that the pulsar is surrounded by a small X-ray emitting nebula ~ 4' in diameter. The characteristic age of this pulsar is 11,000 years, in agreement with the estimated age of the shell. $\dot{E}$ is 1/65 that of the Crab pulsar, partially explaining the low X-ray luminosity of the central nebula, which is 1/5,000 that of the Crab.

## 4. MSH 15-52 and PSR 1509-58

MSH 15-52 has a large patchy radio shell characteristic of middle-aged supernova remnants and, in apparent contradiction to this, an internal pulsar with characteristic age of 1550 years. The pulsar is surrounded by an X-ray nebula with a spectrum characteristic of synchrotron emission. The shell of the supernova remnant is only bright in the Northwest, where it coincides with the H-Alpha Nebula RCW89 (Seward et al, 1983). One small knot in this region is particularly bright in X-rays, H-Alpha emission, and infrared lines from [FeII]. The pulsar has a period of 150 milliseconds and its spindown rate, $\dot{P}$, is the highest measured for any pulsar, three times higher than that of the Crab pulsar. The pulsar loses energy at 1/25 the rate of Crab pulsar and is easily capable of powering the surrounding diffuse nebula.

The pulsar has been detected and monitored in the radio band. It is spinning down smoothly and the second derivative of the period, $\ddot{P}$, has been measured. This is important because with $\ddot{P}$ the breaking index, N, can be calculated. It is customary to assume that the torque slowing the spinning neutron star is proportional to rotational frequency raised to the power N. This is only known for two pulsars, the Crab Pulsar and PSR 1509-58 for which $N = 2.515 \pm .005$ and $2.83 \pm .03$, respectively. We expect $N = 3.00$ if the torque is due to radiation reaction alone.

The surrounding nebula is much larger than that surrounding the Crab pulsar and not as luminous. The X-ray surface brightness is only $2 \times 10^{-5}$ that of the synchrotron emission of the Crab and the total X-ray luminosity is 1/100 that of the Crab Nebula. The surface brightness of the nebula around PSR 1509-58 is too small to detect in radio or optical bands. The nebular magnetic field calculated from the X-ray data is about $10^{-5}$ Gauss, a factor of 20 below that within the Crab, which characterizes the great difference in X-ray surface brightness within the two remnants.

## 5. SNR 0540-69.3

This remnant is in the Large Magellanic Cloud and is closer in character to the Crab than any other remnant so far observed.

The pulsar has a period of 50 ms and a spindown rate 13% higher than that of the Crab pulsar, so $\dot{E}$ is 1/3 that of the Crab pulsar. Optical pulsations have been observed by Middleditch and Pennypacker (1985) and the X-ray continuum spectrum is a bit harder than that from the Crab Nebula. The remnant is in the Large Magellanic Cloud, so its distance and luminosity are known, a happy situation in the study of SNR. Deep optical photographs by Chanan, Helfand, & Reynolds (1985) reveal an 8" diameter oxygen-rich shell surrounding the pulsar (2/3 the size of the filamentary part of the Crab Nebula) and an optical continuum, presumably synchrotron radiation, within this shell. The X-ray emission is 23% pulsed, so the pulsed X-ray fraction is higher than the 4% of the Crab. Since existing southern hemisphere radio telescopes do not have the capability of detecting arcsecond radio structure, the dimensions of the outer part of the remnant are unknown. There is some evidence for a radio outer shell $\sim$ 2" in diameter. Unlike the Crab, the remnant has an oxygen-rich inner shell. It seems likely that the precursor star evolved to the point where oxygen rich material existed in the outer layers of the star.

## 6. CTB 109 and 1E2259+586

This remnant is only 3° from Cas A, the brightest radio source in the sky and was not well mapped in the radio band until after its serendipitous discovery by Gregory and Fahlman (1980) with the Einstein IPC. On the eastern side, an X-ray shell surrounds a bright central unresolved source. There is no radio or X-ray shell on the western side because of a molecular cloud in this region (Gregory et al, 1983). In this dense cloud the shock has evolved more rapidly and is no longer radiating. X-rays from the central source have a period of 6.985 s, confirming the identification as a neutron star. Since there is evidence for orbital motion, the pulsar is probably powered by accretion. The optical counterpart is very faint (B=23.5) and cannot be massive. There are inner regions of radio and X-ray emission which seem to connect the central source to the shell and suggest to some the existence of an energetic jet originating in the binary system. Information about the age of the neutron star has been lost through torques generated by the accretion process. The precursor star cannot have been very massive or this close binary system would have been disrupted by the explosion. Mass ejected by the SN explosion is estimated as only a few tenths of a solar mass.

Table 1. Pulsars in SNR

| Pulsar Name | P (s) | $\dot{P}$ ($10^{-15}$s/s) | $\dot{E}$ ($10^{38}$ erg/s) | Age ($10^3$ yr) | Remnant Name |
|---|---|---|---|---|---|
| Crab | 0.033 | 423 | 4.7 | 1.24 | Crab |
| 0540-69 | 0.050 | 479 | 1.5 | 1.67 | 0540-64.3 |
| 1509-58 | 0.150 | 1540 | 0.18 | 1.55 | MSH 15-52 |
| Vela | 0.089 | 125 | 0.071 | 11.3 | Vela xyz |
| 1E2259+586 | 6.98 | | -in binary system- | | CTB 109 |

## 7. The Four Young Isolated Pulsars

Thus, we know of four isolated pulsars within SNR, and the regular pulsations give positive identification as rotating neutron stars.

The waveform of the pulses emitted differs greatly among the individual objects. The Crab Pulsar shows two sharp pulses separated by 140° of phase from radio frequencies to $\gamma$-ray energies. The pulsar in the LMC has an X-ray and optical waveform which is almost sinusoidal but with structure at the maximum. No radio pulsations have been observed, but the pulsar is so distant that it would have to be abnormally strong to be above the detection threshold. The pulsar within MSH 15-52 also has a sinusoidal waveform in X-rays. It has been detected in the radio band, but not optically. The radio duty cycle is $\sim$ 10% and the X-ray duty cycle $\sim$ 50%. The pulse shape for the Vela pulsar varies with frequency. It progresses from a single sharp radio pulse to a crab-pulsar-like double pulse at $\gamma$-ray energies where most of the pulsed energy is emitted. No pulsations have been detected in X-rays down to a level of $\sim$ 1%.

These variations might be due, at least partially, to variations in geometry. The pulse shape we see is dependent on the orientation of the line of sight, the spin axis, and the magnetic axis of the pulsar. The mechanism by which pulsars emit radiation is not well understood.

## 8. Unresolved Sources within Supernova Remnants

Four remnants contain unresolved sources likely to be neutron stars. High resolution Einstein observations show the sources clearly but do not contain enough counts to make a decent search for pulsations possible. So, no pulsations, X-ray or radio, have been detected from these faint objects.

3C58, a radio source full of faint optical filaments and at a distance of 2.6 kpc, is probably the most interesting. The linear size of the remnant is 50% larger than the Crab Nebula, but it is 1700 times less luminous. The interior of the remnant shows diffuse X-ray emission with a faint unresolved source at the center (Becker et al, 1982). This is commonly accepted as the remnant of SN 1181. It is thus a young remnant with an interior neutron star 2,000 times less luminous than the one within the Crab Nebula, implying a much lower rate of energy loss. Here is the importance of historical identification. If correct, we know that neutron stars can be born with low luminosity and that all young neutron stars are not easy to detect.

CTB80 is, in some ways, similar to 3C58, and has been suggested by some to be the remnant of SN 1408. In X-rays it appears as a faint diffuse source of extent $\sim$ 5' surrounding a weak unresolved object, probably a neutron star (Wang & Seward, 1984). The area around the neutron star contains faint optical filaments and is a fairly strong radio source. The radio appearance of this remnant, however, is unique. Three faint radio arms extend a distance of ~ 30' and no shell-like structure is observed. If the remnant is young, these arms require material to move at ~ 0.15 c. Another possibility is the superposition of a young SNR and an older one.

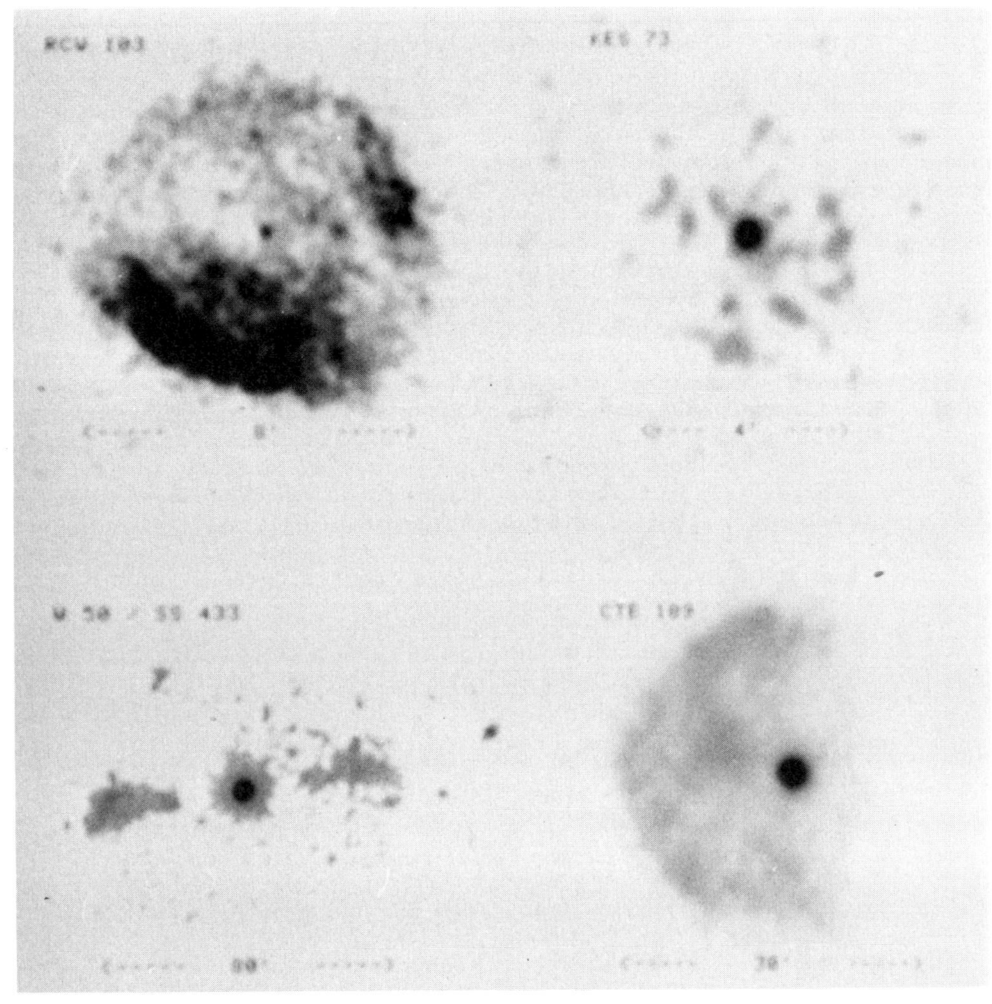

Fig. 2: Einstein pictures of SNR with internal unresolved sources.

RCW 103 is more conventional. It has an X-ray and radio shell of diameter 8', the same as the diameter of Tycho's SNR. At precisely the center, there is an unresolved X-ray source just above detection threshold. No diffuse emission obviously connected with this compact source is seen, but this could easily be obscured by thermal emission from the shell. Touhy, et al (1983) have searched for an optical counterpart and found nothing brighter than V=22 mag. They conclude this is an isolated neutron star and the X-rays are thermal emission from the hot surface. A longer X-ray observation is needed to collect enough counts to do a definitive search for pulsations.

At the other extreme, the internal source within Kes 73 is moderately bright, but the surrounding diffuse emission is ill-defined. Kriss et al (1985) consider it most likely that this source is an accretion powered binary, a reasonable hypothesis since its luminosity is high compared to the surrounding synchrotron emission. Once again more data are needed to determine its nature.

SS433 within W50 is an accretion powered binary, consisting of a massive star and compact object. The jets emanating from the central source with speed 0.26 c contain enough energy to modify the appearance of the surrounding remnant, and it is difficult to derive the age and properties of the SNR. There is some speculation that the compact object is a black hole and not a neutron star.

## 9. Two Remnants with Synchrotron Emission (and unseen pulsars?)

Both X-ray and radio images of G21.5-0.9 show a plerionic (filled-center) morphology with diameter ~ 1' (Becker & Szymkowiak, 1981). The Einstein HRI observation shows only diffuse emission. An unresolved source, if present, contributes 5% or less of the total X-ray flux, which is about the ratio of pulsed to diffuse X-rays from the Crab Nebula. Indications are, therefore, that this remnant is like the Crab, but a factor of 80 less luminous.

The remnant Kes 75 consists of a partial radio shell with diameter 3' half surrounding a central radio component with extent ~ 30" (Becker et al, 1982). This central source, as seen with the Einstein HRI is extended with radius ~ 15" and the presumed point source cannot be distinguished. The distance is thought to be 20 kpc. If so, the X-ray luminosity is ~ 1/5 that of the Crab and the X-ray size of the source is slightly greater than that of the Crab.

These two remnants, then, have strong internal emission indicative of energy generated by isolated but unseen neutron stars.

## 10. Information about Neutron Stars from Synchrotron Nebulae

Figure 3 shows the X-ray surface brightness of nebulae surrounding pulsars or unresolved sources in these SNR. These curves were obtained by subtracting the point response function of the Einstein telescope from the average surface brightness measured radially from the unresolved source. The distances listed by Seward (1985) have been used to set the linear scale. If the distance is not correct the radial scale will shift but the value of the surface brightness will remain constant. The calculated surface brightness does, however, depend critically on assumed X-ray absorption in the ISM.

The pulsar in the Large Magellanic Cloud is too distant for us to separate any surrounding synchrotron nebula from the pulsar. We know that 77% of the emission from this region is not pulsed and we can speculate on the strength of the nebula around the pulsar, but we have no measure of its dimensions.

Figure 3 implies that Kes 75 and G21.5-0.9 contain pulsars with high rates of energy loss. Most other remnants cluster about MSH 15-52. These data can be used to estimate properties of the neutron stars. We arbitrarily use the measured surface brightness 0.3 pc from the central source to characterize the emission. Using the three known pulsars as calibration, we find the surface brightness varies as $(\dot{E})^{2.5}$, and use this to estimate $\dot{E}$ for other remnants.

If $\dot{E}$ and age are known, the pulsar period may be derived, since

$$P(s) = .055\ A^{-1/2}\ (\mathrm{Kyrs})\ \dot{E}^{-1/2}\ (10^{37}\ \mathrm{erg/s}).$$

For example: if 3C58 is the remnant of SN1181, its age is 800 years; and with $\dot{E}$ estimated from the nebular surface brightness, we calculate a pulsar period of 210 ms. Likewise, if CTB 80 is the remnant of SN 1408, the pulsar period is 280 ms. If G21.5-0.9 is 1,000 years old, the pulsar period is 125 ms.

Let us also consider Cas A, an oxygen rich remnant thought to be the result of an unseen Type 2 supernova explosion in ~ 1670. With age of 300 years, it is the youngest SNR yet observed and the 3' diameter shell is very bright. There is no evidence for a neutron star in the interior but a moderately strong synchrotron nebula could be masked by the bright shell. The surface brightness of a knot located close to the center of Cas A is plotted as a dashed line in Figure 3. This can be taken as an upper limit to emission from a synchrotron nebula which could be bright compared to the other remnants (except for the Crab). We estimate a pulsar period of 145 ms or greater is possible. Thus, Cas A could contain a neutron star and synchrotron nebula of moderate luminosity, undetected at present because of bright surrounding material.

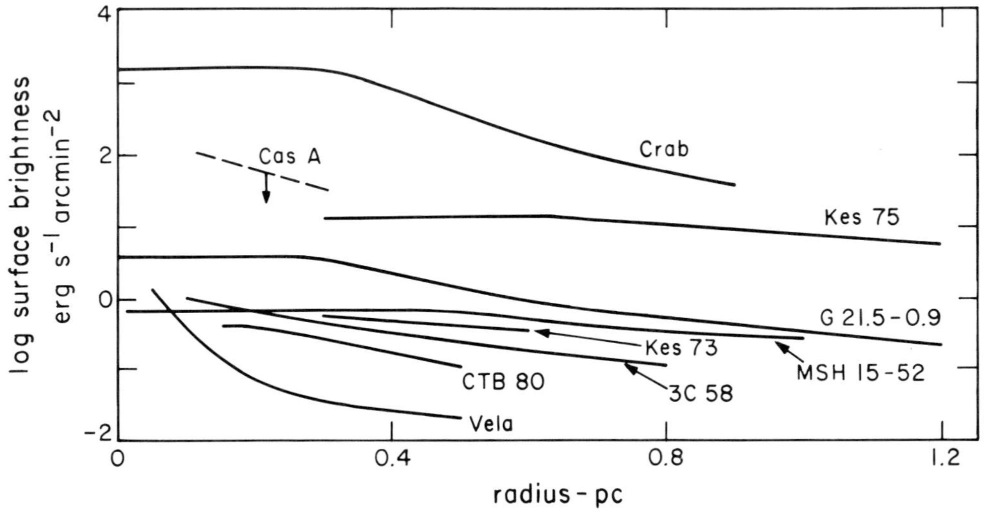


Fig. 3: Surface brightness of X-ray synchrotron nebulae.

## 11. Conclusions

The Einstein observations were more sensitive than radio and optical techniques in the detection of low-surface-brightness nebulae, and in some cases, the pulsed radiation from young pulsars.

We have reviewed twelve remnants. Four contain isolated neutron stars, two contain compact objects in binary systems, and six show evidence for compact objects or energy deposited by them in the central region. The surrounding remnants are all different. Some have shells some do not. Some have oxygen rich material some do not. Some perhaps have jets emanating from the central object.

Perhaps these differences in the appearance of SNR shells reflect differing conditions during gravitational collapse and formation of the neutron star; and neutron stars can be formed from a great variety of progenitor stars. Although variations due to age and the surrounding interstellar medium are important, the young remnants should reflect the characteristics of the progenitor stars.

It seems clear that some young neutron stars have low luminosity. So far we have only detected the brightest. It is likely that some are born in an "inactive" state, and have never been strong X-ray emitters. They might be formed with long rotation periods or they may spin rapidly but have weak initial magnetic fields. Present observations indicate that there is no shortage of neutron stars within SNR. Instruments of greater sensitivity should lead to more identifications and to a fuller understanding of neutron star formation.

This work was supported by NASA contract NAS 8-30731

## References

Becker,R., Helfand, D., and Symkowiak, A., 1982 Ap.J. 255, 537
Becker, R.H. and Symkowiak, A.E, 1981 Ap.J. Letters 284 L23
Chanan, G., Helfand, D., and Reynolds, S., 1985 Ap.J. Lett. 287,L23
Gregory, P.C., Braun, R., Fahlman, G.G., and Gull, S.F., Supernova Remnants and Their X-ray Emission, eds. J. Danziger and P. Gorenstein, D. Reidel, Dordrecht, 1983, 437.
Gregory, P.C., and Fahlman, G.G., 1980, Nature, 287, 805
Harnden, F.R., Jr. and Seward, F.D., 1984 Ap.J. 283, 274.
Harnden, F.R., Jr., Grant, P.D., Seward, F.D., and Kahn, S.M., 1985, Ap.J., 299, 828
Henry, R.G.C., and McAlpine, G.M., 1982, Ap.J. 258, 11
Kriss, G.A., Becker, R.H., Helfand, D.J., and Canizares, C.R., 1985, Ap.J., 288, 703
Middleditch, J., and Pennypacker, C., 1985, Nature 313, 659
Seward, F.D., 1985, Comments on Astrophy. 11, 15
Seward, F.D., Harnden, F.R., Jr., Murdin, P., and Clark, D.H., 1983, Ap.J., 267, 698
Touhy, I.R., Garmire, G.P., Manchester, R.N., and Dopita, M.A., 1980, Ap.J. Lett. 239 L107
Wang, Z., and Seward, F.D., 1984, Ap.J. 285, 607

# THE PROGENITORS OF PULSARS

G. Srinivasan and D. Bhattacharya
Raman Research Institute, Bangalore 560 080, India
Joint Astronomy Program, Department of Physics, Indian Institute of Science, Bangalore 560 012

ABSTRACT. The progenitors of single as well as binary pulsars are discussed with special emphasis on binary pulsars with low mass companions. Several predictions are made concerning millisecond pulsars.

## 1. INTRODUCTION

Till quite recently it was believed that formation of all neutron stars was associated with type II supernovae: in sufficiently massive stars the mass of the degenerate core will eventually reach the Chandrasekhar limit and become dynamically unstable, resulting in the formation of a neutron star and accompanied by a supernova explosion. For a variety of reasons, one has now been forced to entertain additional scenarios for the formation of neutron stars. It may be worth listing these before proceeding further.

The second route for the formation of neutron stars is through white dwarfs. There are three possibilities.

(1) A white dwarf more massive than the Chandrasekhar limit could in principle be stabilized by thermal energy or rotation. Eventually due to cooling or magnetic braking it will become unstable (Shklovsky 1978; Ostriker 1971). But this scenario must be so rare that we shall not discuss it further.

(2) An extremely tight binary consisting of two white dwarfs will eventually coalesce, perhaps resulting in the formation of a neutron star. Since this mechanism is likely to be discussed in detail later in this conference, we shall not discuss it here.

(3) The third possibility is accretion induced collapse of a white dwarf in a binary (Miyaji et al 1980). Very few people dared to speak about it in public a few years ago, and yet this seems to be popular now and, indeed, taken for granted !



*D. J. Helfand and J.-H. Huang (eds.), The Origin and Evolution of Neutron Stars, 109–119.*

## 2. THE OBSERVED POPULATION OF PULSARS

Keeping these theoretical scenarios at the back of our mind, let us turn to the observed population of pulsars.

The B-P plot: Figure 1 shows approximately 300 pulsars with their measured periods and derived surface magnetic fields. Pulsars in binaries are indicated by dots inside open circles. There are seven binary pulsars known so far (J.H. Taylor, this volume). The first thing that strikes one is that the single pulsars have fairly high magnetic fields and predominantly in the range $10^{12}$ - $10^{13}$ gauss. The binary pulsars, however, have a wide spread in their fields, ranging from $10^{12}$ to $4x10^8$ gauss. There may be other differences between the single pulsars and the binary pulsars, such as their space velocities and scale heights, but we shall not pursue them further at the moment. But it *is* worth noting that the seven binary pulsars may be further subdivided into two classes - one in which the companion has mass comparable to that of the neutron star itself and the other group having very low mass companions. The companion of PSR 1913+16 is almost certainly another neutron star; this is probably the case for PSR 2303+46 also. The mass of the white dwarf companion of PSR 0655+64 is again probably $\sim 1\ M_\odot$. On the other hand, the two *millisecond* pulsars in binaries, as well as PSR 0820+02 and PSR 1831-00 have rather low mass companions.

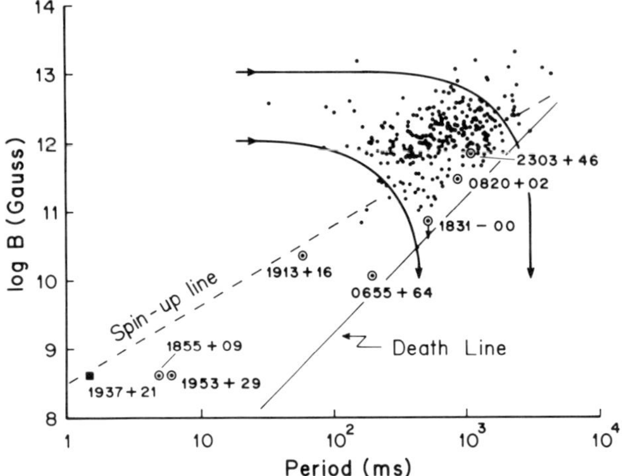


*Fig. 1: The periods and derived magnetic fields for about 300 pulsars are shown. The dots with open circles are pulsars in binaries.*

Given this grouping of pulsars into several famililities, it is natural to ask if they can be traced back to the different progenitor systems outlined before.

## 3. THE ORIGIN AND EVOLUTION OF SINGLE PULSARS

Let us first turn to a discussion of the progenitors of single pulsars and their evolution as pulsars. The standard picture is that they were formed by the collapse of degenerate cores of massive stars. The progenitors could have been single stars or members of binary systems. From the statistics of pulsars one can deduce their birthrate (see e.g. Lyne et al 1985; R. Narayan, this volume). Although there is some uncertainty in this number, for the purpose of definiteness I shall assume a birthrate of 1 in 40 years.

The question to be asked now is the following: Are there enough massive stars to give a pulsar birthrate of 1 in 40 years and what is the mass range which makes the dominant contribution to the pulsar population? To answer this question one takes recourse to some initial mass function (IMF), and a smoothed lifetime of stars as a function of their mass. For example, figure 2a shows the integrated deathrate of stars using the IMF

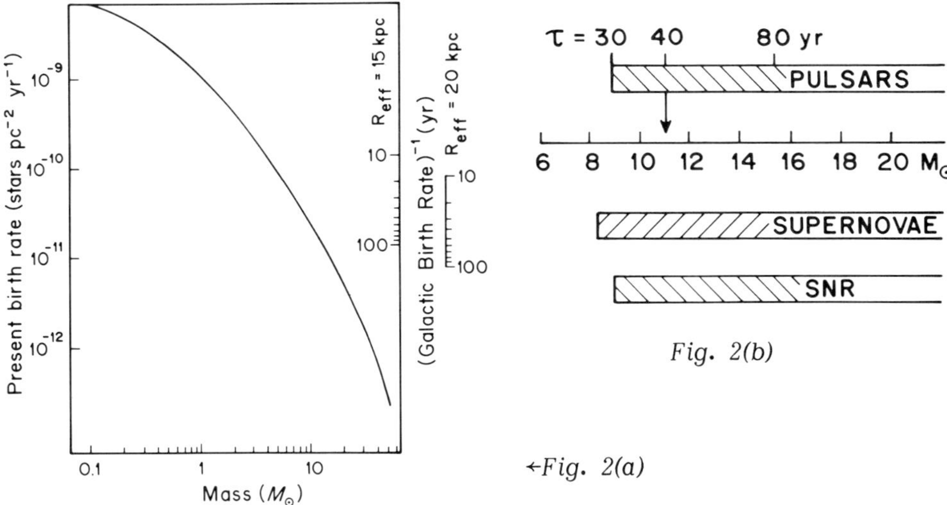


*Fig. 2(b)*

←*Fig. 2(a)*

*Fig. 2: (a) The integrated birth rate of stars derived by Miller and Scalo (1979). This can be used to derive the limiting mass of progenitors of pulsars given their galactic birthrate and an effective radius for the galaxy. This is indicated on the right.*
*(b) The lower limit for the mass of the progenitors implied by the pulsar birthrate, the frequency of supernovae and the birthrate of supernova remnants is shown. The uncertainties in these rates are indicated by hatched regions. ($R_{eff}$ = 20 kpc has been assumed)*

of Miller and Scalo (1979). In order to use this deathrate function, which is derived from a study of stars in the local neighbourhood, one has to convert the *galactic* birthrate of pulsars to a *local* rate. For this one needs to assume a value for the effective radius of the galaxy. If the distribution of pulsars were uniform in the galaxy, then one would simply take a standard number like 15 kpc for the radius. However, it is believed that the radial distribution of pulsars is not uniform; there are more pulsars around ∿ 5 kpc from the centre of the galaxy than in the solar neighbourhood. Therefore, in going from a galactic rate to a local rate one must assume a larger effective radius for the galaxy than its actual dimensions. Figure 2b summarizes the conclusions assuming an effective radius of 20 kpc. We see that a pulsar birthrate of 1 in 40 years implies that all stars above ∿ 11 $M_{\odot}$ should produce neutron stars. The hatched region represents the uncertainty in the pulsar birthrate. Also shown in the figure are the birthrate of supernova remnants and the frequency of supernovae. Although there is considerable uncertainty in each of these rates, it is interesting to note that there is significant overlap between them.

Recently Blaauw (1985) has looked into this question from a different point of view, and has arrived at some important conclusions. His approach may be summed up as follows. He restricts himself to the observed sample of pulsars whose distances projected on to the galactic plane are within 0.5 kpc. From the information about the scale height of pulsars, and their measured proper motions, he concludes that the *local* population of pulsars must be replenished every 5 million years by *local* progenitors. After examining the deathrate of stars in the nearby OB associations he arrives at the conclusion that the local pulsar population *cannot* be replenished by OB associations alone. The field stars must therefore make an important contribution. Blaauw's conclusion is that the progenitors of the overwhelming majority of the local pulsar population must be the relatively old population of field stars with masses in the range 6-10 $M_\odot$. There simply aren't enough stars more massive than 10 $M_\odot$ to replenish the local population of pulsars every 5 million years or so. If one accepts this, then an equally important corollary is that 'Pulsars are, on a galactic scale, tracers of regions of past spiral structure (20-50 Myr) rather than of active spiral structure' (Blaauw 1985).

It should be remembered that all these conclusions regarding progenitors are based on star counts and there are considerable uncertainties involved in going from that to an integrated deathrate function, as has been repeatedly emphasized in the literature. However, if one accepts these conclusions for the moment, the next logical question to ask is whether stars in these mass ranges are likely to produce neutron stars. I shall leave this question to later speakers. Assuming that the answer is positive, let us ask if the observed characteristics of the single pulsars is consistent with what one expects from such progenitors.

As already mentioned one of the most significant characteristics of the population of single pulsars is that they have high field. Indeed, even before neutron stars were discovered it was surmised that flux conservation during core collapse will result in fields $\gtrsim 10^{12}$ gauss (e.g. Woltjer 1964). Therefore there may not be much mystery in that. It was also conjectured that the newly born neutron star will be spinning very rapidly, perhaps close to the limiting period $\sim$ 1 millisecond. The discovery of a fast pulsar in the Crab Nebula seemed to confirm this expectation. However, both the statistics of pulsars and the absence of a large number of bright plerions in the galaxy strongly suggest that the majority of pulsars must be born spinning rather slowly with periods $\gtrsim$ 100 ms (Vivekanand and Narayan 1981; Srinivasan, Bhattacharya and Dwarakanath 1984; Chevalier and Emmering 1986). As the pulsars age their periods will lengthen; thus their evolutionary tracks will be horizontal in the B-P plane (see fig. 1). After a few million years the magnetic field will begin to decay and the trajectory will become vertical. Although it is generally agreed that magnetic fields of neutron stars do decay, there is no consensus on either the mechanism or the time scale over which the field decays significantly. Observations suggest a timescale $\sim$ 4-8 million years (Chevalier and Emmering 1986; Lyne, Manchester and Taylor 1985). If we admit field decay in this sort of time scale then the *distribution of fields at birth will be narrower than the observed spread.*

As the period lengthens, and the field weakens, the pulsar will eventually die. Even though the details of the emission mechanism are poorly understood, there is strong observational evidence and theoretical justification for the existence of a 'death line' (Ruderman and Sutherland 1975; Rawley, Taylor and Davis 1986).

## 4. PULSARS FROM MASSIVE BINARIES

We now turn to the progenitors of pulsars like 1913+16 and 2303+46 which have massive companions. In sufficiently massive binaries two neutron stars will be born. In very rare cases the system will remain bound after the second supernova explosion which signals the birth of the second neutron star. The first born beutron star will be *recycled* during the X-ray phase. The rotational history of a neutron star in a mass transfer binary has been reviewed in several places (see. e.g. van den Heuvel 1984) and therefore I shall summarize it rather briefly (fig. 3a). Initially the

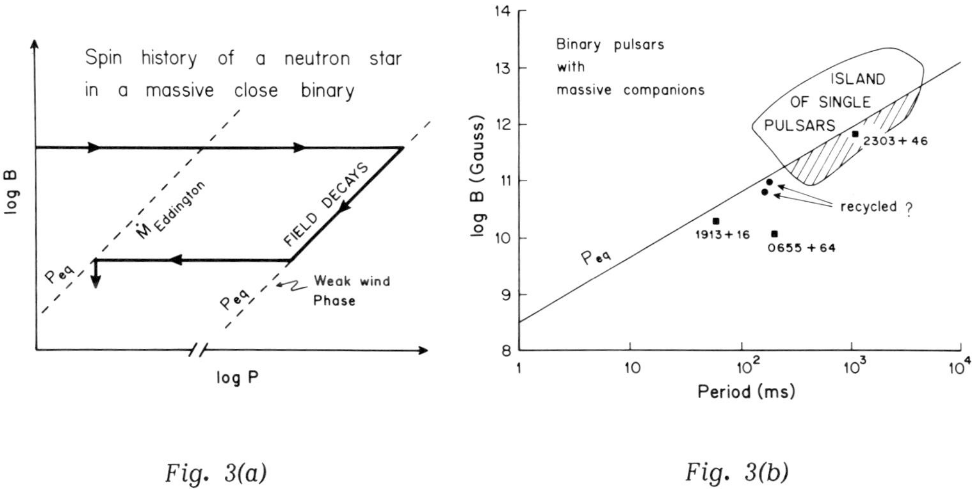


*Fig. 3(a)* *Fig. 3(b)*

neutron star will experience a braking due to its pulsar activity. During the weak stellar wind phase, the braking will be much more rapid, and soon it will cease to function as a pulsar. Eventually the period of rotation will settle down to an equilibrium value determined by its magnetic field and accretion rate. Before the secondary evolves and fills its roche lobe, the magnetic field of the neutron star can decay significantly, and it will move down along this line. When the neutron star becomes a power X-ray source, it will be spun-up to its *critical equilibrium period.* The period after such a recycling will depend upon the time between the first and the second explosion, which will determine the extent of field decay. The anomalous combination of small B and small P of 1913+16 clearly indicates that we are seeing such a recycled pulsar (fig. 3b). PSR 2303+46 is probably the second born pulsar, but it could also be the recycled one if the field had not significantly decayed.

In the majority of cases, however, the system will disrupt during the second supernova. The second born pulsar can end up anywhere in the pulsar island (fig. 3b). The only definite thing we can say about its partner, the recycled member, is that it will be *below* the critical equilibrium period line. If its field had decayed significantly before the onset of accretion then it will stand out from the population of single pulsars, as does PSR 1913+16. PSR 1804-08 and PSR 1541-52, which are just outside the island, may also be recycled pulsars.

What is the formation rate of such recycled pulsars? The only thing certain about the answer is that it is uncertain!

The statistics of the standard massive X-ray binaries gives a birthrate of only one in $\sim$ 1000 years (see van den Heuvel and Habets 1985). Thus they will make only 3-5% contribution to the population of pulsars. van den Heuvel and Habets (1985) have suggested that B-emission X-ray binaries may make substantial contribution, but this is controversial.

The picture is not much clearer if we try to estimate the frequency of occurrence of binaries while both the stars are unevolved. The typical statement one finds in the literature is that $\sim$ 30% of massive stars are in binaries (see e.g. Boland and van Woerden 1985). Blaauw (1985) has also addressed this question. From a study of 135 stars of spectral types B2, B2.5, B3 in the nearby associations he concludes that only 18% are well established binaries, and there is no reason to think that the frequency of binaries among the B stars of the field population will be very different.

This question is of considerable importance. As we have heard, pulsars have a typical velocity $\sim$ 100 kms$^{-1}$. The prevalent opinion about the origin of this velocity is that the neutron star receives a kick at birth due to a small asymmetry in the supernova explosion. However, Radhakrishnan and Shukre (1985) have argued that both the observed velocities, as well as their correlation with magnetic field of the pulsars, will find a natural explanation if most, if not all, pulsars came from binary systems. The idea is that pulsars acquire their velocities from the disruption of the binary during the second supernova. In fact, pursuing this idea further they have been able to put fairly stringent constraints on the parameters of the progenitor binary systems. It should be emphasized that this mechanism is particularly attractive since there is, as yet, no logical basis for believing in an asymmetric explosion. Blaauw (1985) on the other hand, has argued that the high velocities of pulsars cannot be due to their binary origin since only a very small fraction of binaries are *close enough* to be of interest in this connection. Clearly this important question deserves further study.

## 5. PULSARS WITH LOW MASS COMPANIONS

Let us finally turn to a discussion of the progenitors of the binary pulsars with low mass companions, and their formation rates. Since this question will be discussed in detail by others, here we shall confine ourselves to one or two intriguing aspects.

It has been aruged by several people, in particular by Joss and Rappaport (1983), that the progenitor of the 6 ms pulsar must have been a Low Mass X-ray Binary (LMXB). In such systems, given the right parameters, there is no difficulty in maintaining mass transfer at close to the Eddington rate for as long as $10^8$ years or more. Turning to the recently discovered 5 ms pulsar, this must have also evolved from an LMXB. An initial orbital period $\sim$ 1 day would lead to the presently observed period of 12.3 days. Although the absence of a companion for the 1.5 ms pulsar complicates matters this, too, may have come from a LMXB.

The real puzzle concerning the millisecond pulsars is not regarding their progenitors, but their *location*. It is an extraordinary fact that all three of them are very close to the galactic plane. Normal pulsars have a scale height $\sim$ 350 pc, which is much larger than the scale height of their progenitors, which is $\sim$ 60 pc. This is easily understood if the neutron stars acquire substantial velocities at birth. Even if millisecond pulsars are not created with such velocities, one would expect them to have a scale height at least comparable to that of their progenitors. The scale height of LMXBs is $\sim$ 300 pc, consistent with their belonging to old disk population. It is, therefore, extraordinary that all three millisecond pulsars discovered are within 25 pc of the galactic plane!

This remarkable 'coincidence' could in principle be rationalized under two exotic circumstances:

(i) It may be that the scale height of the very old neutron stars associated with LMXBs is indeed large, but for some reason they do not function as pulsars unless they are in an environment that obtains only close to the galactic plane!

(ii) For some unknown reason the velocity dispersion of these old pulsars *decreases* with time, and they 'settle down' near the plane. In other words, the population of millisecond pulsars 'cools' with time.

Since we have no suggestions at present for either of these two exotic possiblities, we shall offer a more modest alternative, viz, the proximity of the millisecond pulsars to the galactic plane is possibly due to selection effects. If so, an immediate implication of this is that there must be many more potentially observable millisecond pulsars.

If the scale height of these pulsars is $\sim$ 300 pc, like that of the LMXBs from which they have come, then one expects only $\lesssim$ 10% of them to be within $\sim$ 30 pc from the galactic plane. A simple scaling which takes into account this factor, beaming and the fact that all the three millisecond pulsars have been discovered in a small sector of the galaxy, suggests that there should be >300 binary pulsars with low mass companions within $\sim$ 4 kpc, implying a total number in the galaxy greater than two thousand.

Unfortunately, this number is much larger than the number of LMXBs in the galaxy. Presently about 30 LMXBs are known, and it is very unlikely that there are more than 100 of them (McClintock and Rappaport 1985). If our basic premise, namely that millisecond pulsars evolve from LMXBs, is correct then the lifetimes of these pulsars must be 20-30 times longer than the X-ray phase of these systems, which are believed to last for $\sim 10^8$ years. We thus conclude that millisecond pulsars must live for $\gtrsim 10^9$ years.

An immediate conclusion one can draw from this is that the magnetic field of a neutron star, after an initial decay in the timescale of a few million years settles down to an asymptotic value. This conclusion has also been arrived at, for essentially the same reason, by van den Heuvel et al (1986). Motivated by the very large apparent age of the white dwarf companion of the binary pulsar 0655+64 Kulkarni (1986) has also concluded this.

Till now the intriguing thing was why the field decays at all. Now we have to understand why the decay stops! Perhaps the correct picture is that the field decay is not a simple exponential process, but a power law. The next question is what is the limiting value of the field, and is this the same for all neutron stars?

Even before the latest millisecond pulsar 1855+09 was discovered, we noticed that the 1.5 ms pulsar and the 6 ms pulsar have almost identical fields. Considering that the 6 ms pulsar is far from the spin up line in the vertical direction (see fig. 1), we felt that this could not be an accident. This led us to speculate that the magnetic fields of very old neutron stars will probably be $\sim 5\times10^8$ gauss. Soon after the discovery of the latest millisecond pulsar 1855+09 we predicted (Bhattacharya and Srinivasan 1986) that its field will be very close to this value. Since then the $\dot{P}$ of this pulsar has been measured (J.H. Taylor, this volume) and its magnetic field is almost exactly the same as that of the other two millisecond pulsars, confirming our conjecture.

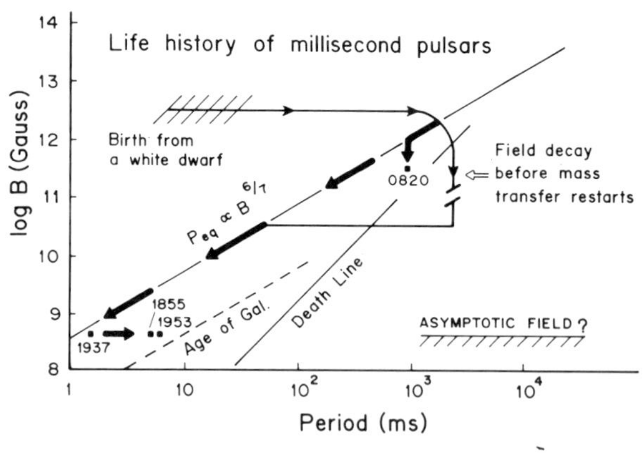


*Fig. 4*

If one accepts all this, then one can summarize the evolution of pulsars born and processed in LMXBs as follows (see fig. 4). Let us begin with the birth of the neutron star due to accretion-induced collapse of the white dwarf. Presumably it had a high field $\sim 10^{12}$ gauss. One cannot be so certain about its initial period, but most likely it would have been spinning rather rapidly and functioning as a pulsar (we are grateful to W.H.G. Lewin for pointing out that accretion on to the newly born neutron star may not commence till several million years have elapsed since its birth and its magnetic field can decay somehwat during this interval). With the onset of accretion it will be spun up to its equilibrium period. It will then dribble down along the equilibrium period line till the mass transfer phase is over. If the initial orbital period is rather large, for example, a couple of 100 days, then the mass transfer will last for less than 10 million years and the field would not have decayed significantly. Then the pulsar will 'peel off' from the 'spin-up' line. The evolutionary history of PSR 0820+02 was probably this; the present orbital period of $\sim$ 1200 days is fully consistent with this picture. But if the mass transfer lasts longer the neutron star will continue to dribble down the 'spin-up' line till the field reaches its asymptotic value of $\sim 5\times10^8$ gauss. After the debris clear away the recycled pulsar will move along a horizontal trajectory. An important implication of this constant-field evolution is that since $\dot{P} \propto P^{-1}$, more pulsars will be found at longer periods. However, given the age of the galaxy, the maximum spin period reached will be only $\sim$ 10 ms. One will, therefore, expect a concentration of millisecond pulsars near $P \lesssim 10$ ms. The periods of the observed millisecond pulsars is consistent with this expectation.

Summary:

1. The fact that all three known millisecond pulsars are so close to the galactic plane can only be a 'selection' effect. Millisecond pulsars must have a scale height comparable to that of LMXBs from which they probably evolve and we predict that there should be $\gtrsim$ 100 potentially observable millisecond pulsars within $\sim$ 4 kpc.

2. Such a large number of millisecond pulsars is consistent with the birthrate of LMXBs only if the lifetime of these pulsars is more than $\sim 10^9$ years. This, in turn, would require that their magnetic fields do not decay indefinitely.

3. The fact that all three known millisecond pulsars have fields very close to each other indicates that the limiting field is around $\sim 5\times10^8$ gauss.

4. The corollary of this is that the majority of ultrafast pulsars will be in binaries and will have periods in the interval 6-10 ms.

REFERENCES

Bhattacharya, D., Srinivasan, G. 1986, Curr. Sci., **55**, 327.

Boland, W., van woerden, H. (eds) 1985, 'Birth and evolution of massive stars and stellar groups', D. Reidel.

Blaauw, A. 1985, in Boland and van Woerden (1985), p.211.

Chevalier, R.A., Emmering, R.T. 1986, Astrophys, J., **304**, 140.

Joss, P.C. Rappaport, S.A. 1983, Nature, **304**, 419.

Kulkarni, S.R. 1986, Astrophys. J. (Lett), in press.
Lyne, A.G., Manchester, R.N., Taylor, J.H. 1985, Mon. Not. R. Astr. Soc., **213**, 613.
McClintock, J.E., Rappaport, S.A. 1985, in 'Cataclysmic Variables and Low Mass X-ray Binaries', eds. D.Q. Lamb and J. Patterson, D. Reidel, p.61.
Miller, G.E., Scalo, J.M. 1979, Astrophys. J. Suppl., **41**, 513.
Miyaji, S., Nomoto, K., Yokoi, K., Sugimoto, D. 1980, Publ. Astr. Soc. Japan, **32**, 303.
Ostriker, J.P. 1971, Ann. Rev. Astr. Astrophys., **9**, 353.
Radhakrishnan, V., Shukre, C.S. 1985, in 'Supernovae, their progenitors and Remnants', eds. G. Srinivasan and V. Radhakrishnan, Indian Acad. Sci., p.155.
Rawley, L.A., Taylor, J.H., Davis, M.M. 1986, Nature, **319**, 383.
Ruderman, M.A., Sutherland, P.G. 1975, Astrophys. J., **196**, 51.
Shklovsky, I.S. 1978, Sov. Astr., **22**, 413.
Srinivasan, G., Bhattacharya, D., Dwarakanath, K.S. 1984, J. Astrophys. Astr., **5**, 403.
van den Heuvel, E.P.J. 1984, J. Astrophys. Astr., **5**, 209.
van den Heuvel, E.P.J., Habets, G.M.H.J. 1985, in 'Supernovae, their progenitors and Remnants', eds. G. Srinivasan and V. Radhakrishnan, Indian Acad. Sci., p.129.
van den Heuvel, E.P.J., van Paradijs, J., Taam, R.E. 1986, Nature, in press.
Vivekanand, M., Narayan, R. 1981, J. Astrophys. Astr., **2**, 315.
Woltjer, L. 1964, Astrophys. J., **140**, 1309.

## DISCUSSION

**J. Shaham:** First a comment: In a forthcoming Phys. Rev. D paper, Harvey, Ruderman and myself calculate the lifetime of neutron star core fields (due to superfluid proton vortices) and find them to be $\gtrsim 10^9$ years. Core fields may, thus, be the slowly relaxing component. Now a question: Do the ages of the 3 millisecond pulsars, measured via their deviation from the zero-age curve for spun-up pulsars, conform statistically to your assumed age of $\gg 10^9$ years?

**G. Srinivasan:** Yes.

**W. Lewin:** When a white dwarf is pushed over the Chandrasekhar limit (due to accretion), its binding energy increases by $\sim 0.2\ M_\odot$. As a result the system will detach (mass transfer will stop). Mass transfer will resume when the Roche geometry again exists. This, according to Ed van den Huevel can take between a few $10^6$ to a few $10^7$ years. Therefore the magnetic field decay could be substantial before the neutron star begins to accrete and is spun up due to the accretion torques.

**J. Taylor:** First a comment: The period derivative of PSR 1855+09 has now been more accurately measured. The new value corresponds to a magnetic field almost exactly the same as those of PSRs 1937+21 and 1953+29, $B = 5 \times 10^8$ G. Now a question: Is it not true that the numbers you quoted for the frequency of binaries among massive stars should be viewed as lower limits?

**G. Srinivasan:** Thank you. I am pleased to hear that the magnetic field of PSR 1855+09 is exactly what we had predicted. Turning to your question concerning the frequency of massive stars in binaries, let me say this. Perhaps it is best if I first quoted Blaauw's remarks pertaining to this question (Blaauw 1985). ". . . 18% are well established (single-line) spectroscopic binaries with determined orbits. The semi-amplitudes of the velocity variations of their primaries range from 14 to 131 km/sec with a median value of 50 km/sec. For the remaining 82%, we are certain that the semi-amplitude of the primary is below 30 km/sec for 13%, below 20 km/sec for 38%, and below 10 km/sec for 31%. These numbers do not leave more than a few percent room for objects of which the explosion of either of the two components could result in a runaway star with a velocity of at least 100 km/sec and a mass of at least the mass of a neutron star." There is of course a possibility that many of the low-velocity binaries may contain a neutron star in which case they would be of interest in this context. However such a postulate is not consistent with the statistics of B-emission X-ray binaries or that of their progenitors.

**D. Eichler:** If the decaying magnetic field "settles down" to a value of $\sim 10^9$ gauss, as you have suggested, in dipolar form, why do accreting X-ray bulge sources not display spin periodicity? A $10^9$ gauss field is enough to funnel the flow needed to produce a typical luminosity.

**G. Srinivasan:** This will be discussed Thursday.

# DISTRIBUTION OF TWO TYPES OF PULSARS IN COMPARISON WITH THAT OF SNRS

H. Sun, and Z.W. Li
Astronomy Departmnet, Beijing Normal University
Beijing, China

ABSTRACT. Recently, pulsars are classified into two types. On the basis of these data, it is possible to discuss the relation between these two types of pulsars and supernova remnants, respectively. In this paper, the statistical K-S test is used to judge whether the two samples of distributions may come from one original distribution or not.

According to the results of classification, 128 Type I and 174 Type II pulsars are presented. The distributions of Type I and II pulsars are compared with the longitudinal one of 160 galactic supernova remnants. The conclusions are that we can not say that the distribution of either Type I or II is similar to that of SNR.

Since there are four PSR being associated with SNR, the reasons why distribution of large sample gives an opposite conclusion may be as follows:

a) Since large number of pulsars observed are around the sun, the selection effect may affect the conclusion strongly.

b) Pulsars may exist in SNR but be undetectable, owing to the limitation of luminosity or the beaming effect.

c) Supernova may occur without the formation of neutron stars.

d) Because pulsars may be no active in the very early stage and its lifetime is about 100 times that of SNR, SNR may be disappeared when PSR become enough active to be detected.

e) High z-distribution of PSR will also affect this conclusion.



*D. J. Helfand and J.-H. Huang (eds.), The Origin and Evolution of Neutron Stars, 121.*

# THE STRUCTURE OF PULSAR NEBULAE

R. A. Chevalier and R. T. Emmering
Department of Astronomy, University of Virginia
P. O. Box 3818
Charlottesville, Virginia
U.S.A.

ABSTRACT. We have found solutions to the problem of a relativistic pulsar wind interacting with a nebula moving at constant velocity $v_n$. The wind is assumed to contain a toroidal magnetic field; the ratio of Poynting flux to particle flux in the wind is a constant $\sigma$. Under these assumptions, the wind is shocked at a radius $r_s$ that moves out at constant velocity. Solutions for the shocked wind are only possible if $\sigma < \sigma_c = 1/(-1 + c/v_n)$. For a given value of $v_n$, $r_s \to 0$ as $\sigma \to \sigma_c$. Kennel and Coroniti (1984) have calculated similar models for the Crab Nebula but they assume that $r_s$ is a constant; i.e. they calculate steady-state models. The steady-state approximation is expected to be good close to the shock wave, but it breaks down in the outer parts of the nebula. For $v_n = 2000$ km $s^{-1}$ and $r_n/r_s = 20$, Kennel and Coroniti find $\sigma = 0.003$. For the same Crab Nebula parameters, our time-dependent model yields $\sigma = 0.0016$. There may be times in the evolution of Crab Nebula or of other pulsar nebulae when a model with a shocked relativistic wind and a toroidal magnetic field cannot apply. It appears that a wide variety of structures are possible for pulsar nebulae.

This work was supported by NSF grant AST-8413138 and NASA grant NAGW-764.

## 1. REFERENCE

1.1 Kennel, C. F. and Coroniti, F. V. 1984, Ap. J., **283**, 694.



*D. J. Helfand and J.-H. Huang (eds.), The Origin and Evolution of Neutron Stars, 122.*

# HIGH-RESOLUTION RADIO OBSERVATIONS OF THE CRABLIKE SUPERNOVA REMNANT 3C 58

Stephen P. Reynolds
Department of Physics
North Carolina State University
Box 8202
Raleigh, North Carolina 27695

H. D. Aller
Department of Astronomy
University of Michigan
Ann Arbor, Michigan 48109

We report Very Large Array[1] observations of 3C 58, the Crab Nebula's closest relative at radio wavelengths. We combined A, B, C, and D configurations (baselines ranging from 0.08 to 35 km) at frequencies of 1446 MHz and 4886 MHz, achieving resolutions of $1''.9$ and $2''.5$, respectively, and a sensitivity at both frequencies of about 70 μJy per beam. We deconvolved the point-source response from the images using a maximum-entropy technique. The 1446 MHz total-intensity image shows that filamentation not only dominates the spatial distribution of flux in the bright inner regions, but appears to continue into the faint outer envelope as well. This envelope fades smoothly into the noise over about half the circumference of the remnant; the lowest contours of our map show maximum extents in RA and $\delta$ of $10'.3$ and $6'.3$ respectively. However, in some places the edge is relatively well-defined, suggesting confinement of some kind. A remarkably jet-like extension can be seen protruding from the south central part of the remnant, with a position angle of about 150° with respect to the radio peak. Unlike the Crab's jet, this is wedge-shaped and somewhat resembles the brighter extension to the north. The X-ray point source (Becker, Helfand, and Szymkowiak 1982, Ap. J. **255**, 557) lies about 30" east of the peak in the radio map. No obvious radio feature coincides with it. We have made spectral-index maps of the remnant, and find only small variations, with little apparent correlation of spectral index with structure. The rms deviation from the remnant mean spectral index is 0.18, but profiles of filaments show less than 0.1 difference in spectral index between peaks and adjacent valleys. Thus the bright filaments are neither flatter nor steeper in spectrum than the remnant mean, ruling out a picture in which bright filaments are locations where shock acceleration produces a steeper particle energy spectrum. Filaments have brightness contrasts of factors of 1.5 to 2, steep edges (<10", sometimes unresolved) and widths of 10" - 30"; if they have similar thicknesses along the line of sight, their emissivity ranges up to 40 - 50 times the remnant mean.

[1]The National Radio Astronomy Observatory is operated by Associated Universities, Inc., under contract with the National Science Foundation.



*D. J. Helfand and J.-H. Huang (eds.), The Origin and Evolution of Neutron Stars, 123.*

# SEARCH FOR PLERIONS IN THE DIRECTION OF TWO YOUNG PULSARS

S. Krishnamohan*, D.K. Mohanty*, A.R. Patnaik* and T. Velusamy+

Radio Astronomy Centre, Tata Institute of Fundamental Research
* P.O. Box 1234, Bangalore 560012, India
+ P.O. Box 8, Udhagamandalam 643001, India

Two of the recently discovered pulsars PSR 1800-21 and 1823-13 have characteristics ages of 17,000 and 22,000 yr respectively and all the three known pulsars that are younger than these two lie within the known supernova remnants (Clifton and Lyne, 1986). These two pulsars are expected to have, by scaling from the Crab nebula, plerions of ~1 Jy each associated with them at 327 MHz. We mapped a field of 1°.5 x 1°.5 around both the pulsars with the Ooty Synthesis Radio Telescope (Swarup, 1984). As the fields are on the galactic plane having complex large scale emission and as the plerions are expected to be compact, we have made maps by excluding baselines less than 500 $\lambda$. This would make our maps insensitive to emission regions larger than ~7 arc min. The synthesised beam is 96 x 36 arc sec in PA 0°. No source with a surface brightness greater than 60 mJy/beam was detected in the direction of PSR 1823-13. An unresolved source of ~150 mJy was detected, in the positional error box of PSR 1800-21, as is shown in the figure. No pulsed emission with an average flux density greater than 10 mJy was detected from this continuum source. It is possible that the pulse is so highly scatter broadened that it becomes undetectable at 327 MHz and the detected source is the scatter broadened pulsar. But, such a possibility seems unlikely as the pulsar's dispersion measure is only 230 $cm^{-3}$ pc, leaving the interesting possibility that the detected source is a plerion associated with the pulsar.

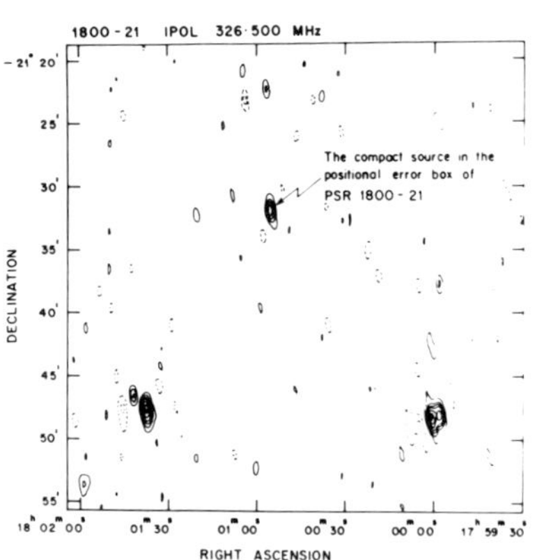


CONTOUR LEVELS = -80,-40, 40,80,120,160,240,320, 400 mJy/beam

## REFERENCES

Clifton, T.R. and Lyne, A.G., 1986. *Nature*, **320**, 43.
Swarup, G., 1984. *J. Astrophys. Astron.*, **5**, 139.



*D. J. Helfand and J.-H. Huang (eds.), The Origin and Evolution of Neutron Stars, 124.*

SUPERNOVA REMNANTS WITH RADIO JETS

M.J. Kesteven, J.L. Caswell, R.S. Roger, D.K. Milne,
R.F. Haynes and K.J. Wellington
Division of Radiophysics, CSIRO
PO Box 76, Epping, NSW 2121
Australia

Two examples are given of probable radio jet/supernova remnant association, G332.4+0.1 and G315.8-0.0. In both cases the jet length is larger than the radius of the remnant's shell; the jet diameter is barely resolved, and is substantially less than the observed shell thickness. The jet luminosity is 5-10% that of the shell. The G332.4+0.1 jet terminates in an extended plume whose luminosity is ~50% that of the shell.

Observations were made at the Molonglo Observatory Synthesis Telescope, at a frequency of 843 MHz and a synthesized beam of 44"x44"/sin (declination).

There are two problem areas:

1. What produces such well-collimated jets? The initial opening angle is less than 10°.
2. The plume of G332. Its lateral extent is comparable to the shell, which implies high velocities in the region where the jet ostensibly has slowed down. Or does it mean that the jet precedes the supernova outburst and remains visible for several thousand years after its source is extinguished?

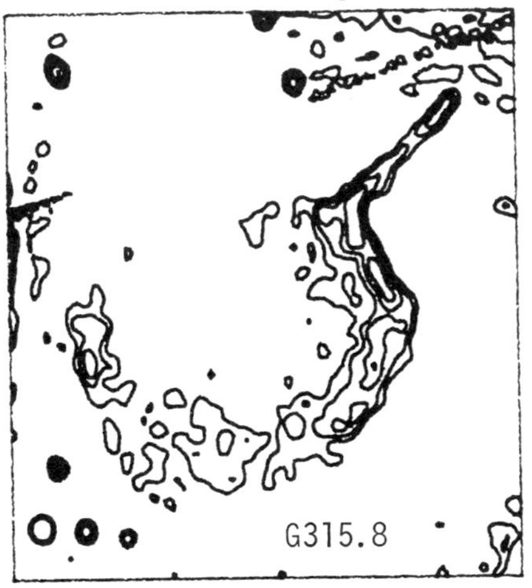


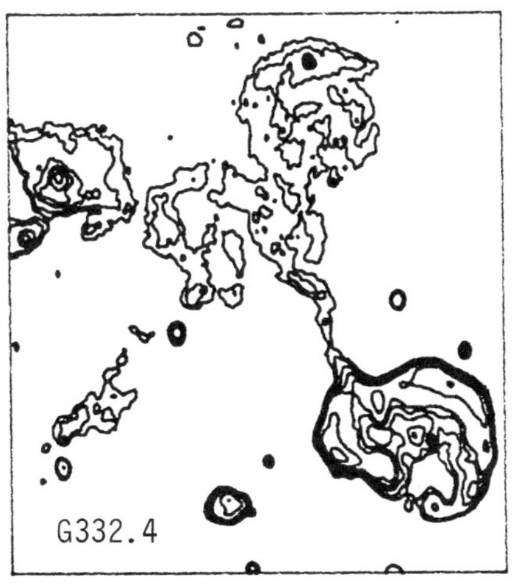




*D. J. Helfand and J.-H. Huang (eds.), The Origin and Evolution of Neutron Stars, 125.*

THE EVOLUTION OF YOUNG SUPERNOVA REMNANTS

John R. Dickel, University of Illinois , Eric M. Jones, Los Alamos National Laboratory, and Jean A. Eilek, New Mexico Tech.

The remnants of the Kepler's and Tycho's supernovae show a thick shell structure with a sharp outer edge and significant brightness irregularities. Models of their expansion have been constructed using a one-dimensional spherical hydrodynamics code, which includes both a leading shock and a reverse shock moving back into the ejectum. The dynamics are controlled by conditions between these shocks. Synchrotron radio emission is produced by acceleration and amplification of initial relativistic particles and fields by eddy motion at the interface between the ejected and swept-up material and at the boundaries of clumps in the surroundings with a mean separation of $5 \times 10^{17}$ cm, Gaussian sizes of $1 \times 10^{17}$ cm and peak densities of $3.6 \times 10^{-24}$ cm$^{-3}$. The latter was typically 10 times the mean density between clumps. A net radial orientation of the magnetic fields is attributed to stretching by Rayleigh-Taylor instabilities at the contact surfaces. To simulate true three-dimensional structure with the one-dimensional model, random contributions from four runs with varying spacings for the clumps were summed along the line of sight. The results are shown in the figure. Without clumps the observed shell is much too narrow and steep on the inside. Whether the clumpiness is a result of presupernova mass loss or a general property of the interstellar medium is not known.

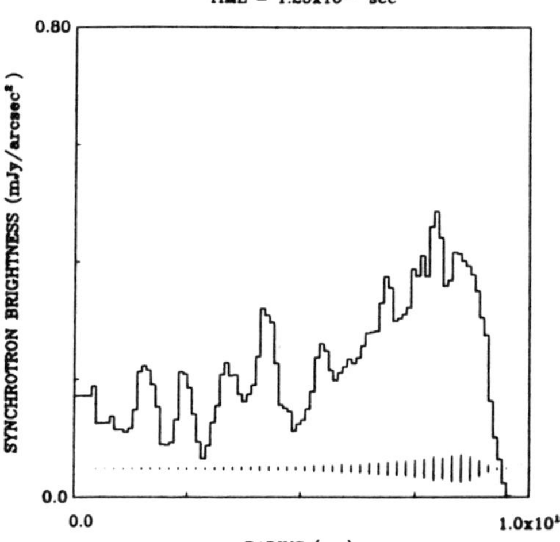


Predicted radio synchrotron emission from the model type I SNR expanding into a clumpy medium. The vertical vectors near the bottom represent the polarized power in a direction corresponding to a radial magnetic field.

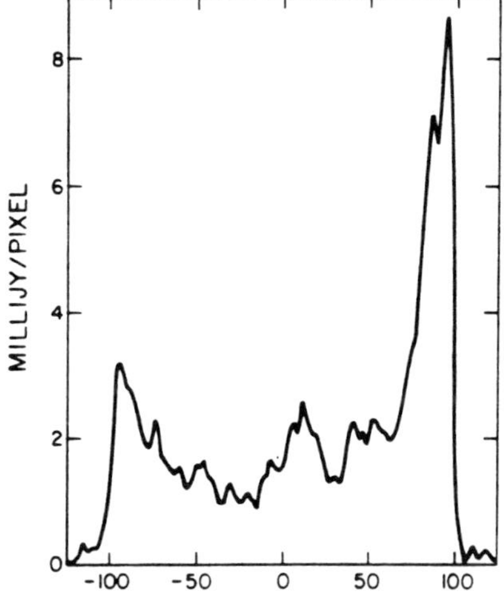


South-north slice through the radio emission from Kepler's SNR. A pixel represents a 1".75 by 2".75 Gaussian beam. The left side can be compared with the model.



D. J. Helfand and J.-H. Huang (eds.), The Origin and Evolution of Neutron Stars, 126.

# ON THE ORIGIN OF KEPLER'S SUPERNOVA REMNANT

R. Bandiera
Osservatorio Astrofisico di Arcetri
Largo E. Fermi 5
I-50125 Firenze
Italy

Aim of my poster is to show that Kepler's supernova progenitor was a massive star, rather than an old population object, as commonly accepted. The star was subject to strong mass loss, and the present remnant is due to the interaction of the blast wave with dense circumstellar matter.

Speculations on the nature of this remnant must account both for the density of the emitting gas (>1 $cm^{-3}$ from X-ray data; $\simeq 10^3$ $cm^{-3}$ from optical spectroscopy) and the velocities of optical knots (only a few hundred km/s, without evidence of expansion, but with an appreciable common motion, 350 km/s for a distance of 5 kpc). The knots cannot be ejecta, as they are not expanding; the interstellar nature can also be ruled out: gas is much denser and clumpier than expected at z=600 pc (the density should be only $3 \cdot 10^{-3}$ $cm^{-3}$), and no systemic motion should be measured. A circumstellar origin looks more natural: the average knots velocity reflects then that of the progenitor. If the star started from the galactic plane, a purely ballistic consideration gives an age of 3 million years.

The wind of a runaway star can be deflected by the interstellar flow. As in a comet, a bow shock forms in the direction of the motion; there matter stagnates and, due to instabilities, is likely to clump. If the blast wave is crossing now the bow shock, a brighter limb should appear on that side: this is actually observed. The size of the bow shock gives $\dot{M} \cdot w$; $\dot{M}/w$ comes from the assumption that the blast wave is in Sedov phase (supported by the broad X-ray radial profile on the side opposite to the bow shock); then mass loss ($\dot{M} \simeq 5 \cdot 10^{-5}$ $M_\odot$/yr) and wind speed ($w \simeq 10$ km/s) are derived. A lower limit of 10 $M_\odot$ can be put to the mass lost in a wind.

An alternative distance estimate is obtained by fitting the pattern of optical knots, in the assumption that they are confined to the intersection of the blast wave with the bow shock: I found a distance of 6 kpc.

Candidates for the progenitor are massive, runaway stars: I suggest Wolf-Rayet (WR) runaway. Such stars can be associated with ring nebulae. These nebulae are very clumpy, show a brighter limb, possibly in the direction of the star motion, and expand at a few tens of km/s: they probably formed during a red supergiant phase. WR runaway stars are suspected to have a neutron star companion. If a neutron star is left also after the WR explosion, one expects to find two neutron stars in eccentric orbits, exactly as for the binary pulsars PSR1913+16 and PSR2303+46.



*D. J. Helfand and J.-H. Huang (eds.), The Origin and Evolution of Neutron Stars, 127.*

# THE X-RAY SPECTRA OF TYCHO AND SN1006

K.Koyama
Institute of Space and Astronautical Science
Komaba 4-6-1 Meguro-ku, Tokyo, 153, Japan

The supernova remnants (SNR) SN1006 and Tycho are considered to be originating from type I supernovae (SN). Therefore one may expect a large amount of heavy elements, particularly iron, in the SNRs. In fact, our recent analysis of new data on the X-ray spectrum Tycho shows strong line emission associated with ionized silicon, sulfur and iron. This result contradicts the results of previous analyses which did not require an overabundance of iron.

Another puzzling has been the X-ray spectra and morphologies of Tycho and SN1006. In the case of SN1006, the spectrum between 1.5 - 20 keV is well-represented by a featureless power law. This spectrum is in sharp contrast to the X-ray emission from Tycho. If one compares the radio and X-ray emission from SN1006 with that from the Tycho, however, one finds very similar morphologies. We observed these two SNRs with the Tenma gas scintillation proportional counters (GSPC) in an attempt to better define the emission line features and the continuum of both the remnants in the X-ray spectral range 1.5-20 keV. We also attempted to resolve some of the confusion associated with these SNRs.

The resulting spectra are displayed in figure 1 along with the best-fit nonequilibrium ionization model spectra for a plasma which has t (sec) after the initial shock-heating to a single (constant) temperature (kT) and density (n). Both spectra can be well-expalined by similar parameter values; $nt=10^{10}$ sec.cc$^{-3}$, kT=2.1 keV. The abundances of heavy elements ( Si - Fe) for Tycho are about 10 times solar values, while those for SN1006 are consistent with solar values.

A possible explanation of the difference in abundances between Tycho and SN1006 is that the X-ray emission of Tycho may come from the ejecta( reverse shock) and that of SN1006 may come from the interstellar matter (blast wave).

Fig.1

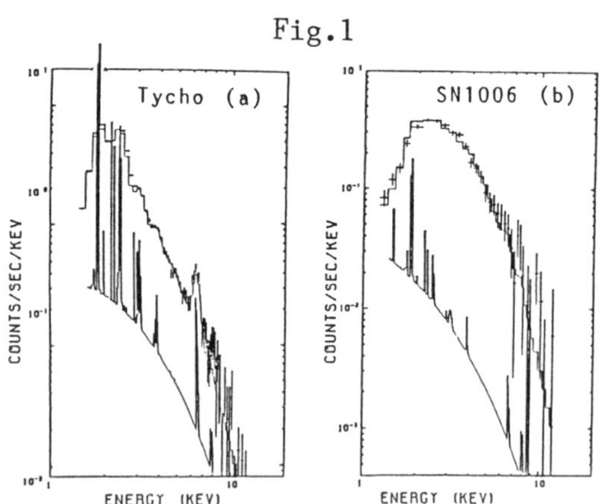


Figure 1
The energy spectra of Tycho (a) and SN1006 (b) along with the best-fit model of a thin thermal plasma in an ionizing state.

## References

Koyama,K.,Tsunemi,H,Becker,R.,Hughes,J.P.1986 submitted to P.A.S.J.
Tsunemi,H., Yamashita,K., Masai,K., Hayakawa,S., and Koyama,K. 1986, Ap. J., in press.



*D. J. Helfand and J.-H. Huang (eds.), The Origin and Evolution of Neutron Stars, 128.*

# THE GALACTIC SOURCES G5.4-1.2 AND G5.27-0.90

J.L. Caswell, M.J. Kesteven, R.F. Haynes, D.K. Milne,
M.M. Komesaroff, R.T. Stewart, and S.G. Wilson
Division of Radiophysics, CSIRO
PO Box 76, Epping, NSW 2121, Australia

Long after a supernova event, the stellar core (neutron star) may continue to excite an extended remnant of ejecta surrounding it, as in the case of the Crab nebula. In contrast, the more common shell supernova remnants (SNRs) appear unaffected by any embedded neutron star.

Recently, Becker & Helfand (1985, Nature 313, 115; hereafter BH85) postulated a new variety of nebula (which they stress is not an SNR but claim nonetheless is the product of a parent stellar-remnant/ binary-system) to account for the radio source G5.3-1.0. The source had hitherto been regarded as an SNR, and we suggest that this earlier, more prosaic interpretation is adequate, albeit with some novel aspects.

We used the MOST* to obtain a new map of the region at 843 MHz with beamsize 42"x110" arc. Our map shows not only the strong arc of emission noted by BH85 at RA $17^h58^m30^s$ but also a similar though weaker arc to the east at RA $18^h00^m30^s$. We suggest that these two arcs are simply the limb-brightened periphery of a roughly circular shell SNR with diameter ~35' arc. We designate it G5.4-1.2, corresponding to the galactic coordinates of its centre. The eastern arc was missed by BH85 because it lay outside the small field of view of their VLA map. Our interpretion of the two arcs as a single shell SNR is borne out by earlier low-resolution maps (e.g. Haynes et al. 1978, Aust. J. Phys. Astrophys. Suppl. No. 45, 1) showing a disk of emission which, although fainter to the east, does exhibit a well-defined eastern boundary.

Just outside the shell lies the source G5.27-0.90, which is quite compact (size ~70" to half-intensity). We find that it has a flat spectrum but believe it is unlikely to be thermal since there is no catalogued IRAS counterpart and we detected no radio recombination line. We suggest it is non-thermal, resembling the Crab nebula, and excited by the short-period pulsar PSR 1758-24 at $\ell=5\overset{\circ}{.}27$, $b=-0\overset{\circ}{.}96\pm0\overset{\circ}{.}1$ (Manchester et al. 1985, MNRAS 212, 975). The compact source G5.27-0.90 shows a weak tail towards the shell remnant G5.4-1.2. This may indicate ejection from the shell remnant like the proposed ejection of Cir X-1 from the shell SNR G321.9-0.3 (Clark et al. 1975, Nature 254, 675).

*Molonglo Observatory Synthesis Telescope, operated by the Physics Dept. of the University of Sydney (Mills 1981, Proc. Astron. Soc. Aust. 4, 156).



*D. J. Helfand and J.-H. Huang (eds.), The Origin and Evolution of Neutron Stars, 129.*

# SUPERNOVA REMNANTS IN THE MAGELLANIC CLOUDS OBSERVED AT THE MOLONGLO OBSERVATORY

A.J. Turtle
School of Physics
University of Sydney
New South Wales 2006
Australia

ABSTRACT. Radio surveys of the Magellanic Clouds at 843 MHz have been made with the Molonglo Observatory Synthesis Telescope. An initial catalogue (Mills et al. 1984) presented details of 38 supernova remnants detected by a combination of X-ray, optical and radio observations. The subsequent completion of the radio survey has revealed at least a further 17 remnants, mainly of large diameter and undetected by the Einstein X-ray Observatory. Though a few remnants have their radio emission concentrated towards their centre there is no evidence in the Magellanic Clouds for a Crab-like plerion (without an associated shell). The two well-established plerions 0540-693 (with optical and X-ray pulsar) and 0538-691 (N157B) appear to be connected with partial shells of strong radio emission which are relatively weak in optical emission lines (0538-691 is superposed on an HII region). Four sources which have a central concentration show larger optical and X-ray shells. The optical spectra of three of these (0505-679, 0509-675 and 0519-690) are dominated by the emission lines of hydrogen and Tuohy et al. (1982) argue that they are remnants of Type I supernovae. The fourth source (0453-685) which has detectable [OIII] emission may be an older but similar type of remnant. A further remnant 0509-687 (N103B) is compact but its radio diameter of 6 pc agrees with the published X-ray result.

## References

Mills, B.Y., Turtle, A.J., Little, A.G. and Durdin, J.M. (1984). Aust. J. Phys. 37, 321.

Tuohy, I.R., Dopita, M.A., Mathewson, D.S., Long, K.S. and Helfand, D.J. (1982). Astrophys. J. 261, 473.



*D. J. Helfand and J.-H. Huang (eds.), The Origin and Evolution of Neutron Stars, 130.*

# A STATISTICAL STUDY OF THE CORRELATION BETWEEN GALACTIC SNRS AND SPIRAL ARMS

Z.W. Li
Astronomy Department, Beijing Normal University, Beijing,China
J.C. Wheeler, F.N. Bash
University of Texas at Austin, USA

ABSTRACT. Based on the statistical studies of external galaxies, our Galaxy is expected to produce Type I and Type II supernovae in closely equal number. There are as of this writing 162 known galactic SNRs.

The basis for our investigation is the observations that SN II are tightly correlated with spiral galaxies and that SN I do not correlate with spiral arms, but are roughly of old disk population.

We seek a method by which a correlation of SNR with spiral arms can be determined. For this preliminary exercise we use only information on the positions of the SNR in galactic coordinates.

As a test, we compared the angular distribution of SNR and giant HII regions which are presumed to define the location of the spiral arms. To the eye, there did seem to have a correlation.

We thus faced a host of questions.

Do SN I still occur in spiral arms, implying that they have massive progenitors?

Is the rate of formation of SN I in Galaxy( and M31 ) considerably less than that of SN II?

Do SN I not produce long lived radio SNRs?

We have developed a quantitative approach to investigate these questions. Using the observed angular distributions of SNRs and giant HII region, we form a cumulative distribution with respect to angle. Two observed distributions can be compared then, and the Kolmogrov-Smirnov statistics was used to determine the probability that the two samples are not drawn from the same distribution.

In addition, we have constructed Monte-Carlo models in which sample objects are distributed in a prescribed pattern in the spiral arms and the galactic disk. The parameters of the Monte-Carlo simulation, such as the opening angle of the spiral arms and the fraction of objects in the disk vs arm population, can be adjusted to obtain the best fit to the observed distribution.

We also study the influence of the selection effects, if the surface brightness of SNR falls below a threshold for detectability ( as $1/r^2$ ).



*D. J. Helfand and J.-H. Huang (eds.), The Origin and Evolution of Neutron Stars, 131.*

# *IIa. ACCRETION POWERED PULSARS*

## *The Population*

*CHAIR: D. Helfand*

# ACCRETING NEUTRON STARS

N.E. White[1]
EXOSAT Observatory/ESOC
Robert Bosch Str 5
6100 Darmstadt
West Germany

1) Affiliated to the Astrophysics Division of the Space Science Dept of ESA.

ABSTRACT.

This paper reviews accreting neutron stars in X-ray binaries, with particular emphasis on how variations in magnetic field strength may be responsible for explaining the spectral and temporal properties observed from the various systems. This includes a review of X-ray pulsars in both low and high mass systems, and a discussion of the spectral properties of the low mass X-ray binaries.

## 1. INTRODUCTION

The strength of the magnetic field on the surface of an accreting neutron star in a binary system critically determines the fundemental X-ray properties of that system. If the field is strong then the accretion flow will be disrupted and funnelled onto the magnetic pole and an X-ray pulsar will be seen. If the field is weak ($<5\times10^8$ Gauss) and the flow directly interacts with the neutron star surface then X-ray emission will be expected not only from the neutron star, but also from the inner regions of the accretion disk.

The fact that X-ray pulsars are found predominantly in the young OB star systems and not in the older low mass X-ray binaries, LMXRB, is consistent with the view that the magnetic dipole moment of a neutron star decays on a timescale of $10^6$ to $10^7$ years (Flowers and Ruderman 1977). The radio pulsars with periods between 1 and 10 milli-seconds, some of which have degenerate binary companions, are most probably the remnants of now exhausted low mass X-ray binary systems, LMXRB (see e.g. Alpar et al 1982; Fabian et al 1983; Joss and Rappaport 1983). The measurement of the slow down of these radio pulsars indicates a range of magnetic field strength of $4\times10^8$ Gauss to $10^9$ Gauss (Taylor 1986). This result raises the possibility that the magnetic field decay timescale increases dramatically when the field has decayed to $\sim10^9$ Gauss and that there may still be a significant field present in many LMXRB (van den Heuvel, van Paradijs and Taam 1986).



D. J. Helfand and J.-H. Huang (eds.), The Origin and Evolution of Neutron Stars, 135–148.

This paper concentrates on how the properties of X-ray pulsars and the spectral properties of LMXRB are influenced by variations in the magnetic dipole field strength. For a discussion of the recent developments arising from the discovery by EXOSAT of quasi-periodic oscillations, QPO, from several of the most luminous LMXRB, the reader is refered to other articles in these proceedings; however some of the conclusions in this review are relevant to models for QPO.

## 2. THE POPULATIONS

In Figure 1 the galactic distribution of X-ray pulsars is compared with those of the other X-ray sources believed to contain neutron stars. Most of the X-ray pulsars lie along the galactic plane, typical of a population I distribution. The exceptions to this are the sources in the LMC and SMC, X Per (which is only ~300 pc away) and the LMXRB systems Her X-1 and 4U1626-67. The X-ray pulsars within our galaxy with OB star companions have an average distance from the galactic plane of 68 pc, comparable with the scale height of OB stars. The galactic distribution of the non-pulsing LMXRB is more concentrated towards the galactic center with a larger average height above the plane of 540 pc, suggestive of an older population.

The LMXRB come in two flavors. The X-ray burst sources typically have orbital periods of between 40 minutes and 7 hours with X-ray luminosities of $\sim 10^{37}$ erg/s. In addition to these ~30 sources there are 11 high luminosity sources that have luminosities typically a factor of ten larger i.e. close to the Eddington limit for accretion onto a neutron star. Optical studies of the optical counterparts of two of these systems, Cyg X-2 and Sco X-1, have revealed periods of 9 and 0.7 day. The other highly luminous sources lie in the heavily obscured region at the galactic center. The failure to detect X-ray orbital modulations in these systems may be because they have very long orbital periods of order tens of days and/or because of the obscuring effects of an accretion disk corona (White 1986).

The lower luminosity X-ray burst sources are driven by the loss of angular momentum from the binary system causing the orbit to decay forcing the companion to fill its Roche lobe. In the high luminosity systems the reverse occurs with the evolution (expansion) of a gaint companion, driving the components apart (c.f. Webbink, Rappaport and Savonije 1983; Rappaport, Verbunt and Joss 1983). The life time of the former sub-group may be $>10^9$yr, whereas for the latter case it is less, ranges between $10^7$ to $10^8$ yr.

The OB star pulsing X-ray binaries can also be divided into two sub-groups depending on whether the companion is a super-giant or a Be star (usually class III-V). The supergiant systems have orbital periods from a few days up to 40 days with, in the longer orbital period systems, eccentricities of up to 0.4. In the systems with orbital periods of a few days quasi-Roche lobe overflow may provide a significant fraction of the accretion flow to the neutron star. In the longer orbital period systems stellar wind accretion is sufficient to power the observed X-ray luminosities . The Be star systems tend to have longer orbital periods than their supergiant counterparts, ranging from ~16 day up to several hundred days. Many of these are only seen as transient X-ray pulsars whose outbursts are caused by

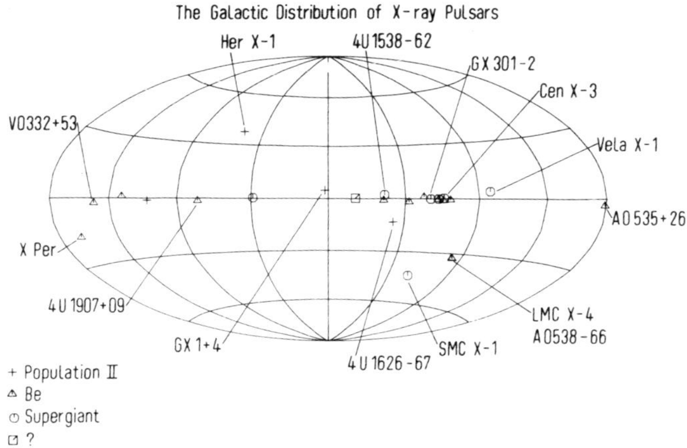


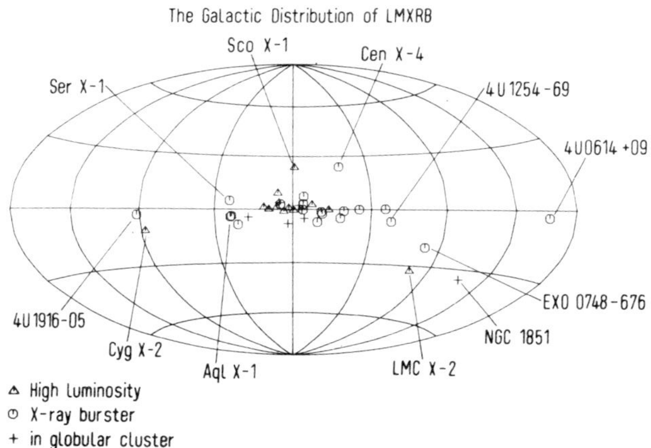


Figure 1. The galactic distribution of X-ray pulsars (top) and low mass X-ray binaries, LMXRB, (bottom). Various sub-populations are indicated by different symbols.

mass ejection episodes from the companion star (c.f. Stella, White and Rosner 1986 and refs therein).

## 3. X-RAY PULSARS

There are now 25 confirmed X-ray pulsars with periods ranging from 0.069 to 835 seconds. The pulse period distributions are shown in Figure 2 for the whole group of pulsars and for the various sub-groups outlined above. When all the pulsars are considered there is a possible bi-modal distribution with a clustering around 10 sec and around 200 sec, although its not a strong effect.

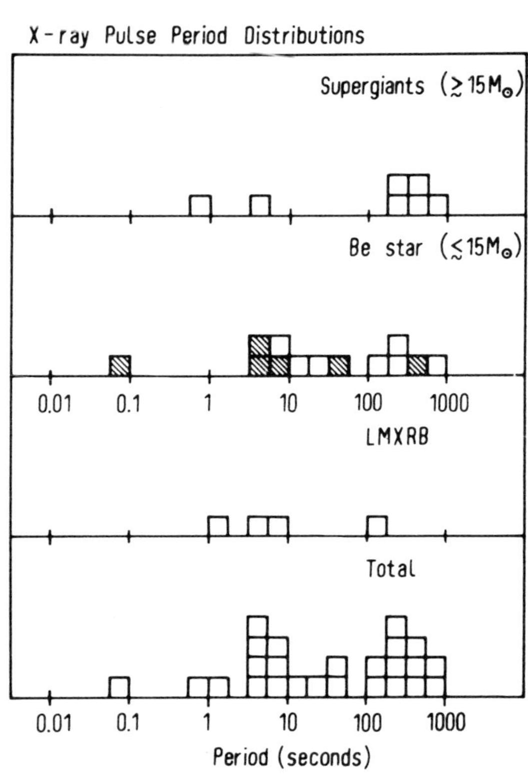


Figure 2. The pulse period distribution of X-ray pulsars. The hatched areas represent transient pulsars in Be star systems.

Looking at the various sub-groups does reveal some possible trends. Three of the pulsars in the LMXRB have periods between 1 and 10 secs, while only one has a period of 100 sec. The reverse trend seems to be the case for the supergiant systems where five out of seven have pulse periods between 100 and 1000 sec. In the Be star systems the distribution is much more evenly divided across the period range. The X-ray transient Be star pulsars are marked with shaded boxes. The pulse periods of all but one of the six are less than 100 sec, with one as short as 69 ms.

## 4. CENTRIFUGAL INHIBITION OF ACCRETION

When a neutron star is born in a binary system it will probably start out spinning rapidly with a period of tens of milli-seconds and then be spun down by various processes to rotate at the periods we now observe (c.f. Henrichs 1983 and refs therein). The spin down process will cease when the rotation period is equal to the Keplerian period at the inner edge of the accretion disk; at periods shorter than this equilibrium value the accretion flow will be centrifugally ejected by centrifugal forces.

In the systems where there is sufficient angular momentum for an accretion disk to form (usually those with Roche lobe overflow) the period is seen to be increasing with a timescale of 100-10,000 yr, depending on the luminosity (c.f. Joss and Rappaport 1984). The spinup timescales are typically much faster than the evolutionary timescale of either the companion or the neutron star magnetic field. This means that the neutron star spin must be re-adjusting to a recent increase in the mass accretion rate. In most of the wind driven systems the pulse periods undergo positive and negative variations in period without any obvious long term trend. In these systems the pulsar is being torqued in random directions by inhomogeneities in the captured stellar wind (Boynton et al 1984).

A centrifugal barrier to accretion will occur when the rotation of the magnetosphere exceeds that of the local Keplerian disk. This sets a lower limit to the X-ray luminosity that can be observed from an accreting neutron star that is given by

$$L_{min}=2\times10^{37}.R_x^{-1}.M_x^{-2/3}.\mu_{30}^2.P_s^{-7/3} \text{ erg/s}$$

where $R_x$, $M_x$, $\mu$ and $P_s$ are the neutron radius, mass, magnetic dipole moment and rotation period in units of 10 km, 1.4 $M_o$, $10^{30}$ Gauss/cm$^3$ and sec. In Figure 3 a plot of luminosity verses pulse period is given for all the pulsars with reasonable distance estimates (from Stella, White and Rosner 1986). Lines of constant $\mu$ defined by the above equation are given for various field strengths. An accreting neutron star can only be a bright X-ray source if it lies above one of these lines i.e. the position of a pulsar in this diagram defines a maximum magnetic dipole moment. It is notable that most pulsars lie above the $\mu=10^{30}$ Gauss/cm$^3$ line (corresponding to $10^{12}$ Gauss).

In plotting Figure 3 the maximum observed luminosity was used because this is the best value documented in the literature. A better restriction on $\mu$ comes from using the minimum observed luminosity and in Figure 3 the vertical lines indicate the range of luminosity seen from a selection of these objects. In most cases the range is not sufficiently large to change the fact that most of the long period

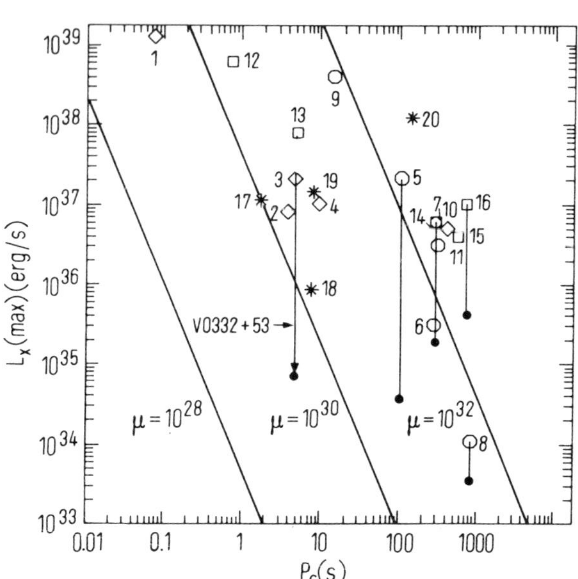


Figure 3. The pulse period vs. maximum luminosity plot for X-ray pulsars. The numbers denote the following sources:

1. A0538-66
2. X0115+63
3. V0332+53
4. X1553-54
5. A0535+26
6. GX301-2
7. X1145-61
8. X Per
9. LMC X-4
10. A1118-61
11. X1145-616
12. SMC X-1
13. Cen X-3
14. Vela X-1
15. X1538-52
16. X1223-62
17. Her X-1
18. 1E2259+59
19. X1626-67
20. GX1+4

The diamonds indicate transients, the boxes supergiants, circles Be systems and stars LMXRB.

pulsars lie well above the $10^{12}$ Gauss line. This means that these systems are either well away from their equilibrium values, or the dipole moments are in excess of $10^{14}$ Gauss. There is no evidence from radio pulsar spin down measurements that such large dipole fields are common and it seems more likely that the equilibrium period of these long period pulsars was orginally much longer.

Many of the transient X-ray pulsars lie close to the $10^{12}$ Gauss line and only a small decrease in luminosity will cause the centrifugal barrier to act. The transient nature of these systems is explained by a small increase in mass accretion rate causing the centrifugal barrier to be overcome and the pulsar to suddenly turn on as a bright X-ray pulsar. These Be X-ray transients may represent a large population of dormant systems that are still in the spin down phase, that will in the future become much longer period pulsars.

In 1973 the transient source V0332+53 underwent an exceedingly luminous outburst during which the flux rose smoothly over an interval of a month, then decayed again over the same time interval (Terrell and Priedhorsky 1984). More recently in 1983 a series of mini outbursts with $L_{max}/L_{min}>500$ were seen which occured every 34 day at the time of periastron passage of the neutron star (Stella et al 1985); no obvious orbital modulation was evident during the 1973 outburst. The overall range of luminosity seen from V0332+53 is indicated in Figure 3. Below the minimum of this range the source was not detected, with an upper limit a factor of 40 below the minimum detected luminosity. The minimum detected luminosity gives an estimate of the dipole field of $\sim 3.5 \times 10^{11}.d^2$ Gauss, where d is the distance in units of 1.5 kpc. This is quite a low value for the dipole field, although uncertainties in the distance estimate could probably bring it closer to the canonical value. The interesting point about V0332+53 is that the mass loss rate from the companion in 1983/4 was such that the equilibrium point straddled the eccentric orbit, such that only close to perigee was the accretion rate high enough to overcome the centrifugal barrier. This explains why a large orbital modulation of the X-ray flux was seen in 1983/4, but not in 1973 when the mass transfer rate was much larger.

A transient X-ray pulsar where a similarly large X-ray orbital modulation is seen is A0538-66. This source has an exceptionally short spin period of 69 milliseconds and is seen at luminosities that are a factor of ten above the Eddington limit. In order to overcome the centrifugal barrier at all, even at these high accretion rates, a lower than average dipole field strength is required. Taking the minimum observed luminosity observed from this source of $\sim 10^{38}$ erg/s gives a dipole field of $\sim 10^{11}$ Gauss. This is a factor of ten below the canonical value.

The position of the LMXRB systems in Figure 3 are also indicated (as * symbols). Her X-1 and 1E2259+586 lie on the $10^{12}$ Gauss line. For Her X-1 this is somewhat lower than the surface field mesurement obtained from the cyclotron line measurement of $4 \times 10^{12}$ Gauss (Truemper et al 1978). This is probably not inconsistent because the cyclotron measurement reflects the surface field, whereas the centrifugal barrier limit reflects the average field at the magnetospheric boundary; any multi-pole components will tend to make the latter lower.

The binary X-ray pulsar 1E2259+586 is located at the center of the supernova remnant G109.1-1.0 which has an age of $\sim 10^4$ years (Gregory and Fahlman 1980). It seems very likely that the remnant is the remains of the supernova which created the X-ray pulsar, which allows the age of the neutron star to established. The faintness of the optical counterpart indicates that this is a LMXRB, with some evidence from X-ray timing and infra-red observations that the orbital period is 38 min (Fahlman and Gregory 1983; Middleditch et al 1983). The observed pulse period of 7 sec is close to the equilibrium period, such that if the pulsar was born spinning rapidly it must have been slowed down to its current value in only 10,000 years, which is several orders of magnitude quicker than current spin down theories predict (c.f. Henrichs 1983). Lipunov and Postnov (1985) suggest that this system evolved from an AM Her system where the white dwarf was pushed over the Chandrasekar limit. Scaling, respectively, the ratio of the moment of inertia and radius of a neutron star to that of a white dwarf gives an orginal rotation period and magnetic field strength for the white dwarf projenitor of 100 min and $10^8$ Gauss, comparable to the parameters of an AM Her system. The alternate explanantion by van den Heuvel and Bonsema (1984) that this is a post common envelope system does not explain the slow spin of the pulsar.

An AM Her system may also have been the projenitor to the 7 sec X-ray pulsar 4U1626-67, which has a 42 min orbital period (Middleditch et al 1981). The high spinup indicates a readjustment of the pulsar spin to a recent increase in accretion rate (Joss, Avni and Rappaport 1978). For Her X-1 the high magnetic field of the neutron star is incompatable with the expected lifetime of the companion star, unless the neutron star was also formed by white dwarf collapse (Sutantyo et al 1986). The same is probably true for GX1+4 where the companion is a symbiotic star in an orbit of one hundred or more days (Taam and van den Heuvel 1986).

5. SPECTRAL PROPERTIES

While a considerable effort has been made to understand the spectral properties of the X-ray pulsars, it is only comparatively recently

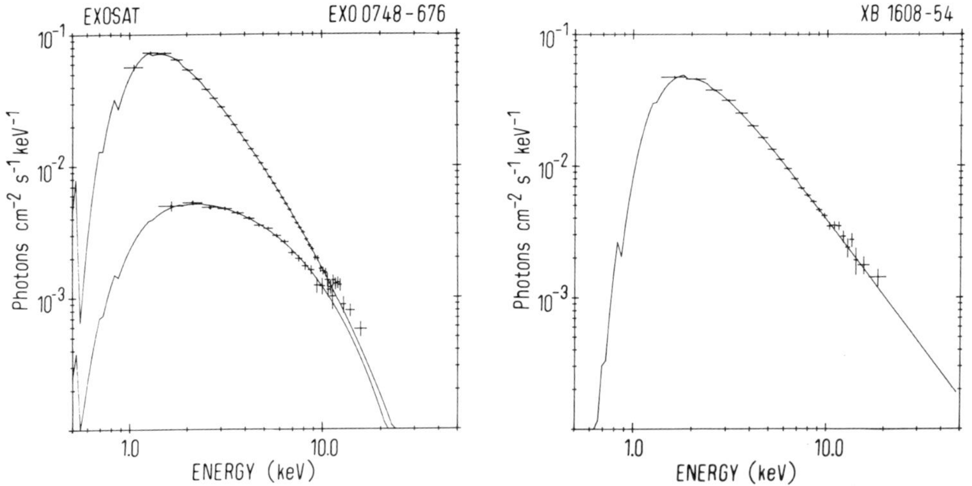


Figure 4. The X-ray spectra from the persistent emission from two typical burst sources. For EXO0748-676 spectra taken at two different intensity levels are shown.

that any serious attempt has been made to model the spectra of the non-pulsing LMXRB in terms of disk accretion onto a weakly magnetised neutron star. The spectral properties of the X-ray pulsars are quite different from those of the non-pulsing LMXRB. The former typically have a power law spectrum with an energy index of 0 to 1, with a high energy cutoff above 10-20 kev (White, Swank and Holt 1983). The non-pulsing LMXRB spectra are typically softer and can be modelled in two ways depending on whether the luminosity is low or high.

For the X-ray burst sources where the luminosities tend to be lower, the spectra can be modelled by a simple power law that, in some cases, is attenuated at high energies by an exponential cutoff. Two examples of these spectra are shown in Figure 4 taken from EXOSAT observations. The power law energy index is typically of order unity and the high energy exponential cutoff varies from ~5Kev, to not being detectable (>50 Kev). For EXO0748-676 the spectrum changed when the source luminosity declined by a factor 5. In the low state the power law index became considerably less steep with a value of ~0.

In the more luminous Eddington limited LMXRB the spectra can be empirically modelled by the same power law with a high energy exponential cutoff, but with an additional blackbody component. A typical example of this, Cyg X-2, is shown in Figure 5. The blackbody luminosity varies from source to source from ~10% to 50% the total. The temperature is between 1 and 2 kev and the infered radius of a spherical emitter ranges between a few and 10 km. The characteristic Sco X-1 like flaring seen from several of these systems is usually caused by increases in the luminosity of the blackbody component.

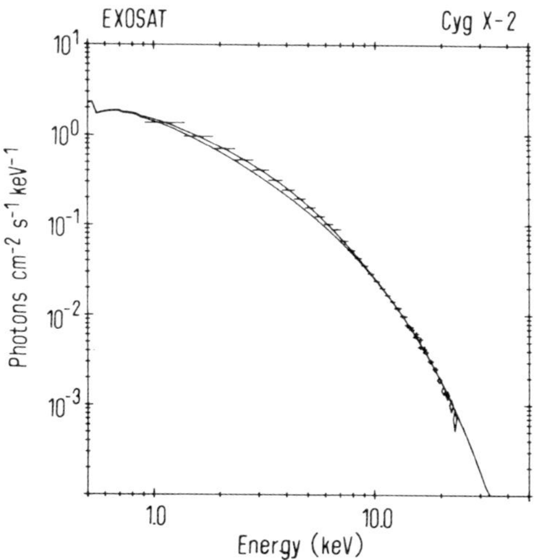


Figure 5. The spectrum of the high luminosity LMXRB Cyg X-2. The lower and upper lines indicate the best fit model with and without the blackbody component.

equal luminosities to the total, unless the neutron star is rotating close to its limiting break up period. Angular momentum conderations show that only ~0.1 $M_o$ of material is required to spin a neutron star up to its break up rotation period, a number which is attained within only $10^7$ years at an accretion rate of $10^{-8}$ $M_o$/yr (Alpar et al 1982). The minimum rotation period of a neutron star will be ultimately limited by rotationally induced instabilities causing distortion of the neutron star and the emission of gravitational radiation. The limiting period first estimated by Papaloizou and Pringle (1978) was 1.5 ms, a factor of two longer than break up. This has now been superceded by more recent calculations that indicate that it may be possible to spinup the neutron star much closer to its limiting breakup value (Friedman, Ipser and Parker 1984). For periods less than about twice the break up value the boundary layer luminosity will asymptotically approach zero.

In the inner region of the accretion disk radiation pressure will dominate the disk structure. If the viscous stress tensor scales as the total gas pressure (Shakura and Sunyaev 1973) this inner region is unstable to secular and thermal instabilites and it may swell up into a quasi-optically thin region (Shapiro, Lightman and Eardley 1976). If the stress tensor scales only with the gas pressure then the disk is stable and optically thick (Stella and Rosner 1984). Most of the calculations have assumed that the disk is optically thick up to some inner radius and then becomes optically thin. Most models assume two components, one from the inner disk and another from the boundary layer, with the relative contribution of both as a free parameter.

In modelling Tenma spectra Mitsuda et al (1984) assume that the disk radiates as a blackbody up to an inner radius where it becomes optically thin. The fits to this model give a radius for the optically thin region of two or more neutron star radii and blackbody luminosities comparable to that from the disk. As noted by Lamb (1986) this model predicts enhanced emission from the boundary layer because

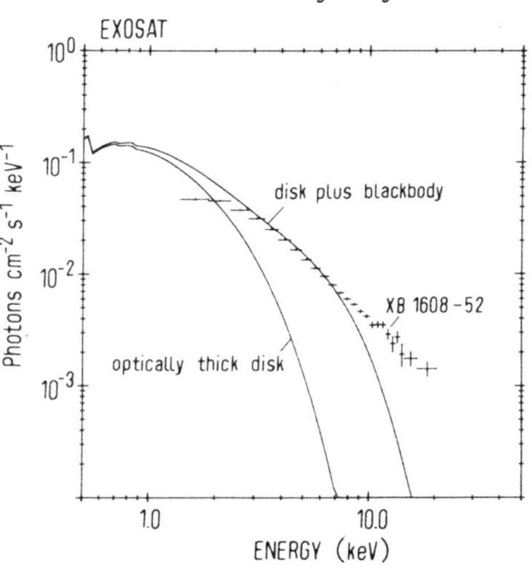


Figure 6. The spectrum of the burst source XB1608-52 fit to the disk model proposed by Mitsuda et al (1984). The two plots represent the model with and without the blackbody emission from the boundary layer.

the gravitational potential energy from the inner optically thin region must be released at the boundary layer; this is not seen. Another problem is that while reasonable fits can be obtained to the spectra of the bright bulge sources, this model fails to explain the power law spectra of the lower luminosity sources. In Figure 5 the Mitsuda et al model is compared to the spectrum of the persistent emission from the burst source XB1608-52, assuming that the luminosity in the blackbody component is equal to that in the disk. At high energies the observed spectrum does not decay in the manner predicted.

In an alternate model White et al (1986) assume the spectrum from the inner accretion disk is dominated by Comptonization (c.f. Shapiro, Lightman and Eardley 1976). The resulting spectrum in the unsaturated case can be approximated by a power law with an exponential cutoff. The key parameters are the optical depth and temperature of the scattering region and are typically ~10 and 3 kev respectively. Variations in the spectrum like those seen fom the burst source EXO0748-676 are explained in terms of changes in these two parameters. The interesting point is that for many burst sources the inclusion of a blackbody component is not required (>10% the total luminosity) indicating the boundary layer emission to be small i.e. the neutron star rotates close to breakup.

The spectra of the high luminosity systems require a blackbody component at the 10-50% level. This may indicate that in these systems the boundary layer contributes a significant fraction of the luminosity. A more slowly rotating neutron star in these systems may be consistent with the shorter lifetime of these systems expected from evolutionary consierations. The presence of a magnetic field will tend to enhance the blackbody component to more than 50% of the total because the flow of the latter will be interupted by the magnetic field pressure; there is no evidence for such an enhancement in any of these systems.

## 6. CONCLUSIONS

The overall properties of X-ray pulsars are strongly dependent on how close they spin to the equilibrium point where the centrifugal radius equals the magnetospheric boundary. Most of the long rotation period (>100s) neutron stars lie far from this boundary, unless their surface fields exceed $10^{14}$ Gauss which seems unlikely. Most of the transient X-ray pulsars in Be star systems do, however, lie close to the critical equilibrium point (when they are in outburst). These sources undergo sharp turn offs during the decline phase when the centrifugal radius moves in side the magnetosphere.

The four (young) X-ray pulsars that are located in the old population of LMXRB most likely resulted from the collapse of an accreting white dwarf. The spectral properties of the remaining LMXRB are dependent on the luminosity (and possibly by the evolutionary history) of the system. In both the high and the low luminosity cases the properties can only be explained if part of the emission originates from Comptonization in the inner accretion disk. Optically thick disk models cannot explain the persistent emission from X-ray burst sources. Many of the burst sources do not show evidence for any blackbody component in the persistent flux which may indicate that the neutron stars have been spun up to close to their breakup period (~0.6 milli seconds). The spectra of the high luminosity systems do show evidence for blackbody components, suggesting that the rotation periods may be longer. The spectra provide no evidence for any residual magnetic field on the neutron star as required by some of the QPO models.

REFERENCES

Alpar, M.A. et al 1982. Nature, 300, 728.
Boynton, P. et al 1984, Ap.J. 283, L53.
Fabian, A.C. et al 1983. Nature, 301, 222.
Fahlman, G.G. and Gregory, P.C. 1983. in 'Supernova Remnants and their X-ray Emission IAU Symp.101', eds: Danziger and Gorenstein], p.445.
Flowers, E. and Ruderman, M.A. 1977. Ap.J., 215, 302.
Friedman, J.L., Ipser, J.R. and Parker, L. 1984. Nature, 312, 255.
Gregory, P.C. and Fahlman, G.G. 1980, Nature, 287, 805.
Henrichs, H.F. 1983. in 'Accretion Driven Stellar X-ray Sources' [Ed: Lewin and van den Heuvel], p.393.
Joss. P.C., Avni, Y. and Rappaport, S. 1978. Ap.J., 221, 645.
Joss. P.C. and Rappaport, S.A. 1984. Ann. Rev. Astr. Ap., 22, 537.
Joss. P.C. and Rappaport, S.A. 1983. Nature, 304, 419.
Lamb, F.K. 1986, in 'the Evolution of Galactic X-ray Binaries', [Ed: Truemper, Lewin and Brinkmann], p.227.
Lipunov, V.M. and Postnov, K.A. 1985. Astr.Ap., 144, L13.
Mitsuda, K. et al 1984, P.A.S.J., 36, 741.
Middleditch, J. et al 1981. Ap.J., 244, 1001.
Middleditch, J. et al 1983. Ap.J., 274, 791.
Papaloizou, J. and Pringle, J.E. 1978. M.N.R.A.S., 184, 501.
Rappaport, S.A., Verbunt, F. and Joss, P.C. 1983. Ap.J. 275, 713.
Shapiro, S.L., Lightman, A.P. and Eardley, D.M. 1976, Ap.J., 204,187.
Shakura, N.I. and Sunyaev, R.A. 1973, Astr.Ap., 24, 337.
Stella, L. and Rosner, R. 1984. Ap.J., 277, 312.
Stella, L. et al 1985. Ap.J. 288, L45.
Stella, L., White, N.E. and Rosner R. 1986. Ap.J., in press.
Sutantyo, W. et al 1986. in 'The Evolution of Galactic X-ray Binaries', [Ed:Truemper, Lewin and Brinkmann], p.261.
Taam, R.E. and van den Heuvel, E.P.J. 1986. Ap.J., 305, 235.
Taylor, J. these proceedings.
Terrell, J. and Priedhorsky, W.C. 1984. Ap.J., 285, L15.
Truemper,J. et al 1978. Ap.J., 219, L105.
van den Heuvel, E.P.J. and Bonsema, P.T. 1984. Astr.Ap., 139, L16.

van den Heuvel, E.P.J., van Paradijs, J.A. and Taam, R.E. 1986. Nature, 322, 153.
Webbink, R.F., Rappaport, S.A., and Savonije, G.J. 1983. Ap.J. 270, 678.
White, N.E., Swank,J.H. and Holt, S.S., 1983, 270, 711.
White, N.E. et al 1986. M.N.R.A.S., 218, 129.
White, N.E. 1986. in 'the Evolution of Galactic X-ray Binaries', [Ed: Truemper, Lewin and Brinkmann], p.227.

## DISCUSSION

**W. Lewin:** I have two comments: You explained in some detail a model that elegantly explained the transient phenomenon of some systems. There are, however, systems (e.g., A0620-00) for which this scenario is not operative, and for which there are not yet attractive models. Comment 2: In calculating the radius of the magnetosphere you used the dipole field approximation. For small values of this radius (e.g. < 3 stellar radii) the results may not be accurate. Higher order fields can then become important. Therefore, the magnetic dipole field strength for which one finds a magnetospheric radius of ~10 km, using your equation, is probably much higher than the actual value. Fred Lamb mentioned to me once that the dipole field may have to be as low as $\sim 10^7$ G before the disk can touch the surface of the neutron star.

**N. White:** No comment.

**J. Dolan:** Certain galactic X-ray binaries with neutron star secondaries (e.g. 4U1700-J7) exhibit nearly circular orbits ($e \approx 0$) but the optical counterpart is not co-rotating. Theory predicts that co-rotation should be established by tidal forces ~10 times faster than the orbit is circularized. If all neutron stars are the result of a SN explosion, then any binary neutron star should be formed with a highly eccentric orbit. Now, co-rotating primaries in X-ray binaries with nearly circular orbits indicate that not all neutron stars are formed in SN explosions, i.e., their orbits have low eccentricity at formation.

**S. Kulkarni:** First a comment: The X-ray source 1E2259+586 is undoubtedly a low mass binary system <u>now</u>. However, the progenitor system need not necessarily be a low-mass system e.g. Savonige <u>et al.</u> (<u>A&A</u> **155**, 51, 1986) suggest that the initial configuration for 1E2259+586 consists of a 5 $M_\odot$ and 8 $M_\odot$ binary. If their scenario is correct it may be inappropriate to classify 1E2259+586 and perhaps 4U1626-67 as low-mass X-ray pulsars. Now a question: I notice that you quote a $Z_{rms} \sim 580$ pc for all LMXB. The $Z_{rms}$ for the 8 bright, bulge X-ray sources is only $\lesssim 300$ pc. Are we being fooled by small number statistics or are there two classes of sources?

**N. White:** The scale height maybe different for the bright bulge sources, but with only eight sources we are dealing with small numbers and the uncertainties are difficult to estimate.

**J. Grindlay:** I would like to make two brief comments: 1) Concerning your discussion of theoretical models for X-ray spectra, I would like to draw attention to work by (Czerny, Czerny and Grindlay 1986) which predicts the emission from the disk vs. boundary layer components which actually fit the data for burst sources. 2) In regard to Kulkarni's question, one should keep in mind that the brightest bulge sources may have had a different evolutionary origin than the lower luminosity systems such as bursters and globular cluster sources. It is unfortunate that there are not more of the bright luminosity galactic bulge sources to better determine their scale height to test whether it is indeed smaller than for the bursters.

**V. Trimble:** Recent models of rapidly rotating neutron stars indicate that for most reasonable equations of state, the m=2 and m=3 (gravitational radiation) modes are not exited before you reach break up (rotational KE > gravitational BE). Presumably this doesn't affect the evolution much since spin up still has to stop near 1 msec.

**N. White:** It is very relevant. If the neutron star rotates close to break up then the energy released in the boundary layer will be small. This would be consistent with the fact that we do not detect blackbody components > 10% of the total luminosity in the X-ray burst sources and would suggest that they have been completely spun up.

**E. van den Heuvel:** The B-star in A0538-66 has a mass of about 8 or 9 solar masses according to the work of Hutchings *et al.*, and is, according to its spectral type at quiescence, already evolved away from the main sequence (luminosity class III). This indicates an age $\gtrsim$ 15 million years for the neutron star. Also, the magnetic field may be larger than the $3\times10^{10}$ G you mention. If I take the period to be the equilibrium period corresponding to maximum X-ray luminosity, of > $10^{39}$ ergs, I get B ~ $3\times10^{11}$ G. Taking this together with the age of 15 million years, it could still have been born with a strong field $\gtrsim 10^{12}$ G.

**N. White:** You are correct about the magnetic field, there was a transcription error on my viewgraph. This is still rather low however. With regard to the evolution timescale we need a good estimate of the eccentricity first to see if the circularization timescale is consistent with the model.

# POPULATION OF ACCRETING NEUTRON STARS IN EXTERNAL GALAXIES.

Ginevra Trinchieri
Osservatorio Astrofisico di Arcetri
Largo E. Fermi 5
50125 FIRENZE

The X-ray observations of the Milky Way galaxy with non-imaging X-ray satellites (e.g. UHURU, HEAO1) has revealed a number of discrete, point -like bright sources clustered around the Galaxy's center (in the bulge region) and on the plane of the Galaxy (see for example Tanambaum and Tucker, 1984). The brightermost ones have been associated with close accreting binary systems, containing an evolved object (a white dwarf, a neutron star or a black hole) and a companion visible star. For sources with X-ray luminosities $L_X \leqq 10^{38}$ erg $s^{-1}$, the compact object needs to be a neutron star or a black hole.

The bright X-ray sources in the Milky Way are divided into two broad main classes (see the review by van den Heuvel, 1980), depending on the mass of the companion star: massive binaries or Type I sources and low mass systems or Type II sources.

There are several problems in studying X-ray sources in the Milky Way: in most cases it is difficult to assign the proper location of a source in the disk, arms or bulge region, the distances of these objects are poorly known and a large fraction of their emission is absorbed by the intervening material in the galactic plane. Part of these difficulties can be solved by observing these sources in other galaxies. This will improve our knowledge of other systems similar to our own galaxy as well as give us new insights on the properties of the galaxy itself.

The instruments on board the EINSTEIN Observatory (Giacconi et al, 1979) made possible the study of the X-ray emission of normal galaxies, and the nature, distribution and characteristics of the evolved stellar population in other galaxies through its X-ray emission. Since this meeting is focused on the origin and the evolution of neutron stars, I will concentrate on the results regarding X-ray sources in binary systems and disregard all other phenomena that have come out as a result of the study of normal galaxies in X-rays (for details, see Fabbiano and Trinchieri, 1985, 1986; Trinchieri and Fabbiano, 1985; Trinchieri, Fabbiano and Palumbo, 1985; Palumbo et al. 1985; and references therein).



*D. J. Helfand and J.-H. Huang (eds.), The Origin and Evolution of Neutron Stars, 149–159.*

## LOCAL GROUP GALAXIES

The first obvious targets for the EINSTEIN Observatory were the nearby galaxies M31 and the two Magellanic Clouds. The observations were then extended to other galaxies in the Local Group and to even farther away galaxies.

The X-ray images of the nearby galaxies have shown that, as in the case of the Milky Way, the X-ray emission is dominated by point-like, bright individual sources (see M31, van Speybroeck et al, 1979; LMC, Long, Helfand and Grabelsky, 1981; SMC, Seward and Mitchell, 1980, and many others), with X-ray luminosities ranging from $\sim 10^{34}$-$10^{36}$ erg $s^{-1}$ (typical limiting sensitivities for the Magellanic Clouds and M31 respectively) to $<10^{39}$ erg $s^{-1}$ in the EINSTEIN band pass (0.2-4.0 kev). Such high luminosities suggest that in most cases these sources are close accreting binary systems, with a neutron star or a black hole as the compact accreting object.

### Massive Binary Systems

A large number of the bright point-like sources detected in these galaxies are found associated with the young Pop I stellar component. Figure 1 (from van Speybroech and Becktold 1980) shows the distribution of the X-ray sources in the M31 galaxy compared to neutral hydrogen emission. Their positional coincidence with regions of star formation would suggest that these sources are binary systems with a young, massive companion. Although secure optical identification of these sources is difficult (see Crampton et al 1984), $\sim$ 13 sources have been associated with blue stars. The average X-ray luminosity of these sources is $2.2 \times 10^{37}$ erg $s^{-1}$ (van Speybroeck et al, 1979), and if the identifications are correct $L_x/L_{opt} \sim 1$ (Crampton et al 1984), typical of the brightest supergiant OB systems in the Galaxy (Bradt and McClintock, 1983).

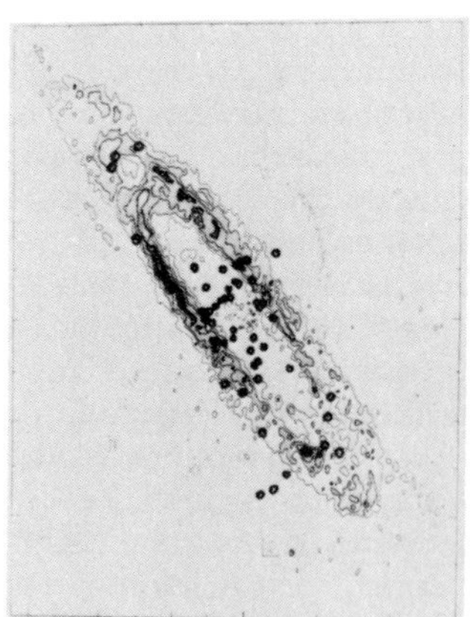

Figure 1: The X-ray sources detected in M31 superposed on the neutral hydrogen emission (from Van Speybroeck and Bechtold, 1980).

Of the ~ 60 sources detected in the LMC, the largest class of identified objects are Supernova Remnants. About 4 sources (LMC X-1, LMX C-3, LMC X-4, A0538-66) and three other possible candidates have been identified with early type stars in massive binary systems (Cowley et al. 1984). The SMC hosts one of the most luminous non transient X-ray source in the Local Group, SMC X-1, with $L_x > 10^{38}$ erg $s^{-1}$, identified with a B star system (Webster et al. 1972). A number of sources in M33 are also identified (or candidate) massive binary systems (see Helfand, 1984 and references therein).

## Low Mass Binary Systems

In the Galaxy, low mass binary systems are located in the bulge region and in globular clusters. In M31, ~ 20 X-ray sources have been identified with globular clusters. Their average X-ray luminosity is $L_x \sim 5 \times 10^{37}$ erg $s^{-1}$, higher than for Milky Way globular clusters (van Speybroeck and Bechtold, 1980). However, their optical luminosities seem also to be higher so that the $L_x/L_o$ ratio is comparable (Crampton et al, 1984). No globular cluster sources have been detected in the Magellanic Clouds or in M33 (see Helfand 1984 and references therein).

The bulge region of M31 is seen as a bright extended X-ray source in the lower resolution IPC instrument, and it is resolved in several, bright, point-like sources by the higher resolution HRI. There seem to be two different "bulge" source populations: about 20 sources with average $L_x \sim 4.5 \times 10^{37}$ erg $s^{-1}$ are clustered in an inner region of ~400 pc radius. Outside this inner X-ray bulge but still inside the optical bulge (~ 1 kpc, Morton, Andereck and Bernard, 1977) ~16 more X-ray sources are detected with average $L_x \sim 2 \times 10^{37}$ erg $s^{-1}$ (van Speybroeck and Bechtold, 1980). The integrated spectrum of the bulge region from the IPC data (Fabbiano, Trinchieri and Van Speybroeck, 1986) shows that the X-ray sources in this region have on average a hard (kT>5 kev) spectrum with small low energy cut-off, consistent with typical spectra of low mass binaries in the Galaxy (Jones, 1977).

The detection of "inner" and "outer" bulge sources in M31 is interesting for the problem of the formation of these systems. It is reasonable to assume that all the sources within ~1 Kpc (the optical bulge) are low mass binary systems, given the expected stellar population in the bulge region. However, while formation by capture can be efficient in the high density globular clusters' environments and possibly in the inner bulge region (see van den Heuvel, 1980), it cannot work in the low density regions of the outer bulge. An alternative formation mechanism must be studied. Moreover, the luminosities of the outer bulge sources are indistinguishable from those of the sources in the disk and arms of the galaxy, suggesting perhaps that their nature may be similar. I will come back to this point later.

## BEYOND THE LOCAL GROUP

As discussed above, in the very nearby galaxies the X-ray emission is due to the same population responsible for the emission in the Milky

Way. However a larger sample of objects needs to be studied to extend the above results to spiral galaxies in general.

Given the limited sensitivity and resolution of the Einstein instruments, only in the Local Group galaxies can the individual sources be resolved and detected singularly. At larger distances, the X-ray emission appears like an extended source cohextensive with the optical emission with only the very high luminosity end of the galactic X-ray sources detected above it. The total observed X-ray luminosity of normal spiral galaxies (between $\sim 10^{39}$ erg $s^{-1}$ and $\sim 10^{41}$ erg $s^{-1}$) can in most cases be totally accounted for by a collection of galactic-type sources similar to those found in the Milky Way, and their X-ray morphology can be similarly explained (see Fabbiano and Trinchieri, 1985; 1986; and references therein). The spectral signature of their emission also confirms the above result, as will be discussed later in more detail.

Several examples of X-ray maps of spiral galaxies are now available. Figure 2 shows the X-ray map of NGC 253 (HRI data, from Fabbiano and Trinchieri, 1984). It is evident that the X-ray emission is dominated by two main components: a diffuse complex emission from the nuclear, inner disk region and a number of bright sources inside the optical image of the galaxy. At the distance of 3.4 Mpc, their luminosity is $L_X \geqq 10^{38}$ erg $s^{-1}$ (Fabbiano and Trinchieri, 1984).

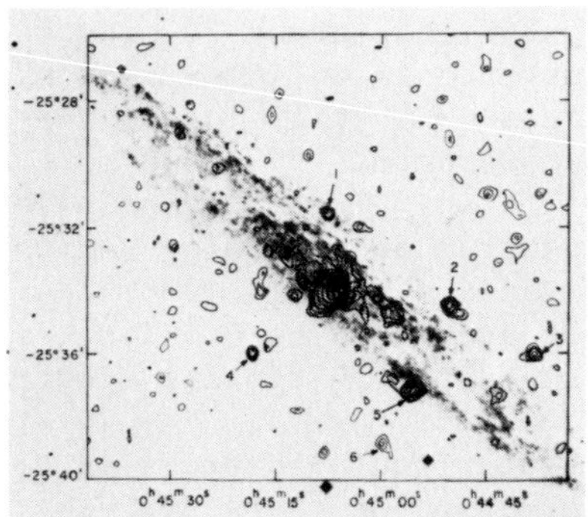


Figure 2: X-ray iso-intensity contours (HRI data) superposed on the optical image of NGC 253 (from Fabbiano and Trinchieri, 1984).

In the X-ray maps of M83, 1 source with luminosity above $10^{38}$ erg $s^{-1}$ (D=3.5 Mpc) is detected on a spiral arm and two more near the nuclear region (see Figures 3 and 4 in Trinchieri, Fabbiano and Palumbo, 1985) above a low surface brightness emission from the plane. Similarly, in M51 there are 3 bright sources with $L_X \geqq 10^{38}$ erg $s^{-1}$ on the spiral arms (see Figure 2 in Palumbo et al. 1985). Other examples of bright sources are found in M101 (see Long and van Speybroeck, 1983; McCammon and Sanders, 1984), IC 342, NGC 2403, NGC 4631 and NGC 6946 (Figure 3, see also Fabbiano and Trinchieri, 1986). Most of these sources have no obvious optical counterpart, although in most cases they are located on the spiral arms and are close to HII regions or radio sources (see Palumbo et al, 1985; Fabbiano and Trinchieri, 1986).

What is the nature of these bright sources in spiral galaxies? The

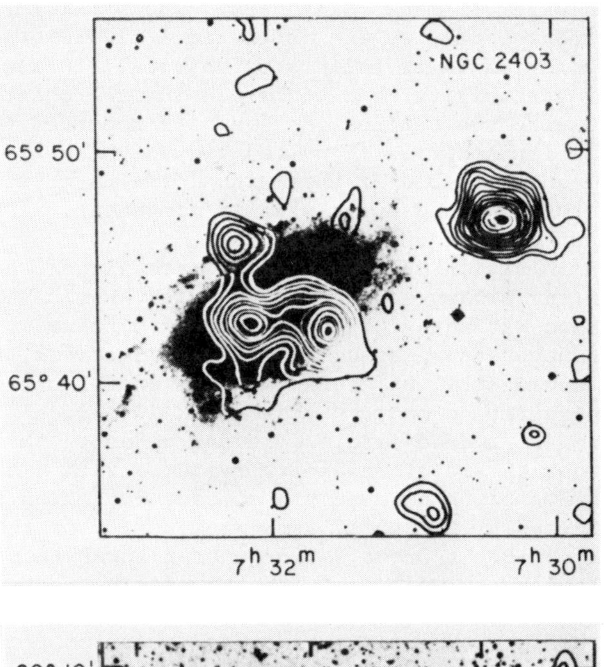


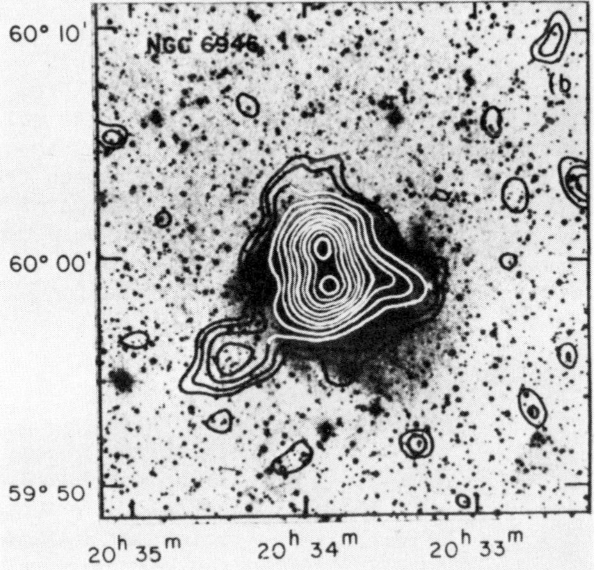


Figure 3: Iso-intensity IPC contours superimposed on the optical plates of NGC 2403 and NGC 6946. The SE source on the spiral arm of NGC 6946 is the supernova SN 1980k (from Fabbiano and Trinchieri, 1986).

quality of the available X-ray data does not allow us to make a detailed study of the nature of these sources. Their X-ray luminosity exceeds ~ $10^{38}$ erg $s^{-1}$, at the galaxies' distances. Their point-like appearance does not guarantee that we are observing a single bright object rather than a complex of lower luminosity objects. In most cases there is no evidence of variability that would confirm the point-like nature of these sources, although variability has been reported for sources in M101 of comparable luminosities (Long and van Speybroeck, 1983). The positional coincidence of some of them with spiral arms would suggest that they belong to the young Pop I stellar component, although projection effects cannot be excluded.

The X-ray luminosities of these sources are higher than typical luminosities of X-ray binary sources in the Milky Way. This could be due in part to wrong estimates of the galaxies' distances, uncertain by large factors. However, even assuming smaller distances, the luminosity of these sources would still be close to or in excess of the Eddington luminosity for an accreting neutron star (~2 X $10^{38}$erg $s^{-1}$). Since at least some of them are indeed single emitting objects, this would suggest that these sources are close binary systems, with a neutron star or possibly a massive black hole as the compact accreting object.

The contributions to the total X-ray luminosity of these galaxies from the extremely bright sources is in general lower than the contribution of lower luminosity objects, with the possible exception of NGC 253 and of the LMC (Fabbiano and Trinchieri, 1986). This component is seen as a low surface brightness, extended emission from the plane. By analogy with the Galaxy and M31, we expect that individual sources, in particular X-ray binaries in the luminosity range $10^{37}$-$10^{38}$ erg $s^{-1}$, dominate this component. Support to this comes also from the spectral signature of the integrated X-ray emission, on average well represented by a thermal bremsstrahlung with typical temperatures above 2 kev and no intrinsic low energy cut-off above the absorption along the line of sight. These spectra are consistent with a population of binary systems dominating the emission (SNR have typically a more complex spectrum with a softer component). By studying the radial distribution of this component, we can therefore study the distribution of the X-ray binary source component in other galaxies.

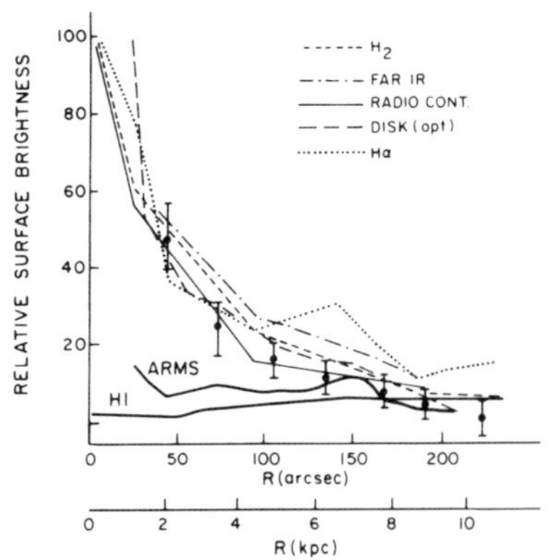


Figure 4: Comparison between the X-ray, Hα, blue light (disk and arms), far-infrared, neutral hydrogen, molecular hydrogen and 21 cm radio continuum surface brightness profiles in M51. X-ray points for R<2kpc are off scale. All curves are arbitrarily normalized.

The spatial distribution of the X-ray emission is remarkably similar to the optical one. In the two best studied cases, M83 (see Trinchieri, Fabbiano and Palumbo, 1985) and M51 (see Figure 4 , from Palumbo et al., 1985), the X-ray and optical radial profiles are consistent with each other out to the last observed point, with the exception of a relative excess X-ray emission in the inner ~ 2 kpc. In NGC 6946 (Figure 5, from Fabbiano and Trinchieri, 1986), the relative increase in the X-ray emission in the inner region is larger and is observed on a larger scale than in the previous examples (Fabbiano and Trinchieri, 1986); however, the presence of a source associated with the starbust nucleus, indistinguishable from the emission from the inner disk, could be responsible for at least part of this phenomenon.

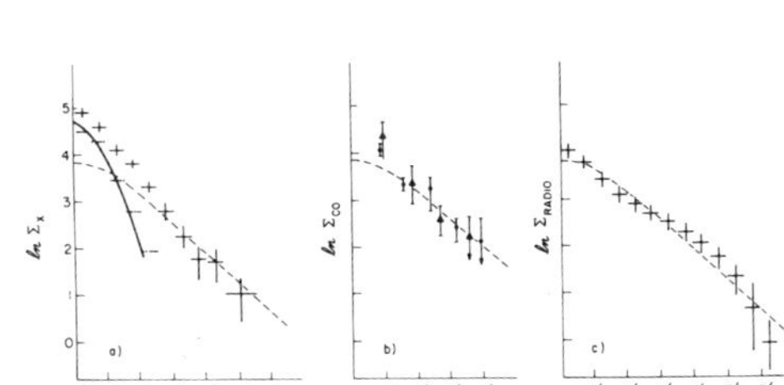

Figure 5: The radial distribution of the emission of NGC 6946 at different wavelengths. The dashed line common to all three figures shows the optical (B-band) profile: a) The crosses represent the X-ray data; b) CO data; c) radio continuum emission. All data are at the same resolution of ~ 1' (from Fabbiano and Trinchieri, 1986).

The X-ray sources are distributed like the exponential disk, rather than the much flatter distribution of the light from the arms or of the neutral hydrogen HI (see M51, Figure 4), although a contribution from the massive binaries associated with the spiral arm population is allowed by the data. This result suggests that the majority of the X-ray binary systems are related to the older ( ~ $10^9$ yr) Population I disk component, rather than to the extremely young ( ~ $10^7$ yr) Population I stellar component of the spiral arms. The relative excess X-ray emission in the inner disk is observed in a region comparable in extent to the "Galactic Bulge" region of the Milky Way. This X-ray excess could be indicative of the presence of a similar population of low mass binary sources, which have on average a higher X-ray luminosity than more massive sources. However these would probably not be associated with the "spheroidal bulge" region (in M83 for example it extends only for $r < 500$ pc) but with a region of the inner disk.

The above results suggest that a significant fraction of the X-ray binary sources in these galaxies are low mass disk systems. This is certainly a new result that has bearing upon the theories for the formation and evolution of low mass binary systems in our own galaxy as well, since the Milky Way is morphologically similar to the above galaxies (Hodge, [1983] classifies it a Sbc or Sc galaxy). A substantial fraction of low mass binary sources could therefore originate from the evolution of disk systems (Fabbiano and Trinchieri, 1985; Fabbiano, 1985).

## STATISTICAL RESULTS

In order to analyze a larger number of normal galaxies, we can study the average properties of a representative sample and compare the integrated luminosities of the sample objects at different wavelengths. The results are shown in Figure 6 and can be summarized as follows (see Fabbiano and Trinchieri, 1985, for details):

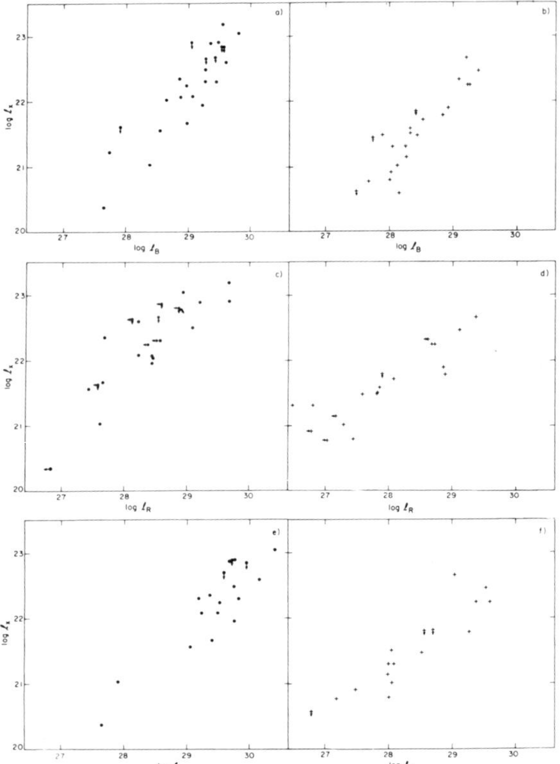


Figure 6: Plot of the total X-ray luminosity ($\ell_X$) versus the optical ($\ell_B$), radio continuum ($\ell_R$) and near infrared ($\ell_H$) luminosities for a sample of normal spiral and irregular galaxies (from Fabbiano and Trinchieri, 1985).

- There is a very tight, linear correlation between the X-ray and the optical (blue band) luminosity of spirals of all morphological types. The X-ray luminosity also correlates with the near-IR luminosity (H band). The slope of the correlation is ~0.7. However, this latter correlation is likely to be the result of stronger links between X-ray and B emission and B and H emission.

The number of X-ray sources scales linearly with the stellar population dominating the B-band light. This is a further indication of a strong association between the X-ray sources and the relatively young disk population, rather than with the galaxy mass and the old stellar component dominating in the near infrared.

- The X-ray and radio continuum emission are well correlated, with a dependence of the kind $\ell_X \propto \ell_R^{\sim 0.5-0.6}$. This correlation is statistically more significant than any of the correlations between radio continuum luminosity and optical or near-infrared luminosities or B-H colour.

The close link between the X-ray and the radio emission is a new and rather puzzling result. The radio emission in spiral galaxies is the result of the interaction between the Cosmic Ray relativistic electrons and the galactic Magnetic Field (see for example the review by Ekers, 1980). If the X-ray emission is dominated by the contribution from low mass binary systems, the above result points to a close link between low mass binary sources and cosmic ray production (see Fabbiano and Trinchieri, 1985, for a more detailed discussion).

## BINARY SYSTEMS IN EARLY TYPE GALAXIES

So far the discussion has focused on the binary systems in spiral galaxies. What do we know about this population in elliptical galaxies? Extended X-ray emission has been detected in many elliptical galaxies, but individual sources have not been seen (except possibly in M32, van Speybroeck and Bechtold, 1980). However, since elliptical galaxies and spiral bulges are similar in their properties, stellar population and colours, we can use the observations of M31 to derive an estimate of the expected X-ray binary population in early type galaxies, and compare the results with the observations (the details of how this estimate has been derived are in Trinchieri and Fabbiano, 1985). Figure 7 shows the results of this comparison: the line labelled $L_{dscr}$ indicates the expected integrated X-ray luminosity due to binary systems as a function of the optical luminosity.

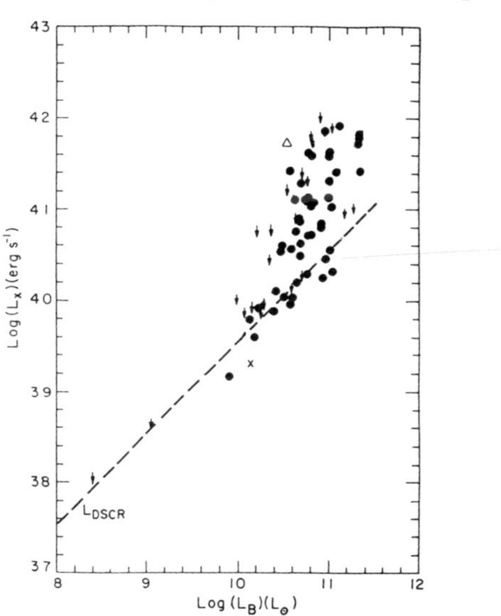


Figure 7: Plot of the total X-ray luminosity $\ell_X$ versus the optical ($\ell_B$) luminosity for a sample of normal elliptical and SO galaxies (from Canizares, Fabbiano and Trinchieri, 1986).

It is clear that for the brightermost objects, the X-ray luminosities are well above the expected contribution from discrete sources. For these objects, the X-ray emission is due to the thermal emission from hot ( ~ 1 Kev) gas (see Forman, Jones and Tucker, 1985; Trinchieri and Fabbiano, 1985; Trinchieri, Fabbiano and Canizares, 1986; and references therein). However, for the low luminosity objects, the

X-ray emission could indeed be dominated by the integrated contribution of X-ray binary systems. Unfortunately, with the available data it is not possible to investigate further whether the discrete source component is indeed dominant in the low luminosity objects. More definitive conclusions will have to wait for new X-ray data from the planned future missions, such as the German X-ray satellite ROSAT and the AXAF project.

REFERENCES


Bradt, H.V. and McClintock, J.E., 1982, Ann.Rev. Astron. Astrophys., Vol. 21, p. 13.

Canizares, C.R., Fabbiano, G. and Trinchieri, G., 1986, Ap. J. submitted.

Cowley, A.P., et al., 1984, Ap. J. 286, 196.

Crampton, D., Cowley, A.P., Hutchings, J.B., Shade, D.J. and van Speybroeck, L.P., 1984, Ap. J. 284, 663.

Ekers, R.D., 1980, in *"Structure and Evolution of Normal Galaxies"*, ed. S.M. Fall and D. Lynden-Bell (Cambridge:Cambridge Univ. Press), p.169.

Fabbiano, G., 1985, in *"Galactic and Extragalactic Compact X-ray Sources"*, ed. Y. Tanaka and W.H.G. Lewin (Japan: Publ. of Institute of Space and Astronomical Science), p. 233.

Fabbiano, G. and Trinchieri, G., 1984, Ap. J. 286, 491.

Fabbiano, G. and Trinchieri, G., 1985, Ap. J. 296, 430.

Fabbiano, G. and Trinchieri, G., 1986, Ap. J. submitted.

Fabbiano, G., Trinchieri, G. and van Speybroeck, L., 1986, in preparation.

Forman, W., Jones, C. and Tucker, W., 1985, Ap. J. 293, 102.

Giacconi, R. et al., 1979, Ap. J. 230, 540.

Helfand, D.J., 1984, P.A.S.P., 96, 913.

Hodge, P., 1983, P.A.S.P., 95, 721.

Jones, C., 1977, Ap. J. 214, 856.

Long, K.S., Helfand, D.J. and Grabelsky, D.A., 1981, Ap. J. 248, 925.

Long, K.S. and van Speybroeck, L., 1983, in *"Accretion driven X-ray sources"*, eds. W. Lewin and E.P.J. van den Heuvel (Cambridge: Cambridge University Press), p. 41.

McCammon, D. and Sanders, W.T., 1984, Ap. J. 287, 167.

Morton, D.C., Andereck, C.D. and Bernard, D.A., 1977, Ap. J. 212, 13.

Palumbo, G.G.C., Fabbiano, G., Fransson, C. and Trinchieri, G., 1985, Ap. J. 298, 259.

Seward, F.D. and Mitchell, M., 1980, Ap. J. 243, 736.

Tanambaum, H. and Tucker, W.H., 1974, in *"X-ray Astronomy"*, ed. R. Giacconi and H. Gursky (Dordrecht: Reidel), p. 169.

Trinchieri, G. and Fabbiano, G., 1985, Ap. J. 296, 447.

Trinchieri, G., Fabbiano, G. and Palumbo, G.G.C., 1985, Ap. J. 290, 96.

Trinchieri, G., Fabbiano, G. and Canizares, C.R., 1986, Ap. J. in press.

van Speybroeck, L., et al., 1979, Ap. J. Lett. 234, L45.

van Speybroeck, L.P. and Bechtold, J., 1980, in *"Proceedings of HEAD-AAS Meeting"*, ed. R. Giacconi (Dordrecht: Reidel), p. 153.

van den Heuvel, E.P.J., 1980, in *"X-ray Astronomy"*, ed. R. Giacconi and G. Setti (Dordrecht: Reidel), p. 199.

Webster, B.L., Martin, W.L., Feast, M.W., Andrews, P.J., 1972, Nature Phys. Sci., 240, 183.

## DISCUSSION

**E. van den Heuvel:** Your results on M51 are extremely important. The fact that the X-ray source distribution scales with the stellar luminosity (which is dominated by the intermediate-age stellar population) indicates, as you point out, that the low-mass X-ray binaries dominate the X-ray luminosity also at larger distances from the center. This is very important for our understanding of the formation of millisecond binary radio pulsars in the outer regions of spiral galaxies like our own.

**G. Trinchieri:** Of course I agree. I should also point out that we are analyzing more data on a few other spiral galaxies in the hopes of providing theoreticians with more convincing examples of the above results.

# OBSERVATIONS OF X-RAY BURST SOURCES

Y. Tanaka
Institute of Space and Astronautical Science
4-6-1 Komaba, Meguro-ku, Tokyo 153
Japan

ABSTRACT. Recent observational results on X-ray bursts and burst sources are reviewed. Two distinct types of bursts, Type I and Type II bursts, are discussed in relation to the mass and radius of neutron stars and to the problems of unusual mass accretion in some burst sources.

## 1. INTRODUCTION

This paper reviews the observational aspects of X-ray bursts. There exist two distinctly different types of X-ray bursts, Type I bursts and Type II bursts according to the designation by Hoffman et al. (1978). Examples of Type I and Type II bursts are shown in Fig. 1, respectively. Type I bursts are characterized by a significant cooling with time; harder X-rays decay faster than softer X-rays. Type I bursts are interpreted as nuclear energy release by thermonuclear shell flashes that occur in the envelope of a neutron star, and are therefore taken as a signature of a neutron star. On the other hand, Type II bursts do not exhibit a significant cooling in the decay; burst profiles in different energies appear essentially the same. Type II bursts are considered to be gravitational

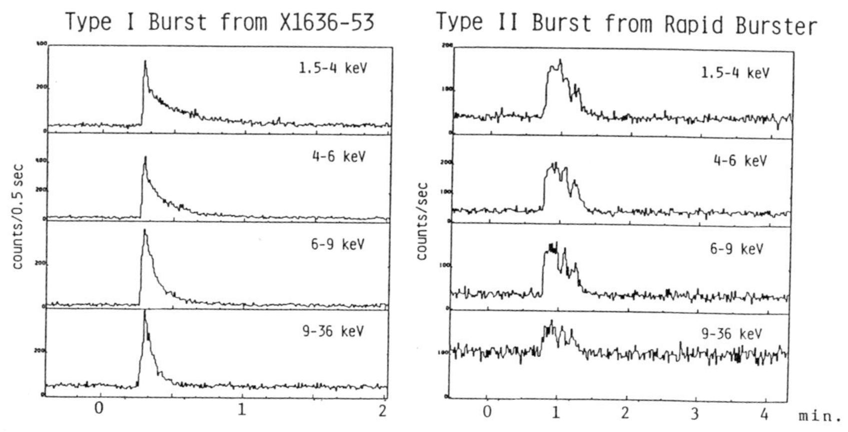


Fig.1. Examples of Type I and Type II bursts (Tenma).



D. J. Helfand and J.-H. Huang (eds.), The Origin and Evolution of Neutron Stars, 161–172.

energy release of accreting matter, when the accretion is not steady but sporadic due probably to certain instabilities. The most remarkable case is the Rapid Burster (Lewin et al. 1976a) of which the X-ray emission is almost totally in the form of Type II bursts. Type II bursts, according to their definition, are not necessarilly limited to neutron stars. However, we shall deal here with only those bursts from neutron stars.

There are about 35 X-ray sources known to date to produce Type I X-ray bursts in our galaxy. These sources are strongly concentrated towards the galactic center, half of them confined within only 10° of the galactic center. Approximately 10 bursters were found to be located within globular clusters. These facts clearly indicate the Pop II nature of the bursters. In fact, there are enough evidences to support that the bursters are low-mass binary systems (White, this Symposium). Furthermore, no burster is an X-ray pulsar, which indicates that the neutron stars in the bursters do not possess a strong magnetic field.

## 2. TYPE I X-RAY BURST

Basic characteristics of Type I bursts are qualitatively but convincingly interpreted by nuclear shell flash models (e.g., Lewin and Joss 1983, and references therein), although there are problems yet to be explained (see 2.5). Since Type I X-ray bursts are an abrupt heating of a geometrically thin envelope of neutron stars, this phenomenon provides means to measure neutron stars themselves. Significant progress along this line has been achieved on the observational side as discussed below. It is possible to determine uniquely the mass M and radius $r_o$ of a neutron star as well as its distance, from observations of Type I X-ray bursts. The methodology and the related observational results are given here, and discussed in more detail by Inoue (this Symposium).

### 2.1. Energy Spectrum

Energy spectrum of the burst emission is well expressed by a blackbody spectrum at any instance during a burst (e.g., Tanaka 1985), except for a significant excess over a blackbody spectrum in the range above 10 keV. This excess is probably due to the effect of Comptonization in a neighboring hotter region. Incidentally, the persistent flux may not necessarily be constant during a burst. Therefore, a special consideration is needed when the burst flux becomes comparable to the persistent level (van Paradijs and Lewin 1986). Observational evidences support that a burst covers the entire neutron star surface and that the burst emission is regarded to be spherical (see 2.3). Then,

$$L = 4\pi d^2 f = 4\pi r_b^2 \sigma T_b^4 = 4\pi r_o^2 \sigma T_b^4 \varepsilon g^{-2} \quad (1),$$

$$g^2 = 1 - (2GM/c^2 r_o) \quad (2),$$

where f is the measured flux with bolometric correction, d the distance, $T_b$ the measured blackbody temperature, $r_b$ the apparent blackbody radius. It is important to note that $T_b$ is a "color" temperature and that, in an electron-scattering dominated envelope during a burst, "color" temperature is significantly higher than effective temperature. The factor ε represents the emissivity correction for this effect. Theoretical studies of the effect of electron scatterings including the Compton effect are well in order, and the correction factor ε has become available (London et al. 1984; Ebisuzaki and Nomoto 1986).

## 2.2. Absorption Line

Significant absorption features were discovered in the spectra of four X-ray bursts observed from X1636-53 (Waki et al. 1984) as shown in Fig. 2a. These features are consistent with a common absorption line at about 4.1 keV. Later, similar absorption lines were found in three bursts from X1608-52 also at about 4.1 keV (Nakamura et al. 1986) as shown in Fig. 2b. If we assume that this absorption line is the gravitationally red-shifted line of iron which would be the most abundant heavy element in the neutron star envelope, we obtain g = 0.61. However, this g-value is uncomfortably small for any stable neutron star model (Waki et al. 1984). Alternative interpretaions of this absorption line have also been proposed (Fujimoto 1985; Ebisuzaki 1986; see Inoue, this Symposium). At present, the origin of this absorption line is still an unresolved issue. However, this absorption line should eventually give us the g-value.

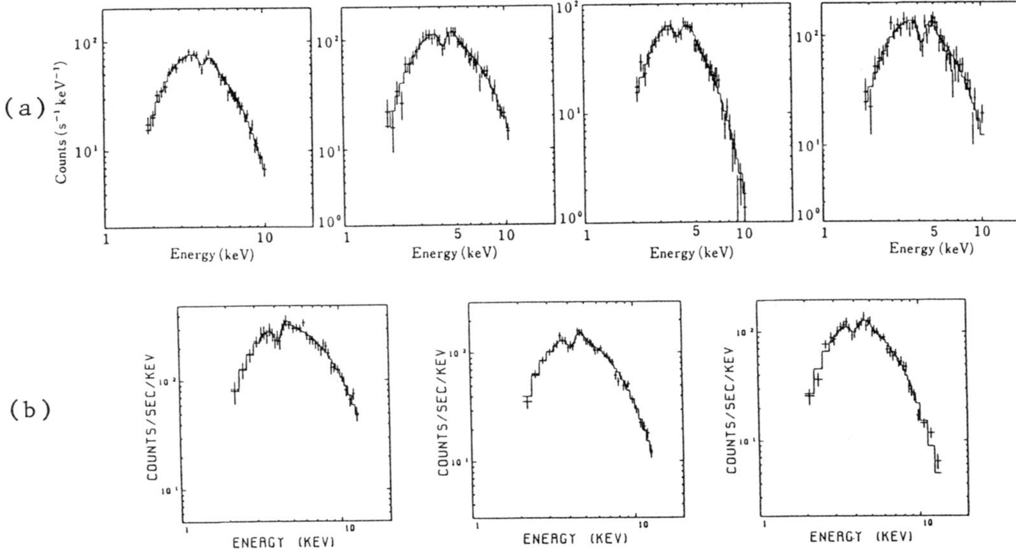


Fig. 2. (a) Absorption lines observed in four bursts from X1636-53, and (b) those in three bursts from X1608-52 (Tenma). Histograms are the best-fit blackbody spectra including the absorption line.

## 2.3. Peak Luminosity Saturation

It was suspected for some time that the burst peak luminosity could largely exceed the Eddington limit for a 1.4 M⊙ neutron star. However, it has become convincing from observational evidences as well as theoretical studies that the burst peak luminosity indeed saturates at the Eddington limit. For example, Fig. 3 shows three bursts with the largest peak fluxes among twelve bursts observed from X1636-53 (Inoue et al. 1984b). These peak fluxes are found to be the same within statistical errors and remain constant for a few seconds. These bursts commonly exhibit the features just as expected when the luminosity reaches the Eddington limit. When it occurs, the radiation pressure starts to expand the envelope, and temperature drops accordingly. As the radiation pressure decreases, the envelope starts to resettle and consequently temperature rises. Whereas, the luminosity is constant during the expansion and contraction of photosphere. The same feature is also observed for the bursts from other sources; when a photospheric expansion is observed, the peak flux saturates at a fixed level for the source, and vice versa. Very long bursts accompanied by a precursor are interpreted likewise (Tawara et al. 1984; Lewin et al. 1984). For these energetic bursts, the temperature drop associated with photospheric expansion was so large that most emission once went away to the longer wavelength band, thus forming a precursor. The luminosity remained constant for more than 100 sec.

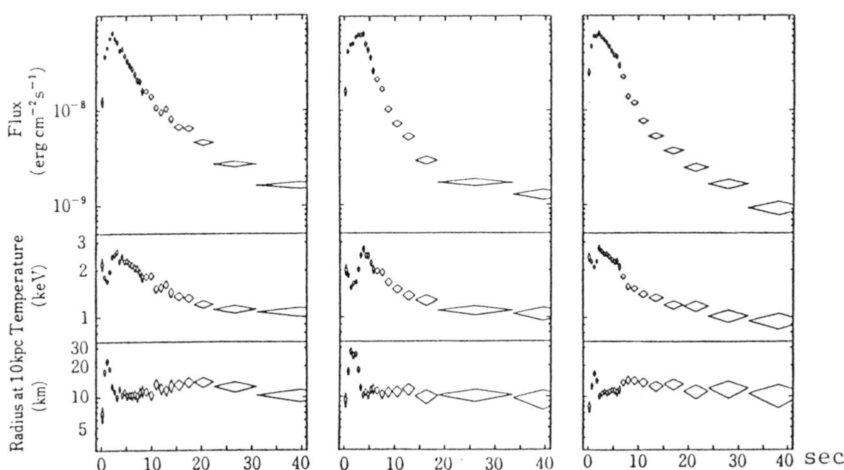


Fig. 3 Three bursts with the largest peak flux from X1636-53 (Tenma). Bolometric flux, color temperature and apparent blackbody radius are shown against time, respectively.

The peak luminosity saturation at the Eddington limit also supports the sphericity of emission, provided the surface magnetic field is less than $10^9$ G. Thus we obtain the third equation,

$$L_E = 4\pi d^2 f_E = 4\pi cGMg/\kappa_o(1 + X) \qquad (3),$$

where $L_E$ and $f_E$ are the Eddington luminosity and the corresponding flux observed, respectively, and $\kappa_o(1 + X)$ is the electron scattering opacity with the hydrogen mass fraction X. For an energetic helium flash, the outer hydrogen-rich envelope would be ejected away by the radiation pressure at the Eddington limit for the inner helium-rich layer (Sugimoto et al. 1984). Ebisuzaki (this Symposium) studied theoretically the luminosity vs. color-temperature relation, following burst evolutions for different models of atmosphere, and showed that the observed relation for those bursts associated with a photospheric expansion agrees well with that expected for an exposed helium-rich envelope, hence X=0. Thus, Eq. 1 for $L=L_E$ with the corresponding maximum temperature $T_b(max)$, together with Eqs. 2 and 3 will give M, $r_o$ and D, when the g-value is fixed.

## 2.4. Distance to the Galactic Center

Use of the burst peak luminosity as a standard candle was originally proposed by van Paradijs (1978). The burst peak luminosity associated with a photospheric expansion, the evidence for saturation at the Eddington limit, provides a good standard candle, if neutron stars in the bursters are alike. Indeed, the blackbody radii derived for the bursters clustering around the galactic center are found to be about the same (Inoue et al. 1981), supporting these neutron stars are similar in size.

Table 1 lists the reported peak fluxes of bursts associated with a photospheric expansion and the distances estimated from the Eddington luminosity for a helium-rich envelope of a neutron star with 1.4 M⊙ and 10 km radius. The average distance for the bursters within 10° of the galactic center turns out to be about 6.5 kpc. Previously, Ebisuzaki et al. (1984) showed that the center of gravity of the similarly estimated burster locations was about 6 kpc away. Because of the strong concentration of bursters to the galactic center, the average distance of bursters should be nearly equal to the distance to the galactic center. Thus, from the burst results, the galactic center distance is likely to be 6 to 7 kpc, unless the Eddington limit is much greater than the value assumed here. Vacca et al. (1986) recently reassessed the globular cluster NGC 6624, the only source for which a fair distance estimate was available, and concluded that even 6 kpc was not inconsistent with its optical data.

Table 1. X-Ray Bursts Showing Photospheric Expansion

| Source Name | l | b | θ | L | $L/L_E$ | d | Ref. |
|---|---|---|---|---|---|---|---|
| one of GCX sources | | | <0.5 | 4.3 | 1.6 | 7.8 | (1) |
| X1744-265 | 2.3 | 0.8 | 2.4 | 7.0 | 2.6 | 6.2 | (1) |
| X1724-307 (Tz 2) | 356.3 | 2.3 | 4.4 | 5 | 1.9 | 7.3 | (2) |
| X1728-337 (Gr 1) | 354.3 | -0.2 | 5.7 | 10 | 3.7 | 5.2 | (3) |
| X1715-321 | 354.1 | 3.1 | 6.7 | 8.1 | 3.0 | 5.8 | (4) |
| X1820-303 (NGC6624) | 2.8 | -7.9 | 8.4 | 5.1 | 1.9 | 7.2 | (5) |
| X1735-444 | 346.1 | -7.0 | 15.6 | 3.6 | 1.3 | 8.6 | (1) |
| X1636-536 | 332.9 | -4.8 | 27.5 | 7.4 | 2.8 | 6.0 | (6) |

l,b: Galactic longitude and latitude (°)
θ : Angular distance from the galactic center (°)
L : Peak luminosity ($10^{38}$ erg/s) of bursts showing a photospheric expansion, assuming d=10 kpc.
$L_E$ : Eddington luminosity of a neutron star of M=1.4M⊙, $r_o$=10km, for X=0.
d : Source distance (kpc), assuming $L=L_E$
Ref. (1)Inoue et al. 1986 (2)Grindlay et al. 1980 (3)Tawara et al. 1986 (4)Tawara et al. 1984 (5)Vacca et al. 1986 (6)Inoue et al. 1984

## 2.5. Puzzles

Although nuclear shell flash models successfully explain the basic characteristics of the burst phenomenon, there remain puzzling problems. Since the same accreting matter powers the persistent emission and also provides the fuel for Type I bursts, a direct relationship is expected between persistent luminosity and burst activity. Observationally, however, once a source becomes burst active, a fairly stable activity in terms of burst size and frequency often persists for some time, apparently unrelated to short term ($\sim$ hours) changes in persistent luminosity (eg., Ohashi et al. 1982). It looks as though there is a buffer against changes in the accretion rate.

More drastic cases are those in which two successive bursts occur within about 10 minutes (Lewin et al. 1976b; Murakami et al. 1980; Inoue et al. 1984a; Gottwald et al. 1986; Nakamura et al. 1986). An example of such events is shown in Fig. 4. If each burst exhausts the available nuclear fuel, such a time interval is too short to replenish the fuel for the second burst. Even if some fuel were left unused, it is not clear how the condition for triggering the second burst is reestablished in

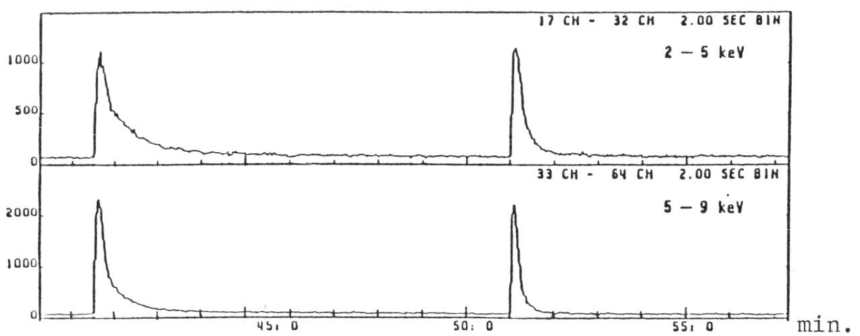


Fig. 4. Two bursts from X1608-52 with a 10-minutes separation (Tenma).

such a short time. As a possible hint, there is a common feature to such pairs of bursts with a short interval, as noticed in Fig. 4: Generally, the first of the two bursts has a long tail in the decay, and the second burst is significantly smaller in size and lacks long tail.

## 3. TYPE II BURST

We shall first discuss Type II bursts from the Rapid Burster. The Rapid Burster is a recurrent transient source discovered in 1976 (Lewin et al. 1976a) which produces rapidly repetitive bursts. It is located in the globular cluster Liller 1. It also produces Type I bursts, and hence believed to be an accreting neutron star. The Rapid Burster shows many distinct characteristics and has been a single source of its kind in our galaxy (e.g., Lewin and Joss 1983). Fig. 5 shows various patterns of the rapid burst activity. One extreme is a rapid, quasi-periodic repetition of short, spiky bursts. The other extreme is a train of long flat-topped bursts lasting as long as 10 minutes. However, this flat top is not the saturation at the Eddington limit, since the level of the flat top varies from burst to burst. No significant softening is observed during the decay of the rapid bursts; these bursts are Type II bursts. One of the outstanding features of the Rapid Burster is that its emission is almost totally in the form of Type II bursts. However, a low-level (∿10% of burst peak flux) persistent emission is observed occasionally. It is usually absent in the burst-active phase, except for some intervals longer than a few minutes between bursts. When the persistent emission appears, it starts to show up about a minute after the end of a burst, and disappears about a minute before the onset of the next burst. Quasi-periodic oscillations of 2-4 Hz were detected on flat-topped bursts as well as the low-level emission between bursts (Tawara et al. 1982; Stella et al. 1986; also discussions by Lewin, this Symposium).

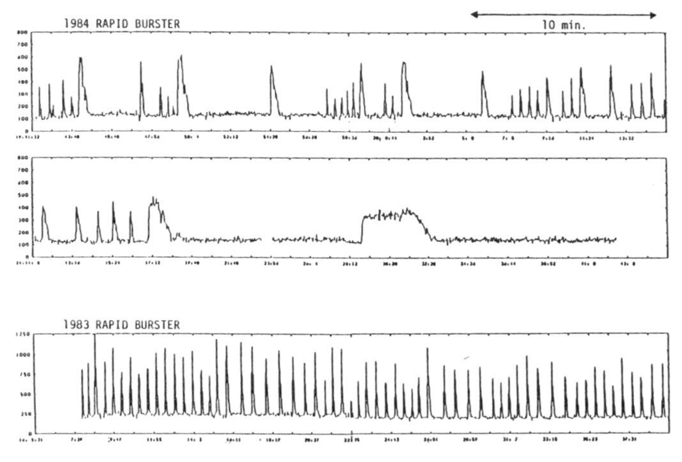


Fig. 5. Patterns of rapid bursts (Tenma).

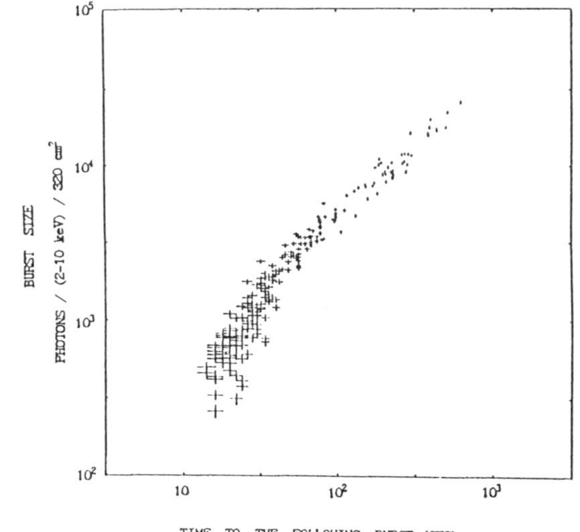


Fig. 6. Size vs. waiting time relation of rapid bursts, July 4-9, 1984 (Tenma)

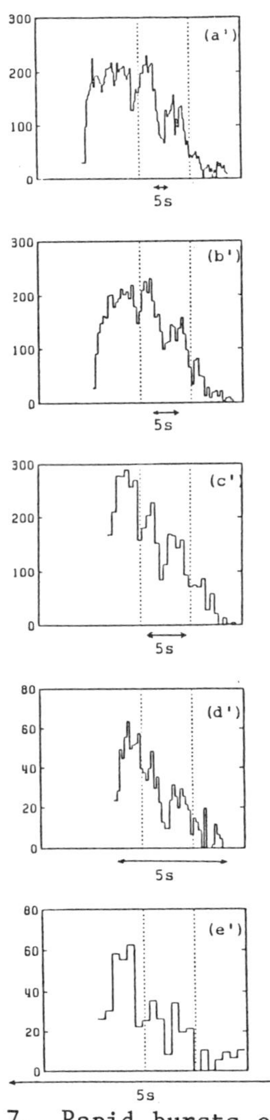


Fig. 7. Rapid bursts over a wide range of duration (Tenma)

### 3.1. Energy Spectrum

Energy spectrum of the rapid bursts is well expressed by a blackbody spectrum of temperature in the range 1.5–1.8 keV, with a significant hard tail above 10 keV (Kawai 1985). In contrast to Type I bursts, no cooling occurs throughout a burst, but rather a slight warming in the decay is observed (Kawai 1985). The burst decay is therefore due to a decrease of the emitting area with time. The peak blackbody radius of individual bursts is not constant but varies in the range 8–15 km for the assumed distance of 10 kpc , depending on the peak luminosity. Whereas, the apparent blackbody radius of Type I bursts from the same source is constant at 7 km, representing the neutron star radius. Thus, Type II bursts are emitted from an opticlly thick region which is more extended than the size of the neutron star. For the high temperature observed, the emission region of Type II bursts must be very near the neutron star.

### 3.2. Relation between Burst Size and Waiting Time

One of the most distinct characteristics of Type II bursts from the Rapid Burster is an approximately linear relation between the burst size (integrated energy, E) and the time interval to the next burst (waiting time, $t_w$); $E = L_p \times \delta t = \langle L \rangle \times t_w$, where $L_p$ is the burst peak luminosity, $\delta t$ the effective burst duration, and $\langle L \rangle$ the time-averaged burst luminosity. Observed burst size vs. waiting time relation is shown in Fig. 6. The linear relation holds over a wide range of E, except for small-size bursts which tend to violate the linear relation. Small-size bursts often repeat quasi-periodically, and in such cases the mean interval appears to saturate at about 16 sec. (Kunieda et al. 1984). The observed linear relation is characteristic of a relaxation oscillator, which makes one suspect the presence of a reservior with a fixed capacity. However, we do not know what serves for the reservoir in reality, nor are we sure if the interpretation by means of a reservoir is right. In this connection, it is important to note that the duty ratio of Type II bursts ( $\delta t/t_w$) remains constant, roughly at 10%, against changes in $\langle L \rangle$ (mean accretion rate). Change in the mean accretion rate causes change in the burst peak luminosity, but does not influence duty ratio.

### 3.3. Time-Scale Invariant Burst Structure

It was earlier noted that Type II bursts from the Rapid Burster exhibit significant structures in the decay, in contrast to Type I bursts which generally show little structure in the decay. We noticed a striking similarity in the multi-peaked decay structure between each other, except for differences in time scale. In fact, if the time scale for each burst is properly adjusted, the detailed decay structures of all Type II bursts, except for flat-topped bursts longer than 100 sec. for which the decay structures tend to be smeared out, are found to be nearly identical over a wide range of burst duration (Tawara et al. 1985), as shown in Fig. 7. In other words, the decay structure of all bursts is expressed by a single function $F(t/\tau)$, where $\tau$ is the characteristic time for each burst. Moreover, relative phases and heights of individual peaks in the decay are found to follow a simple arithmetic rule (Tawara et al. 1985).

This decay structure is almost independent of changes as much as a factor of three in the peak luminosity. The luminosity at a given time during a Type II burst is considered to be proportional to the instantaneous accretion rate to the neutron star. If so, the time-scale invariant structure implies a yet unexplained, complex flow-modulation mechanism.

### 3.4. Bursts from Cir X-1

Cir X-1 has been one of the black hole candidates because of its fast time variabilities and the high-low transitions, similar to the case of Cyg X-1. However, recent discovery of Type I bursts from Cir X-1 (Tennant et al. 1986b) concludes that Cir X-1 is a neutron star source. Earlier, bursts which showed no cooling or very slight cooling were detected from Cir X-1 during its low state (Dotani et al. 1985; Tennant et al. 1986a), which would most probably be Type II bursts. A Type I burst and a composite of Type II bursts from Cir X-1 are shown in Fig. 8.

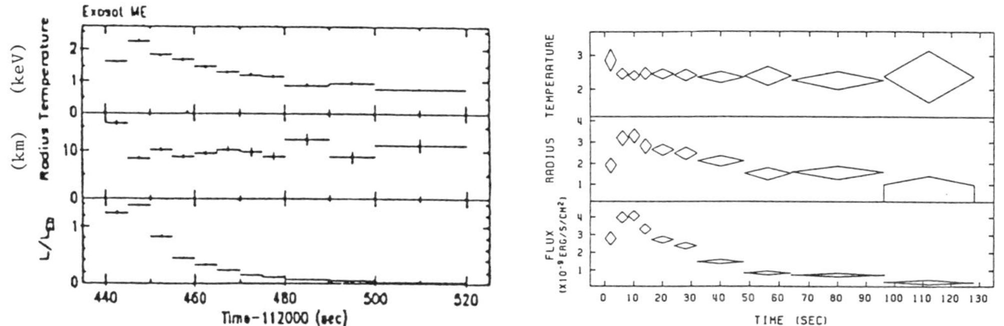


Fig. 8. A Type I burst (EXOSAT)(left), and a composite of Type II bursts observed during a low state (Tenma)(right).

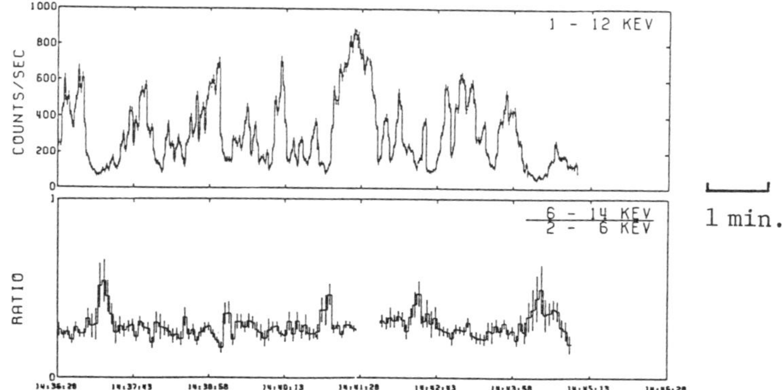


Fig. 9. Bursts from Cir X-1 in its high state (Tenma). Light curve (top) and the hardness ratio (bottom) are shown.

Cir X-1 has several characteristics which are similar to those of the Rapid Burster (Ikegami 1986). It produces both Type I and Type II bursts. Cir X-1 often shows, in its high state, repetitive bursts as shown in Fig. 9. with a large modulation factor of 80-90 %. The energy spectrum of these bursts is of a blackbody of roughly 1 keV without any change against intensity, hence these bursts could also be regarded as Type II bursts. The time scale of repetition on the order of 1 min. is often noted, as seen in Fig. 9, which is the same order of magnitude as that of the Type II bursts from the Rapid Burster. The profiles of these bursts appear similar, if not identical, to each other. An essential difference from the Rapid Burster is that there is little gap between bursts; duty ratio is 100 %. In view of the above results, it may be interesting to examine Cir X-1 based on the working hypothesis that Cir X-1 may be a similar system to the Rapid Burster.

## REFERENCES

Dotani,T. et al. 1985, Proc. ISAS Conf. on Space Astrophys.(in Japanese)
Ebisuzaki,T. 1986, submitted to P.A.S.J.
Ebisuzaki,T.,Hanawa,T. and Sugimoto,D. 1984, P.A.S.J., **36**, 551.
Ebisuzaki,T. and Nomoto,K. 1986, Ap.J., in press.
Fujimoto,M.Y. 1985, Ap.J.(Letters), **293**, L19.
Gottwald,M.,Haberl,F.,Parmer,A.N. and White,N.E. 1986, Ap.J., in press.
Grindlay,J.,Marshall,H.,Hertz,P. et al. 1980, Ap.J.(Letters), **240**, L121.
Hoffman,J.A.,Marshall,H.L. and Lewin,W.H.G. 1978, Nature, **271**, 630.
Ikegami,T. 1986, private communication.
Inoue,H. et al. 1986, in preparation.
Inoue,H.,Koyama,K.,Makino,F. et al. 1984a, P.A.S.J., **36**, 855.
Inoue,H.,Koyama,K.,Makishima,K. et al. 1981, Ap.J.(Letters), **250**, L71.
Inoue,H.,Waki,I.,Koyama,K. et al. 1984b, P.A.S.J., **36**, 831.
Kawai,N. 1985, Ph.D. Thesis, Univ. of Tokyo (ISAS RN. No.302)
Kunieda,H.,Tawara,Y.,Hayakawa,S. et al. 1984, P.A.S.J., **36**, 807.
Lewin,W.H.G.,Doty,J.,Clark,G.W. et al. 1976a, Ap.J.(Letters), **207**, L95.
Lewin,W.H.G.,Hoffman,J.A.,Doty,J. et al. 1976b, M.N.R.A.S., **179**, 83.
Lewin,W.H.G. and Joss,P.C. 1983, "Accretion-Driven Stellar X-Ray Sources", ed. Lewin and van den Heuvel(Cambridge Univ. Press), p.41.
Lewin,W.H.G.,Vacca,W.D.and Basinska,E.M. 1984, Ap.J.(Letters), **277**, L57.
London,R.A.,Taam,R.E. and Howard,W.M. 1984, Ap.J.(Letters), **287**, L27.
Murakami,T.,Inoue,H.,Koyama,K. et al. 1980, P.A.S.J., **32**, 543.
Nakamura,N. et al. 1986, in preparation.
Ohashi,T.,Inoue,H.,Koyama,K. et al. 1982, Ap.J., **258**, 254.
Stella,L.,Parmer,A.N.,White,N.E. et al. 1985, IAU Circular No.4110.
Sugimoto,D.,Ebisuzaki,T. and Hanawa,T. 1984, P.A.S.J., **36**, 839.
Tanaka,Y. 1985, Twelfth Texas Symposium on Relativistic Astrophys., p163.
Tawara,Y. et al. 1986, in preparation.
Tawara,Y.,Hayakawa,S.,Kunieda,H. et al. 1982, Nature, **299**, 38.
Tawara,Y.,Kawai,N.,Tanaka,Y. et al. 1985, Nature, **318**, 545.
Tawara,Y.,Kii,T.,Hayakawa,S. et al. 1984, Ap.J.(Letters), **276**, L41.
Tennant,A.F.,Fabian,A.C. and Shafer,P.A. 1986a, M.N.R.A.S., **219**, 871.
Tennant,A.F.,Fabian,A.C. and Shafer,P.A. 1986b, to appear in M.N.R.A.S.
Vacca,W.D.,Lewin,W.H.G. and van Paradijs,J. 1985, to appear in M.N.R.A.S.
van Paradijs,J. 1978, Nature, **274**, 650.
van Paradijs,J. and Lewin,W.H.G. 1986, A.& Ap., **157**, L10 .
Waki,I.,Inoue,H.,Koyama,K. et al. 1984, P.A.S.J., **36**, 819.

## DISCUSSION

**J.H. You:** Have you observed the X-ray spectrum in Type II burst?

**Y. Tanaka:** Yes, it is also a black body spectrum.

**F. Verbunt:** To get the maximum burst luminosity equal to the Eddington luminosity, you place the galactic center a little bit closer than hitherto assumed (6-7 kpc instead of 8.5 kpc). Do you also move the globular cluster sources closer in?

**Y. Tanaka:** Yes. There are several globular cluster sources whose bursts show evidence for reaching the Eddington limit (see Table I in the text). No optical data are in conflict with these globular clusters being closer than 8.5 kpc (see Vacca et al. 1986).

**D. Backer:** What is the distribution of neutron star masses derived from the analysis of your observation?

**Y. Tanaka:** We are not yet able to determine neutron star mass uniquely from burst observations, because of the remaining ambiguity in the g-value. See paper by Inoue (this Symposium).

**A. Burrows:** Would you care to comment on the recently observed triple-peaked bursts?

**Y. Tanaka:** You refer to the result by van Paradijs et al. (Nature 1986). This is a low-luminosity burst from 1636-53. We do observe occasionally low-luminosity bursts with complex structures, which cannot be explained by a single flash. Perhaps, we would have to consider a possibility of unstready burning for such a burst whose peak luminosity is much lower than the Eddington limit. However, this is not understood yet.

**S. Woosley:** If, as we have heard in other talks, there might be a minimum magnetic field for neutron stars of near $10^9$ guass, the Eddington luminosity might need redefining as the field would inhibit mass loss. Thus, the ordinary Eddington limit might be excluded.

**Y. Tanaka:** Observational results support that mass loss indeed occurs in energetic bursts (Ebisuzaki, this Symposium). In addition, observed evidence for sphericity of the burst emission suggests that the magnetic pressure is perhaps not significant.

**S. Miyaji:** It sounds to me that there is a conflict in your argument on Type II bursts of the Rapid Burster. Is it true that there are two kinds of Type II bursts, i.e., flat-topped bursts showing QPO and time-scale invariant bursts? Or, do you mean that we find higher frequency QPO in short-duration Type II bursts?

**Y. Tanaka:** The time-scale invariant structure and QPO are entirely different phenomena. The time-scale invariant structure is a solid form of Type II bursts of the Rapid Burster, and is not QPO.

**F. Verbunt:** You interpret the Cir X-1 bursts as Type I and Type II. Tennant interprets all as Type I, noting that the bursts from EXO 0748 (Gottwald et al. Ap.J. 1986) show a similar range in cooling curves; some cool more, others less. Can you comment on this?

**Y. Tanaka:** As regards the EXOSAT results by Tennant et al., the eight bursts observed first and the three bursts observed later are clearly different in the burst peak flux and the cooling curve. The last three are convincingly identified as Type I bursts, whereas we suspect that the first eight are Type II bursts. The first eight bursts of Tennant et al. look very similar to those bursts we observed from Cir X-1 (Fig. 8), and we do not find a significant cooling.

**W. Lewin:** There are several sources which exhibit hic-ups in their accretion (e.g., Cyg x-1, GX301-2). They lead to bursts which are generally called Type II bursts to distinguish them from the thermonuclear flashes (Type I bursts). Your data of Cir X-1 clearly also show accretion hic-ups, thus Type II bursts. However, to call the Cir X-1 burst behavior similar to that of the Rapid Burster goes perhaps too far. In the Rapid Burster, the Type II bursts are the result of a relaxation oscillator; there is a unique relation between the energy in a Type II burst and the waiting time to the next Tupe II burst. If that is not observed in Cir X-1, I suspect that the burst mechanism is very different from that in Cir X-1 (and Cyg X-1, GX301-2, etc.).

**Y. Tanaka:** You are right in that Cir X-1 does not exhibit a clean behavior of a relaxation oscillator. However, it is important to mention that we do not know for sure if the rapid bursts are the result of a relaxation oscillator. As a matter of fact, we do not have a picture of the physical mechanims which works analogous to a relaxation oscillator. A mechanism, similar to the as yet unknown mechanism of the Rapid Burster, might also produce bursts with 100% duty ratio as in the case of Cir X-1. This is worth further investigations.

# ON THE ORIGIN OF NEUTRON STARS IN GLOBULAR CLUSTERS

Jonathan E. Grindlay
Harvard Smithsonian Center for Astrophysics
60 Garden St., Cambridge, MA 02138

ABSTRACT. The formation of neutron stars in globular clusters is discussed in light of a number of recent results and, in particular, studies of the origin and evolution of the high luminosity x-ray binaries found in globular clusters. We argue that the neutron stars most probably arise from the accretion-induced collapse of white dwarfs in compact binary systems, themselves detectable as low luminosity cluster x-ray sources. The white dwarfs which can collapse are probably the remnants of relatively more massive stars than those presently found in globulars. This can account for the predominant occurrence of the high luminosity cluster sources in clusters of relatively high metallicity, since those clusters have recently been found to probably have flatter mass functions of their component stars.

## 1. INTRODUCTION

The existence of neutron stars (neutron star is hereafter abbreviated NS; plural is NSs) in globular clusters has long been suspected on the basis of the existence of luminous x-ray sources in globulars (e.g., Clark 1975) as well as the central surface brightness peaks in clusters such as M15 (e.g., Illingworth and King 1977). With the determination of the approximate mass of the the high luminosity ($L_x \geq 10^{36}$ erg/sec) x-ray sources as approximately 1.5 ±0.5 $M_\odot$ (Grindlay et al. 1984, Grindlay 1985a), it was confirmed that these were indeed NSs accreting from low mass companions in compact binaries, as suspected from the fact that the majority of these sources are x-ray bursters as well as other arguments (cf., Lewin and Joss 1983). The alternative model that the luminous x-ray sources in globulars are massive black holes was ruled out (Grindlay 1981), although it has been claimed that globulars could contain a significant fraction of their "dark" mass in stellar mass black holes of approximately 3 $M_\odot$ (Larson 1984). It has been clear from stellar evolution, as well as recent observational evidence (discussed below), that globular clusters must also contain an appreciable fraction of their total mass in the form of white dwarfs (white dwarf is hereafter abbreviated WD; plural is WDs), and indeed the apparently distinct population of low luminosity (with $L_x \leq 10^{34.5}$ erg/sec) x-ray sources discovered in globular clusters (Hertz and Grindlay 1983) has been interpreted as being largely due to WDs accreting in compact binary systems. We return to this point in section 4.



*D. J. Helfand and J.-H. Huang (eds.), The Origin and Evolution of Neutron Stars, 173–185.*

In this paper we consider the origin of NSs in globulars. After reviewing the arguments for the required density and number of NSs, we point out the special constraint on their origin imposed by the relatively low escape velocity from the cluster. We then outline the three principal schemes suggested for the production of NSs: as a consequence of the evolution of single, massive stars with or without "normal" supernova production (referred to below as IMF models 1 and 0, respectively), and as a consequence of the evolution of WDs in compact binaries (referred to here, and by Van den Heuvel 1986, as the accretion induced collapse, or AIC, model). In each case we discuss the assumptions and limitations. Specific predictions for on-going observational programs are made, and the indirect consequences of each of the three models are discussed. We conclude from a variety of arguments that the AIC model is preferred, and that the formation of low mass x-ray binaries in the galactic bulge apparently outside of globulars may be related to globulars.

## 2. STATEMENT OF THE PROBLEM

### 2.1 Direct Evidence for NSs and Constraints on NS Masses and Numbers

The best evidence for the existence of NSs, as opposed to WDs or stellar mass black holes, in globular clusters is the population of accreting x- ray binaries found in globulars. The properties of these sources as well as the clusters they are found in have been recently reviewed (Grindlay 1985b), and a suggested scheme for the origin and evolution of these compact binaries has been proposed (Grindlay 1986). In brief, some 10 clusters contain x-ray sources with persistent luminosities (usually) above $10^{36}$ erg/sec and all of these sources but one (the source in M15) are known to be x-ray burst sources. This virtually requires that the compact objects be NSs, given the overall success of the thermonuclear flash models (cf., Joss and Rappaport 1984 for a recent review). As mentioned above, the statistical determination of the mass of these systems from their distribution of radial offsets from their respective cluster centers, assuming isothermal King models for the cluster potentials, also leads to a 90% confidence interval for their mass of 0.8-2.5 $M_\odot$ (Grindlay 1985a), or consistent with them being NS binary systems (Lewin and Joss 1983, Grindlay et al. 1984).

The companion stars in the compact x-ray binaries in globulars are almost certainly of lower mass than the NSs, since the latter are expected (see discussion in sections 3 and 4 below) to have masses near 1.4 $M_\odot$ (or greater) and the former are limited to be less than the present turnoff mass of 0.8 $M_\odot$ in globulars. Therefore the mass transfer on to the NS will widen the orbit, and the mass transfer will cease unless either the donor star expands to keep its Roche lobe filled or angular momentum can be removed from the system to counteract the orbital expansion. In the first case, however, the resulting mass transfer rate and x-ray luminosity will be too high, since precisely this mechanism of evolution of the companion star up the giant branch to maintain mass transfer in a low mass x-ray binary was considered by Webbink, Rappaport, and Savonije (1983) for the more luminous (Verbunt et al. 1984, Grindlay 1985b) galactic bulge x- ray sources outside globular clusters. Therefore, the binary companions in the globular cluster sources are most probably main sequence stars, since the x-ray luminosities (Grindlay 1985b) imply accretion rates which are generally $10^{-10}$ - $10^{-9}$ $M_{\odot}/yr$, and this can be supplied by gravitational radiation and magnetic braking angular momentum losses in a system with a lower mass secondary star overflowing its Roche lobe (Rappaport, Joss and Webbink 1982).

It is perhaps even more reasonable, and more interesting, to invert the arguments given above and conclude that the observed x-ray luminosities of the high luminosity cluster sources as well as the upper limit on the mass of the mass-losing stars in these binaries implies a corresponding lower limit of 0.8 $M_{\odot}$ for the masses of NSs in globulars. Although the initial analysis for the most likely mass of the high luminosity x-ray sources in globulars suggested (Grindlay et al. 1984) that perhaps the NSs were formed with masses less than the 1.4 $M_{\odot}$ value found for NSs in the field (cf., Rappaport and Joss 1984), the more complete analysis of Grindlay (1985a) showed this is by no means required since a somewhat broader mass range (90% confidence) is indicated. We note also that the one claim for a NS mass significantly lower than 1.4 $M_{\odot}$ (for the NS in the 5 hour binary 4U2129+47), possibly as low as 0.6 $M_{\odot}$ (Horne, Verbunt and Schneider 1986), is subject to possible systematic errors since it was derived in part from the velocity variations in the He II emission line which could be subject to uncertain streaming motions of the gas in the system. A velocity study of this system is currently being attempted by M. Garcia which should avoid this problem, since the x-ray source is currently in an extended low state and no emission lines are visible.

Independent of their mass, the required number of NSs in the cluster core (where the more massive NSs will accumulate in an isothermal cluster) is about 3% by <u>number</u> to produce the observed number of clusters with high luminosity x-ray sources currently detected (Lightman and Grindlay 1982). This estimate is based on the tidal capture model (Fabian, Pringle and Rees 1975; Press and Teukolsky 1977) for the formation of the x-ray binaries, which continues to be the favored mechanism for their production (c.f. Grindlay 1986). Since the NSs are more massive than both the visible cluster stars (i.e., their potential binary companions) by the argument above and the cluster WDs, which are expected to be currently formed at about 0.5 $M_{\odot}$ (but whose mass spectrum probably also extends to above 1 $M_{\odot}$--cf., discussion below), their total contribution to the mass in the cluster core is in the range 10%. Adopting a core mass for a typical globular cluster of about 1% of the total cluster mass, the fractional mass in NSs in a typical cluster implied by the presence of the high luminosity x-ray binaries is only about $10^{-3}$. Thus, only about 100 NSs are required in the cluster (core). We note that this is a much smaller total number, or mass fraction, in the total cluster than that attributed to dark remnants in clusters such as 47 Tuc, where DaCosta and Freeman (1985) find a dark mass fraction of 35%. The comparison with 47 Tuc is appropriate since the (high luminosity) X-ray globular clusters are predominately concentrated to the galactic center and metal rich. The bulk of dark matter in cluster cores, therefore, is probably in the form of WDs, not NSs (or more massive stellar black holes).

## 2.2 Retention of NSs in Globulars

The central problem in obtaining this apparently modest total number of NSs is not only to form the NSs but to keep them in the cluster during the formation process. Globular clusters are both robust and fragile structures (cf., discussion following the paper by Grindlay 1985a), since their escape velocities are typically only 10-30 km/sec. Thus any model for their formation must have a "channel" for low velocity production of the NS. For models where the NS results from the supernova of a massive star, the supernova and collapse event must be able to be very symmetric (to limits much less than 0.1%) to avoid escape. For models involving the NS formation in a binary system, the mass lost from the supposed supernova event leading to the creation of the NS must not exceed half the binary mass or else the binary will become unbound and the NS will be given the orbital velocity of its progenitor star. In the case of a progenitor binary system in a globular cluster, where the only binaries that survive are those with orbital velocities greater than the cluster velocity dispersion (i.e so-called "hard binaries"), this means the NS would be ejected from the cluster in the case when the binaries are unbound by mass loss in the supernova event.

## 2.3 Formation Models to be Considered

Both formation and retention in the cluster are the topics we address in the remainder of this paper. We consider two major scenarios: the IMF model, that the NSs formed directly from the evolution of a prior generation of massive stars, and the AIC model, that they are forming even now from the accretion-induced collapse (AIC) of WDs. IMF models are discussed in section 3, where we review recent work on the measurement of the mass function, and by inference constraints on the initial mass function (IMF), in globular clusters. We consider this in light of the NS formation and retention problems and discuss two possibilities: model IMF- 1, where the NS is produced as a result of a supernova (Type I or, more probably, Type II) explosion vs. model IMF-0, where the NS is produced in a "quiet" collapse (a Type 0 "supernova" event ?) as suggested by Katz(1983). The AIC model is then considered in section 4, where we review the recent evidence for populations and masses of WDs in globulars and the recent theoretical work on the fate of an accreting WD which may lead to either disruption in a Type I supernova or collapse to a NS.

# 3. IMF MODELS: NS FORMATION FROM SINGLE, MASSIVE STARS

## 3.1 IMFs in Globular Clusters

The IMF in globular clusters has long been the subject of much interest as it obviously must constrain the conditions in the early history of star formation in the Galaxy. Numerous attempts have been made to derive IMFs, but it is with the advent of new color magnitude data from the much more sensitive and linear CCD detectors that the greatest progress has come and is yet to come. Recently, McClure et al. (1986) have presented results which, if confirmed, are of fundamental importance: the IMF slope in globulars appears to show marked variations from cluster to cluster (in their sample of seven) and appears to be strongly correlated with the metallicity of the cluster. The derived power law index, x, for the usual mass function formula,

$$\frac{dN}{dm} = \phi(m) \propto m^{-(1+x)} ,$$

where m is the stellar mass and N(m) is the number of stars with mass m, ranges from -0.5 for the most metal rich cluster in the sample (47 Tuc) to 2.5 for the most metal poor cluster (M15). [Recall that the Salpeter mass function index is 1.35, though even this value is uncertain and may be as flat as 0.85 in the solar neighborhood.] The uncertainties in these derived indices are estimated as ±0.5, but in fact the results are subject to a variety of systematic effects which must be considered (and are now being considered in a series of follow-up papers).

First, the measurements used by McClure et al. are of present day luminosity functions and have been converted to mass functions using theoretical isochrones of VandenBerg and Bell (1985). In addition to the dependence on these models, a cluster age (~15 Gyr) and helium abundance must be assumed. Second, and more important, the measurements were conducted at a variety of radial offsets from the cluster centers so that the effects of mass segregation, if the clusters are indeed relaxed, must be considered. Pryor, Smith and McClure (1986) have done this and show that the derived mass function indices are indeed affected somewhat but that an apparently strong correlation with metallicity still remains. Finally, the effects of both the dynamical and chemical enrichment in the clusters themselves (those with more massive stars being expected to have more self-enrichment) must be considered, as has been done by Smith and McClure (1986). They show that the variation in x, while large, may require that the most metal rich clusters with the flattest indices have sharply defined upper mass limits.

It should be noted that these new results must be checked by other groups since they are both consistent and inconsistent with previous results, or inferences, for (in some cases) even the same cluster. For example, the authors cited above point out that the trend of x with metallicity is as expected given the dynamical evidence reported by DaCosta and Freeman (1985) for a fraction of mass in dark remnants, primarily WDs, of 35% in 47 Tuc. Yet DaCosta (1982) found a much steeper mass function for 47 Tuc, though his (photographic) results were probably more contaminated by the non-constant background of stars from the SMC.

We point out that if the results for the IMF and its metallicity dependence are confirmed in future studies, a major puzzle concerning the distribution in the Galaxy of luminous x-ray sources in globular clusters (and perhaps also the galactic bulge) may be elucidated. That is, the strong correlation of the high luminosity sources with galactocentric distance (Lightman and Grindlay 1982), coupled with the strong correlation of metallicity with R, suggests that the primary correlation for the presence of a high luminosity x-ray binary (i.e., a significant NS population) is the cluster metallicity. If so, this can now be understood, since these clusters have flatter IMFs and thus had larger populations of more massive stars from which the NSs were produced. Unfortunately, this does not tell us how the NSs were produced, since the same population of more massive stars will produce both more massive WDs, which might undergo AIC (see section 4) as well as more higher mass stars, which might produce NSs through either the IMF-1 or IMF-0 models. We turn to these next.

### 3.2 Model IMF-1

In this model we assume that the NSs are formed as remnants in the supernova events which presumably end the evolution of the most massive stars formed in the cluster IMF. The lower limit for formation of a NS in an isolated supernova is probably in the range 8 - 12 $M_\odot$, so this scenario depends on the number of stars initially in the cluster in this mass range and above. Since stars in this mass range have lifetimes only of at most about 3 x $10^7$ years, all the NSs are formed while the cluster is still very young. As Smith and McClure (1986) also point out, the gas liberated by this much supernova activity in metal rich globulars in a time comparable with (or shorter than) the cluster dynamical timescales would have both important chemical and dynamical effects on the subsequent evolution of the cluster. That is, a large mass loss in stellar winds (pre-supernova) and supernova ejecta would tend to expand the cluster and also modify the chemical enrichment of the remaining stars. Since the metal rich globulars in the galactic bulge are also typically more centrally condensed, and those containing the high luminosity x-ray sources certainly are (Lightman and Grindlay 1982), this does not seem to have occurred. Thus we agree with Smith and McClure that the high metallicity clusters with apparently flat IMFs must have sharp cutoffs or steepenings of their IMFs above some critical mass. If this critical mass is as low as only 2 $M_\odot$ as they suggest for 47 Tuc, then the flat IMFs will not produce an overabundance of NS remnants (indeed none would be produced for such a low critical mass), but rather would produce more WD remnants.

Whatever the number of initial cluster stars sufficiently massive to form NS remnants in isolation, only a fraction of those formed will be retained in the cluster if the formation process is at all similar to that apparently operative for Pop I stars today. That is, from the observed distribution of radio pulsar velocities (see, e.g., the paper by Cordes in these proceedings), it appears that most pulsars are given a large kick of some 100-200 km/sec at their birth. This, of course, would eject them from the globular cluster. The published velocities of Anderson and Lyne (1983) indicate that at most about 10% of the neutron stars so produced would have velocities $\leq$30 km/sec and so would be retained in the cluster. Thus instead of the ~100 NSs now present in a "typical" cluster, more than 1000 would have to have been produced. This would seem to further complicate the problems of cluster enrichment and expansion mentioned above.

Finally, NS formation in the IMF-1 model would lead to NS ages as old as the cluster itself, or some 15 Gyr. However, if the magnetospheric models are accepted for the quasi-periodic oscillations (QPOs) now known to be present from at least two of the high luminosity x-ray binaries in globulars (the "Rapid Burster" source in the cluster

Liller 1 and the source in NGC 6624), then the magnetic fields on their NSs are probably in excess of the now-presumed base magnetic field of $\sim 10^9$ gauss (Taam and Van den Heuvel 1986). This, together with the strong evidence that magnetic fields on NSs decay down to a base value with a time constant of $\sim 10^7$ years (see paper by Taylor in these proceedings), suggests these NSs are relatively young. In this case, they cannot have been formed by the IMF-1 model, and either a combination of processes or primarily other ones must be operative for NS formation.

### 3.3 Model IMF-0

This model, though not so named, was proposed by Katz (1983) to circumvent the NS escape problem as well as to account for the possibly discrepant age of a NS in an apparently old supernova remnant. It presupposes a mechanism for the formation of NS in a "quiet collapse" event. As such, it would not be subject to the difficulties of mass loss associated with the IMF-1 model. However there is as yet no compelling observational or theoretical reason to believe that such events occur in nature. This formation mechanism, if it also relied on massive stars with short lifetimes, would also be subject to the magnetic field problem mentioned above. It remains an interesting possibility in search of a unique application.

## 4. AIC MODEL: NEUTRON STAR FORMATION FROM WHITE DWARFS IN BINARIES

We therefore turn to the AIC model for the formation of NSs in globulars. Arguments in favor of this model have been given by Grindlay (1986), Van den Heuvel (1986) and Taam and Van den Heuvel (1986). Here we review these and relate them to the new IMF results mentioned above.

### 4.1 Evidence for WDs in Globulars

By analogy with the evidence for NSs in globulars, the most direct evidence for WDs are both observations which directly suggest the presence of WDs in binaries (e.g., the low luminosity x-ray sources in globulars-- Hertz and Grindlay 1983) as well as the studies referred to above which suggest substantial dark matter contributions in cluster cores. Before turning to the arguments for WDs in binaries, which form the basis for the AIC model, we review briefly the arguments for a substantial number of isolated WDs in clusters. Recall that essentially all of the cluster mass originally present in the IMF at stellar masses above the present turnoff mass of $\sim 0.8\ M_\odot$ will have now evolved into post-AGB stars, WDs, NSs and liberated gas. The total fraction of the original cluster mass in this evolved form is obviously dependent on the index x of the IMF as well as the upper mass limit, $m_u$, in the cluster IMF. As shown by Smith and McClure (1986), the ratio of mass in dark remnants (i.e., the WD and NS masses assumed for a given model of stellar evolution, which relates their masses to their progenitor stellar masses) to the still-visible stellar mass (i.e., the mass in stars below the turnoff mass) is between $\sim 0.9$ - 0.1 for x in the range 0 -2 and a maximum initial mass $m_u = 2\ M_\odot$. This is consistent with the dark mass fraction of 35% (of the total) estimated for 47 Tuc (DaCosta and Freeman 1985) and indeed is what fixes $m_u$ given the observed value of $x \simeq 0$ for this cluster (although Pryor et al. 1986 find that x should be ~0.3 flatter still to account for mass segregation; this would reduce $m_u$ still more). Alternatively, Smith and McClure show that the dark mass fraction in 47 Tuc could be produced by a more complicated mass function (and IMF) such as one with a break (at the present day turnoff mass) from $x = 0$ to $x = 1.83$ and $m_u = 20\ M_\odot$.

In either case, a substantial fraction of the dark mass must be in the form of WDs (which would account for <u>all</u> of the dark mass in the first case). However, only in the broken power law case would there be an appreciable range of WD masses (as well as NSs, fractionally retained, from the IMF models of section 3 above) produced from the range of stellar masses above the turnoff mass. We shall see below that this may be necessary for the production of NSs from the AIC model.

The observations of historical novae and, recently, two dwarf nova systems (spectroscopically) in globular clusters (cf., Shara et al. 1985 for a brief review) provide direct "proof" that WD-binaries exist in globulars. However, as Shara et al. also report, no direct optical detection of cataclysmic variables (CVs) in globulars have yet been made down to absolute magnitudes $M_B \simeq 6$ (in the cluster M3). CVs are expected on the basis of the expected number of WDs that should tidal capture main sequence stars in cluster cores (Hut and Verbunt 1983). Indeed this is what led Hertz and Grindlay (1983) to (independently) propose this process for the explanation of the new class of low luminosity x-ray sources they found in globulars with the Einstein Observatory. The fact that the luminosity distribution of compact x-ray sources in globulars (i.e., not including the additional diffuse x-ray emission from hot cluster gas--cf., Grindlay 1985a) has a significant gap between $\sim 10^{34.5}$ and $10^{36}$ erg/sec (Hertz and Grindlay 1983, Hertz and Wood 1985) provides strong evidence that the low luminosity sources are by and large a separate class from the high luminosity neutron star systems. The factor of $\sim 10^3$ in maximum x-ray luminosity for the two distributions points to WDs as the compact objects in the low luminosity systems, although undoubtedly some of the low luminosity sources (e.g., the soft x-ray transient in NGC 6440) are NS systems with their mass transfer essentially turned off.

From the observed distribution of x-ray luminosities for the low luminosity sources as well as the upper limits for the entire cluster sample surveyed with both the Einstein and HEAO-1 x-ray satellites, Hertz and Wood (1985) extended the calculations of Hertz and Grindlay (1983) to derive the rate of tidal capture of main sequence stars on WDs and thus the number of x-ray emitting WD binaries in a "typical" cluster. They concluded that the "typical" globular cluster (with central density $\sim 10^4$ pc$^{-3}$) would contain approximately 10 low luminosity sources for an assumed mass fraction in WDs of 15%, an IMF index $x = 2$, and a fixed WD mass of $0.6\ M_\odot$. This is approximately 100 times the number of high luminosity sources (NS binaries) per cluster (cf., Lightman and Grindlay 1982) and indicates that the ratio of WDs to NSs in the cluster core is at least this ratio. In fact the ratio must be even larger since the NSs, being more massive than the WDs, will be more centrally concentrated in the core and will have a larger gravitational focusing contribution to the tidal capture cross section.

4.2 Requirements for AIC

Clearly the first requirement for the AIC model is met: a globular cluster (core) contains a substantial fraction of its mass in WDs and a relatively large number of these are probably in (tidal capture) binary systems. The next major requirement is for the WD to be able to accrete enough mass to exceed the Chandrasekhar limit without losing the mass in (a series of) nova explosions. Finally, the WD must be able to collapse without being disrupted in a supernova Type I explosion. Recent work by a number of investigators suggests that each of these conditions can be met in different regions of the two-dimensional space of accretion rate vs. "seed" WD mass.

A particularly interesting summary of the possible NS formation regimes in the $M_{WD}$ vs. accretion rate plane is given by Nomoto 1986. A O-Ne-Mg WD can collapse to a NS if its mass is $\geq 1.05$ $M_\odot$ and it accretes at a rate $\geq 10^{-7.3}$ $M_\odot$/yr or at a rate of only $\leq 10^{-9.2}$ $M_\odot$/yr if its mass is above about 1.12 $M_\odot$. Thus these relatively massive WDs, which would arise from the $\sim 8$ $M_\odot$ stars of the cluster IMF or might arise from the mergers of less massive WDs (see discussion below) are very likely progenitor systems for the AIC production of NSs. The more numerous C-O WDs, with masses in the approximate range 0.8-1.15 $M_\odot$, will collapse to form O-NeMg WDs and then NSs if their accretion rate is maintained above about 0.2 of the Eddington value ($\sim 10^{-5}$ $M_\odot$/yr). These results are reviewed by Nomoto but have been obtained previously. However two new accretion rate vs WD mass regimes for C-O WD collapse are also reported: for C-O WD masses $\geq 1.2$ $M_\odot$ and accretion rates $\geq 10^{-7.4}$ $M_\odot$/yr and for $\geq 1.13$ $M_\odot$ and accretion rates $\leq 10^{-9}$ $M_\odot$/yr. At intermediate accretion rates in this same mass range Nomoto finds the possibility of "dim" supernovae (Type Is); these should not be confused with the IMF-0 "quiet collapse" model discussed in section 3 since accretion from a binary companion is required.

The low accretion rate regime for collapse of C-O WDs is particularly relevant to the globular case considered here, since the accretion rates needed ($\leq 10^{-9}$ $M_\odot$/yr) are just those expected for WDs which tidally capture main sequence companions. These are in fact the typical accretion rates indicated for the low luminosity cluster x-ray sources. Thus AIC and not disruption in a supernova appears possible for a range of WD masses and accretion rates provided the WD masses are initially above about 1 $M_\odot$. These C-O WDs will arise "naturally" from the evolution of single cluster stars with masses above about 4 $M_\odot$ so that the AIC production process will be enhanced in accordance with the IMF numbers of stars in this mass range. This, in turn, would suggest that the metal rich clusters, provided they do indeed have flatter IMFs (or at least flatter IMFs up to a cutoff mass, as discussed above for 47 Tuc) will indeed produce more NSs. Thus the significant tendency for the high luminosity cluster x-ray sources to be in clusters relatively near the galactic center, and thus (statistically at least) more metal rich, could be understood.

The presence of NSs in all clusters, and particularly metal poor clusters such as M15, could also be understood if there were a "natural" way to enhance the numbers of C-O (and, ultimately, O-Ne-Mg) WDs in these clusters. A mechanism may be provided by the merger process considered by Nomoto and Sugimoto (1977) for (field) WD binaries, which themselves presumably arise from the evolution of a giant binary pair. In a high central density globular cluster, many such binaries will be formed by tidal capture between various pairings of main sequence, giants and WD stars. All of these will eventually lead to WD-WD binaries, where the typical component WD members will be $\sim 0.6\ M_\odot$ He WDs (now being formed in the clusters). The merger of these will be a C-O WD. Thus these more massive WDs can be "grown" in the dense cores of globulars, and a correlation of NS production rate and central density of the cluster is expected.

## 5. RELATION TO NS ORIGIN IN FIELD LMXRBS

The AIC process seems very likely to occur and to be self-sustaining in globular clusters where the stellar densities are very high and binary evolution is important. The low mass x-ray binaries (LMXRBs) in the field, however, are more difficult to understand. Certainly the bursters in the field, which are now increasingly identified with systems with orbital periods of ~3-5 hours and main sequence companions, are not readily understood by either the IMF or AIC models discussed in sections 3 and 4 above: the IMF models would be expected to disrupt the (pre- existing) binary as the more massive star explodes, while the AIC model doesn't naturally lead to a C-O or O-Ne-Mg WD in the (pre-existing) binary system for the same reason (disruption). Thus the considerations for the formation of the NSs in these field LMXRBs, as well as the evidence for binary evolution and hierarchical triple systems, continue to support the suggestions (Grindlay 1985a,b, and 1986) that these compact binaries were born in globular clusters (by the tidal capture and NS evolution processes discussed above) and either ejected or, more likely, liberated from globulars undergoing disruption in the galactic bulge.

## 6. CONCLUSIONS

At first sight, NSs were (are) not expected in globular clusters, as witnessed by the general surprise and interest in the discovery of the luminous x-ray sources and bursters in globulars in the mid-1970s. The existence of NSs in globulars is now certain beyond a doubt but their origin and implications for both stellar and binary evolution in globulars remains a challenge. We have reviewed a variety of recent results and contributed arguments which suggest that AIC is the dominant mode of formation of NSs in dense globular cluster cores. The arguments will be tested by continuing searches for the optical counterparts of both the high and, especially, the low luminosity x-ray sources. Optical identifications of the low luminosity systems are especially critical to confirm that they are indeed the CVs expected (cf., discussion of on-going searches and preliminary results in Grindlay 1986). NS production should be maximum in clusters which are both metal rich and of the highest central densities $n_c$. Although the distribution of clusters in the [m/H] vs. $n_c$ plane will be studied in a forthcoming paper, it is interesting that the clusters containing high luminosity x-ray sources are generally both metal rich and with high central densities, whereas clusters with comparably high central densities but lower metals (e.g., NGC 5824) generally do not contain high luminosity sources. The clusters with the largest values of $n_c$ and [m/H] should also then be the best prospects for finding the probable evolutionary end-products of NS-binaries in globulars: binary radio pulsars with spin periods of a few milliseconds and orbital periods of ~10 days (though some will have been scattered out of their parent binaries by interactions with both single stars and other binaries in the cluster core). In this regard the steep spectrum radio source in the core of the globular cluster M28=NGC 6626 (Hamilton, Helfand, and Becker 1985) is particularly intriguing and searches for millisecond pulsars in more metal rich globulars should be continued.

I thank Jim Hesser for providing early copies of the exciting new papers by the DAO group on IMFs in globulars. This work was supported in part by NSF grant AST-84-17846.

REFERENCES

Anderson, B. and Lyne, A., 1983, Nature, **303**, 597.
Clark, G.W., 1975, Ap.J. (Letters), **199**, L43.
DaCosta, G., 1982, Astron.J., **87**, 990.
DaCosta, G. and Freeman, K., 1985, in Dynamics of Star Clusters, IAU Symp. 113, (J. Goodman and P. Hut, eds.), Reidel:Dordrecht, p. 69.
Fabian, A., Pringle, J. and Rees, M., 1975, MNRAS, **172**, 15P.
Grindlay, J.E., 1981, in X-ray Astronomy With the Einstein Satellite, (R. Giacconi, ed.), Reidel:Dordrecht, p. 79.
Grindlay, J.E., 1985a, in Dynamics of Star Clusters, IAU Symp. 113, (J. Goodman and P. Hut, eds.), Reidel:Dordrecht, p. 43.
Grindlay, J.E., 1985b, in Proc. US-Japan Seminar on Galactic and Extragalactic Compact X-ray Sources, (Y. Tanaka and W. Lewin, eds.), ISAS, p. 215.
Grindlay, J.E., 1986, in The Evolution of Galactic X-ray Binaries, (J. Trumper, W. Lewin and W. Brinkman, eds.), NATO ASI Series, Vol. 167, p. 25.
Grindlay, J.E. et al., 1984, Ap.J. (Letters), **282**, L13.
Hamilton, T., Helfand, D. and Becker, R., 1985, Astron.J., **90**, 607.
Hertz, P. and Grindlay, J., 1983, Ap.J., **275**, 105.
Hertz, P. and Wood, K., 1985, Ap.J., **290**, 171.
Horne, K., Verbunt, F. and Schneider, D., 1986, MNRAS, **218**, 63.
Hut, P. and Verbunt, F., 1983, Nature, **301**, 587.
Illingworth, G. and King, I., 1977, Ap.J. (Letters), **218**, L109.
Joss, P. and Rappaport, S., 1984, Ann.Rev.Astron. and Astrophys., **22**, 537.
Katz, J., 1983, Astron. and Astrophys., **128**, L1.
Larson, R., 1984, MNRAS, **210**, 763.

Lewin, W. and Joss, P., 1983 in Accretion Driven X-ray Sources, (W. Lewin and E. Van Den Heuvel, eds.), Cambridge Univ. Press, p. 41.
Lightman, A.P. and Grindlay, J.E., 1982, Ap.J., **262**, 145.
McClure, R. et al., 1986, Ap.J. (Letters), **307**, L49.
Nomoto, K., 1986, in Proc. VIth Astrophysics Meeting, Moriond, in press.
Nomoto, K. and Sugimoto, D., 1977, Publ.Astr.Soc.Japan, **29**, 765.
Press, W. and Teukolsky, S., 1977, Ap.J., **213**, 183.
Pryor, C., Smith, G., and McClure, R., 1986, preprint.
Rappaport, S., Joss, P. and Webbink, R., 1982, Ap.J.,**254**, 616.
Shara, M., Moffat, A., and Hanes, D., 1985, in Dynamics of Star Clusters, IAU Symp. 113, (J. Goodman and P. Hut, eds.), Reidel:Dordrecht, p. 103.
Smith, G. and McClure, R., 1986, preprint.
Taam, R. and Van den Heuvel, E., 1986, Ap.J., **305**, 235.
Van den Heuvel, E.P.J., 1986, in The Evolution of Galactic X-ray Binaries, (J. Trumper, W. Lewin and W. Brinkman, eds.), NATO ASI Series, **167**, p. 107.
Webbink, R., Rappaport, S., and Savonije, G., 1983, Ap.J., **276**, 678.
Verbunt, F., Van Paradijs, J. and Elson, R., 1984, MNRAS, **210**, 899.

## DISCUSSION

**R. Dewey:** Are the white dwarfs in the giant-driven sources likely to be of the right composition to collapse rather than completely disrupt?

**J. Grindlay:** That is an excellent question. If massive white dwarfs (O - Ne - Mg) are produced from an early generation of massive stars in globulars, then they should not disrupt. If the more numerous CO white dwarfs have cooled to have a gradient in their composition (with oxygen settling to their core), as suggested by Canal et al., they they should not disrupt.

**S. Woosley:** If the neutron star in your binaries comes from white dwarf collapse the mass should not be much less (or much greater for that matter) than 1.1 $M_\odot$. You start with 1.4 $M_\odot$ subtract no more than 0.2 $M_\odot$ in the explosion (nucleosynthetic limits will likely reduce this to $\leq$ 0.1 $M_\odot$) and about 0.1 to 0.2 $M_\odot$ for the neutron star binding energy.

**J. Grindlay:** I agree; a mass loss of no more than ~0.2 $M_\odot$ is also implied by the requirement to keep the resulting neutron star binary in the globular cluster, where it may subsequently interact with (predominantly) main sequence stars in the cluster core to produce (by exchange collisions) the neutron star - main sequence star binaries predominantly observed.

**F. Verbunt:** I am pleased to see that you have an indication that neutron stars in globular clusters are less massive than 1.4 $M_\odot$. There is one low-mass X-ray binary in which the mass of the neutron star has been estimated, 4U2129+47 (Horne et al. 1986, MNRAS **218**, 63) and the mass of this neutron star is 0.6±0.2 $M_\odot$.

**J. Grindlay:** I mentioned that lower mass neutron stars are possible (i.e, consistent with) but not yet required by the X-ray positions and mass determination. Lower mass neutron stars would be required only to the extent that the accreted mass is a significant fraction of the current total mass, and to the extent required by improved models of the cluster potential to be derived from our optical studies of cluster cores. As for the evidence for a reduced mass in the reference you cite, this is subject to the highly uncertain kinematics of the orbit when derived from emission, rather than absorption lines (i.e., the effects of gas streams on the derived mass are very uncertain).

**S. Kulkarni:** In response to the previous question asked by Verbunt to Grindlay, is it not true that we expect the neutron stars to be more massive than 1.4 $M_{\odot}$ since neutron stars in globular clusters always end up with a companion and may have had more than one episode of extended accretion phase?

**J. Grindlay:** The neutron stars are probably formed in globular clusters with masses of ~ 1.1 - 1.2 $M_{\odot}$ if they are formed from white dwarf collpase. If they accrete for $\sim 10^9$ yrs, they may indeed gain an additional several tenths of a solar mass.

**S. Kulkarni:** Dissipation of a globular cluster is a very inefficient way to create an LMXB. The formation rate of galactic LMXBs is $\sim 10^{-6} - 10^{-7}$ $yr^{-1}$. In your scenario we would need a very large number of galactic clusters in the past. Could you please comment on this point?

**J. Grindlay:** The formation rate of $\sim 10^{-7} - 10^{-8}$/yr is that for the brightest luminosity systems, with mass transfer supplied by a giant. The globular cluster disruption scenario, on the other hand, attempts to account for the lower luminosity LMXBs in which mass transfer is supplied by a main sequence star (typically a K or M dwarf). The lifetime of these systems is much longer ($\sim 10 \times 10^9$ yrs) so that the required rate and number of disrupted globulars is reasonable as I have shown in several papers.

# THE GLOBULAR CLUSTER POPULATION OF X-RAY BINARIES

Frank Verbunt
Max Planck Institut für
Extraterrestrische Physik
8046 Garching bei München
Federal Republic of Germany

Piet Hut *
Institute for Advanced Study
Princeton, New Jersey 08540
U.S.A.

**Abstract.** We discuss formation mechanisms for low-mass X-ray binaries in globular clusters. We apply the most efficient mechanism, tidal capture in close two-body encounters between neutron and main-sequence stars, to the clusters of our galaxy. The observed number of X-ray sources in these can be explained if the birth velocities of neutron stars are higher than estimated from velocity measurements of radiopulsars, or if the initial mass function steepens at high masses. We perform a statistical test on the distribution of X-ray sources with respect to the number of close encounters in globular clusters, and find satisfactory agreement between the tidal capture theory and observation, apart from the presence of low-mass X-ray binaries in four clusters with a very low encounter rate: Ter 1, Ter 2, Gr 1 and NGC 6712.

*EXOSAT* observations indicate that some dim globular cluster sources may be less luminous than hitherto assumed, and support the view that the brighter dim sources may be soft X-ray transients in quiescence.

## 1. Introduction

Globular clusters are a favoured environment for low-mass X-ray binaries: of the $\sim 50$ bright low-mass X-ray binaries in the galaxy, $\sim 10$ are located in globular clusters. With $\sim 10^7 M_\odot$ in globular clusters and $\sim 10^{11} M_\odot$ in the galactic disk, the number of bright low-mass X-ray binaries per unit of mass is $\sim 2000$ times higher in globular clusters than in the galactic disk, or $\sim 200$ times if we consider Population II mass only (Katz 1975). It is thought that this high frequency is the result of the frequent occurrence in globular clusters of a close encounter between a neutron star and an ordinary star, resulting in the formation of a low-mass X-ray binary through tidal dissipation (Sutantyo 1975; Fabian, Pringle & Rees 1975). This mechanism does not operate in the galactic disk because of the low number densities there of stars.

The formation rate of low-mass X-ray binaries in globular clusters depends on the numbers of neutron stars and ordinary stars in the clusters. The observed

*Sloan Foundation Fellowship



*D. J. Helfand and J.-H. Huang (eds.), The Origin and Evolution of Neutron Stars, 187–197.*

number of low-mass X-ray binaries further depends on their lifetime. Thus by combining the observed numbers of low-mass X-ray binaries in globular clusters with our knowledge of their formation mechanisms and of their lifetime we can derive the number of neutron stars in globular clusters, and obtain clues on their origin.

In Section 2 we review the formation processes in globular clusters of low-mass X-ray binaries, in Section 3 the numbers of neutron stars and field stars are derived from the initial mass function, and in Section 4 the lifetime of a low-mass X-ray binary is estimated. Application to observations follows in Section 5. In Section 6 we discuss the nature of the low-luminosity X-ray sources in globular clusters.

## 2. Formation mechanisms of X-ray binaries in globular clusters

### 2.1 Close encounters between neutron stars and other stars.

If a neutron star with mass $m$ passes at closest distance $d$ of a main sequence or giant star with mass $M$ and radius $R$, the tidal bulge on this star will have a maximum height

$$h \simeq \frac{m}{M}\frac{R^4}{d^3}$$

and contain an amount of mass of order

$$m_t \simeq k\frac{h}{R}M \simeq k(\frac{R}{d})^3 m$$

where $k$ is the apsidal motion constant (e.g. Schwarzschild 1958). For a star with deep convection $k \simeq 0.14$ (Motz 1952). The potential energy in the tidal bulge is of order

$$E_t \simeq m_t\frac{GM}{R^2}h \simeq k\frac{Gm^2}{R}(\frac{R}{d})^6$$

This can be compared with the kinetic energy at infinity

$$E_{in} = \frac{1}{2}\frac{mM}{m+M}v_{in}^2$$

where $v_{in}$ is the relative velocity between the stars at infinity. The energy in the bulge exceeds the relative kinetic energy if

$$d \leq 3.2R(\frac{k}{0.14}\frac{m}{M}\frac{m+M}{2M_\odot}\frac{R_\odot}{R}\frac{10km\ s^{-1}}{v_{in}})^{1/6} = yR \quad (1)$$

The small power in this expression, caused by the strong radial dependence of tidal interactions, justifies the rough approximations made before. A binary will be formed if 1) the energy in the tidal bulge exceeds the kinetic energy at infinity, i.e. if eq.(1) is satisfied, and 2) the energy in the tidal bulge is dissipated.

The cross section for passage within distance $d$ of two stars is given by

$$\sigma = \pi d^2 \left(1 + \frac{2G(m+M)}{v_{in}^2 d}\right) \simeq \pi d \, \frac{2G(m+M)}{v_{in}^2} \qquad (2)$$

The second term within brackets gives the effect of gravitational focussing and dominates for the small relative velocities in globular clusters, justifying the approximation in the third member.

The formation rate in globular clusters of low-mass X-ray binaries through tidal interactions follows with eqs.(1) and (2):

$$F = n_{ns} n v_{in} \sigma = 6 \times 10^{-11} \frac{n_{ns}}{10^2 pc^{-3}} \frac{n}{10^4 pc^{-3}} \frac{m+M}{M_\odot} \frac{yR}{R_\odot} \frac{10 km\ s^{-1}}{v_{in}} \, yr^{-1} pc^{-3} \qquad (3)$$

where $n_{ns}$ and $n$ are the number densities of neutron stars and other stars, respectively. The other stars can be main-sequence stars, or giants. The above formula corresponds to the results of Fabian, Pringle & Rees (1975). The different result of Sutantyo (1975) is due to his neglect of gravitational focussing. Press & Teukolsky (1977) estimate the dissipation rate for an n=3 polytrope star (cf. Burke 1967). Lee & Ostriker (1986) correct an error in the earlier estimate for the n=3 polytrope, and also present calculations for the n=1.5 polytrope. Because of the deep convective regions in low-mass stars, the result for the n=1.5 polytrope is more relevant, and it is similar to the result of the simple derivation above.

### 2.2 Exchange collisions

A close encounter of a binary with a single star often leads to the formation of a short-lived triple system, and generally ends with the expulsion of the least massive star of the three, leaving the other two stars in a binary. Thus a close encounter between a neutron star and a binary of two ordinary low-mass stars will usually lead to the replacement in the binary of an ordinary star with the neutron star (Hills 1976). The importance of this process in globular clusters depends on the number density in them of close binaries. Observational evidence on this is lacking at the moment. Even with an optimistic estimate of the number of close binaries in globular clusters the formation of low-mass X-ray binaries through exchange collisions barely competes with two-body encounters (Hut & Verbunt 1983). We will therefore neglect triple-body interactions.

### 2.3 Pushing a white dwarf over the Chandrasekhar limit.

Just as low-mass X-ray binaries are formed by close encounters between neutron stars and field stars, cataclysmic variables are formed by close encounters between white dwarfs and field stars (Hut & Verbunt 1983). Under special circumstances the white dwarf may accrete a sufficient amount of mass from its companion to transgress the Chandrasekhar limit and implode to a neutron star in a sufficiently quiet fashion to keep the binary intact. In the galactic disk the number of cataclysmic variables is $\sim 10^7$, the number of low-mass X-ray binaries $\sim 50$. Thus formation of a low-mass X-ray binary from a cataclysmic variable is an extremely rare event, and in globular clusters this mechanism does not compete with tidal capture in two-body encounters. We will therefore neglect white dwarf implosion.

## 3. The stellar population of globular clusters

### 3.1 The initial mass-function

To estimate the number densities of main-sequence stars, giants and neutron stars in globular clusters, Sutantyo (1975) and Van der Woerd & Van den Heuvel (1984) use the method of Tinsley (1974). It starts from the assumption that all stars in a globular cluster are born simultaneously, in numbers given by $dN = C_o m^{-1-x}\, dm$ with $dN$ the number of stars with mass in the mass interval $dm$ around mass $m$, and $C_o$ a normalization constant. The number of stars originally in the interval between $m_1$ and $m_2$ is given by

$$N(m_1, m_2) = \int_{m_1}^{m_2} C_o m^{-1-x} dm = \frac{C_o}{x}(m_1^{-x} - m_2^{-x})$$

Gravitational interactions between the stars in a globular cluster enhance equipartition in kinetic energy between stars, leading to low velocities of massive stars and high velocities of less massive stars. Thus very light stars will escape from the cluster. The mass of the least massive stars still present in the cluster is called the cutoff mass, $m_{co}$. The time that a star takes to evolve away from the main sequence is smaller for more massive stars. The mass of the most massive stars in a cluster still on the main sequence is called the turnoff mass, $m_{to}$. Thus the number of main sequence stars is given by $N(m_{co}, m_{to})$.

The turnoff mass as a function of the age $\tau$ of the cluster can be written (Rood 1972)

$$log\frac{m_{to}}{M_\odot} = -0.28\ log\frac{\tau}{10^9 yr} + 0.013\ log Z - 0.75\ log Y + 0.453$$

$Z$ and $Y$ are the metal and helium contents of the cluster. The evolution of a giant star is fast with respect to the age of a globular cluster, and all giants will have a mass close to the turnoff mass. If the time that a giant spends at a given locus i of the giant branch is $\Delta t_i$, the number of giants at that locus is

$$N_{G,i} = 0.28\ C_o\ \frac{\Delta t_i}{\tau}\ m_{to}^{-x}$$

The total number of giants is found by summing the numbers at different evolutionary stages along the giant branch. All stars more massive than the turnoff mass will have completed their evolution. If all stars with original mass below $m_a$ evolve into white dwarfs, and all stars with original mass between $m_a$ and $m_b$ into neutron stars, the number of white dwarfs in the cluster is given by $N(m_{to}, m_a)$, and the number of neutron stars by $fN(m_a, m_b)$. Here $f$ is the fraction of neutron stars born with a velocity lower than the cluster escape velocity (see Section 3.2).

In the above we assumed that all stars leave either a white dwarf or a neutron star. It is possible that stars in a given mass interval do not leave remnants. Also, stars in close binaries may exchange mass with their companion, and leave a different remnant than single stars with the same initial mass. Inclusion of such effects is straightforward, but for the sake of clarity we will ignore them.

The constant $C_o$ can be determined in a number of ways, for example:

a. count the number of main sequence stars in a given luminosity interval, and use the mass-luminosity relation to translate this into a mass interval.

b. count the number of red giants.

c. determine the total luminosity of the cluster, and equate this to the sum of the main-sequence and giant luminosities, found with the appropriate mas-luminosity relations.

### 3.2 Birth velocities of neutron stars

The average velocity of radio pulsars is $\sim 200\ km\ s^{-1}$, much higher than the escape velocity of a globular cluster. Katz (1975) argued that the presence of X-ray sources in globular clusters indicates that some compact objects are born with high velocity, and others with low velocity. Hut and Verbunt (1983) interpret this as indicating that single neutron stars are born with low velocity, whereas neutron stars born in a binary retain the high velocity of the orbital motion. The velocity measurements of 26 radio pulsars (Lyne, Anderson & Salter 1982) are compatible with such an interpretation. Consider a globular cluster with escape velocity 25 $km\ s^{-1}$. If the velocity distribution of the newly born neutron stars is identical to that of the 26 radio pulsars, some 16 % of the neutron stars would remain in the cluster ($f \simeq 0.16$). If there are no binaries in globular clusters the fraction remaining could be as high as 40 % ($f \simeq 0.4$).

## 4. Lifetimes of X-ray binaries

Our understanding of the evolution of low-mass X-ray binaries is too limited for exact estimates of the lifetime of these systems. For a system with a main-sequence companion an estimate for the lifetime is found by dividing the available mass by the mass transfer rate: a companion of 0.44 $M_\odot$ transferring at $10^{-9} M_\odot\ yr^{-1}$ is exhausted after $T_i = 0.44 \times 10^9 yr$. There is evidence that systems with mass transfer lower than $\sim 10^{-10} M_\odot yr^{-1}$ become transient, i.e. they transfer most of the mass in bursts lasting some months but are quiescent in the years between bursts (White, Kaluzienski & Swank 1984). The observed gap between bright and dim sources in globular clusters offers further evidence for a switch-off of systems at low mass- transfer rates (see Section 6). The theory of the evolution of low-mass X-ray binaries suggests that the mass transfer is drastically reduced when the mass of the main-sequence star is $\sim 0.3\ M_\odot$ (Rappaport, Verbunt & Joss 1983). The lifetime of bright X-ray binaries could then be as short as $T_i \simeq 0.1 \times 10^9 yr$ in globular clusters.

The lifetime of X-ray binaries with a giant is set by the evolution timescale of the giant: once the system passes the top of the giant branch, the system becomes detached and mass transfer stops.

## 5. Expected number of X-ray binaries in globular clusters

### 5.1 One cluster

As an example we consider the globular cluster NGC 6440. For this cluster we use the following parameters: $x = 2$, $m_{co} = 0.3M_\odot$, $m_{to} = 0.8M_\odot$, $\tau = 14\ 10^9 yr$ (cf. the values of Gunn & Griffin (1979) for M 3), $m_a = 8M_\odot$ (cf. Blaauw 1985), $m_b = 80M_\odot$, $m_{wd} = 0.6M_\odot$. The ratio of the numbers of main sequence stars, white dwarfs and neutron stars then is: $N_{ms} : N_{wd} : N_{ns} = 1 : 0.16 : 0.0016\, f$. The average mass of the main sequence stars is $m_{ms} = M_{ms}/N_{ms} = 0.44M_\odot$. For the central density $\rho_o$ we take $5 \times 10^5 M_\odot pc^{-3}$ (cf. Webbink 1985). If we neglect mass separation, it follows that the central number densities are $n_{ms} = 95 \times 10^4 pc^{-3}$, $n_{wd} = 15 \times 10^4 pc^{-3}$, $n_{ns} = 15 f \times 10^2 pc^{-3}$. With the radius approximately proportional to the mass the average radius is $R_{ms} = 0.44 R_\odot$. The average relative velocity of stars in the cluster is somewhat higher than the dispersion velocity, but capture is more efficient at lower velocities. Estimating $v_{in}$ with the dispersion velocity $\sigma$ of the cluster is therefore quite accurate. With an average mass for the neutron stars of $1.4M_\odot$ and a dispersion velocity $\sigma = 13\ km\ s^{-1}$ the central formation rate of low-mass X-ray binaries through tidal capture with main-sequence stars is $F = 1.6 \times 10^{-7} f\ yr^{-1} pc^{-3}$.

To find the ratio of tidal capture by giants with that by main-sequence stars we average the radii of the giants (taken from Rood 1972) weighing with the number of giants at each evolutionary stage:

$$\frac{F_G}{F_{ms}} = \frac{\Sigma\ N_{G,i} R_i}{N_{ms} R_{ms}} \simeq 0.07$$

Of this 0.07, 0.03 are due to subgiants, and 0.04 to giants.

To get the formation rate for the cluster as a whole, one has to integrate the formation rate over the volume of the cluster. In the absence of mass separation, a good estimate of the total formation rate can be made by multiplying the central formation rate with the core volume. To predict the number of observable low-mass

X-ray binaries one must multiply the formation rate with the expected lifetime of the X-ray binary. For systems with main-sequence stars we get

$$N_x = \frac{4}{3}\pi r_c^3 F\ T_i$$

With a core radius of 0.24 pc the predicted number of X-ray binaries is 0.9f to 4.0f, for $T_i = 0.1$ to $0.44 \times 10^9 yr$.

Systems with giants will live longer when the capture takes place at an earlier evolutionary stage. We assign to each system formed at stage i a lifetime $T_{G,i}$ equal to the time the giant takes to reach the top of the giant branch (taken from Rood 1972). Again scaling the number of systems with giants to the number of systems with main-sequence stars, we find for the ratio of the expected numbers:

$$\frac{N_{x,G}}{N_{x,ms}} = \frac{\Sigma\ F_G T_{G,i}}{F_{ms} T_i} = 0.06.$$

0.05 is due to subgiants, and 0.01 to giants on the giant branch.

In the above we have neglected mass separation. As neutron stars are more massive than the average cluster member, they will be strongly concentrated to the cluster core. Since most close encounters occur in or near the core, this will enhance the formation rate and hence the expected number of X-ray binaries. We estimate that the enhancement is approximately one order of magnitude. This would lead to an expected number of X-ray binaries in NGC 6440 of 9 $f$ to 36 $f$. This cluster contains 1 X-ray transient.

## 5.2 The galactic system of globular clusters

From the above discussion follows that the number of expected X-ray binaries in a globular cluster is proportional to $C = \rho_o^2\ r_c^3/\sigma$ (see Eq.(3)). We use the compilation of Webbink (1985) to calculate $C$ for the 147 clusters for which values for these parameters are available. The sum of all values of C is 0.385 $\times 10^{10} M_\odot^2 pc^{-3} s\ km^{-1}$. By comparison with the single cluster we then estimate the number of expected low-mass X-ray binaries in all galactic globular clusters together as $125f$ to $500f$ sources. This assumes that all parameters not entering $C$ are the same in all clusters. Small variations between clusters in the parameters not used in $C$ ($m_{co}, m_{to}, m_a, \tau$, etc) do not affect this estimate much. This means that the observed number of low-mass X-ray binaries in globular clusters can be explained with $f \simeq 0.1$, somewhat lower than our estimate in Section 3.2. Other possibilities are that a sizeable fraction of supernovae does not produce a neutron star remnant, or that the number of stars at higher mass is lower than that predicted from extrapolating the initial mass function (which after all is determined at masses an order of magnitude below the range of supernova predecessors).

Next we order the clusters with respect to their value of $C$, and assign space to them on a line running from 0 to 1 proportional to their value of $C$ (Figure 1).

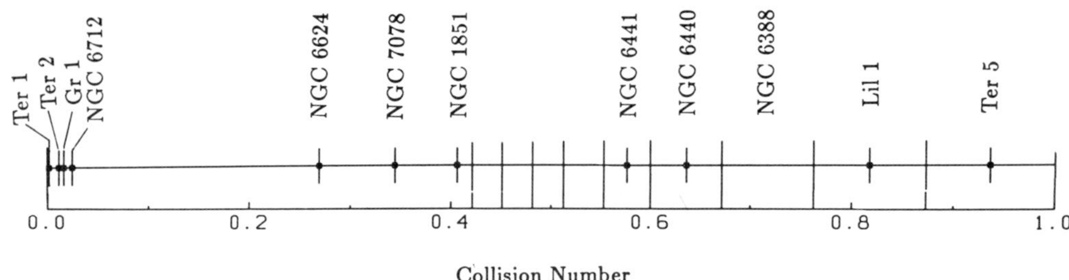


*Figure 1. This Figure tests the hypothesis that low-mass X-ray binaries in globular clusters are formed through close encounters. It is made as follows: we calculate the number of close encounters C for 147 clusters listed by Webbink (1985). We order the clusters with respect to C and assign each cluster a space between 0 and 1 proportional to its value of C. For the clusters with the 9 largest C's this space is indicated in the Figure. The clusters containing low-mass X-ray binaries are marked. The source in NGC 6440 is transient, the source in Lil 1 (the rapid burster) is a recurrent transient. If sources are formed through close encounters they should be evenly distributed between 0 and 1.*

This allows us to test the hypothesis that the number of low-mass X-ray binaries in a cluster is proportional to $C$: if this is the case the sources should be evenly distributed along the line (cf. Lightman & Grindlay 1982). For example, we expect that $\sim$ 13 % of all X-ray sources in clusters are in Ter 5. Also, whereas each individual cluster with small $C$ is predicted to have a small probability of having an X-ray source, many of such clusters together should have some. We see from Figure 1 that 7 of the 11 sources known in globular clusters are indeed following the expected distribution, indicating that they are indeed formed by close encounters. Surprisingly, however, 4 of the sources are in clusters with a very small $C$. It should be noted that the values of $C$ for individual clusters are rather uncertain, perhaps even more so in the five clusters which were studied only after an X-ray source was discovered in them (all of these clusters are located at the extremes of the $C$-distribution). It is therefore hard to asses the significance of the over-representation of low $C$ clusters among the clusters containing an X-ray source. It may be worth while, however, to start looking for formation mechanisms for low-mass X-ray binaries in globular clusters which do not depend on close encounters, and to investigate whether Ter 2, Ter 1, Gri 1 and NGC 6712 are special in any respect.

## 6. The luminosity function and the dim sources

The sources discussed in the previous Sections all have been observed at $L_x >$

$10^{36} erg\ s^{-1}$ at some time, and therefore must contain neutron stars. These sources are all located within 2 $r_c$ of the cluster center. With the sensitive *EINSTEIN* detectors Hertz & Grindlay (1983) discovered sources in a lower luminosity range ($L_x < 10^{34.5} erg\ s^{-1}$). No sources were discovered in the range $10^{34.5} < L_x(erg\ s^{-1} < 10^{36}$. Current theories of the evolution of low-mass X-ray binaries all predict that they have low mass-transfer rates once the mass of the secondary becomes low. The existence of a lower limit to the average luminosity of the bright sources at $L_x \simeq 10^{36} erg\ s^{-1}$ therefore indicates that low-mass X-ray binaries become transient if their mass-transfer rate drops below $\sim 10^{-10} M_{\odot} yr^{-1}$. It is more difficult to explain that the upper limit to the average luminosity of the globular cluster sources at $L_x \simeq 10^{37.3} erg\ s^{-1}$ is lower than the brightest galactic disk sources. The main difficulty is that the sources in the globular clusters of M 31 reach average X-ray luminosities as high as the brightest sources in our galactic disk, although the globular clusters in M 31 are very similar to the ones in our galaxy (Verbunt, Van Paradijs & Elson 1984).

The nature of the dim ($L_x < 10^{34.5} erg\ s^{-1}$) sources is the subject of some debate. Hertz and Grindlay (1983) argue that these sources are cataclysmic variables. Cataclysmic variables are expected to be formed in globular clusters by close encounters between white dwarfs and main-sequence cluster members (Hut & Verbunt 1983). Verbunt, Van Paradijs & Elson (1984) agree that the sources at the low end of the luminosity distribution of the dim sources, some of which are located far outside the cluster core, are presumably cataclysmic variables, but note that the brightest of the dim sources, all located in the cluster cores, seem too bright to be cataclysmic variables, and propose that they are transient low-mass X-ray binaries in quiescence.

Two *EXOSAT* observations shed new light on this debate. An observation of $\omega$ Cen indicates that some dim sources have very soft spectra ($kT \simeq 40eV$). For these sources the luminosities derived by Hertz & Grindlay, who assume $kT = 5keV$, may be an order of magnitude too high (Verbunt et al. 1986). In order to determine whether the brightest of the dim sources are too bright to be cataclysmic variables it is necessary to measure their temperatures.

The other *EXOSAT* observation concerns a source in the galactic disk: the soft X-ray transient Cen X-4. This source was detected at a level of $L_x \simeq 10^{32} - 10^{33} erg\ s^{-1}$, the first unambiguous detection of a soft X-ray transient in quiescence (Van Paradijs, Verbunt & Shafer, in preparation). This shows that transients in quiescence can indeed have luminosities in the range of the dim sources in globular clusters, and strengthens the case for identification of some dim sources with quiescent transients.

## References

Blaauw, A. 1985. In: *Birth and Evolution of Massive Stars and Stellar Groups*, eds. W. Boland & H. van Woerden, Reidel, Dordrecht, p.211.

Burke, J.A. 1967. *Mon. Not. R. astr. Soc.* **136**, 389.

Fabian, A.C., Pringle, J.E. & Rees, M.J. 1975, *Mon. Not. R. astr. Soc.* **172**, 15p.

Gunn, J.E. & Griffin, R.F. 1979, *Astron. J.* **84**, 752.

Hertz, P. & Grindlay, J.E. 1983, *Astrophys. J.* **275**, 105.

Hills, J.G. 1976, *Mon. Not. R. astr. Soc.* **175**, 1p.

Hut,P. & Verbunt, F. 1983, *Nature* **301**, 587.

Katz, J.I. 1975, *Nature* **253**, 698.

Lee, H.M. & Ostriker, J.P. 1986. Preprint.

Lightman, A.P. & Grindlay, J.E. 1982. *Astrophys. J.* **262**, 145.

Lyne, A.G., Anderson, B. & Salter, M.J. 1982, *Mon. Not. R. astr. Soc.* **201**, 503.

Motz, L. 1952, *Astrophys. J.* **115**, 562.

Press, W.H. & Teukolsky, S.A. 1977. *Astrophys. J.* **213**, 183.

Rappaport, S., Verbunt, F. & Joss, P.C. 1983, *Astrophys. J.* **275**, 713.

Rood, R.T. 1972, *Astrophys. J.* **177**, 681.

Schwarzschild, M. 1958, *Structure and Evolution of the Stars*, Princeton University Press.

Sutantyo, W. 1975, *Astron. Astrophys.* **44**, 227.

Tinsley, B.M. 1974, *Publ. Astr. Soc. Pac.* **86**, 554.

Van der Woerd, H. & Van den Heuvel, E.P.J. 1984, *Astron. Astrophys.* **132**, 361.

Verbunt,F., Van Paradijs, J. & Elson, R. 1984, *Mon. Not. R. astr. Soc.* **210**, 899.

Verbunt,F., Shafer, R.A., Jansen, F., Arnaud, K.A. & Van Paradijs, J. 1986, *Astron. Astrophys.* in press.

Webbink, R.F. 1985, in: *Dynamics of star clusters, IAU Symp. No. 113*, eds. J. Goodman & P. Hut, Reidel, Dordrecht, p.541.

White,N.E., Kaluzienski,J.L. & Swank,J.H., 1984. In: *High Energy Transients in Astrophysics*, ed. S.E.Woosley, American Institute of Physics, New York, p.31.

**Note:** Some time after the conference a paper by S. Djorgovksi & I.R. King appeared (Astrophys. J. Letters **305**, L61; 1986) which presents evidence that Ter 1 and Ter 2 have undergone core collapse. This seems to support the suggestion of Grindlay at the conference that the four bright X-ray sources in clusters with low (present) encounter rates have formed during core collapse. On the other hand, NGC 6712 shows no sign of core collapse.

## DISCUSSION

**A. Blaauw:** Your scenario first forms by capture, a binary which becomes the X-ray source. Is it possible that encounters, not close enough to lead to (semi-permanent) binaries, yet rather close, contribute to explaining observed X-ray sources?

**F. Verbunt:** Such sources would live a very short time, a day or a week or so. The observed sources are much more long-lived.

**J. Grindlay:** I wish to clarify what appears to be a misunderstanding of my talk based on the comments of the speaker, who stated that the number of CVs per LMXB in the plane is very different than in globular clusters. In fact the number of CVs per highest luminosity LMXB (i.e. the ~10 GX sources with luminosities $\sim 10^{38}$ erg/s) is $\sim 10^{4}$ both in the field and in globulars. On this basis, I argued that accretion-induced-collapse (A.I.C.) of both CVs in globular and in giant-fed systems in the bulge are occurring. Secondly, it was apparently not clear that the neutron star formation by A.I.C. in globulars produces subsequently "re-cycled" X-ray sources by exchange collisions in the dense cluster cores so that they are predominantly observed as bursters with main sequence companions

**F. Verbunt:** This comment does not alter my opinion that direct close encounters between neutron stars and field stars are by far the most efficient way to make low-mass X-ray binaries in globular clusters. The comment is wrong in the sense that not a single globular cluster contains a source as bright as the bright galactic bulge sources.

**J. Grindlay:** I showed on a viewgraph (but did not mention) that formation of neutron stars in globulars from collapse of single massive stars may lead to the problem of escape from the cluster but also to a problem of "pollution" of the cluster with too many metals (from the many SN-II). Yet neutron stars are formed (in abundance) in metal poor clusters like M15. Globular cluster high luminosity sources (bursters, except M15) have the same luminosity distribution as bursters in the field (cf., Grindlay 1985, Proc. Japan-US Seminary on X-ray Sources). In contrast to your statement that none of the X-ray clusters appear to have undergone core collpase, this is not true for both NGC 6624 and M15, which show central cusps and possibly also NGC 6712, which may show an enhanced number of binaries (produced at core collpase) as indicated in my talk.

**F. Verbunt:** I agree. I don't think it affects any of my arguments. The main statement I make is that there is to my knowledge no evidence that Ter 1, Ter 2, Gr 1 or NGC6712 have undergone core-collapse. Further, there are several clusters that do show evidence of core-collapse that have much higher densities than the four just mentioned, and that do not contain low-mass X-ray binaries.

# HARDNESS RATIO IN EVOLVING LOW-MASS X-RAY BINARY SYSTEMS

J. Shaham and M. Tavani
Columbia Astrophysics Laboratory and Department of Physics
Columbia University
New York, New York 10027

Spectral observations of low-mass X-ray binaries (LMXBs) show that the soft component usually dominates over the hard one. These results provide additional support to an interpretation based on models of LMXBs in which the neutron star while, on the average, spinning up, is also experiencing a spinning down torque. Under these conditions, a fraction of the luminosity associated with the gravitational release of energy on the surface of the accreting neutron star may manifest itself as luminosity originating in the inner part of the accretion disk. It is probably possible to separate the two contributions; the stellar luminosity can be associated with the hard component of the spectrum and the disk luminosity, related to the exchange of energy due to the torque between the rapidly spinning neutron star and the accretion disk, can be associated with the soft spectral component.

We have calculated the evolution of the binary system and of the stellar and disk luminosities for several initial conditions with H and He companion stars. The hardness ratio turned out to be quite sensitive to the evolution of the mass transfer rate, or, in other words, to the nature of the companion star and to the nature of the torques. In particular, for a He companion, a regime exists in which the energy released in the slowing down of the neutron star is in excess of the accretion energy. This slowing down energy is assumed to be deposited not on the neutron star surface but mostly in the disk.

We conclude that in the course of the evolution of LMXBs, from $10^{37}$ ergs $s^{-1}$ down to $10^{36}$ ergs $s^{-1}$, the hard component may become more dominant depending upon the mechanism by which spin down energy is deposited in the disk and on the nature of the companion star. A survey of a sample of such sources can thus provides clues to the actual form of the torques.



*D. J. Helfand and J.-H. Huang (eds.), The Origin and Evolution of Neutron Stars, 199.*

# RESULTS OF THE TIMING ANALYSES OF X-RAY PULSARS OBSERVED BY HAKUCHO AND TENMA

F. Nagase
Department of Astrophysics, Faculty of Science
Nagoya University
Furo-cho, Chikusa-ku, Nagoya 464
Japan

**ABSTRACT.** A dozen of X-ray pulsars have been observed with the Japanese X-ray astronomy satellites *Hakucho* and *Tenma* between 1979 and 1984. The obsetvations revealed remarkable pulse period changes both in disk-fed and wind-fed pulsars. From the histories of the pulse period changes so far measured, we are able to classify the X-ray pulsars into three categories:

(1) Pulsars which exhibit a steady spin-up with a constant rate of period change (e.g., 4U 1626-67 and GX 1+4).

(2) Disk-fed pulsars in which considerable period changes are superposed on the secular trend of spin-up (e.g., Her X-1 and Cen X-3).

(3) Wind-fed pulsars whose pulse periods fluctuate randomly about the average pulse period on time scales from days to years (e.g., Vela X-1 and GX 301-2).

Among the dozen X-ray pulsars observed by *Hakucho* and *Tenma*, Vela X-1 was most extensively studied. Timing studies of Vela X-1 revealed an episode of secular spin-down of the pulsar between 1979 and 1982, and large rates of short-term period fluctuations superimposed on the secular trend. Recent studies of the wind-fed pulsar suggest that the angular momentum transfer expected from the wind accretion is too small to account for the observed rates. Spectroscopic studies of Vela X-1 by *Tenma* observations provided a variety of phenomena related to the binary system and detailed properties of matter surrounding the system, especially using an iron emission line fluoresced thereby as a probe.

Formation of a disk in the wind-fed pulsar is proposed to accommodate a variety of phenomena observed from Vela X-1. A disk is expected to be formed through the interaction of the spherically accreting matter with the rotating magnetosphere, under the assumption that the closed magnetosphere extends to the corotation radius. Application of this model to Vela X-1 requires a large magnetic moment of $\mu \approx 10^{32}$ gauss cm$^3$ ($10^{13} \lesssim B \lesssim 10^{14}$ gauss, depending on the stellar parameters), but once the large value is accepted, this model fits well with the large rates of period change and phenomena which would otherwise be regarded as strange. If this model works for Vela X-1, it will be also applied to some of other wind-fed pulsars.



*D. J. Helfand and J.-H. Huang (eds.), The Origin and Evolution of Neutron Stars, 200.*

# HERCULIS X1: RESULTS AND INTERPRETATION

Ph. Durouchoux
Service d'Astrophysique
DPhG/Ap, CEN Saclay
91191 Gif/s/Yvette
FRANCE

The hard X-ray spectrum of Her X1 was measured for the first time with a high-resolution (1.4 keV FWHM) germanium spectrometer (LEGS: GSFC/Saclay collaboration). The observation was performed near the peak of the OB state in the 35 day cycle and 1.24 s pulsations were observed between the energies of 20 keV and 70 keV. The best fit energies are 35 keV for an absorption line and 39 keV for an emission line. These are significantly lower energies than those derived from previous experiments.

It seems that this line energy variation is correlated with the intensity of the line as well as the continuum luminosity emitted by HER X1. In order to explain such a correlation, we consider the cooling by comptonization on a cold plasma of a black body spectrum, injected at the bottom of the accretion column.

We show that a temperature variation of the injected spectrum could produce the line energy shift as well as its correlation with the line intensity and the continuum luminosity.

The results of satellite observations as well as balloon borne experiments are summarized and compared with our model.



*D. J. Helfand and J.-H. Huang (eds.), The Origin and Evolution of Neutron Stars, 201.*

# A HARD X-RAY OBSERVATION OF CYG X-1 IN 1985

Y.Q.Ma, G.H.Li, C.M.Zhang, Q.Y.Xiao, Y.M.Qian,
Z.G.Li, M.Wu, CJ.Dai, Y.D.Gu, X.Y.Zhang, T.P.Li

Institute of High Energy Physics, Academia Sinica
P.O. Box 918, Beijing, China

On September 22, 1985, a hard X-ray observation of Cyg X-1 was performed by using a balloon-borne CsI-NaI phoswich telescope HAPI-2 at Xianghe Balloon Facility in China. The main detector is CsI(Tl) with a thickness of 0.4cm and an area of $140cm^2$. The energy range is 20-200KeV. The telescope reached a float altitude of about 38Km($4g/cm^2$). The photon's arrival time and energy loss spectrum were measured for both background and active tracking on-source observations.

For deconvoluting an observed energy loss spectrum I', we quoted two commonly used methods: a) Minimize the objective function $\chi^2$ for a given expected incident photon spectrum f(E,c) to get the optimum estimation of the parameters c and relevant confidence regions; b) Solve the matrix equation I'= F*I, then fit the estimation of the incident photon spectrum I with a function f(E,c) by minimizing $\chi^2$. The response function F of the detector was obtained by Monte-Carlo simulation. The incident photon spectrum shown here, which was obtained by the method b) using a simplified diagonal matrix F, lies on the superlow state spectrum suggested by Ling et al.(1983) above 90KeV. The X-ray luminosity in 10-200KeV was estimated as $(1.07\pm0.08)\times10^{37}$ ergs/s from a fitted thermal spectrum $(A/E)\exp(-E/B)$, where A is 0.114 (+0.010, -0.009) and B is 103.0(+10.2,-8.9). These indicate that Cyg X-1 probably was not in the normal low state.

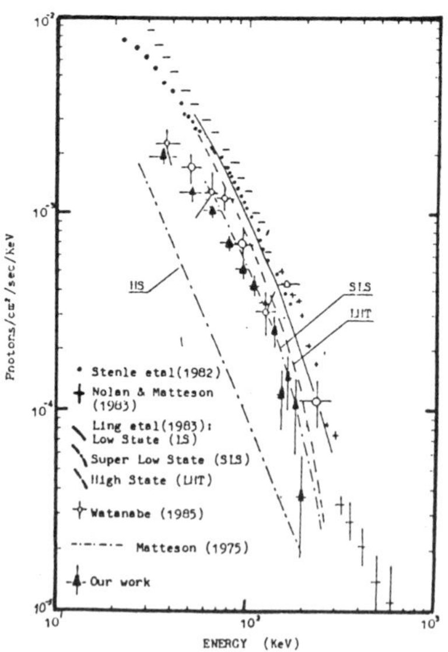


## References

Ling, J.C. et al., Ap.J., 275(1983)307.
Matteson, et al, 1975, NASA SP-389.
Nolan, P. L. et al, Ap. J., 265(1983)389.
Steinle, H. et al. A. Ap., 107(1982)350.
Watanabe, H., RUP 85-2, 1985.



*D. J. Helfand and J.-H. Huang (eds.), The Origin and Evolution of Neutron Stars, 202.*

# THE X-RAY TRANSIENT EXO 2030+375

A.N. Parmar, N.E. White, L. Stella[1] and P. Ferri

EXOSAT Observatory, Affiliated to the Space Science Department of ESA.

[1] On leave from ICRA, Universita di Roma

EXOSAT has observed a bright transient X-ray pulsar EXO 2030+375 that decayed in intensity by a factor ~5000 between 1985 May and August. The variations in 42s pulse period enable an orbital period of 37.9±1.3 days with an eccentricity of 0.31±0.02 to be determined. The spin-up timescale of ~30 years suggests that this is a very large outburst of order a few $10^{38}$ ergs/s. The mass function of $5M_{o}$ is consistent with the unidentified companion being a Be star, similar to many other pulsing X-ray binaries. The position of EXO 2030+375, obtained with the EXOSAT Imaging Telescope, is RA: 20 30 21.25, Decl. +37 27 51 (1950; with an uncertainty radius of 10").

The evolution of the pulse profile and spectrum were studied in detail as the intensity decayed by at least a factor ~100. At the maximum observed luminosity of 4 x $10^{38}$ ergs/s (for an assumed distance of 10 kpc) the 2-10 keV pulse profile showed a broad smooth maximum with a narrow sharp notch. As the source decayed a second feature located ~0.5 $\phi$ earlier than the original maximum became gradually more prominent until by the time the luminosity had fallen to 5 x $10^{37}$ ergs/s the two maxima were of approximately equal intensities. The next observation (2 x $10^{37}$ ergs/s) showed both the original maximum and the sharp notch to have virtually disappeared. These changes might be expected if the radiation beaming mechanism changes from being a fan beam at high accretion rates to a pencil beam at lower rates.

The 1-40 keV phase averaged spectrum of EXO 2030+375 is similar to those of other X-ray pulsars being well represented by a power law continuum with a high energy cut-off and significant low energy absorption. In addition an iron emission feature at ~6.5 keV is required. As the outburst decayed the spectrum became harder.

A second outburst from EXO 2030+375 was seen by EXOSAT some 50 days later. The shape of the X-ray lightcurve had changed dramatically being dominated by a series of flares that occurred every ~5 hours. The flares do not appear to be strictly periodic and were not associated with significant increases in low energy absorption. During the flares the pulse profile changed in a way consistent with the profile changes seen during the earlier decay.



*D. J. Helfand and J.-H. Huang (eds.), The Origin and Evolution of Neutron Stars, 203.*

ANALYSIS OF X-RAY SPECTRUM FOR GLOBULAR CLUSTER M15 X-RAY SOURCE

F.Z.Cheng
Center for Astrophysics, University of Science and Technology of China, Hefei, Anhui, China
J.E.Grindlay
Harvard-Smithsonian Center for Astrophysics, USA

Investigation of all X-ray spectrum obtained with MPC and HRI in HEAO-B for globular cluster M15 X-ray source shows us several interesting results:

(1) If we only consider one radiation component and fit HRI data to MPC spectrum, then the results indicate that bremstrahlung model is a little bit better than exponential one. Maybe, two radaition components should be taken into account. One is from neutron star and the other from the accretion disk around the neutron star.

(2) The column density $N_H$ is inversely proportional to X-ray count rate ( see the figure below ), suggesting that photoionization mechanism plays an important role, and that low energy X-ray absorption in the system does account for some of the reduction in source flux.

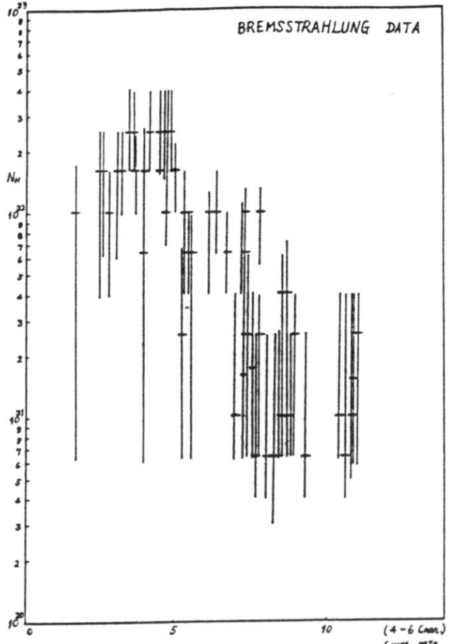


(3) Lowest column density, $N_H \sim$ 6.3E20, from X-ray data is consistent with the results from optical observation.

(4) If M15 X-ray source is a binary system, then the period of X-ray binary, that we can estimate, is at least 1 day.

(5) It seems that there are a few marginal soft X-ray flares during 1979 HEAO-B observation. But more observations to confirm this property are needed.

Fig 1. The column density to the globular cluster M15 X-ray source is obviously in inverse proportion to X-ray count rate.



*D. J. Helfand and J.-H. Huang (eds.), The Origin and Evolution of Neutron Stars, 204.*

# IIb. ACCRETION POWERED PULSARS

## Theoretical Considerations

*CHAIR: K. L. Huang*

# ACCRETION ONTO MAGNETIZED NEUTRON STARS: POLAR CAP FLOW AND CENTRIFUGALLY DRIVEN WINDS

Jonathan Arons
Astronomy Department and Physics Department, University of California at Berkeley, and Institute of Geophysics and Planetary Physics, University of California, Lawrence Livermore Laboratory

ABSTRACT. Some basic concepts of accretion onto the polar caps of magnetized neutron stars are reviewed. Preliminary results of new, multidimensional, time-dependent calculations of polar cap flow are outlined, and are used to suggest the possible observability of fluctuations in the X-ray intensity of accretion powered pulsars on time scales of 10 – 100 msec. The possible relevance of such fluctuations to Quasi-Periodic oscillations is suggested. Basic concepts of the interaction between a disk and the magnetosphere of a neutron star are also discussed. Some recent work on the disk-magnetosphere interaction is outlined, leading to the suggestion that a neutron star can lose angular momentum by driving some or all of the mass in the disk off as a centrifugally driven wind. The relevance of such mass loss to the orbital evolution of the binary is pointed out.

## 1. INTRODUCTION

I was asked to review accretion onto neutron stars. I don't know how to do this without writing a monograph. Therefore, I have chosen to discuss some basic concepts and recent work of immediate interest to me. These are the structure of accretion flow onto the polar caps of magnetized neutron stars, widely thought to be of relevance to the observable emission of X-rays from accreting neutron stars, and the interaction between a thin disk and the rotating magnetosphere of a neutron star, a subject of relevance to the rotational evolution of these objects.

## 2. POLAR CAP ACCRETION ONTO MAGNETIZED NEUTRON STARS

### *2.1. The Eddington limit and Optically Thick Accretion*

I'll begin by discussing a few basic concepts. Most of the discussion in this section is review of well-known ideas, going back to Basko and Sunyaev (1976). Spherical, optically thin flow of a fully ionized plasma with cosmic composition onto a gravitating body cannot occur when the luminosity exceeds the Eddington luminosity $L_{ED} = 4\pi GM_{*}mc/\sigma_T = 1.3 \times 10^{38}$ erg/s, since the radiation pressure then exceeds gravity (eg., Shapiro and Teukolsky



*D. J. Helfand and J.-H. Huang (eds.), The Origin and Evolution of Neutron Stars, 207–225.*

1983). Here $M_*$ is the mass of the star, taken to be 1 $M_\odot$, m is the mean molecular weight, taken to be 61% of a proton's mass, and $\sigma_T$ is the Thomson cross section.

Accretion onto a magnetic pole modifies the magnitude and meaning of this limit. If one imagines all the radiation to be coming out of the stellar surface through a polar cap area $\pi R_*^2 \theta_c^2 \sim 3\ (\theta_c/0.1)^2\ km^2$ and assumes the radiation flux is directed anti-parallel to the plasma flowing in along the magnetic field B, then the Eddington luminosity for that pole is $L_{ED}^{(eff)} = (\theta_c^2 H_{\parallel}/4) L_{ED} \sim 3 \times 10^{35}\ (\theta/0.1)^2 H$ erg/s. $H_{\parallel} \geq 1$ is the ratio of the Thomson cross section to the magnetically modified cross section for coherent scattering of photons travelling along B; for photon energies exceeding $11.6 B_{12}$ keV, $H_{\parallel} = 1$, while $H_{\parallel} \approx 0.06(B_{12}/T_{10})^2$ for $B_{12}/T_{10} \gg 4.5$; $H_{\parallel}$ departs substantially from unity only for $B_{12}/T_{10} > 3$ (Arons, Klein and Lea 1986, hereafter called AKL). Here, $B_{12} = B/10^{12}$ Gauss, $T_{10} = T/10$ keV with T the photon temperature. $L_{ED}^{(eff)}$ is determined mainly by the magnetic requirement that all the radiation come from a smaller area; in very strong fields, the reduction of the opacity pushes $L^{(eff)}$ up again, but in general, reduction of the area is the dominant effect.

If this effective Eddington limit meant the same thing as in spherical free fall, one would wonder how the accretion of plasma onto magnetic polar caps could be the only going model for the X-ray emission from the accretion powered X-ray pulsars, and also for the magnetized stars suspected of being in low mass X-ray binaries, since most of these objects have X-ray luminosities much in excess of $L_{ED}^{(eff)}$. However, one can readily show that the scattering optical depth from the surface to infinity, measured along an almost radial dipole magnetic field through freely falling plasma is $1.5[c/v_{ff}(R_*)][L_{cap}/L_{ED}^{(eff)}] \gg 1$ when the polar cap accretion luminosity exceeds $L_{ED}^{(eff)}$. Thus, locally super-Eddington flow is no longer optically thin, and the correct interpretation of $L_{ED}^{(eff)}$ is as a fiducial luminosity, above which scattering in an optically thick medium causes the radiation to be redirected out of the accretion column into a fan beam emission pattern. When accretion occurs from a thin disk, one suspects the plasma falls in the form of a hollow cone flowing along B. Radiation emitted from the outside of this optically thick cone forms a fan beam, while radiation emitted into the inner parts of the cone flows out as a broad or narrow pencil beam, with the opening angle determined by the distribution of scattering optical depth as a function of magnetic azimuth around the flow cone. The proportion of fan beam to pencil beam emission also depends on this optical depth distribution. For simplicity, I will discuss only the simplest case of optically thick accreting plasma falling in all over the polar cap, as may be a respectable model when plasma enters the magnetosphere with low angular momentum (Arons and Lea 1980; Burnard et al 1983; see also Arons et al 1984, Lamb 1984).

### *2.2. Physical Conditions in Polar Columns*

Over the years, quite of lot of data have been gathered by the satellite experiments equipped with proportional counters. A typical example of the continuum spectral fluxes of these objects is contained in the survey of the accretion powered pulsars published by White et al (1983). The parameters typical of polar cap flow, inferred by assuming densities on the order of free fall, "temperatures" comparable to the quasi-exponential roll-overs seen in the spectra, and photospheric areas and magnetic field strengths characteristic of the magnetospheric models, yield an impression of a very lively part of the universe. Typically, for photospheric emission area $\sim$ 1 km$^2$, one finds "temperatures" $\sim$ 10 keV, plasma densities $\sim 10^{21}$ cm$^{-3}$ (higher, in regions where the flow is decelerated below free fall), free fall dynamic pressures $\sim 10^{17}$ dyne/cm$^2$, LTE radiation pressure $\sim 10^{17.5-18}$ dyne/cm$^2$, plasma pressure $\sim 10^{14}$ dyne/cm$^2$, and magnetic pressure $\sim 10^{22.5} B_{12}^2$ dyne/cm$^2$. Clearly, radiation pressure is enough to decelerate the flow to speeds below free–fall, while plasma pressure is unimportant, at least at the observable photosphere. If pressures everywhere in the column are comparable to these observationally constrained values, then perfect magnetic confinement in the directions across the magnetic field looks like a good idea[1].

I occasionally remind myself of this environment's lively character by looking up the X–ray exposure limits issued by the United States Occupational Health and Safety Administration, which tell me that even a few minutes exposure to the X–rays from these stars at a distance of one parsec would be enough to make me very sick indeed. Put differently, the X–ray flux at the source typical of these objects corresponds to roughly 300,000 nuclear wars[2]/m$^2$–sec. Examples like these remind one of the hostility of some parts of the universe to our rather fragile form of life.

### 2.3. Spectroscopic Problems

When one tries to advance beyond these order of magnitude views of polar columns, an immediate problem is encountered. The observed spectra of the brighter sources are not well represented by the simplest models. No model employing one value of density, temperature, with a single radiation process converting plasma energy into photons, has succeeded in giving a good account (meaning physically self–consistent) of the emergent spectra, with their flat power law form below a quasi–exponential roll–off. Furthermore, the columns clearly form scattering atmospheres, whose photospheric surfaces have radiation number density depleted well below the Planck level appropriate to the plasma temperature – the emergent fluxes might be modified black bodies, if the lines of sight through the column ever become optically thick to absorption at all. Some weak observational evidence that such modifications of the emission, due to the dominance of scattering in the atmosphere, really do play a role, can be gleaned by comparing the exponential roll–overs expected from complete local thermodynamic equilibrium models to the observed rollovers. Because freely falling, optically thick plasma traps the photons emitted from the higher parts of the decelerated mound of plasma contained within the free fall zone, the emission area of the fan beam turns

---

1. Of course, anyone who has followed the efforts to achieve controlled fusion will realize how dangerous and unrealistic such a conclusion can be, based only on a simple comparison of magnetic to total gas pressure. In fact, polar cap flows may admit an interesting class of interchange flows which may allow plasma to "diffuse" anomalously fast across B (Arons et al 1985). For my discussion here, I follow the conventional astronomical discussions and **assume** perfect magnetic confinement.
2. Being a bit of a pessimist, I have assumed one war is a 10,000 megaton "exchange" (1 megaton $\approx 10^{22}$ ergs).

out to be comparable to the area of the polar cap, even when the decelerated column has a height large compared to the polar cap's diameter (Arons and Klein 1987a,b); the emission is never that of a tall skinny cylinder, in high luminosity flows. Then the effective temperature of the photons is expected to be $\sim 4\ (L_{cap}/L_{ED})^{1/4}(\theta_c/0.1)^{1/2}$ keV, corresponding to a quasi–exponential roll–over in the spectrum at energies $\sim 11\ (L_{cap}/L_{ED})^{1/4}(\theta_c/0.1)^{1/2}$ keV. Observed spectra have exponential rollovers usually factors of 2 – 4 higher than this, consistent with the plasma having a radiating surface characteristic of a modified black body [although the continuum absorption opacity isn't bremsstrahlung, but is more likely to be some combination of Comptonized cyclotron emission (AKL) and a resonant form of double Compton scattering (Kirk and Melrose 1986)].

Because the radiation pressure is strong enough to decelerate the flow, and because the flow is necessarily inhomogeneous across as well as along B, one expects a wide range of physical conditions to contribute to the emergent spectra. In particular, in attempting to model the already measured phase resolved spectra of the accretion powered pulsars, one needs to have some means of dynamically determining the range of photon temperature and chemical potential (which measures the depletion of the radiation field below the Planck level) in order to have a hope of using the observations to probe the flow and magnetic field structure. In addition, as we will see, the flows almost certainly impose short time scale ($\sim$ 10–1000 msec) fluctuations on the emergent intensity, even if the accretion flow is free of "raindrops" (Arons and Lea 1976, 1980, Michel 1977, Elsner and Lamb 1977, 1984), "blobs" (Lamb et al 1985) and other semi–quantitative constructs of theorists who have studied the entry of plasma into the magnetosphere.

### *2.4. Theoretical Models*

The inhomogeneities in space and time built into the basic model for polar cap accretion can't be inferred by direct inversion of the data, so far as I know, so a more useful method appears to be the construction of **dynamical** models which allow one to determine the flow structure, within the context of an assumed geometry and profile of mass flux of the flow along **B**. In a completely *a priori* theory, these would be known from the theory of the magnetosphere, and **that** would be known, once one specified the evolutionary state of the binary. Needless to say, one doesn't try to be this general, and instead imposes simplified models of the mass flux profile of freely falling plasma. Various authors have made analytical and numerical models of the polar flow when radiation pressure supports the plasma in the direction along **B** (Davidson 1973, Basko and Sunyaev 1976, Wang and Frank 1981, Braun and Yahel 1984, Kirk 1985). Basko and Sunyaev and Braun and Yahel assumed the scale of transverse gradients is always equal to the width of the flow at each altitude. This allowed them to create an approximate one–dimensional theory by replacing all the differential operators in the directions across **B** by the inverse of their assumed length scale. At high luminosity and high optical depth, this one dimensionalization is misleading, since the actual transverse length scale decreases, in order to allow radiation diffusion across **B** to carry the dynamically determined radiation flux. Davidson, Wang and Frank and Kirk all carried out steady state, two dimensional axisymmetric calculations, in each case dropping all the details of the radiation emission processes and thus leaving out the information needed to determine the photon number density, required if one is to determine the modifications to local black body emission imposed by the scattering character of the atmosphere. In the rest of my discussion, I will concentrate on the most ambitious attempt made so far to model the dynamics of polar flows, being carried out by Richard Klein and myself. For more details, see the poster papers at this meeting (Klein and Arons 1987a, Arons and Klein 1987a).

With suitable restrictions it turns out to be possible to make steady flow, multi–dimensional analytic models (Kirk 1985; Arons and Klein 1987a,b). I will outline the numerical models that we have done so far, since these have the greatest realism. I emphasize that this work is preliminary; we are still in the throes of putting together the last pieces of the physics and numerical methods, but some interesting results are already clear. The program is fully time dependent, and assumes axisymmetry of the flow with respect to the magnetic axis. We include magnetically modified brems–strahlung emission and Coulomb collisional excitation/deexcitation of quantized cyclotron gyration as the photon emission and absorption processes. The cyclotron line emission is treated using a non–LTE, two Landau level "atom" model with complete redistribution for resonant scattering. Because the optical depths are large in the line **and** in the neighboring continuum in the high optical depth models we consider, the problem of cyclotron line transfer is solved "on the spot", with the result that cyclotron line radiation is primarily a source for the continuum[3]. The continuum photons undergo fully saturated Comptonization, using a rate for energy exchange with elec–trons that takes account of the magnetic modifications of scattering. The consequence is that in most of the flow, the radiation spectrum can be safely approximated as having a Bose–Einstein distribution with photon tempera–ture and chemical potential varying in space and time. The emergent spec–trum at each point is characterized by these two numbers.

In order to find them, we include a new theory of the coupled diffu–sion and advection of photon energy density **and** number density, using ap–propriate, flux limited "Rosseland" averages of the angle and frequency de–pendent anisotropic, coherent scattering opacity in the continuum. The elec–trons and ions are allowed to have separate temperatures, described by the energy equations for the plasma. The plasma temperatures are coupled by Coulomb energy exchange. Mass and momentum conservation are described by the continuity and one–fluid momentum equations for flow strictly along **B**. In the momentum conservation equation, we include gravitational acceleration along **B**, radiation force, thermal pressure gradients and the buoyancy force. The assumed geometry is flow along a dipole magnetic field, and in all of our calculations done to date, the mass flux has been assumed to be independent of distance from the magnetic axis, out to the boundary of the flow where the density is assumed to drop abruptly to zero. At the top of the calculational grid (for now, taken to be 20 km above the magnetic pole of a star whose radius is 10 km), we assume the plasma is in free fall.

3. The high optical depth and the associated saturated Comptonization, ap–propriate for sources with high luminosity, is the essential difference between our work and the pure transfer models of Meszaros and Nagel (1986), for example.

We emphasize that this mass profile is chosen to keep the problem as simple as possible in the first calculations; more complex flows, such as hollow cones, will be done in the near future. We also assume no heat flows into the star, as is plausible since the photon opacity of the crust immediately below the piled up mound of plasma is much greater than in the accreted plasma itself, and we assume photons escaping from the sides of the column go off into black sky.

I will show a few of the results from two calculations. In both, the flow was initialized as cold plasma in free fall all the way down to the stellar surface. The first example was run with an accretion luminosity onto the cap of 5 x $10^{37}$ erg/s. The magnetic colatitude of the footprint of the field lines at the edge of the polar cap is $\theta_c$ = 0.4 radians; for self–congratulatory reasons, we used the model of Arons and Lea (1980) for the polar cap size in these preliminary calculations. There is no implication intended that this is the correct way to scale the geometry for all accreting neutron stars, although there is some weak evidence that it is relevant to lower luminosity ($L < 10^{37}$ erg/s) accretion (White *et al* 1983). The neutron star radius is 10 km, and the surface magnetic field strength was chosen to be 3 x $10^{12}$ Gauss (magnetic moment $\mu$ = 1.5 x $10^{30}$ cgs), yielding an energy for the cyclotron line at the surface of 35 keV.

After the passage of a shock through the flow in order to adjust to quasi–stationary conditions, a mound, in approximate hydrostatic equilibrium developed with the subsonic plasma supported by radiation pressure; radiation pressure also provides the full deceleration from free fall. A fan profile of emergent radiation is formed after a couple of photon diffusion times. An example is shown in Figure 1, a snapshot taken at 130 $\mu$sec into the problem (this is about 50 times the dynamical time). To get this far takes about 2.3 hours on a Cray XMP–48 – this type of calculation is not for the pencil and paper or VAX crowd! The vectors represent the photon flux leaving the point at the foot of each vector, whose direction shows where the flux is going and whose length indicates the flux magnitude. The velocity field at this time is rather dull, showing the same exponential stratification along **B** and Gaussian variation across **B** as is found in the analytic solutions. The rest of the structure – density and pressure, as well as velocity – is also well represented by the steady flow analytic models, if one uses the numerical data to properly choose the magnetic modifications of the opacity.

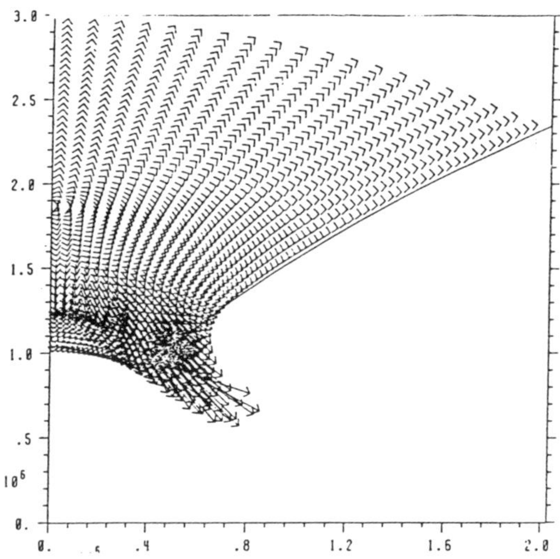


Figure 1: Radiative flux of $10^{37.7}$ erg/s of accretion onto the polar cap. The spatial scales are in units of 10 km.

A bit later, things become more interesting. Along the magnetic axis, the plasma develops a density depression, giving rise to a net buoyancy force.

We believe, although we have not yet conclusively proved, that this is the onset of a form of overstable convection (*e.g.*, Proctor and Weiss 1982). The initial non–linear consequences for the velocity field are shown in Figure 2, a snapshot taken at 600 μsec into the problem. Along the central axis, a lower density, subsonically **rising** "bubble" forms with density lower than the surround–ings; the emergent flux in the fan beam has yet to show a substantial change. The appearance of such structures is not especially surprising, since the polar cap flow is strongly super–Eddington with respect to the **effective** Eddington limit. Many authors have used linear stability arguments (*e.g.*, Hameury *et al.* 1984) to suggest that polar cap flows would form convective motions, and long ago, Hsieh and Spiegel (1976) speculated that the such flows would form "photon bubbles".

The final fate of this struc– ture is still unknown; the lower densities have driven the Courant time down to the point where fur– ther evolution using our explicit treatment of the momentum equa– tion requires prohibitive amounts of computer time (the transfer and energy equations are already treat– ed implicitly). Technical develop– ments are in progress which should allow us to get around this prob– lem. If these succeed, we will be able to evolve models to times as long as 1 second. My sus– picion is that the bubble will break up into finer scaled structures because of the weight of the accreted plasma piling up on top – the magnetic field precludes the falling away of plasma towards the bubble's sides that one expects in ordin– ary convection, or was postulated in Hsieh and Spiegel's photon bubbles. Because radiation diffuses faster through the lower density zones, one expects this "convection" to lead to faster, percolative loss of radiative energy across the columns, which **may** lead to observable, short time scale (10 – 100 μsec) fluctuations in the emergent flux. A prediction of the magnitude of such fluctuations, so far not studied observationally in the rotation powered pul– sars, requires running calculations to times long enough to determine the number and scale of the bubbles contributing to the percolation, a task to be done as soon as the ability to follow the flow to long times is in place.

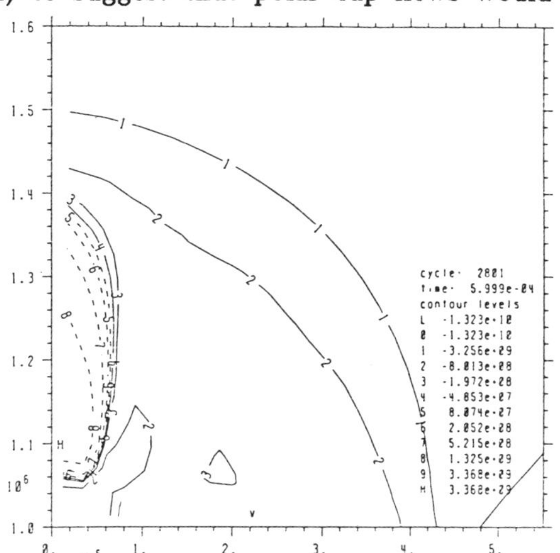


Figure 2: Velocity field of the same model as in Figure 1, but at the later time 600 μsec. The vertical scale is in units of 10 km; the horizontal scale has units of 1 km. Solid curves are contours of down– ward (negative) velocity, in cm/s; dashed curves show of upflow.

The character of these "convective motions clearly changes with vary– ing luminosity. Figure 3 shows the velocity field of a model constructed in the same manner as that of Figure 2, but now for an accretion luminosity onto one cap of 1.5 x $10^{38}$ erg/s. The snapshot is taken at 0.75 msec into the flow. One sees now three "bubbles" – because of the axisymmetry, these bubbles form toruses in the plasma. As time goes on to 0.9 msec, when we quit trying to follow the flow because of the declining time step, the central bubble had grown to fill in on the magnetic axis, and the satellite bubbles had greatly increased their size.

I don't have room to go into the spectroscopy of these models, except to remark that the output spectra are the superposition of the Wien spectra determined by the variable temperature and chemical potential over the surface of the decelerated mound. The beaming pattern is mainly controlled by the trapping of the photons in the optically thick, freely falling plasma above and outside the mound, until at the bottom of the flow, the layer of freely falling plasma becomes optically thin to scattering and photons escape in a narrow ring around the base of the mound, as is shown in Figure 1. One can show that the height of this ring is roughly $\theta_c R_* [v_{ff}(R_*)/c]$, so that the area from which the radiation is beamed is comparable to the area of the polar cap on the stellar surface. The **quality** of the radiation field is determined by the emission from the much larger mound surface, however. So far, the results we have found for the spectrum form a surprisingly good representation of the spectra of some of the rotation powered pulsars – my surprise is a consequence of the simplicity of these calculations' geometric assumptions. One other aspect is that at photon energies well below the cyclotron energy at the surface (35 keV in these models), the emitted radiation is very highly linearly polarized, because the opacity in the extraordinary mode is much smaller than in the ordinary mode, and therefore has an intensity reflecting the Wien intensity of deeper plasma layers. If someone, someday, would finally get the chance to fly an X–ray polarimeter attached to a telescope of decent collecting area, we would be able to observationally probe the geometric structure of the flow, to which the spectroscopy seems to be insensitive.

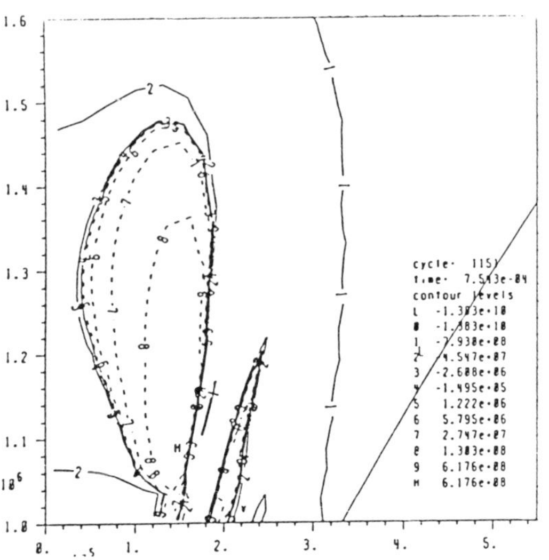


Figure 3: Velocity field of a model with assumptions the same as in Figure 2, except the luminosity is now $10^{38.2}$ erg/s and the polar cap opening angle is 0.33 radians. This snapshot is at 0.75 msec. The meaning and units of the contours are as in Figure 2.

I can't resist leaving this topic without a final speculation, perhaps especially appropriate to this meeting. While we have concentrated on the strongly magnetized, accretion powered pulsars in the galactic disk, the same physics applies, with lower magnetic field strength, to the now popular, more weakly magnetized models for galactic bulge/QPO sources which are likely to be the precursors to the millisecond pulsars. It has not escaped our notice that the likely fluctuation times appearing in the polar flow models are comparable to the fluctuation periods seen in the quasi–periodic oscillators, leading to intriguing thoughts about a model in which the observed fluctuations may not by the direct consequence of orbital phenomena at the magnetopause.

## 3. CENTRIFUGALLY DRIVEN WINDS FROM NEUTRON STARS

### *3.1 Standard Ideas*

The standard ideas on disk–magnetosphere interaction were invented to explain the average behavior of the rotation period P and dP/dt, as observed during the "American" era of X–Ray astronomy in the 1970's. The basic thoughts were first outlined by Pringle and Rees (1972) in advance of substantial observations of dP/dt, and were applied to the system shown in Figure 4. Here $\mu$ is the magnetic moment, $\Omega_*$ is the angular velocity, $R_m$ is the magnetopause radius

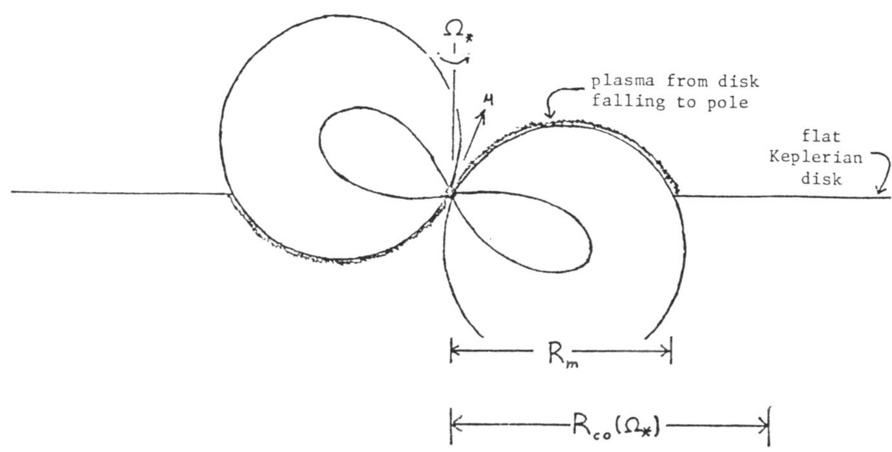


Fig. 4: Cartoon model of a magnetosphere confined by a Keplerian disk, when accretion to the stellar surface is in progress.

in the disk plane, and $R_{c*}$ is the corotation radius, where the Keplerian angular speed $(GM_*/R_{c*}^3)^{1/2}$ equals the stellar angular velocity. This is thought to be relevant to magnetospheric problems since almost all authors have assumed the field lines to have at least one foot on the stellar surface, and to be perfect conductors by virtue of the ideal plasma presumed to completely fill the magnetosphere. If these assumptions are met, the magnetospheric electric field is $\mathbf{E}_{mag} = -(\mathbf{\Omega}_* \times \mathbf{r}) \times \mathbf{B}/c$ and the velocity of plasma across B is $\mathbf{v} = c(\mathbf{E}_{mag} \times \mathbf{B})/B^2 = \mathbf{\Omega}_* \times \mathbf{r}$. Therefore, if $R_{c*}(\Omega_*) < R_m$, net gravity (= real gravity + centrifugal force) at the magnetopause is outward, and "no" accretion to the stellar surface could occur. Numerically, $R_{c*}(\Omega_*) = 1500\ (P/1s)^{2/3}(M_*/M_\odot)^{1/3}$ km. The magnetopause radius in the disk plane has been estimated to be at $R_m = 3000[\mu_{30}^4(M\ /M\ )/L_{37}^2]^{1/7}$ km (Pringle and Rees 1972, Lamb, Pethick and Pines 1973, Ghosh and Lamb 1979, Arons et al 1986), a value close to that determined when the infall lacks angular momentum (Davidson and Ostriker 1973, Arons and Lea 1976, Elsner and Lamb 1977, Michel 1977). Here $\mu_{30}$ is the magnetic moment in units of $10^{30}$ cgs, and $L_{37}$ is the accretion luminosity in units of $10^{37}$ erg/s. Thus, under all circumstances, short period ($P < 1-2$ s) accreting neutron stars with strong fields ($\mu_{30} > 1$) should have trouble accreting mass and angular momentum, as long as the magnetic field lines are rooted in the star and are good conductors. Indeed this limiting period for spin up is one of the ideas behind the probable fact that the shortest period radio pulsars are the consequence of accretion spin up (eg., Damashek et al 1982, Alpar et al 1982).

The simplest, surprisingly successful ideas (Pringle and Rees 1972, Lamb et al 1973, Rappaport and Joss 1977, Mason 1977) of how a pulsar changes its rotational period come from assuming that each bit of plasma enters the magnetosphere from the inner edge of a Keplerian disk, with specific angular momentum $\sim (GM_* R_m)^{1/2}$. This gives a spin up torque $T_+ = \dot{M}_x (GM_* R_m)^{1/2} \propto L^{6/7}\mu^{2/7}$, independent of P, and therefore a spin up rate $\dot{P}/P \propto PL^{6/7}\mu^{2/7}$. This described the spin up data from the 70's for most pulsing X-ray sources in X-ray binaries, if the dispersion in $\mu^{2/7}$ is not too large. Because of the weak dependence on $\mu$, this doesn't prove that these objects have $10^{12}$ Gauss surface fields, nor is this explanation of spin up unique – the angular momentum accreted from a wind, with no disk formed can give as good an explanation for many of these sources (eg., Arons and Lea 1980, Stella et al 1986). Nevertheless, the simplest idea, of matter being transferred from a disk to a strongly magnetized neutron star through attachment of the plasma to a corotating magnetosphere, seemed to be qualitatively correct.

Unfortunately, in Her X-1, where we are certain (Boynton et al 1980) that mass transfer occurs through a disk, $\dot{P}/P$ is $\sim 1/30$ of the value expected. Most people do not think it possible to obtain this reduction simply through reducing $\mu$ by $\sim 2 \times 10^5$, mainly because it is hard to see how to form a well defined polar spot from which the radiation arises without a much stronger surface, **large scale** field. The same argument can be applied if the field is not dipolar but is dominated by some other, higher order multipole. The magnetosphere is smaller (Arons and Lea 1980), with a consequently smaller angular momentum/gram accreted, but the effect is still much too small to explain the reduced spin-up torque observed in this system, while still being consistent with the narrow cyclotron line observed – in particular, the variations of the line shape and centroid energy observed require the presence at the line photosphere of a field of strength $\sim 10^{12.6}$ Gauss, with no magnetic broadening within the instrumental resolution (Voges et al 1982). The rather good fit of simple radiative transfer models of cyclotron line formation in a quasi-homogeneous magnetic field to these data (Meszaros and Nagel 1986) also suggest that the surface field has rather large scale structure. Then one cannot suppose the magnetospheric field to be a very intense ($B > 10^{12}$ Gauss) collection of multipoles, with a very weak field at the magnetopause – the result would be too many different field strengths contributing, which creates an inadmissably broad line. For an alternate view of possible weak field models of Her X-1, see Ruderman (1986).

Therefore, one concludes that some sort of "spin-down" torque operates which is almost in equilibrium with the simple spin-up due to accretion. The first candidate proposed for such a mechanism is the closed, force–free magnetosphere, "resistive" disk theory of Ghosh and Lamb (1979). Since these earlier data and theories, observations carried out primarily on Hakucho and on Tenma have revealed substantially more complex spin–up and spin–down behavior, much of which might be interpretable in terms of spin–down torques competing with the accretion of angular momentum. Our work, as well as that of Anzer and Borner (1980, 1983) and of Ghosh and Lamb (1979), concerns the nature of the interactions between disk and magnetosphere which lead to these torques, as well as models of the torque and the resulting spin behavior of the star, under various assumptions as to the nature of the basic magnetic configuration and approximations to the relevant physics.

### *3.2 Mass Entry: A Thought Experiment*

I now describe some new work I have been doing, in collaboration with C.F. McKee and R. Pudritz. We construct our theory by making thought experiments to answer the question of how a magnetosphere responds to an impinging disk of conducting plasma. We assume the disk has no large scale magnetic field of its own; at present, there is no direct evidence for such fields. In our thought experiments, we first assume ideal MHD describes the interaction, then imagine the effects of dissipation being turned on by the instabilities in the ideal system.

Aly (1980) has found exact analytic solutions to this model problem. The disk is modeled as a perfectly conducting plate with a hole in the center of radius R . The structure of the magnetosphere is outlined in Figure 5, drawn for a rapid rotator so that the corotation radius $R_{c*}$ based on the stellar rotation rate is close to the magnetopause radius – in the jargon of Ghosh and Lamb (1979), the "fastness" $\omega_* = \Omega_*/\Omega_K(R_m) = (R_m/R_{c*})^{3/2}$ is close to unity. A notable feature of Aly's model is the presence of a neutral point at $R_n$. Field lines which would have closed exterior to $R_n$ are now open, because of the surface currents on the (infinite) disk.

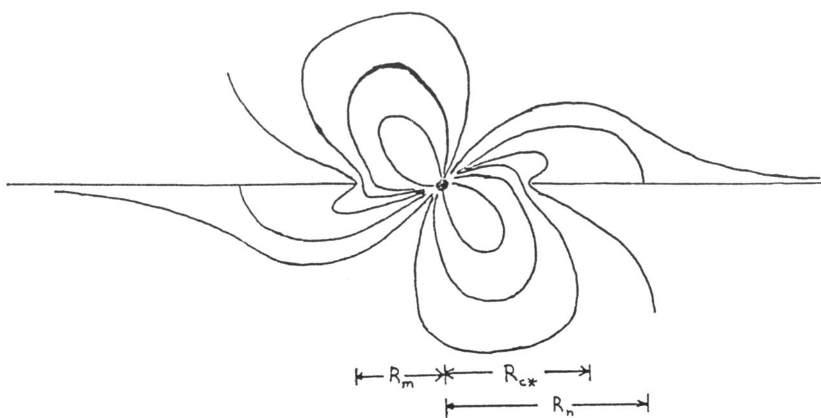


Fig. 5: Field lines of Aly's model. $R_m$ is the magnetopause radius where the disk ends, $R_n$ is the radius of the neutral points, and $R_{c*}$ is the Keplerian corotation radius based on the stellar angular velocity.

For the rest of my discussion, I assume $R_n > R_m$, as well as assuming $R_m$ to be less than all the other radii which turn out to have significant in forming the wind. Roughly speaking, this implies the angle of obliquity i between the magnetic moment and the rotation axis to be less than $\sim 60^o$. Then the topology of the aligned rotator, where $R_n$ is at infinity, is the same as that of the dynamically interesting region near $R_m$, and we can study the aligned rotator with some confidence that it's physics gives a zeroth order approximation to the oblique problem. When $R_n$ of Aly's model is interior to the radii of interest, the topology is like that assumed in the $i = 90^o$ model studied by Anzer and Borner (1980), whose structure is intrinsically 3D. For simplicity, we have studied the aligned case, and have uncovered new physics which we think makes the model relevant to the spin behavior of objects with arbitrary values of the Ghosh–Lamb fastness $\omega_*$. Since the aligned model is relevant to about half the solid angle available to the magnetic moment, we think the aligned case is relevant to the general behavior of accreting magnetized neutron stars.

Interior to $R_m$, the field is close to the undisturbed dipole, with $B \propto r^{-3}$, while for $r > (2-3)R_m$, the field strength is proportional to $r^{-(2+\nu)}$, with $\nu \simeq 2$. Because Aly models the disk as having no thickness, his solutions have infinite magnetic field at the inner edge. This is artificial, of course, and for a disk of finite half thickness H, his solutions suggest the true value at the inner edge should be $\sim \mu(R_m/H)^{1/2}R_m^{-3}$. We have not attempted to calculate an improved magnetic field model including the finite disk thickness, since we expect that Aly's model provides a reasonable zeroth order estimate of the poloidal field strength, even with dissipation included.

In ideal MHD, the magnetopause separating the disk and the magnetic field lines folded around it is a tangential discontinuity, with no plasma transport across the field lines. However, in general, the magnetospheric field and the disk's plasma move at different speeds, with toroidal relative velocity $v_{rel} = [\Omega_B - \Omega_K(r)]r$. If the field lines corotate with the star, $\Omega_B = \Omega_*$, but we will find that $\Omega_B$ can be significantly different than $\Omega_*$. Such relative motion is unstable to the formation of breaking waves (the Kelvin–Helmholtz instability). The mixing induced by such wave breaking is the dominant effect in causing mass to cross the magnetopause boundary, if the disk has no large–scale magnetic field of its own; reconnection between small magnetic loops in the disk and the magnetospheric field acts mainly to trigger the Kelvin–Helmholtz instability by providing a minimum mass density on the magnetospheric side of the interface, if no other mass source is present (Arons et al 1986).

For waves with wavelength along the relative velocity long compared to the thickness of the boundary layer, the linear growth rate is $\gamma \simeq k[\Omega_B - \Omega_K(r)]r(\rho_w/\rho_a)^{1/2}$, where k is the wavenumber, $\rho_a$ is the density in the upper atmosphere of the disk where the boundary layer is formed, and $\rho_w$ is the density in the wind outflow formed within the magnetosphere. From approximate pressure balance, $\rho_a = B^2/8\pi c_s^2(\text{disk}) \sim 10^{-5}$ gm cm$^{-3}$. Because of X–ray heating, $c_s$ (disk), the sound speed in the disk atmosphere, is $\sim 10^3$ km s$^{-1}$. We remark that a mass density above $10^{-7}$ gm cm$^{-3}$ of plasma corotating in the magnetosphere at $R_{c*}$ exerts sufficient stress to burst the magnetosphere, converting closed field to open. Thus, even if only 1% of the disk plasma is scraped off and converted into magnetospheric plasma at radii exceeding $R_{c*}$, the magnetospheric topology is completely al–

tered. The principle mechanism that does this scrape–off is the breaking of the "cat's eyes" formed by the Kelvin–Helmholtz instability. Hydrodynamic experiments and numerical calculations, and MHD simulations (eg., Miura 1984, 1985) show that after $\eta^{-1} \sim 10$ linear growth times, the waves with wavelength $2\pi/k$ break; the characteristic (vertical) velocity at breaking is therefore the wavelength divided by the time to breaking, or $\eta\gamma/k$. Once the waves break, successive overturns rapidly diffuse the momentum and mass (Arons and Lea 1980), entraining the disk plasma with the magnetosphere. The effective diffusion coefficient for this process is $\sim \eta\gamma/k^2 \sim 0.1\gamma/k^2$; the momentum diffusion coefficient found by Miura is $0.098\gamma/k^2$, once proper ac–count is taken of the distributed shear profile used in his calculations. The instability also generates MHD waves (primarily fast modes); because the Alfven speed on the magnetospheric side of the boundary layer is high com–pared to the relative speed in the final, self–consistent models, these waves carry off the disk momentum deposited in the magnetosphere with only small amplitude pertubations of the magnetospheric structure – here, we differ from the assertion made by Anzer and Borner, that the magnetic field in and near the boundary layer must be fully turbulent.

The net entry velocity of disk plasma into the magnetosphere is then $\sim \eta\gamma/k = \eta[\Omega_B - \Omega_K(r)]r(\rho_w/\rho_a)^{1/2}$, with the plasma distributed over a layer whose vertical thickness is on the order of the longest wavelengths which can grow and break. For practical purposes, this turns out to be roughly the scale height of the disk atmosphere, since the large inertia of deeper layers in the disk prevents much growth at smaller values of k. The plasma injected onto the field has sufficient rotational inertia to break open the field and be flung away as a centrifugally accelerated wind; at this stage in the thought experiment, we assume this to be so, and find the assumption to be self–consistent at the end. The resulting configuration is shown in the cartoon in Figure 6. Here, the disk lies in the midplane, with the open field lines wrapped around the disk and now disconnected from the star, held in only by the accretion of plasma from the disk's inner edge. Therefore, the angular velocity of the open field lines and of the wind is $\Omega_B$, with $\Omega_* < \Omega_B < \Omega_K(R_m)$. If it turns out that $\Omega_B = \Omega_K(R_m)$, all rotators are fast with respect to angular momentum loss in the wind.

### *3.3 Angular Momentum Loss and a Quantitative Model*

The consequence of our thought experiment, therefore, is that for radii larger than $R_c(\Omega_B) < R_{c*}$, where the field lines rotate faster than the underlying disk, the Kelvin–Helmholtz instability feeds plasma from the disk into the overlying magnetosphere at a rate sufficient to blow the field lines open and maintain them in the opened configuration. Indeed, our thought experiment suggests that closed models, such as that of Ghosh and Lamb (1979), are unstable to switching over into the opened state described here. Furthermore, the plasma injected into the poloidal field must be rotationally spun up, which requires the field lines to be swept back into a spiral. Therefore, the wind carries off angular momentum. Since the field lines connect through the boundary layer between the inner edge of the disk and the corotating, closed field region of the Poynting fluxes from this boundary layer region. At the very least, therefore, the wind provides a reduction of the specific angular momentum accreted below the Keplerian value first estimated by Pringle and

Rees (1972). The possibility of producing a true spin–down torque is discussed briefly below.

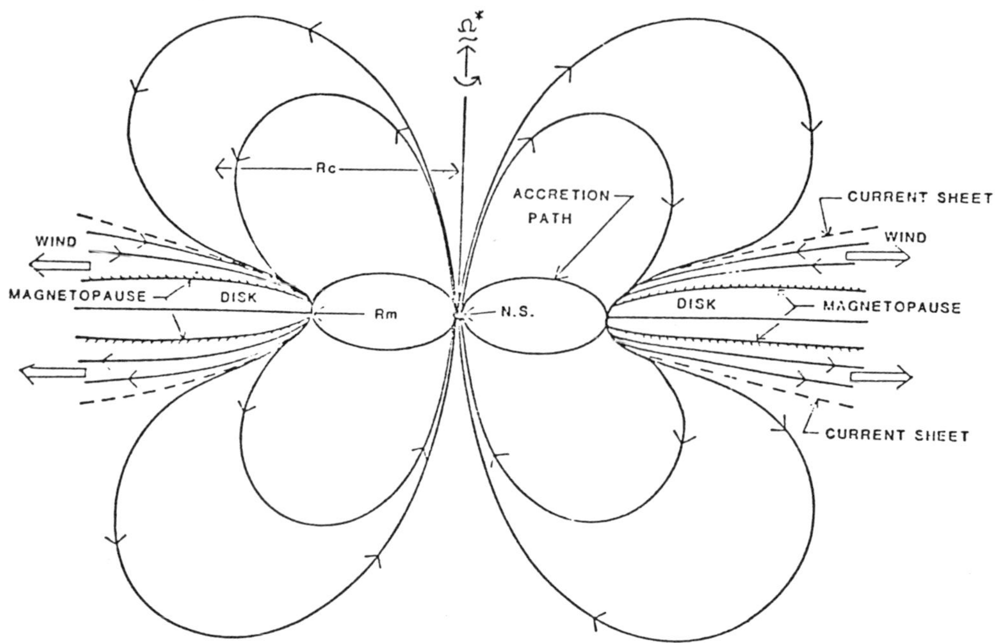


Figure 6: Magnetospheric structure with a wind from a disk.

In our calculations, we assume, and find to be self–consistent, that the region of mass entry is interior to the Alfven radius of the wind $R_A$, where the poloidal component of the wind velocity equals the poloidal Alfven speed. Then the rate at which angular momentum is lost is given by $T_- = \dot{M}_w \Omega_B [R_A^2 - R_c^2(\Omega\ )]$ (Hartmann and MacGregor 1982). The task of a quantitative model is to find $\dot{M}_w$, $\Omega_B$ and $R_A$, all of which depend on $B[R_c(\Omega_B)]$ and on the temperature of the X–ray heated gas, and also to determine if $T_-$ corresponds to a true spin down torque or is limited to a torque reduction alone; this latter task is closely related to finding $\Omega_B$.

The full details of our calculations are reported elsewhere (Arons *et al* 1986). What we did was to use our simple mixing model as a mass source and a momentum sink for the magnetospheric layers bounding the disk, and integrated the equations for flow along the largely poloidal B to find the asymtototic mass loss rate in the wind. This yields

$$\dot{M}_w \approx \begin{cases} 2 \times 10^{17} \left(\frac{\eta}{0.1}\right)^2 \mu_{30}^2 \left(\frac{10^{8}\,\mathrm{K}}{T_a}\right) P_B^{-10/3} \ \mathrm{g\ sec^{-1}} \\ 10^{18} \left(\frac{\eta}{0.1}\right)^2 \left(\mu/10^{27}\right) \left(\frac{10^8\,\mathrm{K}}{T_a}\right) (P_B/10\ \mathrm{msec})^{-10/3} \ \mathrm{g\ sec^{-1}} \end{cases} \quad (1)$$

these values assume the field strength varies as $r^{-3}$. We then solve the MHD wind equations as if all the mass is injected with the rate (1) at the corotation radius $R_{co}(P_B)$. This collapse of the mass entry zone to a single injec–

tion surface works if the Alfven radius is well outside the region of mass entry. The poloidal magnetic field is allowed to vary in proportion to $r^{-(2+\nu)}$. The flow is in approximate corotation with the field lines, at speed $\Omega_B r$, out to the Alfven radius $\approx R_A \approx (\mathrm{flux}^2/\dot{M}_w\Omega_B)^{1/3} \approx 10^{8.2}(0.1/\eta)^{1/3}P_B$ cm ($\nu = 1$). This radius can be guessed by dimensional analysis; all the work of solving the MHD equations yields multiplicative factors nearly equal to unity which depend on $\nu$. The asymptotic wind speed for $\nu = 1$ is $v_\infty = 10^{9.2}\ (0.1/\eta)^{1/2}$ cm s$^{-1}$. The rate of angular momentum loss is the wind is $T_- = \Omega_B \dot{M}_w (R_A^2 - R_c^2)$. Plugging in yields the ratio of the angular momentum loss rate to the rate stellar angular momentum is gained from accretion

$$\frac{T_-}{T_+} = 1.4 \left(\frac{0.1}{\eta}\right)\left(\frac{M_\odot}{M_*}\right)^{23/21} \mu_{30}^{12/7} \left(\frac{10^8 K}{T_a}\right)^{1/2} P_B^{-7/3} L_{37}^{-6/7} . \quad (2)$$

As before, P is the rotation period of the open field lines.

The final step needed is to calculate $\Omega_B$. This requires solving a model of the boundary layer between the disk's inner edge and the co-rotating region of the magnetosphere, which we do by treating the effects of the Kelvin-Helmholtz instability as a mass and momentum diffusion, as well as including the loss of angular momentum from the boundary layer by Poynting fluxes directed along the open field lines which connect to the wind zone. This work is not entirely complete, but present indications are that $\Omega_B$ is closer to $\Omega_K(R_m)$ than it is to $\Omega_*$, so that all rotators with obliquity not too large should be interpreted as "fast" rotators. Furthermore, when the accretion rate is sufficiently small, corresponding to accretion luminosities below $10^{37}$ erg/s, the angular momentum loss into the wind can drive the angular velocity of the boundary layer flow below $\Omega_B R_m$. Since the plasma is brought into co-rotation by viscosity between the co-rotating region and the boundary layer, angular momentum can be extracted from the co-rotating part of the magnetosphere and therefore from the star. When the wind's ability to transport angular momentum is sufficiently large, therefore, the torque $T_-$ **does** correspond to a stellar spindown torque. In its current state of development, the model looks like an interesting candidate to interpret some aspects of the spin behavior of Vela X-1. The quantitative details of these results, and their application to rotational evolution of accreting neutron stars, is described elsewhere (Arons, McKee and Pudritz 1986; Arons and McKee 1986).

### *3.4. Applications and Conclusions*

We have made several preliminary applications of our models to observations, existing or potential, of X-ray pulsars. In the case of Her X-1, the net average spin-up torque is 1/30 of the value expected if no spin-down torques operate. Application of our model to the data yields, assuming $P_B \cong P_*$, a predicted magnetic moment of 1.1 x $10^{30}$ Gauss-cm$^3$, with a dipole component of the surface magnetic field of about 2 x $10^{12}$ Gauss. The result doesn't differ much if the rotation rate of the magnetic field is taken to be the Kepler rate, since in this case $R_{c*}$ is already close to $R_m$.

A more interesting application of the model is to the interpretation of the broad Fe fluoresence line observed in this and other X–ray pulsars. The fluorescing plasma must have temperature well below $10^7$ K, a Thomson optical depth of order unity (eg., Pravdo et al 1979, White et al 1983), and a full width at half–maximum of $\sim 2 \times 10^4$ km/s (Trumper 1986). In fact, our wind model has all these properties, without forcing any of the parameters. The wind plasma is cooled by cyclotron emission to a temperature well below 1 keV. From the mass loss rate (1), the Thomson optical depth is close to unity, and the magnetic moment is inferred by applying the theory to the reduced spin up observed in Her X–1. The predicted velocity full width is then about 15,000 km/s, using the theory without taking proper account of the toroidal fields in the mass entry zone [ie., without worrying about $R_A/R_c(\Omega_B) \sim 1$]. In fact, the theory constructed with proper account of $B_{toroidal}$ gives even better results. Since the basic parameters of our model are already constrained by the dynamical observations (ie., by studying P and dP/dt), the spectroscopic observations provide a test and the possibility of refined modeling of the system.

A third application can be made to the nonconservative evolution of binaries, since the plasma is flung off at $10^4$ km s$^{-1}$ and must leave the system. Then it also carries off the orbital angular momentum characteristic of the neutron star, which leads to evolution of the binary on the time scale $\approx 10^8 \ (0.1/\eta)^2(10^{27}/\mu)^2(P_B/10 \text{ msec})^{10/3}$ years, a time of possible relevance to the evolution of low mass X–ray binaries.

A final application is to the formation of either enhanced mass loss from the outer regions of the disk, or to the formation of "bubbles" in the interstellar medium. Depending on the shape of the disk's outer region, the wind may or may not collide with the disk. If it misses, the wind goes on and blows a bubble in the interstellar medium, of radius $R_{bubble} \sim 5 \ (n_{ISM}/1 \text{ cm}^{-3})^{-1/5}\mu^{2/5}(t/10^5 y)^{3/5}P_B^{-2/3}$ parsecs, rather like the wind from an O star. In this case, a possible test of the theory would be to find bubbles around X–ray sources with B or later optical characteristics, but with bubble properties like those expected for the winds from very early stars (eg., McKee et al 1984). An alternate possibility is that the wind does encounter the disk's outer regions. Then the shock heating might cause the disk plasma to be lifted up above the disk plane, possibly causing the peculiar absorption features seen in some low mass sources whose luminosity is too low to drive the formation of winds from the outer regions by photon heating from the central source.

Our preliminary conclusion is that centrifugally driven winds from the inner edges of disks around magnetized neutron stars are likely to form at a level interesting to the interpretation of dynamical and spectroscopic phenomena observed in these systems.

## ACKNOWLEDGMENTS

The new research outlined here has been supported by grant numbers AST 82–15456 and AST 83–17462 from the US National Science Foundation, by grants numbers 82–04, 85–06 and 86–09 from the Institute of

Geophysics and Planetary Physics, by the Adolph C. and Mary Sprague Miller Foundation, and by the taxpayers of California. We have benefitted from discussions with R. Rosner and S. Kahn.

*References*

Alpar, A., Cheng, A.F., Ruderman, M.A., and Shaham, J. 1982, *Nature*, **300**, 728
Aly, J.–J. 1980, *Astron. and Ap.*, **86**, 192
Anzer, U., and Borner, G. 1980, *Astron. and Ap.*, **83**, 133
_______. 1983, *op. cit.*, **122**, 73
Arons, J., Burnard, D.J., Klein, R.I., McKee, C.F., Pudritz, R.E., and Lea, S.M. 1984, in *Proc. Santa Cruz Workshop 'High Energy Transients in Astrophysics'*, S.E. Woosley, ed. (New York: American Institute of Physics), 215
Arons, J., Hernquist, L., and Klein, R.I. 1985, *Bull. Am. Phys. Soc.*, **30**, 1471
Arons, J., and Klein, R.I. 1987a, in *Proc. IAU Symp. No. 125, 'Origin and Evolution of Neutron Stars'*, D.J. Helfand, ed. (Dordrecht: D.Reidel).
_____. 1987b, *Ap. J.*, submitted
Arons, J., Klein, R.I., and Lea, S.M. 1986, *Ap. J.*, in press
Arons, J., and Lea, S.M. 1976, *Ap.J.*, **207**, 914
_______. 1980, *Ap.J.*, **235**, 1016
Arons, J., and McKee, C.F. 1986, in preparation
Arons, J., McKee, C.F., and Pudritz, R. 1986, in preparation
Basko, M.M., and Sunyaev, R.A. 1976, *MNRAS*, **175**, 395
Boynton, P.E., Crosa, L.M., and Deeter, J.E. 1980, *Ap.J.*, **237**, 169
Braun, A., and Yahel, R.Z. 1984, *Ap. J.*, **278**, 349
Burnard, D.J., Lea, S.M., and Arons, J. 1983, *Ap.J.*, **266**, 175
Damashek, M., Backus, P.R., Taylor, J.H., and Burkhardt, R.K. 1982, *Ap.J.(Lett.)*, **253**, L57
Davidson, K. 1973, *Nature Phys. Sci.*, **246**, 1
Davidson, K., and Ostriker, J.P. 1973, *Ap.J.*, **179**, 585
Elsner, R.F., and Lamb, F.K., 1977, *Ap.J.*, **215**, 897
_____. 1984, *Ap. J.*, **278**, 326
Ghosh, P., and Lamb, F.K. 1979, *Ap.J.*, **232**, 259
Hartmann, L., and MacGregor, K.B. 1982, *Ap.J.*, **259**, 180
Hameury, J.M., Bonnazola, S., Heyvaerts, J., and Lasota, J.P. 1984, *Adv. Space Res.*, No. 10–12, 297
Hsieh, S.–H., and Spiegel, E.A. 1976, *Ap.J.*, **207**, 244
Kirk, J. 1985, *Astron. and Ap.*,**142**, 430
Kirk, J., and Melrose, D.B. 1986, *Astron. and Ap.*, **156**, 257
Klein, R.I., and Arons, J. 1987a, in *Proc. IAU Symp. No. 125, 'Origin and Evolution of Neutron Stars'*, D.J. Helfand, ed. (Dordrecht: D.Reidel).
_____. 1987b, to be submitted to *Ap. J.*
Lamb, F.K. 1984,in *Proc. Santa Cruz Workshop 'High Energy Transients in Astrophysics'*, S.E. Woosley, ed. (New York: American Institute of Physics), 179
Lamb, F.K., Pethick, C.J., and Pines, D. 1973, *Ap.J.*, **184**, 271
Lamb, F.K., Shibazaki, N., Alpar, A. and Shaham, J. 1985, *Nature*,
Mason, K. 1977, *MNRAS*, **178**, 818
McKee, C.F., van Buren, D., and Lazareff, B. 1984, *Ap.J.*, **278**, L115

Meszaros, P., and Nagel, W. 1986, *Ap.J.*, in press
Michel, F.C. 1977, *Ap.J.*, **214**, 261
Miura, A. 1984, *J. Geophys. res.*, **89**, 801
______. 1985, *Geophys. Res. Lett.*, **12**, 635
Pravdo, S.H., *et al.* 1979, *Ap.J.*, **231**, 912
Proctor, M., and Weiss, N. 1982, *Rep. Prog. Phys.*, **45**, 1317
Pringle, J.E., and Rees, M.J. 1972, *Astron. and Ap.*, **21**, 1
Rappaport, S., and Joss, P.C. 1977, *Nature*, **266**, 683
Ruderman, M.A. 1986, *Cargese Lectures (1985)*, to be published
Shapiro, S.L., and Teukolsky, S. 1983, *Black Holes, Neutron Stars and White Dwarfs* (New York: John Wiley)
Stella, L., White, N.E., and Rosner, R. 1986, *Ap.J.*, in press
Trumper, J. 1986, in *Proc. Mediterranean School on Plasma Astrophysics*, N. Kylafis and J. Ventura, eds. (Iraklion: University of Crete), in press
Voges, W., *et al.* 1982, *Ap.J.*, **263**, 803
Wang, Y.–M., and Frank, J. 1981, *Astron. and Ap.*, **93**, 255
White, N.E., Swank, J.H., and Holt, S.S. 1983, *Ap.J.*, **270**, 711

## DISCUSSION

**A. Burrows:** Have you considered energetic particle production in this coupled neutron star/disk context?

**J. Arons:** Yes, I don't think disk electrodynamic leads to much interesting acceleration in the X-ray source context. Dissipative MHD instability at the bottom of the accretion column is more interesting, but the story is too long for discussion here.

**J. Grindlay:** Does your work provide any clues as to why cyclotron line features are apparently so difficult to detect in X-ray pulsars with magnetic fields presumably in the $10^{12}$ - $10^{13}$ gauss range (e.g. from spin-equilibrium arguments)?

**J. Arons:** I don't think spin equilibrium or other large scale aspects of the magnetosphere have much to do with line visibility. The dominant effect is likely to be velocity shear, which is enormous in the models I showed. Unless one looks almost across the magnetic field, the variable Doppler shift from the bulk motion smears out the line. If the field is simple, and if one could look at it broadside, then one has a chance to see a narrow feature. Clearly, the phase space for this is small, so finding line features in only a few objects is consistent with the kind of theory I've described.

**D. Arnett:** What is the ion temperature in the wind?

**J. Arons:** The collisional equilibration time is short, so it's probably about the same as the electrons, $\sim 10^6$ K.

**X.J. Mao:** What kind of material is in the disk of your model? According to the Goldreich-Julain model, the density of plasma is getting lower and lower and your disk doesn't go into the magnetosphere of the pulsar.

**J. Arons:** The disk is made of normal plasma accreted from the normal star in the binary. The Goldreich-Julian charge density occurs in the corotating region of the magnetosphere, and is easily maintained by the accreting plasma which passes through the magnetopause onto closed field lines.

**J.H. You:** If the temperature in the accreting column is $\gtrsim 10^8$ K, pair production will occur. It is so important for cooling, and radiation from pair-annihiliation may be dominant over bremstrahlung and cyclotron radiation. Don't you consider this effect?

**J. Arons:** Yes, maybe it can't be neglected and it should be taken into account.

# LINE RADIATION FROM ACCRETING MAGNETIZED NEUTRON STARS

D.Y. Wang
Purple Mountain Observatory
Academia Sinica
Nanjing, China

ABSTRACT. A new non-thermal model on cyclotron line radiation from accreting magnetized neutron stars have been reviewed. Above the magnetic polar cap of accreting neutron stars, the maser instability may be excited by a part of non-thermal electrons, and the radiation with the frequency near electron cyclotron frequency and its second harmonic may be emitted as well. According to this model,the intensity and energy of line radiation will vary with pulsation phase of X-ray pulsars. These phenomen have been observed from Her X-1.

## 1. INTRODUCTION

The most direct measurement evidence for intense magnetic field of neutrc stars is the discovery of a rather sharp hard X-ray spectral feature fron X-ray pulsar source Her X-1 since 1976 ( Trümper et al, 1978 ). A lot of balloon borne experiments ( Sheepmaker et al, 1981; Polcaro et al,1982; Voges et al,1983; Prantzos et al,1984 ) as well as satellite observations ( Maurer et al,1979;Evans et al,1980; Gruber et al,1980 ) on Her X-1 harc X-ray spectrum have been analysed. An excess around 55 kev ( or 40 kev corresponding absorption line) is interpreted as a cyclotron line, corresponding to a magnetic field of about $5\times10^{12}$ Gauss ( Trümper et al, 1978 ). A second excess at about 110 kev also reported by Trümper (1978) has not been confirmed yet. Recently, some new observational results from Her X-1 hard X-ray spectral analysis have been reported: a) An analysis of phase-dependent spectra shows a significant variation of the emission line centroid of the cyclotron feature during the main pulse ( Gruber et al,1980; Voges et al,1982 ); b) There exists a correlation between the pulsar's hard X-ray luminosity and the intensity, energy of line feature ( Prantzos and Durouchoux, 1984; Durouchoux, 1987 ). Another object that shows clear evidence for cyclotron lines is 4U 0115+63 ( Wheaton et al,1979; White et al,1983 ).

On the other hand, a straightforward explanation of the cyclotron feature observed in the spectra of some binary X-ray pulsars is a very complicated task, because of the effects of the intense magnetic field in accretion column, and the coupling of the plasma flow of the accreted



D. J. Helfand and J.-H. Huang (eds.), The Origin and Evolution of Neutron Stars, 227–232.

matter to the radiative transfer process ( Mészáros,1983 ). No self-consistant exists up to now, but two approximation models have been developed during the past few years. One suggestion is the emission feature to be produced by electron quantum transition between Landau levels in the action of strong magnetic field ( Basko,1975;Qu,1980 ); the other is caused by cyclotron resonsnces absorption of radiation transfer in the accretion column ( Venttura et al,1979; Mészáros et al,1980; Yahel,1980; Nagel,1981; Arons,1987 ). However, the velocity distribution of electron considered in both models is thermal equilibrium.

In fact, above the magnetic polar cap of neutron stars, the velocity distribution of a part of electrons may be a non-thermal distribution. The possibility of non-thermal eletron is caused by rapid electron rotation around the strong magnetic field ( hollow beam distribution ), or by reflected electrons at magnetic mirror point ( loss-cone distribution ), or under the action of rotary electric field. For simplicity, we consider the magnetic mirror effect only. As the strength of magnetic field in the accretion column varies inversely with the cube of distance from neutron star, it form a magnetic mirror; a part of relativistic electrons with large pitch angle will be reflected at magnetic mirror point, and form a non-thermal distribution. In these case, the maser instability can be excited by these non-thermal electrons, and the radiation with the frequency near the cyclotron frequency can be emitted as well. In the case of X-ray pulsars, the accretion matter from companion flow uninterruptedly through accretion disk to the magnetic polar cap, and then a part of electrons with large pitch angle will be pumped continuouesely to upper accretion column. The line radiation with cyclotron frequency are emitted with the excited maser instability. These lines have been observed from Her X-1 and some X-ray pulsars.

## 2. BASIC THEORY OF MASER INSTABILITY FOR LINE RADIATION

We assume the electron density of background plasma $n_o \gg n_s$, where $n_s$ is the non-thermal electron density produced from magnetic mirror effect, and growth rate of instability $\omega_i \ll$ resonance frequency of instability $\omega_r$, thus the growth rate of maser instability might be expressed as ( Wu and Lee, 1979; Melrose et al, 1980; Freund et al,1983 )

$$\left(\frac{\omega_i\, n_o}{\Omega_e\, n_s}\right)_j = \frac{2\pi^2}{G}\left(\frac{\omega_{pe}}{\Omega_e}\right)\left(\frac{\Omega_e}{\omega_r}\right)\sum_{m=j}^{\infty}\int_{-\infty}^{\infty} du_{\parallel}\int_0^{\infty} du_{\perp}\,\frac{u_{\perp}}{\gamma}\frac{m\Omega_e}{k_{\perp}}\left(m\Omega_e\frac{\partial}{\partial u_{\perp}} + k_{\parallel}u_{\perp}\frac{\partial}{\partial u_{\parallel}}\right)F(u_{\parallel},u_{\perp})\,\Psi_m(u_{\perp})\,\delta(\gamma\omega_r - m\Omega_e - k_{\parallel}u_{\parallel}) \tag{1}$$

where $\gamma = (1 - v^2/c^2)^{-1/2} = (1 + u^2/c^2)^{1/2}$ ; $u_{\perp} = \gamma v_{\perp}$ , $u_{\parallel} = \gamma v_{\parallel}$;
$\omega_{pe}$ is the electron plasma frequency; electron cyclotron frequency. G and $\Psi_m$ are expressed in the same formulae as that in our previous paper ( Wang and Mao, 1986 ); $\omega_r$ is obtained from cold plasma approximation for backgorund plasma. The loss-cone velocity distribution with intense gravitational potential has been adopted as the non-thermal electron velocity distribution function $F(u_{\parallel}, u_{\perp})$

$$F(u_{\parallel}, u_{\perp}) = \begin{cases} F_{>} = A\exp(-u^2/\alpha^2)\ ; \ u_{\perp}^2 \geqq \dfrac{u_{\parallel}^2 + \Phi}{d} \\ F_{<} = A\exp(-u^2/\alpha^2)\exp\left(\dfrac{u_{\perp}^2 - (u_{\parallel}^2 + \Phi)/d}{\beta}\right)\ ; \ u_{\perp}^2 < \dfrac{u_{\parallel}^2 + \Phi}{d} \end{cases} \tag{2}$$

where $d = \frac{B_M}{B} - 1$, $\Phi = \frac{2(U - U_1)}{M}$, $U_1$ and $B_M$ are gravitational potential and magnetic field at magnetic mirror point, respectively. A is normalization constant. We obtain

$$\left(\frac{\omega_i n_o}{\Omega_e n_s}\right)_j = \frac{4\pi^2}{G}\left(\frac{c}{\alpha}\right)^2\left(\frac{\omega_{pe}}{\Omega_e}\right)^2\left(\frac{\Omega_e}{\omega_r}\right)^2\sum_{m=j}^{\infty}\Bigg\{\Bigg(\int_{u_m(-)}^{u_{\parallel}(A_m^-)} du_{\parallel} + \int_{u_{\parallel}(A_m^+)}^{u_m(+)} du_{\parallel}\Bigg) m\Big[m\Big(\frac{\alpha^2}{\beta^2} - 1\Big)\frac{\Omega_e}{k\sin\vartheta} - \Big(1 + \frac{\alpha^2}{\alpha\beta^2}\Big)\frac{\cos\vartheta}{\sin\vartheta}u_{\parallel}\Big] u(A_s) F_{<} \Psi_m - \int_{u_{\parallel}(A_m^-)}^{u_{\parallel}(A_m^+)} du_{\parallel} \cdot m\Big(\frac{m\Omega_e}{k\sin\vartheta} + \frac{\cos\vartheta}{\sin\vartheta}u_{\parallel}(A_s) F_{>} \Psi_m\Bigg\}\ ;$$

$$\Delta > 0 \tag{3.a}$$

$$\left(\frac{\omega_i n_o}{\Omega_e n_s}\right)_j = \frac{4\pi}{G}\left(\frac{c}{\alpha}\right)^2\left(\frac{\omega_{pe}}{\Omega_e}\right)^2\left(\frac{\Omega_e}{\omega_r}\right)\sum_{m=j}^{\infty}\int_{u_m(-)}^{u_m(+)} du_{\parallel}\Big[m\Big(\frac{\alpha^2}{\beta^2} - 1\Big)\frac{\Omega_e}{k\sin\vartheta} - \Big(1 + \frac{\alpha^2}{\beta^2}\Big)\frac{\cos\vartheta}{\sin\vartheta}u_{\parallel}\Big] u_{\perp}(A_s) F_{<} \Psi_m$$

$$\Delta \leqq 0 \tag{3.b}$$

where $\Delta \equiv \left(1 + \frac{1}{d}\right)\left(\frac{m\Omega_e}{\omega_r}\right)^2 - \left(1 + \frac{\Phi}{dc^2}\right)\left(1 + \frac{1}{d} - m^2\cos^2\vartheta\right)$

The relation between resonance ellipse in (1) and $\Delta$ has been discussed in detail ( Mao and Wang, 1986 ). In order to compare with line radiation from Her X-1, following parameters have been used in the calculation, $M = 1 M_\odot$, $r_M = 10^6$ cm, $c/\alpha = 1.7$, $c/\beta = 11$, $d = 0.728$, $B = 4.5\times10^{12}$ Gauss and magnetic mirror point $r = 1.2\ r_M$.

## 3. DISCUSSION AND SUMMARY

a) The growth rate of extraordinary mode ( X mode ) $\gg$ the growth rate of ordinary mode ( o mode ). This means that line radiation is produced from X mode maser instability. The polarization of line radiation from X mode is large.

b) The growth rates increase with the electron density ( Fig. 1 ).

Therefore, the line radiation from maser instability would be coherent, comparing with the incoherent radiation from other models.

c) In the region of $\omega_{pe}/\Omega_e \approx 0.01 - 0.1$, $\omega_i(j=2)/\omega_i(j=1) \simeq 0.05$ ( Fig.1). Thus, second harmonic line ( ~ 110 kev ) is quite weak from maser instability, if $n_s \ll n_o$ and $n_o < 10^{27}/cm^3$ are satisfied.

d) The growth rates depend intensely on the angle relative to magnetic field, as shown in Fig.2. This means that the emission line from maser instability can be observed only in a limited angle region, or the intensity of emission line will depend on the pulsation phase. The energy of emission line depends on the angle also ( Wang and Mao,1986 ). These results are consistent with observations by Voges et al ( 1983 ) qualitatively.

e) The resonance condition in the maser instability is ellipse ( $\delta$ function in eq.(1) ). Therefore, it is not only the electron velosity with $u_\perp \geqq u_{\perp c}$ can contribute to line radaition, but $u_\perp \leqq u_{\perp c}$ can also contribute to this line ( $u_{\perp c}$ is a critical perpendicular velocity for single electron emitting cyclotron radiation ). The electron number contributed to resonance emission line will increase vastly as compared with single electron process.

f) The ratio of quantum synchrotron radiation power $p^{(q)}$ and classical synchrotron radiation power $p^{(c)}$ for single electron may be expressed as follows (Sokolov,1968 )

$$\frac{p^{(q)}}{p^{(c)}} = 1 - \frac{55}{16}\frac{B\gamma}{B_c}\gamma + 48\left(\frac{B\gamma}{B_c}\right)^2 + \cdots\cdots ; \quad \text{for } B\gamma \ll B_c$$

where $B_c \simeq 4.4\times10^{13}$ Gauss, in

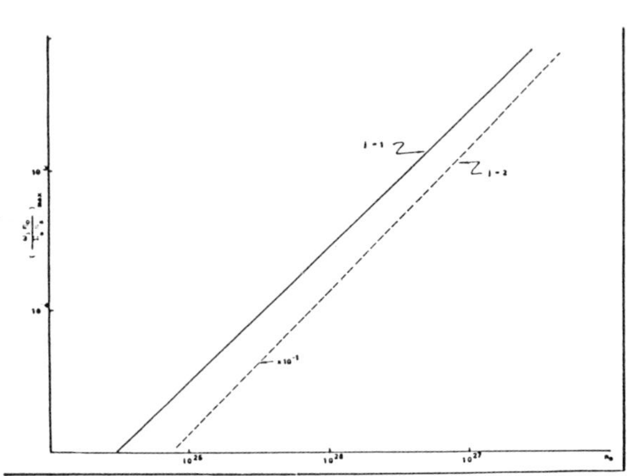

Fig.1 The maximum growth rate of X mode vs the electron desity $n_o$

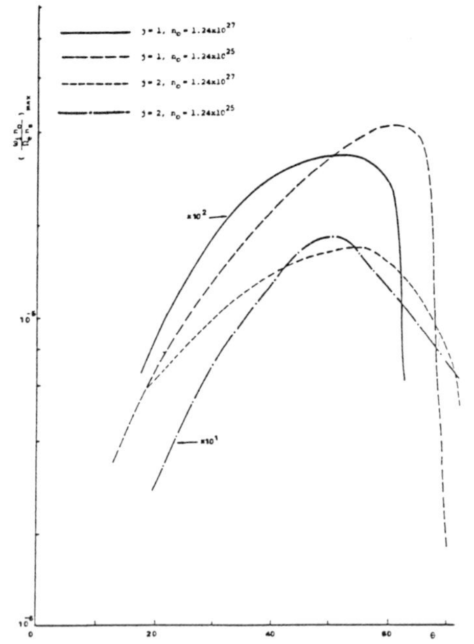


Fig.2 Maximum growth rate as a function of angle respect to magnetic field

the case of Her X-1, $p^{(q)}/p^{(c)} \simeq 1.0$. This result shows distinctly that classical synchrotron radiation process is same important as quantum Landau levels transition process in the X-ray pulsar Her X-1.

g) In order to investigate the relaxation time of non-thermal electron distribution, two effects should be considered. The first one is that the relaxation time of synchrotron radaition of single electron ( include classical and quantum ) may be expressed as below

$$\tau_{rel}^{(R)} \simeq \frac{3}{4}\frac{c}{r_o \Omega_e^2 \gamma} \simeq 0.16 \times 10^{-16} \text{ sec}$$

However, the synchrotron radiation of single electron is limited to influence of the electrons with normal velocity $u_\perp > u_{\perp c}$. The second one is that the relaxation time of electron collision has been studied by Zhong and Hu ( 1986 ) in detail. They have found that the relaxation time of electron collision will be extended by Larmer motion of electron around the intense magnetic field. In the case of $B = 10^{12}$ Gauss, kT = 10 kev, $n_o = 10^{26}/cm^3$; $\tau_{ee}^{(c)} \simeq 0.73 \times 10^{-15}$ sec. Therefore, the maser instability might be excited, if the relaxation time of non-thermal electron, $\tau_M$, $\tau_M \approx \frac{1}{\omega i} \leqq \tau_{ee}^{(c)}$ and/or $n_s/n_o \geqq 0.01$ are satisfied.

consequently, the maser instability ( even if classical theory is considered only ) should play an important role in the line emission from accreting magnetized neutron stars.

References

Arons,J.: 1987, in this Proceedings
Basko,M., and Sunyaev,R.: 1975, Astron.Astrophys., 42, 311
Durouchoux,P., and Prantzos,: 1987, in this Proceedings
Evens,A.,Quenby,J., and Engle,A.: 1980, Astron.Astrophys.Suppl.,41,13
Freund,H.P., Wang,H.K., Wu,C.S., and Xu,M.J.: 1983, Phys. Fluids,26,2263
Gruber,D.E.,Matterson,J.L.: 1980, Astrophys.J.,240, L127
Mao,D.Y., and Wang,D.Y.: 1986, in preparation
Maurer,G.,et al.:1979, Astrophys.J., 231,L906
Melrose,P., Rönnmark,K.G., and Hewitt,R.G.: 1982, J.Geophys.Rev.,87,5140
Mészáros,P.: 1983, Space Sci.Rev.,38, 325
Mészáros ,P., Nagel,W., and Ventura,J.:1980, Astrophys.J.,238,1066
Nagel,W.: 1981, Astrophys.J.,251,278
Polcaro,V.,et al.:1982, Astron.Astrophys.,108,249
Prantzos,N., and Durouchoux,P.: 1984, Astron.Astrophys.,136,363
Qu,Q.Y.,Wang,Z.R., and Zhang,H.Q.: 1980, Acta astronomica Sinica,21,180
Sokolov,A.A., and Ternov,I.M.: 1968, Synchrotron Radiation, Academic-Verlag ( Berlin )
Trümper,J., et al.:1978, Astrophys.J., 219,L105
Ventura,J., Nagel,W., and Meszaros,P.: 1979, Astrophys.J.,233,L125
Voges,W., Pietsch,W., Reppin,C.,Trumper,J., Kendziorra,E., and Staubert, R.: 1983, Astrophys.J.,263,803
Wang,D.Y., and Mao,D.Y.: 1986, Publication of Purple Mountain Obs.,6,1
Wheaton,W.A.,et al.:1979, Nature, 28,240
White,N., Swank,J.H., and Holt,S.S.: 1983, Astrophys.J.,267,318

Wu,C.S., and Lee,L.C.: 1979, Astrophys.J.,230,621
Yahel,R.Z.: 1980, Astron.Astrophys.,90,26
Zhong,Y.S., and Hu,J.M.: 1986, to be published in Chinese Nuclear Fusion and Plasma Physics

## DISCUSSION

**D. Eichler:** The optical depths of the accretion columns of neutron stars are many orders of magnitude larger than the plasma in the earth's magnetosphere. Would the mass emission be absorbed at the nth harmonic ($n \geqslant 2$) on the surfaces where B is 1/n of its value at the maser sight?

**D. Wang:** In the Her X-1 case, we used the parameter $W_{pe}/n_e \cong 0.01 \rightarrow 0.1$. If these values are quite small in the maser instability and in this parameter region the growth rate of the 2nd harmonic resonance is much lower than the first harmonic $(\omega_i, j=2)/(\omega_i, j=1) \cong 0.05$. Thus, I think the absorption at the nth harmonic is negligible.

# X-RAY BURSTING NEUTRON STARS

H. INOUE
Institute of Space and Astronautical Science
Komaba, Meguro-ku, Tokyo 153
Japan

ABSTRACT. The maximum peak luminosity of the X-ray bursts from a burster is most likely interpreted as the Eddington luminosity of a helium-rich envelope surrounding a neutron star. If this interpretation is true, we can obtain a relation between the mass and the radius of the neutron star in terms of the maximum effective temperature of bursts. On the other hand, the most naive understanding of the origin of the 4.1 keV absorption line often detected in X-ray burst spectra gives us another relation of the neutron star mass with its radius. By solving two simultaneous equations, we can determine the values of the mass and the radius of the neutron star, respectively. However, the result is critical to every neutron star model currently considered.

The persistent emissions from X-ray bursters are also discussed.

## 1. MASS AND RADIUS OF X-RAY BURSTING NEUTRON STARS

### 1.1. Luminosity saturation of bursts at the Eddington luminosity

Figure 1 shows the peak flux distribution of bursts from X1636-53 observed from Hakucho (Ohashi et al. 1982) and Tenma (Inoue et al. 1984) as a function of the integrated flux. This figure indicates the presence of an upper limit in the peak flux of bursts from the burster. The peak flux values of eight bursts reaching the upper limit in Fig.1 are the same within errors. After fitting a blackbody spectrum to the time-resolved spectra of the bursts, it is found that the bursts with the maximum peak flux all exhibit a flat top at the same maximum level accompanying a rapid radius change (Inoue et al. 1984). The time variation of energy flux versus apparent radius during the first 6.5 sec of a burst with the maximum peak flux is shown in Fig.2. The apparent radius quickly increases to 20-30 km during the burst rise and returns to about 10 km within a few seconds (assuming 10 kpc distance). These observed features strongly suggest that the peak luminosity of X-ray bursts from a X-ray burster is the Eddington luminosity.

When the huge nuclear energy is released in a very short time, the excess energy flux above the Eddington luminosity is converted into



D. J. Helfand and J.-H. Huang (eds.), The Origin and Evolution of Neutron Stars, 233–243.

kinetic energy of the ambient matter. This results in envelope expansion while the luminosity is kept at the Eddington luminosity. This is qualitatively consistent with the radius change seen during the flux peak. However, as seen from Fig.2, the luminosity has not reached the Eddington luminosity during the envelope expansion. Sugimoto et al.(1984) suggested that since the Eddington luminosity of the hydrogen-rich envelope at the neutron star surface is lower than that of the inner helium-rich envelope, the hydrogen-rich envelope is ejected before the flux reaches its maximum. Furthermore, they pointed out an existence of a gap in the peak flux distribution by a factor 1.7 as indicated in Fig.1, which value corresponds to the difference in the Eddington luminosity of neutron stars having helium or hydrogen-rich envelopes. Thus, the observed maximum flux is consistently interpreted as due to the Eddington luminosity of the helium-rich envelope after the surface hydrogen-rich envelope has been lost.

## 1.2. Maximum effective temperature at the Eddington luminosity

If the maximum peak luminosity is indeed the Eddington luminosity at the neutron star surface, the maximum effective temperature, $T_{e,Max}$, at the Eddington luminosity relates to the mass, M, and the radius, R, of the neutron star as

$$\sigma T_{e,Max}^{4} = cGM(1-(2GM/c^2R))^{3/2}/\kappa_e R^2, \tag{1}$$

where $\sigma$ the Stephan-Boltzman constant, c the velocity of light, G the gravitational constant and $\kappa_e$ the opacity for electron scattering (e.g. Goldmann 1979). Figure 3 shows the mass and radius relations, calculated with Eq.1, in terms of two values of the maximum effective temper-

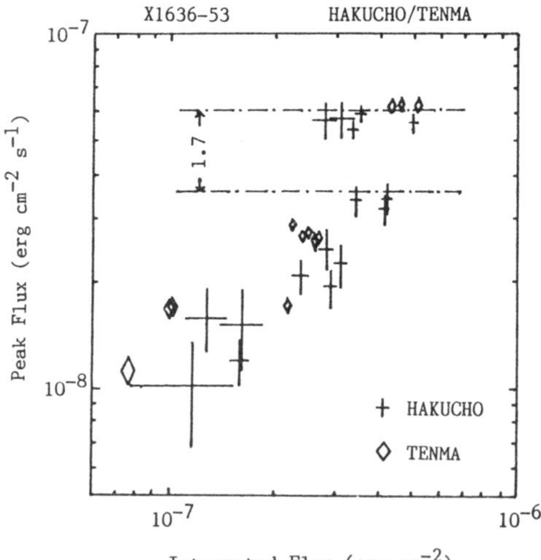


Fig.1 Peak flux distribution of X-ray bursts from X1636-53 as a function of the integrated flux.

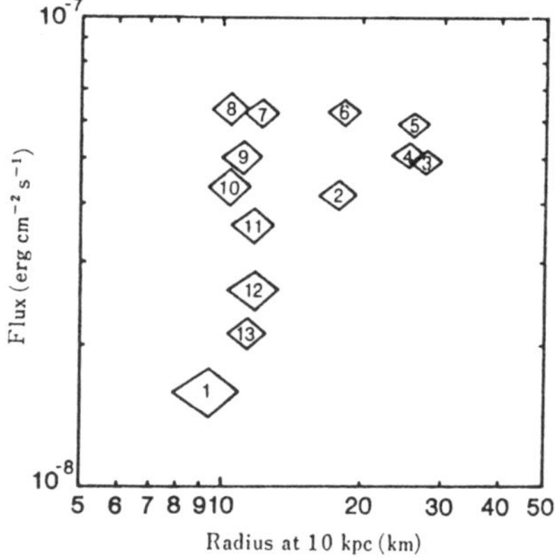


Fig.2 The time variation of flux versus apparent radius during the first 6.5 sec of a burst with the maximum peak flux. Numbers in the diamonds indicate time sequences with 0.5 sec interval.

ature. However, the observed maximum color temperature at the Eddington luminosity is as high as about 3 keV (see Fig.5, as an example), which indicates too small a mass of the neutron star. Then, it was pointed out that the color temperature can be higher than the effective temperature for the radiation from an atmosphere in which the electron scattering opacity is dominant (e.g., van Paradijs 1982).

In order to obtain the quantitative relationship between color temperature and effective temperature, London et al.(1984) and Ebisuzaki and Nomoto (1985) solved the multi-frequency diffusion problem of photons in a neutron star atmosphere. The result from Ebisuzaki and Nomoto (1985) is shown in the temperature and luminosity diagram (identical to the Herzsprung - Russel diagram) in Fig.4. We can see that the color temperature (solid line) is substantially higher than the effective temperature (dashed line). Since the theoretical curve is a function of the maximum effective temperature at the Eddington luminosity, $T_{e,Max}$, we can now determine the value of $T_{e,Max}$ by fitting the curve to the data points on the color temperature and the luminosity plane.

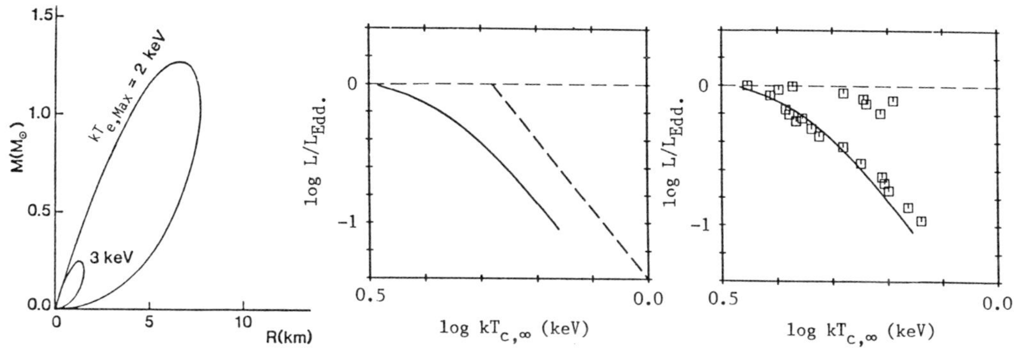


Fig.3 Mass-radius relations for two values of maximum effective temperature.

Fig.4 Relation of color temp. to effective temp. as a function of luminosity.

Fig.5 An example of the fitting of the theoretical curve to the data.

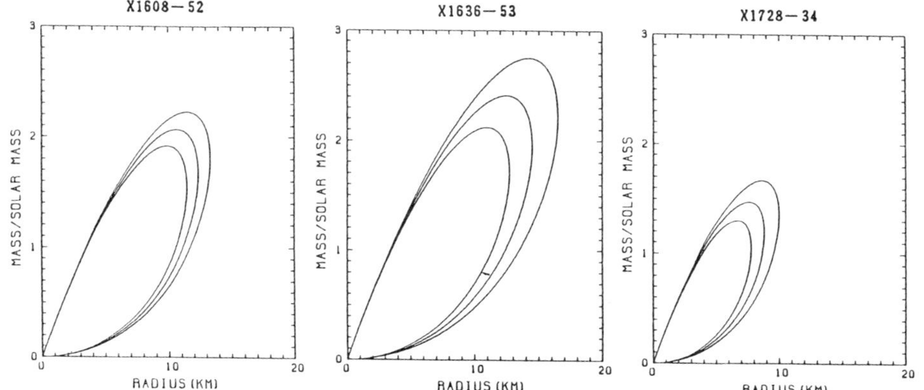


Fig.6 Preliminary results for the mass-radius relations of neutron stars in X-ray bursters; X1608-52, X1636-53 and X1728-34.

Fujimoto and Taam (1986), Ebisuzaki (1986) and Nakamura (1986) performed such fits to the data during the decay part of bursts observed from Tenma. Figure 5 shows an example from Ebisuzaki (1985). Once we have the value of $T_{e,Max}$ from the fits, we can obtain the mass-radius relation of the neutron star by using Eq.1. The preliminary results by Nakamura (1986) for three X-ray bursters; X1608-52, X1636-53 and X1728-34 are plotted in Fig.6. Outer and inner loops represent the error domain at 90% confidence limit derived from the fittings. Note that uncertainties in the theoretical calculations are not included.

## 1.3. Absorption lines in X-ray burst spectra

Tenma discovered significant features in X-ray burst spectra. Among sixteen bursts from X1636-53, four bursts exhibited five cases of significant dips in the spectra, whose nature was consistent with an absorption line (Waki et al. 1984). On the other hand, three bursts among seventeen bursts from X1608-52 also exhibited absorption lines in the spectra (Nakamura et al. 1986). Although Tenma detected as many as twelve bursts from X1728-34 and the data have the same statistical level as those from the above two burst sources, no significant dip was found in the X-ray burst spectra from X1728-34 (Tawara et al. 1986).

A line was observed at 5.7 keV only near the peak of a burst from X1636-53, the luminosity of which reached the Eddington luminosity. On the other hand, the line center energies in the other 7 cases are all consistent with 4.1 ± 0.1 keV, taking account of their errors. The equivalent widths of those absorption lines are about 100 to 200 eV. The observed line profile is well reproduced by the detector's responce function for a single line without significant intrinsic line width and the upper limit for the line width is about 500 eV (FWHM).

It is most likely that the absorption lines are produced by atomic processes in the atmosphere surrounding the neutron star. Then, for the plasma temperature during a burst, the helium-like ions are considered to be most abundant for the heavy element responsible for the absorption. Taking account of the possibility of the energy redshift due to the general relativistic effect on the neutron star surface, the candidate elements forming the absorption lines are those heavier than Ca for the 4.1 keV line and than V for the 5.7 keV line (Waki et al.1984).

The matter producing the absorption line may be either the products of helium burning in the bursts or the accreted matter from the companion star. The accretion rate is estimated to be $10^{16-17}$ g/sec, corresponding to an accumulation of matter at about $10^{3-4}$ electron scattering optical depth/sec over the neutron star surface. If the freshly accreted matter spreads over the neutron star surface during a burst, the most plausible element for the observed absorption would be iron which is most abundant in the accreted matter. On the other hand, when the luminosity reaches the Eddington luminosity, the surface accreted matter can be ejected by the radiation pressure and the following accretion might be hampered by the ram pressure of the ejected matter. In that case, the nuclear-burning products could be exposed on the surface. Then, the possibility of absorption by elements other than iron could not be excluded. However, the peak luminosities of a burst from X1636-

53 and three bursts from X1608-52 among the seven bursts exhibiting the 4.1 keV line are lower than the Eddington luminosity of these sources by factors 3 to 4. Matter ejection is quite unlikely for these bursts. This is in favor of the iron origin of the 4.1 keV absorption line. If so, this gives the redshift factor, $(1+z)^{-1}$, of the 4.1 keV line to be $0.61 \pm 0.02$.

The most naive understanding is that this absorption line is produced in the atmosphere surrounding the neutron star. Then, the atmosphere is considered to be geometrically thin, since the effect of the radiation pressure is negligible when the 4.1 keV line is observed. Thus, this redshift would represent the general relativistic effect on the neutron star, and we have an equation for the mass-radius relation of the neutron star as

$$( 1 - (2GM/c^2R) )^{1/2} = 0.61 \pm 0.02 \ . \tag{2}$$

1.4. Mass and radius of X-ray bursting neutron stars

Now that we have had two independent equations; Eqs.1 and 2, for the mass-radius relation of the neutron star, we can determine each of the values of the mass and the radius. The results are shown on the mass-radius plane in Fig.7 for two X-ray bursters; X1608-52 and X1636-53. The mass and radius values are suggested to be about 2 $M_o$ and about 10 km, respectively, for both of the two neutron stars.

However, these results have a serious consequence. The obtained values for the mass and the radius are critical to every neutron star model currently considered (Baym and Pethick 1979, and references therein; see also Alcock, 1987).

Fujimoto (1984) proposed a model to avoid the difficulty of the too stiff neutron star. He considered the boundary layer between the neutron star and the accretion disk as the line formation region. Since

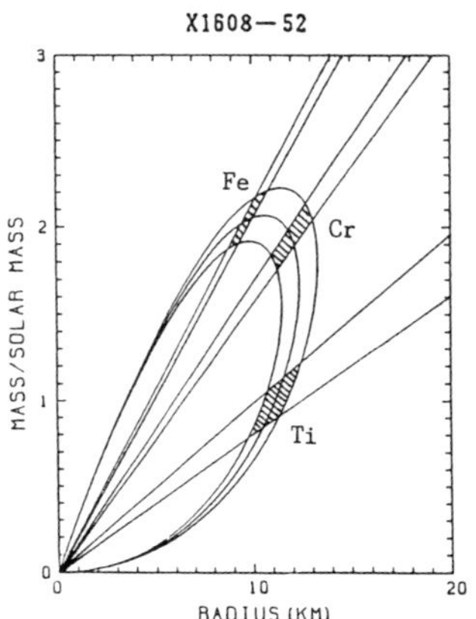


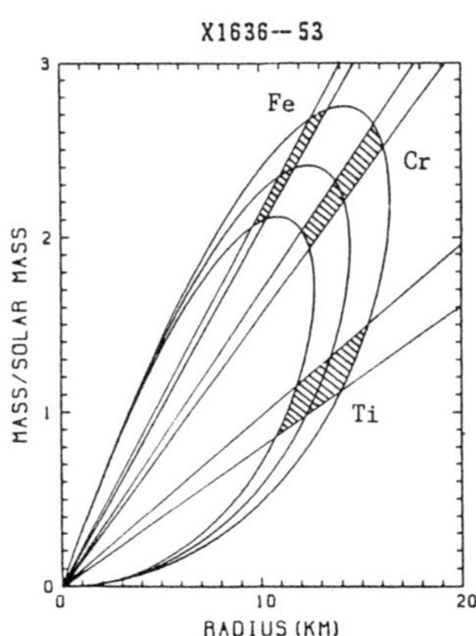


Fig.7 Mass-radius domains satisfying Eqs.1 and 2 simultaneously, for two X-ray bursters; X1608-52 and X1636-53. Results in the cases that the 4.1 keV line is originated from Fe XXV, Cr XXIII and Ti XXI are shown respectively.

the boundary layer is expected to rotate very rapidly with the accretion disk, the transverse Doppler effect due to this rotation is important in evaluating the redshift as well as the general relativistic effect. He calculated the redshift factor expected from the boundary layer, and the mass-radius relation obtained gives a good agreement with some neutron star models. However, this model has a serious difficulty.

The rotation of the line producing matter causes not only the shift of the line center energy but also the broadening of the line profile. In order to accomodate the rotational broadening with the observed upper limit for the intrinsic line width, the inclination angle of the line of sight to the rotaion axis of the line producing matter should be smaller than about 15°. However, the possibility for the line of sight to be within such a small cone is very small. Hence, it seems difficult for this model to explain the Tenma observations that among three bursters whose burst spectra were extensively searched for dip features, two sources exhibited the 4.1 keV line.

If there exists some unknown mechanism for mixing the nuclear burning products with the accreted matter, Cr or Ti are still possible candidates for the line producing element. If so, the corresponding values of mass and radius of the neutron star agree with some neutron star models (see Fig.7). A possible mechanism may be a continuous shear mixing of the rotating accreted matter with the neutron star atmosphere at the boundary layer between the accretion disk and the neutron star surface (Fujimoto et al. 1986; Hanawa 1986). Further study on this mechanism is expected.

## 2. PERSISTENT EMISSIONS FROM X-RAY BURST SOURCES

### 2.1. Spectral hardening with intensity-decrease at the dim phase

Figure 8 shows the change of the spectral hardness associated with the change of the luminosity of an X-ray burst source; X1608-52, observed from Tenma, together with those of three bright low-mass binary X-ray sources; Sco X-1, GX349+2 and GX5-1. The burst activity of X1608-52 in the period of the Tenma observation is also shown in Fig.8 in terms of $\alpha$-value, which is the ratio of the averaged persistent flux to the averaged burst flux. The larger value of $\alpha$ corresponds to the lower activity of bursts. The steep increase of $\alpha$-value with the increase of luminosity and the similarity in the luminosity-spectral hardness relation between X1608-52 at the bright phase and the other three sources suggest that the three bright low-mass binary X-ray sources are probably potential X-ray burst sources, although no X-ray burst has been detected from them.

When the luminosity is in the range from $5 \times 10^{36}$ to $10^{37}$ erg/sec, the hardness increases as the luminosity decreases, as seen in Fig.8. The change of the spectrum with the luminosity decrease is more clearly seen from Fig.9. As the luminosity goes down, the high energy tail becomes more pronounced and consequently the whole spectrum approaches a power-law spectrum. This demonstrates that the spectrum of a X-ray burst source can undergo a dramatic change from a thermal type spectrum to a power-law type spectrum depending on the accretion rate.

After the detailed spectral analysis, Mitsuda and Tanaka (1986) showed that the spectral hardening with the luminosity decrease can be interpreted as due to the increase in the degree of the Comptonization for blackbody photons from the neutron star surface by hot electrons surrounding the neutron star. This may indicate that an optically thin hot region surrounding the neutron star extends when the accretion rate is reduced. The change in the accretion flow near the neutron star will be interpreted as being due to a transition in the accretion disk between an optically thick/geometrically thin state and an optically thin/geometrically thick state, depending on the accretion rate (Inoue and Hoshi 1986).

## 2.2. Spectral hardening with intensity-increase at the bright phase

Contrary to the above phase, X-ray burst sources exhibit a positive correlation of the spectral hardness with the luminosity, when the luminosity is as bright as $10^{37-38}$ erg/sec. This behaviour was investigated in detail for the four bright low-mass binary X-ray sources in Fig.8 by Mitsuda et al.(1984).

These sources exhibit intensity variations of a factor of two to three on time scales of the order of an hour. Mitsuda et al.(1984) compared the spectra when the intensity was high with those in adjacent periods of lower intensity. They found that the difference between high- and low-intensity spectra is always expressed very well by a single blackbody spectrum with kT of approximately 2 KeV for all sources examined. Furthermore, the blackbody temperature is found to be fixed

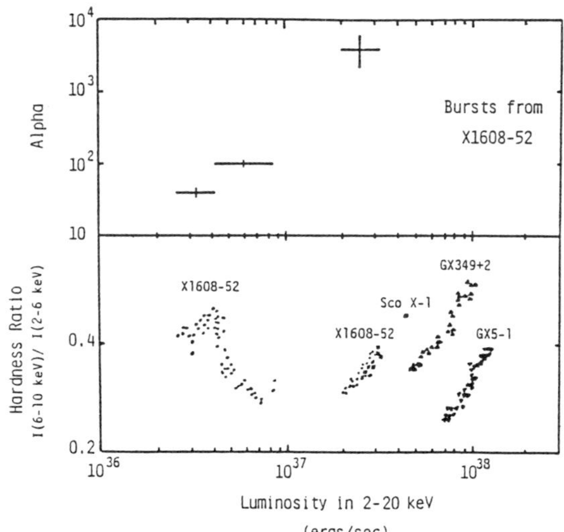


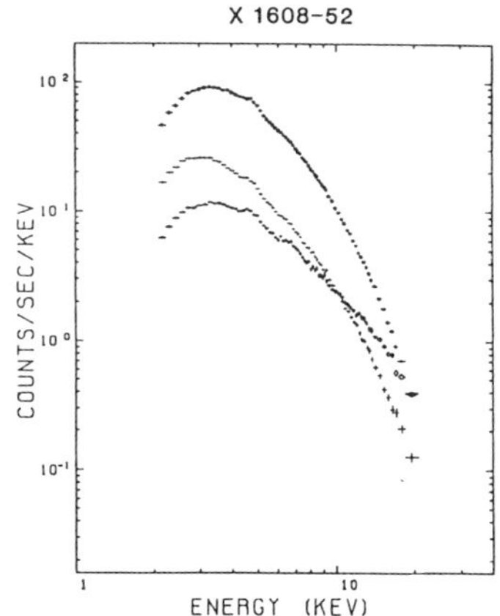


Fig.9 Change of spectrum of X1608-52.

Fig.8 Dependency of alpha on the 2-20 keV luminosity of X1608-52 (upper panel), and those of hardness ratio on the luminosity of four low-mass binary X-ray sources (lower panel). (Assumed distances are 4.4, 6, 6, and 1 kpc for X1608-52, GX349+2, GX5-1 and Sco X-1, respectively.)

for a given source. Thus, the hardening of the spectrum can be interpreted as due to an intensity increase in this blackbody component.

By subtracting the 2-keV blackbody component from the observed spectra, Mitsuda et al.(1984) decomposed the observed spectrum into two spectral components; the "2-keV blackbody" component and a softer component. Following theoretical considerations about the accretion disk around a weakly magnetized neutron star (e.g., Hoshi 1984), Mitsuda et al.(1984) interpreted the softer component as the emission from an optically thick accretion disk around the neutron star and the 2 keV component as that from the neutron star surface, since the decomposed spectrum of the softer component is found to be well expressed by that expected from an optically-thick accretion disk.

Figure 10 shows the intensities of the two components of GX5-1 as a function of time (Mitsuda et al.1984). We clearly see that the 2-keV blackbody component is highly variable. While the flux of the 2-keV blackbody component varies widely, the blackbody temperature remains constant. This implies that the emitting area on the neutron star surface changes with flux. This may be understood in terms of the local Eddington limit. If the radiation flux per unit area of the neutron star surface reaches the local Eddington limit, the temperature cannot increase further. For more energy to be radiated, the surface area of the emission will have to increase. In fact, the temperature 2 keV is close to the maximum effective temperature determined from Eq.1.

However, a question is why this maximum temperature 2 keV is lower than the maximum temperature of X-ray burst emission. As discussed before, the maximum temperature of X-ray bursts goes up to about 3 keV and is interpreted also as due to the local Eddington limit at the neutron star surface. This question will be solved if we take account of the following two factors. One factor is the difference in the hydrogen abundance of the atmosphere from which the flux comes out. The Eddington limit of X-ray burst is probably that in helium rich envelope, while the matter emitting the persistent emission has cosmic element abundances. This gives a factor of about 1.7 for the difference between the two Eddington limits. Another factor will come from the transverse Doppler Effect due to the rotation of accreted matter in the boundary

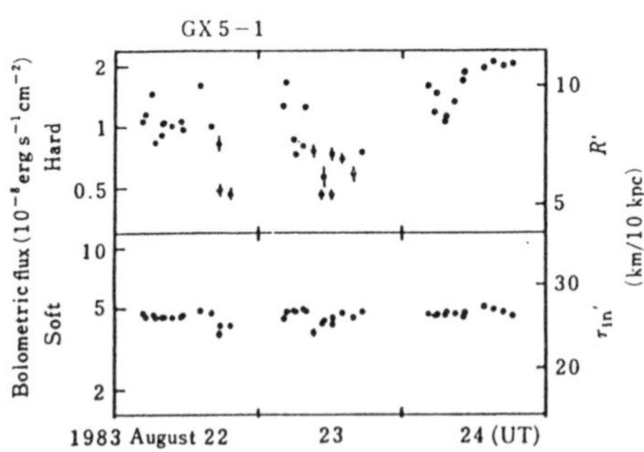


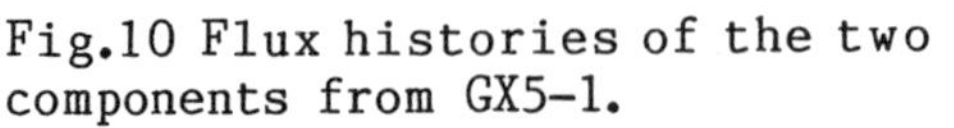
Fig.10 Flux histories of the two components from GX5-1.

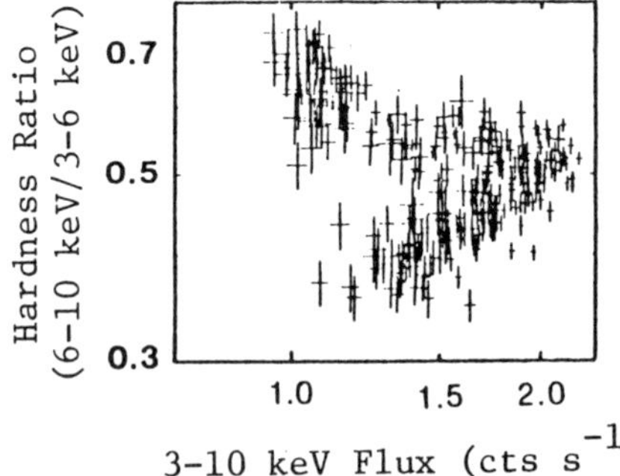


Fig.11 Dependency of hardness ratio on flux of GX5-1 observed from Hakucho.

layer between the accretion disk and the neutron star surface. This will make the observed temperature of persistent emission from the boundary layer lower than that of the X-ray burst emission, even if the two temperatures are the same at the neutron star surface.

## 2.3. The "horizontal branch" at the brightest phase

Hakucho observed GX5-1 for about 40 days in April to June in 1980. In most of the observational period, GX5-1 showed large intensity variations on a time scale of hours, when the spectral hardness correlated positively with the intensity as observed with Tenma ("normal branch"). However, the intensity variation on a time scale of hours occasionally disappeared, when the intensity-spectral hardness diagram reveals the "horizontal branch" as shown in Fig.11 (Mitsuda 1984). The spectral hardness increases as the intensity increases in the "normal branch". However, when the intensity reaches the brightest point, there exists an alternative branch where the spectral hardness is almost constant against the intensity variation within the same range as that of the "normal branch".

As discussed in the preceding subsection, the emission region of the 2-keV blackbody component changes its area with the intensity in the "normal branch". The projected radius of the emission region of the 2-keV blackbody component is indicated at the right hand side of Fig.10, where we find that the emitting area seems close to the whole surface of the neutron star at the brightest point. This suggests the luminosity at the brightest point to be almost the Eddington luminosity, since the 2-keV temperature is considered to be determined by the local Eddington limit on the neutron star surface. In fact, the luminosity estimated from the maximum flux of GX5-1 exceeds $10^{38}$ erg/sec, even if we assume the distance to be as small as 6 kpc. This value is consistent with the Eddington luminosity for about a solar mass neutron star.

These facts may suggest that the transition from the "normal branch" to the "horizontal branch" will take place when the mass accretion rate onto the neutron star exceeds the critical rate corresponding to the Eddington luminosity.

In the case of the accretion rate larger than the critical rate, the radiation pressure will prevent all of the accreted matter from reaching the neutron star surface and part of matter will be blown away. Then, if the "horizontal branch" indeed corresponds to the case of super-critical accretion, it is very interesting to note that the observed flux is smaller than the maximum flux. This may indicate the presence of the focussing of radiation into some direction unobservable by us, which accelerates the matter to form the cosmic jets. Although this suggestion is too speculative presently, the possibility will, at least, stimulate further study about the "horizontal branch".

Finally, we will give a short comment on the quasi-periodic oscillation (QPO) from GX5-1 (for recent observational results of QPOs from various sources, see van der Klis 1987; Hasinger 1987; Lewin 1987).

Van der Klis et al.(IAU Circ. No.4140) reported that the QPO from GX5-1 seems to appear mainly in the period when the source is on the "horizontal branch". If the "horizontal branch" corresponds to a state when the accretion exceeds the critical rate as discussed above, the radiation pressure will be dominant in the neutron star envelope at that state. Then, it is to be pointed out that the following evidences probably indicate the presence of the oscillation in the radiation-pressure dominated atmosphere on the neutron star surface.

High time resolution light curves near the peaks of X-ray bursts from X1728-34 (Hoffman et al.1979) and from X1608-52 (Hayakawa 1981) show the oscillatory behaviour on a time scale of about 0.5 sec during the peak. Since the peak fluxes of these two bursts are consistent with the maximum peak luminosity of the bursts from these sources, which is probably the Eddington luminosity, the oscillatory behaviour can be considered to occur in the radiation pressure dominated atmosphere. Hence, the stability of the radiation pressure dominated atmosphere on the neutron star surface will have to be studied in relation to QPOs and also to the "horizontal branch".

## REFERENCES

Alcock,C. 1987, this conference.
Baym,G. and Pethick,C. 1979, Ann.Rev.A.Ap., **17**, 415.
Ebisuzaki,T. 1986, submitted to P.A.S.Japan.
Ebisuzaki,T. and Nomoto,K. 1986, Ap.J., in press.
Fujimoto,M.Y. 1985, Ap.J.(Letters), **29**?, L19.
Fujimoto,M.Y.,Sztajno,M.,Lewin,W.H.G.and van Paradijs,J. 1986, preprint.
Fujimoto,M.Y. and Taam,R.E. 1986, Ap.J., **305**, 246.
Goldmann,I. 1979, A.Ap., **78**, L15.
Hanawa,T. 1986, preprint.
Hasinger,G. 1987, this conference.
Hayakawa,S. 1981, Sp.Sci.Rev., **29**, 221.
Hoffman,J.A.,Lewin,W.H.G.,Primini,F.A. et al., 1979, Ap.J.(Letters), **233**, L51.
Hoshi,R. 1984, P.A.S.Japan, **36**, 785.
Inoue,H.,Waki,I.,Koyama,K. et al. 1984, P.A.S.Japan, **36**, 831.
Inoue,H. and Hoshi,R. 1986, submitted to Ap.J.(Letters).
Lewin,W.H.G. 1987, this conference.
London,R.A.,Taam,R.E. and Howard,W.M. 1984, Ap.J.(Letters), **287**, L27.
Mitsuda,K. 1984, Ph.D.Thesis, Univ of Tokyo (ISAS RN. 251).
Mitsuda,K.,Inoue,H.,Koyama,K. et al. 1984, P.A.S.Japan, **36**, 741.
Mitsuda,K. and Tanaka,Y. **1986**, Proc. NATO Advanced Workshop on the Evolution of Galactic X-Ray Binaries, Tegernsee.
Nakamura,N. 1986, private communication.
Nakamura,N. et al. 1986, in preparation.
Ohashi,T.,Inoue,H.,Koyama,K. et al. 1982, Ap.J., **258**, 254.
Sugimoto,D.,Ebisuzaki,T. and Hanawa,T. 1984, P.A.S.Japan, **36**, 839.
Tawara,Y. et al. 1986, in preparation.
van der Klis,M. 1987, this conference.
van Paradijs,J. 1982, A.Ap., **107**, 51.
Waki,I.,Inoue,H.,Koyama,K. et al. 1984, P.A.S.Japan, **36**, 819.

## DISCUSSION

**A. Burrows:** The neutron star masses you derive for this group of bursters are rather large. Taking these results at face value can you say why this population is comprised of neutron stars whose masses are so different from those derived for other classes of neutron stars?

**H. Inoue:** The high mass neutron star in the X-ray burst sources can be interpreted in terms of the increase of the mass due to the accretion. The persistent emission of X-ray burst sources requires the mass accretion rate of the order of $10^{-9}$ $M_\odot$/year. If the life time of the X-ray burst sources is of the order of $10^9$ years, the amount of mass accreted by the neutron star during that period will be about a solar mass. Hence, the neutron star mass in X-ray burst sources can be 2 $M_\odot$ or more, even if the initial mass is 1.4 $M_\odot$ or less.

# POLAR CAP ACCRETION ONTO MAGNETIZED NEUTRON STA... AN ANALYTIC SOLUTION

Jonathan Arons* and Richard I. Klein#
Department of Astronomy, University of California at Berkeley, and Institute of Geophysics and Planetary Physics, University of California, Lawrence Livermore Laboratory

This abstract should be read in conjunction with the papers by Arons and by Klein and Arons in these proceedings. In the context of the accretion models described there, one can find an analytic solution for the flow down the polar field lines if a number of simplifying assumptions are made. These are (1) steady flow in the co–rotating frame; (2) radiation pressure large compared to gas pressure; (3) pure scattering for the Rosseland opacity, with the magnetic corrections set equal to constants instead of using the actual functions of temperature; (4) diffusion flux of radiative energy proportional to the gradient of the energy density alone, instead of the correct sum of terms proportional to the photon energy density and the number density gradients; and (5) below a radiative shock, subsonic flow in approximate hydrostatic equilibrium. We assumed dipole geometry, and also assume the mass flux is independent of distance from the magnetic axis. The essential trick is to use (1), (2) and (5) to write the advective contribution to the radiation transfer equation as $\dot{M}g$/area = rate at which gravity does work on a fluid element, and use (3) and (4) to write the nonlinear diffusion flux as the ratio of gradients in the energy density. Then the multidimensional diffusion equation can be cast in a separable, linear form by using the logarithmic radial gradient of the energy density as the basic variable (see also Kirk, J., 1985, Astron. and Astrophys., **142**, 430). The result is exponential stratification of the energy density, velocity and mass density along B with scale height $R_* [L_{ED}^{(eff)}/4L_{cap}]$; the effective Eddington luminosity is discussed by Arons, these proceedings. This result can be understood as the result of almost exact balance between upward diffusion and downward advection of photons in the optically thick medium. The same fluid quantities are stratified in a Gaussian manner across B, with angular half width at half maximum $\Delta\theta = [L_{ED}^{(eff)}/L_{cap}](r/R_*)^{3/2}$. These distributions agree well with more sophisticated computational results, during times when the flow is steady. When used as a basis for calculations of the radiative entropy, the calculated emergent spectra are not dissimilar to the spectra of high luminosity, accretion powered pulsars.

---

*also at Department of Physics, University of California at Berkeley
#also at Lawrence Livermore Laboratory



*D. J. Helfand and J.-H. Huang (eds.), The Origin and Evolution of Neutron Stars, 245.*

# RADIATION GAS DYNAMICS OF POLAR CAP ACCRETION ONTO MAGNETIZED NEUTRON STARS

Richard I. Klein and Jonathan Arons
University of California, Berkeley Department of Astronomy
601 Campbell Hall, Berkeley, California 94720, USA
and Institute for Geophysics and Planetary Physics
Lawrence Livermore National Laboratory
P.O. Box 808, Livermore, California 94550, USA

ABSTRACT. We present results of the first self-consistent, time-dependent, 2-D calculations of the accretion of plasma onto polar caps of high luminosity ($L_{*} > 10^{36}$ erg-s$^{-1}$) magnetized neutron stars. We follow the temporal and spatial evolution of three fluids, electrons, ions and photons in a superstrong ($B=3 \times 10^{12}$ Gauss) dipole magnetic field where radiation pressure dominates plasma pressure by solving coupled 2-D equations of radiation hydrodynamics. We have included several physical processes in the radiation-plasma coupling in superstrong magnetic fields (Klein, et al., 1984, Santa Cruz Workshop on High Energy Transients, and Arons, this conference). We solve the resulting system of coupled 2-D PDEs on a Cray XMP-48 by applying implicit finite-difference techniques with iterative operator splitting methods. We present results for two models of $5 \times 10^{37}$ erg-s and $1.5 \times 10^{38}$ erg-s$^{-1}$ super-Eddington luminosity on one polar cap, each having initial mass flux independent of co-latitude of a field lines footprint. We find (a) Radiation develops a broad transverse fan beam that emerges from an annulus 0.2-0.5km above the polar cap. (b) The beam profile is determined by advective trapping of radiation in optically thick ($\tau_{\parallel}, \tau_{\perp} \sim 10^3$) flow. Here the time for diffusion of radiation up through the accretion column is >> the time for downward advection. (c) There is a three fluid nonequilibrium with $T_i >> T_{\gamma} \gtrsim T_e$. (d) Maximum photon temperature of $\sim$10-20 keV in the fan beam is in the observed range. (e) Cyclotron emission >> bremsstrahlung as a source of photons. (f) At early times (<<1ms) radiation pressure strongly decelerates flow to $10^{-3}$ of freefall in central regions of accretion column resulting in a density mound, but plasma freefalls down the sides of the column. (g) Analytical models have reasonable agreement with numerical calculations; velocity and energy density roughly Gaussian transversally and exponential vertically, until the onset of "photon bubbles" after several dynamical times ($\sim$1ms). (h) Multiple "photon bubbles" rising subsonically in the accretion column form in the high luminosity model. We believe the photon bubbles to be a possible consequence of overstable convection in super-Eddington flows. These photon bubbles could be observable as 10-100μs fluctuations in the emergent flux and, thus, be an important diagnostic for inhomogeneous structure of the column.



*D. J. Helfand and J.-H. Huang (eds.), The Origin and Evolution of Neutron Stars, 246.*

# X-RAY IRRADIATED ACCRETION DISK AND BIMODAL STATES

R. Hoshi [1] and H. Inoue [2]
1 Department of Physics, Rikkyo University,
Nishi-Ikebukuro, Tokyo 171
2 Institute of Space and Astronautical Science,
4-6-1 Komaba, Meguro-ku, Tokyo 153

It is well known that black hole candidates, Cyg X-1 and GX 339-4, have distinct high and low states, known as bimodal states. Detailed spectroscopic studies of these X-ray sources have revealed high and low states corresponding to optically thick and thin states of the surrounding accretion disks.

We have found that bimodal states are the natural consequence of the irradiation of the outer accretion disk by X-rays emanating from the vicinity of the accreting black hole.

It has been found by Hayakawa that the outer accretion disk is heated by Compton scattering of X-rays emerging from the vicinity of the accreting black hole. As a result, even if the outer disk is initially optically thick (geometrically thin), Compton heated hot gas continuously evapolates from the low density surface layer of the optically thick disk and finally forms an optically thin hot outer disk. This optically thin disk is maintained by X-ray heating (external source) if the accretion rate is larger than a certain critical value $\dot{M}cr$. If the accretion rate is lower than $\dot{M}cr$, the disk is maintained by viscous heating (internal source).

For the former case, as we proceed inward, the disk changes its structure to an optically thick-geometrically thin state, since cooling due mainly to free-free emission dominates over X-ray heating. This optically thick disk extends down to the radius of the innermost stable circular orbit. Soft X-rays are exclusively emitted from the optically thick disk surface in accord with the soft X-ray excess observed in the high states of Cyg X-1 and GX 339-4.

When the accretion rate is decreased below $\dot{M}cr$, the disk is maintained by viscous heating. In this case, no cooling process outweighs viscous heating. The disk remains optically thin down to the vicinity of the black hole, which corresponds to the low states characterized by a power law spectrum in the X-ray range. Thus, a high to low state transition occurs when the accretion rate is decreased below $\dot{M}cr$. In order to change to a high state, an increase in the accretion rate of more than 3 $\dot{M}cr$ is required. This hysteretic behavior explains why black hole candidates remain in one state for weeks or months before changing to the other state.



*D. J. Helfand and J.-H. Huang (eds.), The Origin and Evolution of Neutron Stars, 247.*

# AN INVERSE COMPTON SCATTERING MODEL FOR THE SPECTRA OF X-RAY PULSARS

G.J. Qiao, X.J.Wu, H. Chen
Geophysics Department, Peking University, Peking, China
X.Y. Xia
Department of Physics, Tianjin Normal University,Tianjin, China

Many observations have been reported in the field of X-ray pulsars, but the mechanism for X-ray emission is not well understood. The X-ray spectra can not be simply described in terms of blackbody or thermal bremsstralung. The high-energy cutoff could be due to cyclotron absorption in high ( $\geqq 10^{12}$ Gauss ) magnetic fields. For the lower energy it can be fitted by a power law with energy index $\alpha$.

We present a model of inverse Compton scattering in strong magnetic fields, which might have certain application to the X-ray continuum spectra of X-ray pulsars at energies below that of the first cyclotron harmonic. This mechanism is used to get a good fit for X-ray spectrum of Her X-1. The beaming effect of inverse Compton Scattering in strong magnetic fields can give a beaming mechanism naturally for X-ray emission of X-ray pulsars.

When the electrons and ions infall into polar cap region a space-charge limited flow must develop, as discussed for radio pulsars. We perform calculations using the method discussed by Sutherland ( 1979 ). It is shown that this is a very efficient mechanism for acceleration of particles and that the acceleration region is not thick. Using the formulae of Thomson and Compton scattering in strong magnetic fields given by Herold ( 1979 ), via Lorentz transformation, the spectra of inverse Compton scattering in strong magnetic fields of an electron have been obtained, then we consider the collective effect to fit the observational data. The theoretical line fits the lower energy spectra of Her X-1 quite well.

References

Herold,H.:1979, Phys.Rev., D19,2868
Sutherland,P.G.: 1979, Fundamentals of Cosmic Physics, 4,95



*D. J. Helfand and J.-H. Huang (eds.), The Origin and Evolution of Neutron Stars, 248.*

# THE X-RAY RADIATION MECHANISM OF THE COMPACT (NEUTRON) BINARY STARS

Tong Yi, Fang Geng and Mao Xinjie
Department of Astronomy
Beijing Normal University
Beijing, China

Usually we think a X-ray source may be a compact(neutron) binary star on which the X-ray radiation might be generated by gravitational acceleration for the particles coming from the primary and going along magnetic field lines of the compact star to the poles. But, in the past, people don't consider well the problem of particle acceleration. It seems to be simplified for the situation only to consider the gravitation effect, because some electric-magnetic effect in a strong magnetic field could not be neglected. However, it is unreasonable to neglect the plasma turbulent waves in an electric-magnetic field, because strong enough turbulent waves such as Alfven waves, whistlers generated nearby the surface of neutron stars probably contribute energy to accelerate particles, which may be more important than gravitation sometimes. For a binary system with a neutron star if ion number density $N > 10^{17}$ /cm$^3$ in its surface atmosphere, the turbulent wavess will be stimulated that will accelerate the particles reaching a speed over $10^8$cm/s. they strike the atmosphere of the compact star in the system, so that a shock wave is formed which turns part of kinetic energy to heat to form hot spots of about $10^8$ K to emit X-ray.

We place the origin of the coordinates at the centre ot the compact star, then we have following equation, 1. The conservation equation of the wave action density. 2. The state equation. 3. The conservative equation for the momentum. 4. The continuity eqution. 5. The conservation equation of magnetic flux.

Using a computer to solve above 5 equations, we can get the particle velocities of this kind of binaries and their temperatures of hot spots calculated by $T = \alpha mV^2/K_B$, where $\alpha$ indicates the energy transfer efficiency when particles going through the shock wave. The result is as follows:

**Table 1: Temperature of hot spots ( $\alpha = 0.1$)**

| Item | r(cm) | v(km/s) | T(K) |
|---|---|---|---|
| Neutron star | $10^8$ | 15000 | $1.36 \times 10^9$ |



*D. J. Helfand and J.-H. Huang (eds.), The Origin and Evolution of Neutron Stars, 249.*

# X-RAY SPECTRA AND ATMOSPHERIC STRUCTURES OF BURSTING NEUTRON STARS

TOSHIKAZU EBISUZAKI
Department of Earth Science and Astronomy
College of Arts and Sciences
University of Tokyo
Komaba Meguro-ku, Tokyo 153

ABSTRACT. High quality spectra of Japanese X-ray satellite TENMA permit us detailed studies of X-ray burst on color temperature vs. luminosity diagram. Using this diagram, Sugimoto Ebisuzaki and Hanawa(1984) divided bursts from MXB1636-536 into two classes: bright class and faint class. In the decay phase of bright class bursts, surface of neutron star should be covered with pure helium matter processed on the surface of neutron star because hydrogen-rich matter is completely lost due to mass loss. On the other hand, surface of neutron star should be covered by hydrogen-rich matter in the case of faint class bursts because any mass loss is not driven in the faint class bursts. I constructed color temperature vs. luminosity diagrams of X-ray burst sources MXB1636-536 and MXB1608-522 using data of Japanese X-ray satellite TENMA and found systematic deviations between evolutional paths of the bright class and faint class bursts on these diagrams. The deviations are successfully accounted for by the difference in chemical composition between bright class and faint class bursts: evolutional paths of bright class and faint class are respectively in good agreement with theoretical curves for pure helium and hydrogen-rich matter. This is a direct evidence of ejection of hydrogen-rich envelope in the bright class bursts.

## REFERENCE

Sugimoto, D., Ebisuzaki, T., and Hanawa, T 1984, Publ. Astron. Soc. Japan, 36, 839.



*D. J. Helfand and J.-H. Huang (eds.), The Origin and Evolution of Neutron Stars, 250.*

# MODEL ATMOSPHERES FOR X-RAY BURSTING NEUTRON STARS*

Richard A. London, Ronald E. Taam and W. Michael Howard
Physics Department, LLNL
University of California
P.O. Box 808
Livermore, CA 94550 U.S.A.

Self consistent neutron star atmospheric models have been constructed which include the effects of Comptonization, free-free and bound-free absorption. It has been demonstrated that for parameters relevant to x-ray bursting neutron stars the atmosphere does not radiate like a blackbody during any phase of an x-ray burst. In particular, during the initial rise and final decline of the burst the temperature structure of the atmosphere is affected by backwarming associated with the high opacity due to free-free processes at low frequencies to an extent that the radiation spectrum is shifted to higher energies than a blackbody of the same effective temperature. On the other hand, near the peak of the burst, the opacity is more gray-like as the electron scattering opacity dominates; however, in this case thermalizaton of the radiation field occurs at such large optical depths ($\tau \sim 5$) that the spectral temperature is higher than the effective temperature. This result is found despite the importance of Comptonization in the thermalization process. Thus, the super Eddington fluxes implied by the spectral data alone are misleading and result from the improper use of the spectral temperature for the effective temperature. For neutron stars characterized by a soft equation of state and radiating near the Eddington effective temperature, fluxes obtained in this way could be overestimated by a factor of about 5.

Because the spectral hardening factor varies throughout an x-ray burst, deconvolution of the spectrum is required before definitive statements can be made concerning the variation of the size of the emitting region. We may say that because of the spectral hardening effect, radius determinations, based on the spectral temperature are only lower limits.

We have also found that the shape of the continuum can be affected by the presence of Fe in the atmosphere; however, the features from bound-free transitons in partially ionized Fe are not particularly strong in the non-LTE approximation for solar abundances.

*Work performed under the auspices of the U.S. DOE by the LLNL under contract number W-7405-ENG-48.



*D. J. Helfand and J.-H. Huang (eds.), The Origin and Evolution of Neutron Stars, 251.*

# III. NEUTRON STAR FORMATION IN THEORETICAL SUPERNOVAE

*CHAIR: R. Chevalier*

# THE BIRTH OF NEUTRON STARS

S. E. Woosley

Board of Studies in Astronomy and Astrophysics
Lick Observatory, University of California at Santa Cruz
Santa Cruz CA 95064
and
Special Studies Group, Lawrence Livermore National Laboratory
Livermore CA 94550

ABSTRACT. Presupernova models of massive stars are discussed and their explosion by either the "core bounce" or neutrino energy transport mechanism briefly reviewed. Special consideration is given to those attributes of the stellar evolution and explosion that might influence the properties of the neutron star remnant: its mass, rotation rate, magnetic field, and "kick" velocity.

## 1. INTRODUCTION

By now it is recognized that neutron stars are a likely product of the evolution of at least some massive stars. This follows from such diverse considerations as 1) the theoretical certainty that stars more massive than about 8 $M_{\odot}$ must evolve to a state of gravitational collapse, and barring the most massive ($M \gtrsim 100\ M_{\odot}$; §2.4), produce either a black hole or a neutron star; 2) the presence of neutron stars in the remnants of historical supernovae (with records kindly provided by our hosts!) such as the Crab Nebula whose non-solar abundances, especially helium, indicate a relatively massive ($\sim$10 $M_{\odot}$) progenitor; 3) the distribution of pulsars with galactocentric radius (Lyne 1981) consistant with Population I objects; 4) the fact that x-ray pulsars and other forms of accreting neutron stars are frequently, though not always, found in binary systems with massive companions, (that the neutron star formed before the short lived companion died again suggests that the neutron star itself had a massive, short-lived progenitor); and 5) the masses of the binary pulsar PSR 1916+13 which are too large to have been formed by the collapse of any but a massive star (Burrows and Woosley 1986; §4.1). Further, the inferred pulsar birth rate (Lyne 1981; this volume) is *consistent* with the observed rate of Type II supernovae (with conservative error bars of a factor of two on each) and the z-distribution of pulsars is consistent with birth in the Galactic disk, albeit with a large peculiar velocity.

None of these conditions requires, however, that *every* neutron star, or even a majority, be produced by massive stars, nor do they require, nor is it even likely, that *every* massive star end its life in producing a neutron star (see e.g.. Helfand and Becker 1984). The possibility of neutron star production by a massive star is, as we shall see, sensitive to the stellar mass and, unfortunately, to the physics that a given theoretician employs in his or her calculation. It turns out that the more massive the progenitor star. the less likely it is that the remnant will



*D. J. Helfand and J.-H. Huang (eds.), The Origin and Evolution of Neutron Stars, 255–272.*

be small enough to stabilize as a neutron star. Further, it is possible to produce neutron stars by accretion at certain rates on white dwarfs in binary systems and thus bypass the need for a massive stellar parent (§3.2).

## 2. TYPE II SUPERNOVAE

Since this subject has been very recently reviewed (Woosley and Weaver 1985; 1986ab; Woosley 1986; Nomoto 1986; Wilson *et al.* 1986; Hillebrandt 1984, 1985), our treatment here can be terse and will concentrate only upon those aspects of the presupernova evolution and supernova mechanism that directly affect the observable properties of neutron stars, especially their mass, rotation rate, and velocity.

### 2.1 *Presupernova Evolution of Massive Stars*

The theoretical evolution of single massive stars allows their segregation, upon the basis of mass, into three groups, each of which encounters gravitational collapse in its own unique fashion (§2.2, 2.3, 2.4). Though the existence of these three groups is clear, the precise boundaries of each are sensitive to the physics employed by various researchers, especially the initial helium abundance, the rate for the $^{12}C(\alpha,\gamma)^{16}O$ reaction, the degree to which mass loss modifies the star's total mass prior to helium depletion, and the prescription for convection. Based upon a standard set of assumptions, helium abundance near 28%, $^{12}C(\alpha,\gamma)^{16}O$ as given by Caughlan *et al.* (1985), no mass loss, and a moderate amount of convective overshoot mixing (Weaver, Zimmerman, and Woosley 1978), Woosley and Weaver (1986ab) estimate the following mass delineations and behavior:

### 2.2 *Stars in the 8 to 10 $M_\odot$ Range*

Stars above 8 $M_\odot$ will ignite carbon burning in a non-degenerate fashion, but for a relatively small interval of masses above this value the later stages of evolution continue to be complicated by effects of electron degeneracy. Stars in this mass range have been recently studied by Nomoto (1984ab; 1986; this volume), Woosley, Weaver, and Taam (1980), and Wilson *et al.* (1986) and their evolution is best segregated upon the basis of the helium core mass, *i.e.* the mass interior to the hydrogen burning shell at the time the star becomes a supernova. Main sequence stars in the 8 to 10 $M_\odot$ range end up with helium cores of roughly 2 to 3 $M_\odot$. Helium cores between about 2.2 and 2.5 $M_\odot$ develop within themselves oxygen cores that do not ignite either neon or oxygen burning in hydrostatic equilibrium. Rather the metallic cores grow to the Chandrasekhar mass (which may be decreased to $\sim$1.37 $M_\odot$ by electron capture), ignite oxygen burning at a very high central density, $\sim 2 \times 10^{10}$ g cm$^{-3}$, and *implode.* The collapse occurs because degenerate oxygen burning proceeds all the way to nuclear statistical equilibrium, producing abundant iron group nuclei having low electron capture thresholds. The thermal pressure increase obtained as the temperature rises to $\sim 10^{10}$ K is inconsequential compared to the effect of the pressure lost to electron capture, and thus the core collapses even as burning continues. Explosion of these stars has been studied by

Hillebrandt, Nomoto, and Wolff (1984) and Wilson *et al.* (1986) with conflicting results. The former group calculated violent hydrodynamical explosions while the latter obtained only a weaker explosion caused by neutrino energy transport (§2.5).

For larger helium cores between about 2.5 and 3.0 $M_\odot$, neon, oxygen, and silicon burning do occur, albeit in complicated stages characterized by off-center ignition, and an iron core in hydrostatic equilibrium is eventually formed (Woosley, Weaver, and Taam 1980; Nomoto 1984a, 1986). This iron core, which is near 1.40 $M_\odot$, collapses chiefly because of electron capture and, to a lesser extent, photodisintegration. Again the results of computer simulation of the core bounce and explosion are not in agreement. Hillebrandt (1982) gets a marginal explosion by hydrodynamical bounce of a 10 $M_\odot$ star, but Wilson (1985) and Burrows and Lattimer (1985) do not and Wilson *et al.* (1986) get an explosion based instead upon neutrino energy transport.

Stars in the 8 to 10 $M_\odot$ range all have very steep gradients in mass density just above the iron core implying a small gravitational binding energy for the external matter. Thus when an explosion does develop it always ejects all of the mass external to the iron core.

## 2.3 *Stars in the 10 to ~70 $M_\odot$ Range*

Stars in this mass range ignite hydrogen, helium, carbon, neon, oxygen, and silicon burning non-degenerately in hydrostatic equilibrium at their centers. The iron core thus formed collapses owing to a combination of electron capture ($M \lesssim 20$ $M_\odot$) and photodisintegration ($M \gtrsim 20$ $M_\odot$). The size of the iron core at collapse becomes less sensitive to the Chandrasekhar mass as one goes to the larger mass stars characterized by higher values of central entropy. Larger central entropies have two effects that greatly influence the development of the iron core. First they allow core masses to exist in hydrostatic equilibrium that are much larger than the Chandrasekhar value. Second a larger central entropy and the shorter time scales associated with massive stellar evolution imply that the radial gradients in entropy are and remain smaller. It is these gradients which determine the extent of convective shells especially during the critical oxygen and silicon shell burning stages. The extent of the convective shells ultimately sets the mass of the iron core at collapse.

Thus there exists a main sequence mass, currently estimated at ~20 $M_\odot$, above which the final iron core begins to appreciably exceed 1.4 $M_\odot$. The exact value where this occurs is of obvious great interest, all the more so because the possibility of exploding the star is very sensitive to the size of the iron core (§2.5). A calculation of this critical value depends upon proper inclusion of all sources and sinks of entropy, such as neutrino losses, radiation transport, and nuclear burning, and upon the theory of convection. Somewhat surprisingly, though reasonably, the results also depend sensitively upon the carbon abundance that exists in the core following helium burning. If the carbon abundance is low, $\lesssim$10% by mass fraction, then carbon and neon burning never generate enough energy to surpass, even locally, that which is lost to neutrinos. Thus, insofar as entropy is concerned, the star proceeds directly and rapidly from helium exhaustion to oxygen ignition without ever spending time convectively burning carbon and neon in its center. Such rapid evolution allows the entropy to remain high in the inner regions and ultimately leads to large iron cores (Woosley and Weaver 1986ab).

This implies a sensitive dependence of the late stages of stellar evolution upon the carbon nucleosynthesis that occurs during helium burning. Carbon nucleosynthesis, in turn, depends in an obvious fashion upon the uncertain rate for the $^{12}C(\alpha,\gamma)^{16}O$ reaction and, in a more subtle manner, upon the theories of semi-convection and convective overshoot employed in the computer code. Some degree of convective penetration beyond that radius formally unstable to convection by the Ledoux criterion is expected to occur, especially near the outer edge of the helium convective shell at a time when the central helium abundance has dropped to only a few percent and where the temperature gradient is only slightly below the critical adiabatic value. Mixing by convective overshoot during this stage brings new helium into a region in which the oxygen to carbon ratio is already high. Since the formation rate of carbon depends upon the cube of the $\alpha$-particle abundance while rate of its conversion to $^{16}O$ is linear, each new helium nucleus brought in almost invariably converts a carbon nucleus to an oxygen nucleus. Thus the carbon nucleosynthesis is sensitive to how much overshoot mixing is assumed.

The models calculated by Woosley and Weaver (1986b), Wilson *et al.* (1986), and Weaver, Woosley, and Fuller (1985) all assume the overshoot parametrization discussed by Weaver, Zimmerman, and Woosley (1978). That is, single zones bounding regions that are formally unstable to convection by the Ledoux criterion are treated as *semi-convective.* Though no convective *energy* transport occurs, the composition of this boundry zone is slowly mixed into the convective helium core. The diffusion coefficient for this "overshoot zone" is calculated in the same fashion as in an ordinary convective zone, but the temperature gradient used in the convective velocity calculation is some adjustable fraction larger than the adiabatic one. In all calculations thus far the value of this constant is 1%, though sensitivity studies using other smaller values are currently underway. It is believed that any non-trivial value for this parameter will give similar results. The diffusion coefficient for these special "overshoot" zones is further restricted, as it is in all semi-convective zones, to be no greater than 10% of the radiative diffusion coefficient (see Weaver, Zimmerman, and Woosley 1978 for further discussion and justification).

If there were no overshoot mixing, the carbon abundance would be higher and it would be easier to form iron cores of near the Chandrasekhar mass for larger values of main sequence mass. If a greater amount of overshoot mixing were employed, the iron cores would be larger for smaller stars. Calculations to quantify these statements are in progress.

### 2.4 *Stars Heavier Than 70 $M_\odot$*

Stars that have helium core masses larger than 32 $M_\odot$ at the end of helium burning become unstable as they evolve to more advanced burning stages. For the present $^{12}C(\alpha,\gamma)^{16}O$ rate and parametrization of convection, the carbon and neon abundances are too small to provide exoergic burning stages powered by these fuels.

Thus the central nuclear energy generation does not exceed neutrino losses until oxygen begins to burn. Oxygen ignition occurs at relatively high values of central entropy ($S/k \gtrsim 7$) and temperature which favor the production of electron-positron pairs. Production of the pair rest mass temporarily reduces the structural $\Gamma$ of the star below 4/3 and collapse begins (Barkat, Rakavy, and Sack 1967). What follows depends very sensitively upon the initial entropy, and hence mass, of the helium core.

For helium core masses greater than $\sim$50 $M_\odot$, with the precise value still awaiting determination, the instability is violent enough that the core collapses to between 3.5 and $5 \times 10^9$ K, burns a fraction of the oxygen core to silicon and heavier elements, and, for helium cores lighter than about 120 $M_\odot$, produces enough energy to reverse the implosion and completely disrupt the star (Woosley and Weaver 1982; Bond, Arnett, and Carr 1984; Ober, El Eid, and Fricke 1983; Glatzel, El Eid, and Fricke 1985; El Eid and Langer 1986). The light curves from such explosions can be very brilliant, especially towards the heavier end of this mass range, since a great deal of radioactive $^{56}$Ni is produced in the explosion. For helium cores still more massive than $\sim$120 $M_\odot$, the kinetic energy of infall becomes too great to be reversed by a nuclear explosion (unless one includes the effect of rotation which can raise the limit considerably; Stringfellow, Woosley, and Bodenheimer 1983; Glatzel, El Eid, and Fricke 1985). As the nuclear burning passes from oxygen burning to silicon burning and finally nuclear statistical equilibrium, reactions begin to absorb more energy than they release. The star collapses directly to a black hole (Woosley, Wilson, and Mayle 1986). Thus one way or another, helium cores heavier than about 50 $M_\odot$ (*i.e.*, stars heavier than about 110 $M_\odot$ on the main sequence) avoid neutron star production.

For helium cores between 32 and $\sim$50 $M_\odot$, the pair instability manifests itself as an extreme form of pulsational instability and does not lead directly to total stellar disruption. Main sequence stars of 75 $M_\odot$ and 100 $M_\odot$ studied by Woosley and Weaver (1986b) produced helium cores of 36 and 45 $M_\odot$ respectively (Table I) which became violently unstable at oxygen ignition. The 45 $M_\odot$ core, whose instability was followed in some detail, ejected about 3 $M_\odot$ with total kinetic energy $4 \times 10^{50}$ erg over the course of 4 "pulsations." Substantial cooling by neutrino losses during the Kelvin-Helmholtz stage following each outward excursion reduced the entropy sufficiently that stable silicon burning was finally ignited in the core. Eventually an iron core of 2.3 $M_\odot$ was produced which collapsed on the photodisintegration instability (Table I). Whether neutrino energy transport can eject the matter external to a core of this size is still to be determined although Wilson *et al.* (1986) studied a similar core and found growth to over 3 $M_\odot$ without explosion. The possibility of an explosion energized by rotation and nuclear burning (Bodenheimer and Woosley 1983) also needs to be examined for this configuration. But one way or another, it seems likely that stars in the pulsationally unstable regime will also leave black hole remnants, not neutron stars.

Table I. Presupernova Models and Explosions

| Main Seq. Mass | Helium Core Mass | Iron Core Mass | Expl. Energy[a] ($10^{50}$erg) | Residual Baryon Mass[a] | Neutron Star Mass[a] | Heavies Ejected ($Z \geq 6$) |
|---|---|---|---|---|---|---|
| 11 | 2.4 | —[b] | 3.0 | 1.42 | 1.31 | ~0 |
| 12 | 3.1 | 1.31 | 3.8 | 1.35 | 1.26 | 0.96 |
| 15 | 4.2 | 1.33 | 2.0 | 1.42 | 1.31 | 1.24 |
| 20 | 6.2 | 1.70 | — | — | — | 2.53 |
| 25 | 8.5 | 2.05 | 4.0 | 2.44 | 1.96 | 4.31 |
| 35 | 14 | 1.80 | — | — | — | 9.88 |
| 50 | 23 | 2.45 | — | — | — | 17.7 |
| 75 | 36 | —[c] | — | — | BH? | 30? |
| 100 | 45 | ~2.3[c] | $\gtrsim$4 | — | BH? | 39? |

[a] All except for 100 $M_\odot$ determined by Wilson *et al.* (1986)

[b] Never developed iron core in hydrostatic equilibrium

[c] Pulsational pair instability at oxygen ignition

## 2.5 *Core Collapse and Explosion in $M \lesssim 70\ M_\odot$*

For many years the outward propagation of the shock wave generated by the bounce of the collapsing iron core of stars in this mass range has been studied in the hope that it might provide the energy required to eject the stellar mantle and envelope and produce a supernova. Unfortunately the most careful theoretical studies are still at variance regarding the success of this and other possible mechanisms, e.g. neutrino energy transport.

To summarize briefly, the shock wave generated by core bounce in a massive star is born with a certain characteristic energy ( $\sim 7 \times 10^{51}$ erg; Burrows and Lattimer 1983) deep within the interior, but not at the center, of the neutronized core. Typically the mass interior to the point where the shock first develops is ~0.8 $M_\odot$. As it moves outwards this shock experiences energy losses, principally owing to photodisintegration of heavy nuclei and, once the density falls below a few times $10^{11}$ g cm$^{-3}$, to neutrino losses. The larger the mass of the collapsing iron core, the greater are the losses, especially photodisintegration which takes about $1.5 \times 10^{51}$ erg for each 0.1 $M_\odot$ of core material that must be traversed. For iron core masses greater than 1.25 to 1.40 $M_\odot$, with the precise value depending upon choice of nuclear equation of state and the distribution of entropy and electron fraction, the shock wave dies before reaching the edge of the iron core. A supernova powered by hydrodynamical energy transport then becomes impossible for larger cores.

This range marginally intersects smaller core masses given in Table I and, not surprisingly, various researchers have calculated qualitatively different results for the core bounce. Baron, Cooperstein, and Kahana (1985ab) obtain energetic explosions of 12 and 15 $M_\odot$ model stars but employ an equation of state that is softer, hence more favorable to explosion, than traditional in the nuclear physics community (though Baron *et al.* 1985 defend their choice as natural for the special environment appropriate to supernova cores). Other researchers have not yet attempted confirmation of the Baron calculations though work is in progress by Wilson and Mayle. Presently it appears marginally possible that stars in the 8 to $\sim$20 $M_\odot$ range will explode promptly. If so, the time scale for explosion following maximum compression is $\sim$20 ms and the explosion energy, around $10^{51}$ erg.

If these stars do not explode promptly, and certainly the more massive cores in Table I will not, then the outgoing shock stalls and becomes an accretion shock. Hope for exploding the star now hinges upon the efficiency of neutrinos in transporting a sufficient fraction of the core binding energy to a region that is simultaneously sufficiently optically thick to neutrinos and insufficiently bound by gravity that thermal overpressure can lead to its ejection. Wilson and coworkers have presented calculations (Wilson 1985; Wilson *et al.* 1986; Mayle 1986) and arguments (Bethe and Wilson 1985) to show that such a mechanism can indeed produce eventual explosion in all save the most massive stellar cores. The mechanism depends upon the capture over a long period of time, typically several hundred milliseconds, of neutrinos by matter directly beneath the accretion shock. Only a few percent of the neutrinos flowing through are captured. This revives the outward motion of the shock, eventually giving rise to ejection of all external matter and explosion energies in the range $\sim 3 \times 10^{50}$ to $10^{51}$ erg. Two points are of special interest here. First the possibility of a succesful explosion that leaves behind a baryonic mass of over 2.4 $M_\odot$ (Table I) implies that some supernovae could leave black hole (or at least very massive neutron star) remnants. The remnants would spend several seconds, perhaps even a minute before losing enough neutrinos to collapse inside their event horizon. Second, during the relatively long period that the "delayed explosion" mechanism requires for its operation, a substantial quantity of matter may accrete onto the neutronized core. Thus the core that is left behind may be larger than the original iron core that collapsed. For the lower energy explosions some of the inner mantle material may end up moving slower than the escape velocity and fall back onto the neutron star, again increasing the probability of massive neutron star or black hole formation.

The neutrino energy transport model of Wilson has yet to be verified by other researchers and, as Arnett discusses in this volume, remains quite controversial. In order to produce a more energetic and less marginal explosion, one needs to boost the neutrino flux coming out of the core. This might be achieved by either a global overturn of the core (Epstein 1979; Colgate and Petschek 1980) or by a more docile form of convection (Mayle 1986; Arnett 1986, this volume) which would bring neutrinos trapped deeper in the core to the neutrinosphere, thereby increasing the neutrino luminosity of the star at a critical time.

## 3. OTHER SYSTEMS THAT MIGHT LEAVE NEUTRON STARS

### 3.1 *Type Ib and "Type III" Supernovae*

The name "Type II Supernova" refers specifically to a supernova whose spectrum displays lines of hydrogen. It is quite probable that very massive single stars (M $\gtrsim$ 40 $M_\odot$), or some less massive stars in interacting binary systems, end their lives devoid of a hydrogen envelope. Such stars would exhibit bolometric and spectroscopic properties quite distinct from Type II supernovae despite identical explosion mechanisms operating in the core. I have recently calculated (Woosley 1986; Ensman and Woosley 1986) the light curves of the helium cores of the 15 and 25 $M_\odot$ models of Table I, but with the hydrogen envelopes artificially removed. The light curves are totally dominated by energy from radioactive decay ($^{56}$Ni and $^{56}$Co) and are qualitatively similar to those calculated for Type I supernovae (Woosley and Weaver 1986ab). They are, however, broader (FWHM $\sim$ 70 d for the 25 $M_\odot$ star; 8 $M_\odot$ He core) and dimmer at peak light ($L_{max} = 3.4 \times 10^{42}$ erg s$^{-1}$) than a typical Type Ia supernova. These are also properties attributed to a recently discovered subclass of "Type I" supernovae called Type Ib (Wheeler and Leverault 1985; Gaskell *et al.* 1986) or sometimes Type III (Chevalier 1986). Other proposed examples of this class are SN 1985f (the "Filippenko-Sargent object," Filippenko and Sargent 1985, 1986) and Cas A. While I do not at the present wish to join the bulk of the community in attributing Type Ib supernovae to massive stellar cores (because of reservations I have regarding the spectrum at peak light; see also Branch and Nomoto 1986), it is worth noting the probable existence of a class of supernovae that 1) lack hydrogen, 2) are associated with a massive stellar population, 3) may have been surrounded by envelopes from pre-explosive mass loss, 4) have a light curve powered totally by radioactivity, 5) are optically dim compared to Type Ia, and 6) unlike Type Ia, *may leave neutron star or black hole remnants.*

### 3.2 *Accreting White Dwarfs*

It has long been speculated that continued accretion onto a white dwarf star might push that star over the Chandrasekhar mass and cause collapse directly to a neutron star. If this occurred a neutron star would be formed in a binary system (although see §4.3) in an event that might be very difficult to detect. Lacking both hydrogen envelope and substantial ejection of $^{56}$Ni during the explosion, the collapse would be optically dim, hence the term "silent supernova" (Nomoto 1984c).

Historically the problem in producing such events has been avoiding the total disruption of the white dwarf by a degenerate carbon explosion. Recently, however, several scenarios have emerged that circumvent this difficulty. All utilize white dwarfs which, for one reason or another, have zero or trivially small carbon abundances in their centers as they approach the Chandrasekhar mass.

Canal and Isern (Canal and Schatzman 1976; Canal and Isern 1979; Isern *et al.* 1983) have proposed a phase transition leading to a mechanical separation of carbon and oxygen in systems with very low accretion rates. By the time the

Chandrasekhar mass is reached, the center of the star consists almost entirely of oxygen. Ignition of nuclear reactions is thus forestalled until such high densities are reached $\gtrsim 10^{10}$ g cm$^{-3}$ that electron capture leads to collapse (see §2.1) much as in a 2.2 $M_\odot$ helium star.

Saio and Nomoto (1985), Nomoto (1986), Woosley and Weaver (1985), and Nomoto and Iben (1985) have discussed an alternate scheme wherein the merging of two carbon-oxygen white dwarfs accomplished by gravitational radiation leads to such *high* accretion rates that carbon burning is ignited non-degenerately off-center. A diffusive burning front propagates to the center of the star while it remains in hydrostatic equilibrium, depleting the star of carbon and converting the inner regions to oxygen and neon. Continued accretion then leads to oxygen ignition and collapse of the white dwarf as above.

Finally one may just start with a neon-oxygen white dwarf, the product of an 8 to 10 $M_\odot$ star that has lost its envelope in a binary system, and accrete on that object until it collapses. Observational evidence for white dwarfs of this type in accreting binary systems has been recently reviewed by Starrfield, Sparks, and Truran (1986) within the context of classical novae.

## 4. NEUTRON STAR PROPERTIES

### 4.1 *Masses*

From the considerations of the previous sections and references therein it follows that neutron stars should be born with a variety of masses (see *e.g.* Rappaport and Joss 1981). In particular, there is no reason why the preferred mass should be near 1.44 $M_\odot$. In the simplest case imaginable where a white dwarf is pushed over its critical mass (Saio and Nomoto 1985), the collapse starts with a mass closer to 1.42 $M_\odot$ because of previous electron capture and because a central density of only $10^{10}$ g cm$^{-3}$ need be achieved in order to trigger collapse. More importantly, the binding energy of the final neutron star must be subtracted off from this mass (*e.g.* Burrows and Woosley 1986), the difference being the neutrinos emitted during the collapse. The specific amount subtracted depends upon the nuclear equation of state adopted but is near 10%. Thus the gravitational mass of the neutron star would be near 1.3 $M_\odot$, possibly 1.35 $M_\odot$.

Somewhat smaller masses might be achieved from the collapse of massive stellar cores, especially in the 8 to 20 $M_\odot$ range. There, depending upon the pre-explosive electron capture and explosion mechanism, the residual iron core mass might be as small as $\sim$1.35 $M_\odot$, (*e.g.* Baron, Cooperstein, and Kahana 1985ab; Wilson *et al.* 1986), thus yielding a neutron star mass near 1.25 $M_\odot$. General considerations of shock wave strength (Burrows and Lattimer 1983; Hillebrandt, Nomoto, and Wolff 1984; Baron, Cooperstein, and Kahana 1985ab) suggest that if the iron core became smaller than $\sim$1.25 $M_\odot$ the shock wave would begin to blow away a significant part of the neutron star. This is strictly forbidden in all but very rare events by nucleosynthetic considerations (Hartmann, Woosley, and El Eid 1985). It is almost certainly impossible to get a stellar core smaller than this size to collapse anyway. Photodisintegration would be negligible and the central density would be too low to cause collape by electron capture. Subtracting *at most* 10% for the binding energy gives a stringent lower limit to the neutron star mass of about 1.15 $M_{\odot}$. As discussed above a more likely value would be 1.3 $M_{\odot}$.

The *largest* neutron star mass is given entirely by considerations of nuclear equation of state since the evolution of massive stars is capable of providing a continuum of masses up to and above the maximum value. The more massive neutron stars should be rarer owing to the decreasing abundance of stars of higher mass. Adopting as a representative example a critical mass near 20 $M_{\odot}$ as the heaviest star for which the entropy is low enough to allow convergence of the iron core mass to near the Chandrasekhar value, a value of 40 $M_{\odot}$ as the heaviest star that leaves a neutron star remnant rather than a black hole, and 8 $M_{\odot}$ for the lowest mass star that undergoes gravitational collapse, one finds (Miller and Scalo 1979) that about 10% of neutron stars should have masses appreciably in excess of the typical 1.3 $M_{\odot}$ value. This number is insensitive to the assumption of 40 $M_{\odot}$ for those stars that produce black holes (unless the value becomes much smaller), but is very sensitive to the assumed 8 and 20 $M_{\odot}$ values as well as the IMF of Miller and Scalo. This number should therefore be treated as a rough approximation. Of course one must also consider the possibility that not all neutron stars have a massive parent (§3.2).

A case of special interest is, PSR 1913+16, the first discovered of the binary pulsars for which the masses of both components have now been determined to high accuracy (Weisberg and Taylor 1984; Taylor 1985, Taylor, this volume). Recent analyses give 1.444 $M_{\odot}$ and 1.384 $M_{\odot}$ for the two stars, both of which are apparently neutron stars (Srinivasan and Van den Heuvel 1982) and the heavier of which is the pulsar. From the above discussion, neither the pulsar nor its companion could have evolved from a collapsed white dwarf unless accretion continued long after the collpase of the neutron star that became the pulsar. The latter is probably ruled out by the young age of the pulsar and, in any event, one is still left with the large mass of the companion. Burrows and Woosley (1986) present arguments to show that this binary pulsar probably evolved from a system of two massive stars, both having main sequence masses between about 15 and 20 $M_{\odot}$. If so, a "kick velocity" of $\sim$200 km $s^{-1}$ is required if the system is to remain gravitationally bound following the second explosion.

### 4.2 *Rotation Rate and Magnetic Field*

Little in the way of quantitative predictions can presently be made, or are likely to be made by theoreticians in the near future, regarding these important attributes of young neutron stars. Upper main sequence stars earlier than F5 are well known to have high equatorial rotational velocities. Assuming that these stars rotate uniformly, the inferred total angular momentum, $J$, is very large, the dimensionless quantity $cJ/GM^2$ lying in the range 9 to 18 (Kiguchi and Sato 1985). If angular momentum were conserved and not transported, rotation would unavoidably play a major role during the final stages of stellar evolution. The core would evolve to the point where the ratio of rotational energy to gravitational energy $(T/W)$ exceeded about 25% and triaxial deformation would then ensue on a dynamical time scale. Bifurcation and/or rapid transport of angular momentum, perhaps by gravitational radiation, would occur and ultimately a neutron star would be born which, neglecting the effects of magnetic interaction (?), would rotate at a substantial fraction of breakup speed. From the standpoint of stellar evolution there is no apparent fundamental reason why this should not occur.

However, angular momentum transport will occur throughout the stellar lifetime, if by no means other than convection. Endal and Sofia (1976; 1978) using one parametrization of angular momentum transport found that rotation would *not* become a major effect during hydrogen, helium, or carbon burning in a 10 $M_\odot$ star, but extrapolated that rotation would become important during later stages. These later stages will be convective however, at least in stars more massive than $\sim$10 $M_\odot$. The several stages of convective oxygen and silicon shell burning could transport substantial angular momentum out of the inner several solar masses, causing the core to tend towards solid body rotation. A large velocity shear near the edge of this core could lead to turbulent dissipation or magnetic braking.

Beyond these already speculative comments it is difficult to say more. Core contraction from the onset of dynamical instability ($\rho_{cent} \sim 10^9$ to $10^{10}$ g cm$^{-3}$) should proceed rapidly enough that, unless triaxial deformation develops, angular momentum is conserved. In particular the viscosity afforded by escaping neutrinos should be negligible (Lindblom and Detweiler 1979). No reason is readily apparent why the rotation rate of the young neutron star should be any particular number (other than the limiting value required for triaxial deformation) or why the magnetic field should tend to lie in the range $10^{12}$ to $10^{13}$ gauss, though such fields are easily attainable if flux is conserved and the silicon core has a magnetic field similar to that of a magnetic white dwarf (*e.g.* Am Herc).

It is worth noting, however, that different scenarios for producing neutron stars might reasonably lead to different ranges of rotation rates and magnetic fields. A collapsing white dwarf in a merging binary pair might be very rapidly spun up to near breakup (Nomoto 1986). A somewhat more distant binary pair might lead to the collapse of a white dwarf (a presumed rigid rotator) that is tidally locked to the period of the binary, thus a more slowly rotating neutron star. Stars in the 8 to 10 $M_\odot$ range will not experience oxygen and silicon burning in hydrostatic equilibrium. The last convective phase of many of these stars is carbon burning at a central density of $10^5$ g cm$^{-3}$. Yet in more massive stars convection during silicon burning occurs up to central densities of several times $10^8$ g cm$^{-3}$, thus affording the possibility of angular momentum transport out of a core that is already quite compact.

Obviously almost any observed rotation rate or classes of rotation rates can be accommodated.

### 4.3 *"Kick" Velocities*

Pulsars are born having a relatively high peculiar velocity, typically on the order of a few hundred kilometers per second. This could be due to escape from a binary pair or could reflect some asymmetry in the explosion itself. Kick velocities of the same order may be required in order to keep those few binary pulsars that are observed gravitationally bound to their companions. Since no theoretical simulations of core collapse and supernova explosion have been carried out in three dimensions any comments one would make on this subject must presently be of a speculative nature. Still we note several attributes of the explosion mechanism that might easily lead to a recoil. The required effect is only at the level of several percent in the supernova energy.

Most calculations of core bounce carried out so far have used 1D hydrodynamics codes and, of necessity, generate radially symmetric results, but even these studies are useful for showing that a large fraction of the core, that fraction external to the "homologous core" ($\sim$0.6 to 0.8 $M_\odot$) is collapsing supersonically when the center of the star reaches nuclear density (*e.g.* Baron, Cooperstein, and Kahana 1985ab). Thus, during core collapse and bounce, large regions of the star are, for the first time since its formation, out of sonic communication with one another. Any asynchronism between collapse onset or difference in collapse rate between two sides of the core will result in one side bouncing slightly in advance of the other. Then the shock wave that moves out will not have radial symmetry but will be egg-shaped, with its leading edge along the angle that bounced first. This egg-shaped shock could be decomposed into a central radially symmetric shock and a smaller one, originating off-center. It is the small off-center component that gives an asymmetric momentum or "kick" to the explosion.

Alternatively one may consider a "delayed" explosion driven by neutrino energy transport (Wilson *et al.* 1986). In this model the explosion occurs at late times owing to the capture by material behind a stalled accretion shock of a fraction of the neutrino luminosity coming from the mantle and outer layers of the young neutron star. There is no reason that this flux must be precisely the same on both sides of the core. Once again disjoint regions are out of communication. An asymmetry in the neutrino luminosity of a few percent can be envisioned, if not easily calculated.

Thus it seems reasonable, though perhaps not necessary, that the neutronized core acquire some peculiar velocity as a result of small but fundamental asymmetries in the energizing mechanism of the explosion. In the case of the core bounce mechanism, the non-radially symmetric hydrodynamics should ultimately translate into an angular asymmetry of the momentum of the ejecta of comparable magnitude (properly scaled for mass). In the neutrino transport model or during neutrino cooling following a sucessful core bounce, the asymmetry could be contained in the momentum distribution of the neutrinos themselves. The effective energy in the neutrinos is, after all, $\sim$10% of the rest mass of the neutron star. Thus a non-radial asymmetry in neutrino momentum distribution of only 1% would give the neutron star a recoil velocity of a few hundred kilometers per second. Such an asymmetry in total neutrino momentum could be a consequence of inhomogeneous composition and neutrino opacity or, especially, of large scale convective motions in the core during the first few seconds of its existence (Epstein 1979; Bruenn, Buchler, and Livio 1979; Colgate and Petschek 1980; Livio, Buchler and Colgate 1980; Smarr *et al.* 1981; Wilson *et al.* 1986).

The overall lesson that stellar evolution brings to those studying neutron stars is one of great possible diversity in the initial properties.

This work has been supported at UCSC by the National Science Foundation (AST-84-18185) and, at LLNL, by the Department of Energy through contract number W-7405-ENG-48. The author is also grateful for a faculty research grant form UCSC to cover a portion of the expenses incurred in attending the meeting.

## REFERENCES

Arnett, W. D. 1986, Preprint Enrico Fermi Inst., Univ. Chicago, submitted to *Ap. J. Lettr.*

Barkat, Z., Rakavy, G., and Sack, N. 1967, *Phys. Rev. Lettr.* **19**, 379.

Baron, E., Cooperstein, J., and Kahana, S. 1985a, *Phys. Rev. Lettr.*, **55**, 126.

_____. 1985b, *Nucl. Phys.*, **A440**, 744.

Baron, E., Brown, G. E., Cooperstein, J., and Prakash, M. 1985, Preprint.

Bethe, H. A., and Wilson, J. R. 1985, *Ap. J.*, **295**, 14.

Bisnovatyi-Kogan, G. S. 1971, *Sov. Astron. A. J.*, **14**, 652.

_____. 1980, *Ann. N. Y. Acad. Sci.*, **336**, 389.

Bisnovatyi-Kogan, G. S. , Popov, Y. P., and Samochin, A. A. 1976, *Ap. and Spac. Sci*,. **41**, 287.

Bodenheimer, P., and Woosley, S. E. 1983, *Ap. J.*, **269**, 281.

Bond, J. R., Arnett, W. D., and Carr, B. J. 1984, *Ap. J.*, **280**, 825.

Branch, D. and Nomoto, K. 1986, preprint submitted to *Astron. and Ap. Letters.*

Bruenn, S. W., Buchler, R. J., and Livio, M. 1979, *Ap. J. Lettr.*, **234**, L183.

Burrows, A. and Lattimer, J. M. 1983, *Ap. J.*, **270**, 735.

_____. 1985, *Ap. J.*, **299**, L19.

Burrows, A. and Woosley, S. 1986, *Ap. J.*, in press.

Canal, R., and Schatzman, E. 1976, *Astron. and Ap.*, **46**, 229.

Canal, R., and Isern, J. 1979, in *White Dwarfs and Variable Degenerate Stars*, Proc. IAU Colloq. 53, ed. H. M. Van Horn and V. Weidemann, Univ. of Rochester, p. 52.

Caughlan, G. R., Fowler, W. A., Harris, M. J., and Zimmerman, B. A. 1985, *Atomic Data and Nuclear Data Tables*, **32**, 197.

Chevalier, R. A. 1986, in *Highlights of Astronomy*, in press.

Colgate, S. A., and Petschek, A. G. 1980, *Ap. J. Lettr.*, **236**, L115.

El Eid, M. F., and Langer, N. 1986, *Astron. and Ap.*, in press.

Endal, A. S., and Sofia, S. 1976, *Ap. J.*, **210**, 184.

____. 1978, *Ap. J.*, **220**, 279.

Ensman, L. and Woosley, S. E. 1986, in preparation for *Ap. J.*

Epstein, R. 1979, *M. N. R. A. S.*, **188**, 305.

Filippenko, A. V., and Sargent W. L. W. 1985, *Nature*, **316**, 407.

____. 1986, *Astron. J.*, in press.

Gaskell, C. M., Cappellaro, E., Dinerstein, H. L., Garnett, D., Harkness, R. P., and Wheeler, J. C. 1986, preprint Univ. Texas, Austin, submitted to *Ap. J.*

Glatzel, W., El Eid, M. F., and Fricke, K. J. 1985, *Astron. and Ap.*, **149**, 413.

Hartmann, D. H., Woosley, S. E., and El Eid, M. 1985, *Ap. J.*, **297**, 837.

Helfand, D. J., and Becker, R. H. 1984, *Nature*, **307**, 215.

Isern, J., Labay, J., Hernanz, M. and Canal, J. R. 1983, *Ap. J.*, **273**, 320.

Hillebrandt, W. 1982, *Astron. and Ap. Lettr.*, **110** L3.

____. 1984, *Ann. N. Y. Acad. Sci*, **422**, 197.

____. 1985, To appear in *Cosmogonical Processes*, proceedings of A. G. W. Cameron's 60th Birthday symposium, Boulder, Colorado.

Hillebrandt, W., Nomoto, K., and Wolff, R. G. 1984, *Astron. Ap.* **133**, 175.

Lindblom, L., and Detweiler, S. 1979, *Ap. J. Lettr.*, **232**, L101.

Livio, M. J. R., Buchler, R. A. and Colgate, S. A. 1980, *Ap. J. Lett.*, **238**, L139.

Lyne, A. G. 1981, in *Supernovae: A Survey of Current Research*, ed. M. J. Rees and R. J. Stoneham, (D. Reidel: Dordrecht), p. 405.

Kiguchi, M., and Sato, K. 1985, preprint, Univ. Tokyo, UTAP-30.

LeBlanc, J. M., and Wilson, J. R. 1970, *Ap. J.*, **161**, 541.

Mayle, R. 1986, Ph.D. thesis, UC Berkeley.

Meier, D. L., Epstein, R. I., Arnett, W. D., and Schramm, D. N. 1976, *Ap. J.*, **204**, 869.

Miller, G. E., and Scalo, J. M. 1979, *Ap. J. Suppl.*, **41**, 513.

Müller, E., and Hillebrandt, W. 1979, *Astron. and Ap.*, **80**, 147.

Nomoto, K. 1984a, *Ap. J.*, **277**, 791.

____. 1984b, *The Crab Nebula and Related Supernova Remnants*, ed. M. Kafatos and R. B. C. Henry, Cambridge Univ. Press: Cambridge. In press.

_____. 1984c, *Problems of Collapse and General Relativity*, ed. D. Bancel and M. Signore, (D. Reidel: Dordrecht), p. 89.

_____. 1986, *Proceedings of the 12th Texas Symposium on Relativistic Astrophysics*, *Ann. N. Y. Acad. Sci.*, in press.

Nomoto, K., and Iben, I. Jr. 1985, *Ap. J.*, **297**, 531.

Ober, W., El Eid, M. F., and Fricke, K. J. 1983, *Astron. and Ap.*, **119**, 61.

Ohnishi, T. 1983, *Technical Reports of the Institute of Atomic Energy*, Kyoto Univ., No. 198.

Rappaport, S., and Joss, P. C. 1981, in *X-Ray Astronomy With the Einstein Satellite*,ed. R. Giaconni (D. Reidel: Dordrecht), p. 123.

Saio, H., and Nomoto, K. 1985, *Astron. and Ap.*, **150**, L21.

Smarr, L., Wilson, J. R., Barton, R. T., and Bowers, R. L. 1981, *Ap. J.*, **246**, 515.

Srinivasan, G., and Van den Heuvel, E. P. J. 1985, *Astron. and Ap.*, **108**, 143.

Stringfellow, G. S., Woosley, S. E., and Bodenheimer, P. 1983, *Bul. Am. Astron. Soc.*, **15**, 955 and in preparation.

Starrfield, S., Sparks, W. M., and Truran, J. W. 1986, Univ. of Illinois preprint IAP-86-30, *Ap. J. Lettr.*, in press.

Taylor, J. H. 1985, private communication.

Weaver, T. A., Zimmerman, G . B., and Woosley, S. E., 1978, *Ap. J.*, **225**, 1021.

Weaver, T. A., Woosley, S. E., and Fuller, G. M. 1985, *Numerical Astrophysics*, ed. J. Centrella, J. LeBlanc, and R. Bowers. Jones and Bartlett: Boston. p. 374

Weisberg, J. M., and Taylor, J. H. 1984, *Phys. Rev. Lettr.*, **52**, 1348.

Wheeler, J. C., and Leverault, R. 1985, *Ap. J. Lettr.*, **294**, L17.

Wilson, J. R. 1985, *Numerical Astrophysics*, ed. J. M. Centrella, J. M. LeBlanc, and R. L. Bowers, (Jones and Bartlett: Boston), p. 422.

Wilson, J. R., Mayle, R., Woosley, S. E., and Weaver, T. A. 1986, *Proc. of Twelfth Texas Symp. on Rel. Ap.*, *Ann. N. Y. Acad. Sci.*, in press

Woosley, S. E. 1986, in *Nucleosynthesis and Chemical Evolution*, Proc. 16th Advanced Course, Swiss Society of Astronomy and Astrophysics, Saas-Fee 1986, ed. B. Hauck and A. Maeder, (Geneva Observatory: Switzerland), in press.

Woosley, S. E., and Weaver, T. A. 1982, *Supernovae: A Survey of Current Research*, ed. M. J. Rees and R. J. Stoneham, D. Reidel: Dordrecht, p. 79

_____. 1985, *Nucleosynthesis and Its Implications On Nuclear and Particle Physics*, Proc. Fifth Moriand Astrophysics Conference, ed. J. Audouze and N. Mathieu, (D. Reidel: Dordrecht), p. 145.

_____. 1986a, *Ann. Rev. Astron. and Ap.*, in press.

_____. 1986b, in *Radiation Hydrodynamics in Stars and Compact Objects*, Proc. IAU Colloq. 89, Copenhagen 1985, ed. D. Mihalas and K. H. Winkler.

Woosley, S. E., Weaver, T. A., and Taam, R. E. 1980, *Type I Supernovae*, ed. J. C. Wheeler, Austin: Univ. of Texas, p. 96

Woosley, S. E., Wilson, J. R., and Mayle, R. 1986, *Ap. J.*, **302**, 19.

## DISCUSSION:

**F. Seward:** What is the minimum mass for a neutron star made during the collapse you have described?

**Woosley:** As I have discussed in § 4.1, the minimum mass for the core is $\sim 1.25$ $M_\odot$. After subtracting off the binding energy, the smallest neutron star mass is $\sim$ 1.15 $M_\odot$. More typical values should be near 1.3 $M_\odot$.

**Z. W. Li:** What is the effect of the [nuclear] equation of state on the calculation of SN II? Is it important?

**Woosley:** For those stars having sufficiently small iron cores that a hydrodynamical explosion might be possible, the nuclear equation of state is important in giving the explosion energy for a given core mass and for determining the maximum core mass that can explode by the mechanism. As Baron, Cooperstein, and Kahana (1985ab) have shown, an equation of state that is "softer" at high density is more favorable to explosion.

**A. Burrows:** [A comment and a question] I believe that the fatter iron cores that are destined to form black holes will do so in seconds, not minutes. The mass accretion rate through the shock should be large enough to quickly put the core over the critical mass. Can you estimate the amount of $^{56}$Ni produced in Type II's as a function of progenitor mass?

**Woosley:** Bodenheimer and I (1983) studied the continued collapse of the core in a 25 $M_\odot$ model which had failed to explode. We found that the collapsed remnant grew to 3.8 $M_\odot$ in about 20 seconds. Thus I agree that, excluding large amounts of rotation, the mass will increase above any value allowing a final stable neutron star within much less than one minute as you say. The real issue however, is the time scale for removing so much entropy from the remnant that it cannot be supported by thermal pressure. Those calculations remain to be done in an object that is becoming a black hole and including convection, though I know you are working on just such a study. All in all, a time scale of 10 seconds (to a factor of several) seems reasonable.

The $^{56}$Ni synthesis depends upon the explosion energy and, more importantly on the pre-explosive density gradient surrounding the core. The shallower the gradient the more material stands near the "bomb" and gets heated above the $5 \times 10^9$ K required to produce $^{56}$Ni. Lower mass stars have steeper density gradients near the core. The largest $^{56}$Ni mass should be $\sim$0.5 $M_\odot$ and should characterize the most energetic explosions (say a few times $10^{51}$ erg) in the most massive stars that explode by core bounce or neutrino energy transport ($25 \lesssim M/M_{\odot} \lesssim 70$). In a 15 $M_{\odot}$ model we have studied, $^{56}$Ni synthesis ranges from 0 to 0.1 $M_{\odot}$ depending upon the energy of the explosion. For explosion energies $\lesssim 5 \times 10^{50}$ erg, the inner nickel-rich layers may fall back onto the neutron star. For still lighter stars, especially 8 to 10 $M_{\odot}$, the $^{56}$Ni synthesis is still less, perhaps a hundredth of a solar mass, though this has yet to be calculated. Of special interest will be the $^{56}$Ni produced in the collapse of an accreting white dwarf (§3.2). Wilson and Mayle are working on that.

**T. Ebisuzaki:** I would guess that newly born neutron stars should evolve along the equivalent of the Hayashi track due to convection, I guess. What do you think about this idea?

**Woosley:** This is very much in line with Burrow's question above and the work he is doing on the early evolution of neutron stars. I think that you are right, *i.e.*, that a surface convection zone of increasing depth will play an important role in removing the entropy of a young neutron star (or black hole). Such convection is also the essential aspect of Arnett's recently proposed model for boosting the neutrino flux *during the explosion itself.* Obvious implications loom for the surface magnetic field, configuration and strength in a young neutron star.

**G. S. Bisnovatyi-Kogan:** A remark about two important physical processes in supernova explosions. First, the interaction of neutrino flux from the core with falling envelope is very important and strongly reduces the falling velocity and bounce, as was shown by Nadëzhin in 1977. So the results of Hillebrandt *et al.* [1982, 1984] have not only been not confirmed later, but also earlier. Second, the magnetic field in rapidly rotating cores may transform the rotational energy into the energy of explosion and give SN explosion (magnetorotational) even when there is no explosion at the end of initial collapse. We are now working on this problem.

**Woosley:** Hillebrandt is not here to defend his work, but it is my understanding that his calculations include a very detailed and physical treatment of neutrino momentum and energy transport throughout the core implosion and explosion. The initial models that he starts with are provided by our group at a time when the *central* density has first reached $10^{10}$ g cm$^{-3}$, thus neutrino momentum transport at an earlier stage is unlikely. I do not think that the braking of infalling matter by neutrinos is the cause for descrepancy with more recent calculations. Rather such differences are attribable to disparate treatments of nuclear equation of state, nuclear burning, neutrino energy transport in and behind the shock, and slight differences in the pre-explosive models (*e.g.* Hillebrandt *et al.* 1984; Wilson *et al.* 1986).

The possible importance of magnetic energy in obtaining a supernova explosion has stressed in a number of calculations (Le Blanc and Wilson 1970; Bisnovatyi-Kogan 1971, 1980; Bisnovatyi-Kogan *et al.* 1976; Meier *et al.* 1976; Müller and Hillebrandt 1979; Ohnishi 1983). In general, the results are quite sensitive to uncertain assumptions that are made, but show that large magnetic fields *might* be instrumental in obtaining explosions (though the works listed above come to differing conclusions in this regard). Unfortunately the calculations are some of the most difficult in all of astrophysics.

**S. R. Kulkarni:** You described two different mechanisms to get velocity kicks during collapse to a neutron star. Are both these mechanisms also operative in neutron stars formed by accreting white dwarfs going over the Chandrashekar mass limit? Observationally one may have to invoke kicks for neutron stars formed from massive stars to explain observed pulsar velocity distributions. [But] if the kick mechanism(s) is also operative in neutron stars born in low-mass systems then virtually none of the low-mass binaries (radio pulsars and X-ray binaries) can be formed.

**Woosley:** Yes, both mechanisms would operate to some extent in collapsing white dwarfs as well, especially the neutrino momentum asymmetry. But lacking quantitative estimates it is difficult to know how "robust" these kick mechanisms are. I envision them as stochastic. Thus large peculiar velocities may be obtained sometimes and smaller ones at others. It also makes a big difference whether the kick occurs along the orbital velocity vector or in opposition. As Burrows and I (1986) have described, a kick may sometimes be necessary to keep a binary *bound*.

**S. Reynolds:** Could you comment on the possible formation, in a rotating star, of a metastable core that takes a year or so to settle to a neutron star? Could you make more massive neutron stars that way?

**Woosley:** Maintaining a neutron star above its critical mass by supporting the mass, in part, with rotation is regarded as difficult. Tri-axial deformation would lead to gravitational radiation and angular momentum loss. The possibility of "fizzlers" (T. Gold) has been most recently reviewed by Kiguchi and Sato (1986) with negative conclusions.

# NUMERICAL EXPERIMENTS AND NEUTRON STAR FORMATION

W. David Arnett*
Enrico Fermi Institute
University of Chicago
Chicago, IL 60637
USA

ABSTRACT. Theory and numerical experiments on neutron star formation are critically reviewed. Several new numerical experiments are summarized, and the importance of advection is discussed.

## 1. THEORETICAL CONTEXT

### 1.1. Neutrino Deposition, Diffusion and Shocks

In this paper we will emphasize the difference between theory (a set of quantitative ideas) and numerical experiment ("computer theory"). Because some of the older theoretical ideas are reappearing in the current models for the explosion mechanism, a brief review is in order. The idea of neutrino deposition is due to Colgate and White (1966). Neutrinos, escaping from a collapsed core, heat the infalling mantle so much that an outgoing shock is formed, which causes the explosion. While at low densities the neutrino flux is simply attenuated, at higher densities the mean free paths are so short that diffusion occurs. Arnett (1966, 1967) found that although the mechanism worked if the neutrino opacities were low enough so that diffusion could carry a lot of energy, higher opacity choked down the neutrino luminosity, trapping the neutrinos in the infalling matter. Mazurek (1975) argued that the neutrinos from electron capture were directly trapped, giving a large excess of neutrinos over antineutrinos, and that these neutrinos would be degenerate. With the discovery of neutral currents and the realization that this gave an increased opacity, it became clear that electron-type neutrinos were indeed strongly trapped and degenerate (Sato 1975; Arnett 1977a). Using these results, and emphasizing that the entropy change and lepton loss in the collapsing core would be small, Bethe, Brown, Applegate and Lattimer (1979) developed a theoretical model in which the core "bounce" at nuclear densities caused an outward shock. This was confirmed by numerical experiments (Hillebrandt 1982), but despite initial euphoria this shock did not prove to be the solution to the supernova problem. Extensive work on the equation of state (cf. review by Lattimer 1981), the electron-capture process (Fuller, Fowler

* A. von Humboldt Awardee, Max-Planck-Institut für Astrophysik, Garching bei München, FRG



D. J. Helfand and J.-H. Huang (eds.), The Origin and Evolution of Neutron Stars, 273–280.

and Newman 1982), and the initial models (Arnett 1977b, Weaver, Zimmerman and Woosley 1978, Nomoto 1982) restricted the plausible parameter space available for numerical experiments, and most (all?) shocks died as they propagated to lower density. The killers were energy losses by (1) photo dissociation of nuclei (8 MeV/nucleon) and (2) neutrino escape (Arnett 1983, Lattimer and Burrows 1984). Ironically the role of neutrinos in the numerical experiments changed, from driving to damping the explosion. Considerable effort is being expended to see if this "prompt" mechanism works for any acceptable choice of parameters.

### 1.2. Late Mechanisms

If the initial collapse does not give rise to an explosion directly, what happens then? Wilson (1985) found a weak explosion at "late" times (about 1 second rather than tens of milliseconds) in numerical experiments. Bethe and Wilson (1985) developed a theory of the process which involves neutrino diffusion from the core heating the infalling mantle. Because of the large opacity, this process is slow, and produces weak explosions. Other numerical experiments (Arnett, see below) do not confirm even weakly explosive behavior.

The bottleneck is the slow diffusion rate. This occurs in a region which has high entropy (due to strong shock heating). It is below a region of lower entropy (due to the weaker heating of the dying shock). This region of the mantle is hydrodynamically unstable to overturn, which transports energy and leptons to lower densities where diffusion can take over. This idea is distinct from previous discussion of overturn which involved the collapsed core (which is stable). While the theory is clear enough, numerical experiments run to such late times raise questions about the validity of the techniques used.

## 2. VALIDITY OF NUMERICAL EXPERIMENTS

### 2.1. Correctness

A fundamental problem for the scientist using a computer is that of validation of results. How do we know they are correct? An obvious and useless test is that of prejudice: the result seems "reasonable". While "unreasonable" results may be a useful clue for finding errors, the fact that results agree with our preconceptions does not imply that they are correct.

Test problems are somewhat more persuasive. Even so, we do not do research on problems to which we know the answer, so test problems are only a partial aid. Analytic solutions are seldom available for interesting new problems, if they were we would not need the computer. Discretization errors may be tested by repeating a calculation with different sizes of zone and time step; this is difficult if the job to be tested already pushes the practical limits of computer and budget. For some problems there are integral constraints which may be monitored to test the quality of computation (for example, energy conservation or entropy "conservation"; lepton conservation). These are only necessary but not sufficient for correctness. I think the best approach is to treat computer results as numerical experiments, and use strategies analogous to those of laboratory experimenters to establish reliability. Correspondingly, publication of the results of at least the most important of these tests is necessary for establishing a numerical experiment as more than a scenario.

### 2.2. Some Types of Errors

In the literature dealing with gravitational collapse one can find examples of many types of errors: (1) choice of physics (omitting crucial physical processes, using the wrong theory or too crude a version of the right theory), (2) choice of algorithms (e.g., putting in the desired answer by choice of method of calculation), (3) programming blunders ("the only correct code is an obsolete one"; this is especially difficult for those who must compute on different machines or who do not exploit modern techniques for software developement), and (4) discretization (truncation, roundoff, level of convergence, etc.).

## 3. BIRTH OF NEUTRON STARS

### 3.1. Error Estimates

#### 3.1.1. Energy

The rest energy of a Chandrasekhar mass ($M \simeq 1.5\ M_\odot$) is $3 \cdot 10^{54}$ erg; the gravitational binding energy B of a neutron star is about 0.1 of this. A weak explosion involves as little as $10^{50}$ ergs (e.g., Crab Nebula?), which is only $3 \cdot 10^{-4}$ of the binding energy released by core collapse. A typical computational sequence takes of order $10^4$ steps in time. Suppose we want an accuracy of 1% in explosion energy for this weak explosion, and that we make some small error in energy at each time step. If these errors are random (they grow as $\sqrt{n}$), then the fractional error per step is

$$|\Delta E|/B \simeq 3 \cdot 10^{-8}$$

For systematic errors (they grow as n), we have

$$|\Delta E|/B \simeq 3 \cdot 10^{-10}$$

These goals are much more stringent than usually required in computational work.

#### 3.1.2. Rates

At present, "late times" mean about 1 second. If the weak explosion mentioned above were driven by some form of energy transfer, then the energy transfer rate would be $L \simeq 10^{50}$ erg/s. Now, the average over the collapsed core of the neutrino energy emissivity need to do this is

$$\langle\epsilon\rangle \simeq L/M \simeq 0.7\cdot 10^{17} \text{ erg/g s.}$$

In the center, the true emissivity is

$$\epsilon \simeq 10^{27} \text{ erg/g s,}$$

which is closely balanced by the neutrino absorption rate. Numerically speaking, to avoid errors of 1% in $\langle\epsilon\rangle$, we need 12 significant figures for $\epsilon$ to cancel the corresponding absorption.

These estimates give a sense of the nature of the problem. Consider that at a given time, the state of the star is represented by an abstract vector whose components represent temperature, composition, neutrino group energy density, velocity, etc. at each selected point in space (each "zone"). This state vector is subjected to an evolution operator which updates each element to its value at the next time step. It is believed (without much proof) that crude state vectors (e.g., a dozen energy groups to represent the neutrino distribution) are adequate. The errors discussed above are related to the evolution operator, and the problem posed is this: can we show that the cumulative errors in this evolution operator are smaller than the effect we seek?

### 3.2. Numerical Experiment: $Y_e$ fixed

A simple limiting case for which errors can be acceptably small is a collapse during which $Y_e$, the electron abundance per nucleon, is held fixed. As no neutrinos are formed, there is no neutrino cooling. The lepton pressure is a bit larger if all leptons are electrons. A third flaw is that $Y_e$ is frozen somewhat arbitrarily at a point after the core collapse begins ($\rho_c \simeq 2\cdot 10^{10}$ g/cc). Such a model gives a strong bounce, with the shock propagating all the way to the surface of the 4 $M_\odot$ helium star. Using a mild modification of Van Riper' method for general relativistic hydrodynamics, the energy check showed a cumulative error of less than $10^{49}$ ergs, about 0.1% of the explosion energy ($3\cdot 10^{51}$ ergs). An implicit, second order hydrodynamic scheme was used, so that the calculation could be continued to 3 minutes, at which point the envelope density was $\rho \simeq 10^{-2}$ g/cc, the shock had reached the surface, the outflow velocities were above escape velocity, and a neutron star of 1.1 $M_\odot$ had formed. Dynamic rezoning was necessary to resolve (for the first time) the steep density gradient between the neutron star and the ejecta ($\rho$ changes by a factor of $10^{12}$ in a tiny fraction of the mass). Some thermonuclear burning occurred during the explosion, with 0.2 $M_\odot$ of $^{56}$Ni being formed. This object would SNI-like light curve. Fig. 1 shows snapshots of the density structure at several different times, beginning with a well-evolved shock and continuing until after the shock reaches the structure of the helium star.

The experiment fails on two predictions: (1) the <u>gravitational</u> mass (10% reduction for gravitational binding) is 1.0 $M_\odot$, so that such an event could not explain the masses in the binary pulsar PSR 1913+16 (which are 1.45 and 1.38 $M_\odot$), and (2) about 0.3 $M_\odot$ of

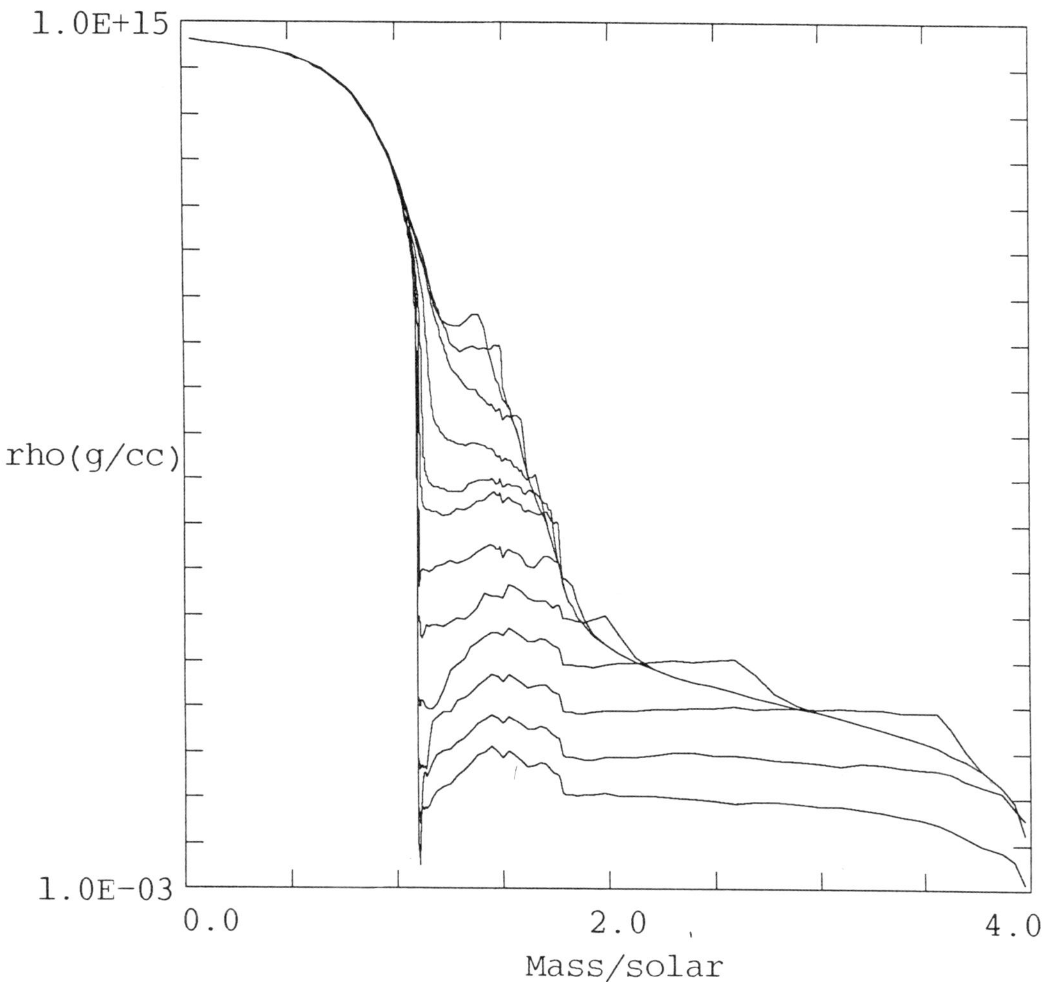


Figure 1. The formation of a neutron star. Density (in g $cm^{-3}$) is plotted versus Lagrangian mass coordinate (in solar masses); snapshots at several different times are shown. In the first snapshot the shock has propagated down to a density of about $10^8$g $cm^{-3}$, prior to the oxygen burning shell. Some explosive burning occurs; see the region from mass 1.4 to 1.8 solar units. Recombination of dissociated nuclei occurs in the mass above 1.1 solar units, and affects the hydrodynamics. Resolution of the steep density gradient at a mass of 1.1 requires special techniques; an implicit hydrodynamic scheme (second order, Lagrangian) with dynamic rezoning was used. The masses shown represent nucleon number, a conserved quantity; the gravitational mass of the neutron star is about 10% less.

$^{54}Fe$ are ejected. This matter is already neutron rich when the collapse begins, is dissociated to nucleons, and recombines to give the shock a significant kick at late times. As $^{56}Fe$ is the most abundant isotope observed, this high yield of the rarer $^{54}Fe$ is an embarrassment.

### 3.3. Numerical Experiment: Death of the Shock

We repeat the previous experiment with weak interactions turned on. The core bounce is now weaker, but a strong shock begins to travel out to lower densities. It is weakened by nuclear dissociation, but is killed at low densities ($\rho \simeq 10^{10}$ g/cc) by neutrino transport. As mentioned above, the mantle develops an unstable gradient, but if the code is not told about two dimensional phenomena, it happily ignores them. The calculation was continued to 1 second, with no hint of a late mechanism of the Wilson type. No further continuation was made because the energy check grew to be of the same order as the energy of Wilson's explosions (several times $10^{50}$ ergs).

### 3.4. Advection and Explosion

A direct integration of the buoyancy force gives turnover velocities of the order of 20% of sound speed, which is enough to mix the mantle in 0.1 seconds. It appears that the crucial hydrodynamics is simply a large scale overturn rather than the more complex development of a turbulent cascade. It is easy to get explosions of the sort I found a long time ago, but hard to do the neutrino transport accurately in both the thick core and tenuous (spherical) mantle. That will be necessary to predict accurately the neutron star mass, and the energy and compositon of the ejecta.

Although we started with a spherical system, we were lead to a discussion of significant nonspherical motion. It is unlikely that an asymptotic calculation will lead us back to symmetry. Whether such phenomena can give new pulsars a kick of the magnitude suggested by observation remains to be seen.

## 4. PROSPECTS

The frontier of collapse research has now progressed to times of order seconds to minutes after the bounce, which is considerably later than the 20 milliseconds discussed a few years ago. There are a number of prospective developments.

Radiation hydrodynamics of neutrinos at low depths is crucial for the late mechanisms, and for the possible survival of the shock with early mechanisms. Fu and Arnett (in preparation) have developed a method which should be an improvement over the flux-limiting technique now in use, with no computational penalty. Also, greater care in the consistent treatment of neutrino-matter coupling at high and low density will help with the energy check problem.

With the new hydrodynamic methods of Woodward and Colella, and van Leer, two and three dimension computations become feasible; the discussion above suggests that they are necessary. The question of a pulsar kick at birth is of prime importance; numerical experiments can be brought to bear on some mechanisms.

Observed neutron star masses (especially PSR 1913+16) have direct overlap with the results of late-time collapse experiments. Similarly, estimates of the nucleosynthesis yields are better if the ejection process actually falls out of the experiment. We may be approaching the point of seriously connecting young supernova remnants with evolutionary histories of their progenitor stars. Remnant energies, masses and compositions, as well as light curve and spectral data, are providing vital empirical information to which we can compare numerical experiment.

## REFERENCES


Arnett, W.D. 1966 Can. J. Phys. **44**, 2553.
Arnett, W.D. 1967 Can. J. Phys. **45**, 1621.
Arnett, W.D. 1977a Ap. J. **218**, 815.
Arnett, W.D. 1977b Ap. J. Suppl. **35**, 145.
Arnett, W.D. 1983 Ap. J. (Letters) **263**, L55.
Bethe, H.A., Brown, G.E., Applegate, J., and Lattimer, J.M. 1979 Nucl. Phys. **A324**, 487.
Bethe, H.A. and Wilson, J.R. 1985 Ap. J. **295**, 14.
Colgate, S.A. and White, R. 1966 Ap. J. **143**, 626.
Fuller, G.M., Fowler, W.A. and Newman, M.J. 1982 Ap. J. **252**, 715.
Hillebrandt, W. 1982 in Supernovae: A Survey of Current Research, ed. M.J. Rees and R.J. Stoneham (D. Reidel: Dordrecht).
Lattimer, J.M. 1981 Ann. Rev. Nucl. Particle Sci., **31**, 537.
Lattimer, J.M. and Burrows, A. 1984 in Problems of Collapse and Numerical Relativity, ed. D. Banal and M. Signore (D. Reidel: Dordrecht).
Mazurek, T. 1975 Ap. and Space Sci. **35**, 117.
Nomoto, K. 1982 in Supernovae: A Survey of Current Research, ed. M.J. Rees and R.J. Stoneham (D. Reidel: Dordrecht).
Sato, K. 1975 Prog. Theor. Phys. **53**, 595.
Weaver, T.A., Zimmerman, G.B. and Woosley, S.A. 1978 Ap. J. **225**, 1021.
Wilson, J.R. 1985 in Numerical Astrophysics, ed. J.M. Centrella, J.M. LeBlanc, and R.L. Bowers (Jones and Bartlett: Boston).


## DISCUSSION

**A. Burrows:** You don't reproduce Wilson's results. How do your effective neutrinospheric temperature and neutrino luminosities compare with his?

**W. Arnett:** As you would guess, my temperatures and luminosities are less than Wilson's. The problem is not that the numerical experiments produce results which are obviously unphysical, but that they produce different, seemingly self-consistent results.

**T. Weaver:** With regard to numerical errors, do you get convergence results when you reduce zone size, time steps, convergence criteria, etc.?

**W. Arnett:** It is difficult to summarize briefly two years of an extensive variety of collapse calculations, but basically the answer is yes, I do get convergent results. The errors which I worry most about are more subtle; over long times small effects can build, giving systematic differences. To the present level of zone size and time step, the calculations seem to converge; the most "accurate" ones do not explode by the Wilson mechanism.

**S. Colgate:** The large neutrino crossection due to neutral currents led to trapping during collapse of electron-capture neutrinos. The lepton-trapped neutron star core has only ~ 1/10 the binding energy of the lepton-depleted core. If neutrino oscillations $\nu_c \leftrightarrow \nu_t$ similar to but with a larger $\Delta m$ than the $\nu_c \leftrightarrow \nu_u$ applied to the solar neutrino problem by Bethe occurred, we would have a lepton-free core, a strong accretion shock, and neutrino emission for neutrino ejection as in Colgate and White.

**W. Arnett:** These collapsing cores may be the only place in the Universe where the assumption of lepton conservation is of crucial importance. My strategy (which I have often questioned) is to get the standard case sorted out before exploring more exotic ones.

# NEUTRON STAR FORMATION IN THEORETICAL SUPERNOVAE — LOW MASS STARS AND WHITE DWARFS —

Ken'ichi Nomoto
Physics Department, Brookhaven National Laboratory; on leave from the Department of Earth Science and Astronomy, University of Tokyo

ABSTRACT: The presupernova evolution of stars that form semi-degenerate or strongly degenerate O+Ne+Mg cores is discussed. For the 10 - 13 $M_{\odot}$ stars, behavior of off-center neon flashes is crucial. The 8 - 10 $M_{\odot}$ stars do not ignite neon and eventually collapse due to electron captures. Properties of supernova explosions and neutron stars expected from these *low mass progenitors* are compared with the Crab nebula. We also examine the conditions for which neutron stars form from accretion-induced collapse of white dwarfs in close binary systems.

## 1. INTRODUCTION

### 1.1 Neutron Star Formation in 8 - 13 $M_{\odot}$ Stars

*Low mass stars* are defined here as stars that form an electron degenerate core during the cource of evolution. The evolution and the final fate of these low mass stars are completely different from more massive stars becasue of the difference in the gravothermal nature of the stars as will be described in §2. Through the study of such *low mass stars*, we can determine the lowest mass of stars that form neutron stars. The stars that form a degenerate C+O core, i.e., 0.5 - 8 $M_{\odot}$ stars, will become white dwarfs or be completely disrupted by carbon deflagration and, hence, will not form a neutron star. What about slightly more massive, 8 - 13 $M_{\odot}$, stars? These stars form a strongly degenerate or semi-degenerate O+Ne+Mg core and their evolution will be discussed in §3 - §5. Supernova explosion from the progenitors in this mass range is interesting because it may be the origin of the Crab nebula in view of elemental abundances. The white dwarf is also a potentail neutron star progenitor if it belongs to a close binary system. This possibility will be the subject of §6 - §9 with the following motivation.

### 1.2 Neutron Star Formation from Accretion-Induced Collapse of White Dwarfs

The final fate of accreting white dwarfs will be either thermonuclear explosion or collapse to form a neutron star, if the white dwarf mass grows to the Chandrasekhar mass. The exploding white dwarf model, in particular, the carbon deflagration



*D. J. Helfand and J.-H. Huang (eds.), The Origin and Evolution of Neutron Stars, 281–303.*

model is in good agreement with many of the observed features of Type Ia supernovae (Nomoto 1986a; Woosley and Weaver 1986b). For Type Ib supernovae, though currently most popular models are explosions of Wolf-Rayet stars, there is a suggestion that the maximum light spectrum can better be explained by off-center explosions of white dwarfs. Recent observations of several interesting binary systems, low mass X-ray binaries, QPOs, and binary radio pulsars have suggested that in these systems a neutron star has formed from accretion-induced collapse of a white dwarf (van den Heuvel 1984; Taam and van den Heuvel 1986).

These variations in the final fate of accreting white dwarfs originate from the differences in the parameters of the binary system, i.e., composition, mass, and age of the white dwarf, its companion star, and, in particular, mass accretion rate. It is important to know under what condition white dwarfs will collapse to form neutron stars.

Possible models for the white dwarf collapse involve solid C+O white dwarfs, in which carbon and oxygen may or may not have chemically separated (Canal and Isern 1979; Isern et al. 1983) and O+Ne+Mg white dwarfs (Nomoto et al. 1979). For a wide range of mass accretion rates and initial white dwarf masses, the O+Ne+Mg white dwarfs collapse due to electron capture on $^{24}$Mg and $^{20}$Ne (Nomoto 1980; Miyaji et al. 1980). On the other hand, depending on the conditions of the white dwarfs and binary systems in which they are formed, the C+O white dwarfs could either explode or collapse. Chemical separation in such objects is still hypothetical and in any case could not be complete before carbon burning starts (Mochkovici 1980). It takes a carbon fraction of only a few percent to sustain a deflagration. Therefore, it is worth determining the critical condition for which a carbon deflagration induces the collapse of a C+O white dwarf rather than its explosion.

In §6, the fate of accreting white dwarfs are summarized in the parameter space of accretion rate and the initial mass of the white dwarfs. In §7, models for Type Ia and Ib supernovae are briefly presented. In §8, possibilities of white dwarf collapse is examined. The companion stars, the fate of merging white dwarfs, etc. are discussed in §9.

## 2. EVOLUTION AS A FUNCTION OF STELLAR MASS

### 2.1 Effects of Electron Degeneracy on Stellar Evolution

The hydrostatic stellar structure gives a relation between $P_\mathrm{c}$, $\rho_\mathrm{c}$, and stellar mass, $M$, as $P_\mathrm{c}^3/\rho_\mathrm{c}^4 \sim M^2$, (Sugimoto and Nomoto 1980; Arnett 1978). For an ideal gas, this gives $T_\mathrm{c}^3/\rho_\mathrm{c} \sim M^2$. The specific entropy, $s$, is then given as

$$(\mu H/k)s_\mathrm{c} = \ln(T_\mathrm{c}^{3/2}/\rho_\mathrm{c}) + C_1 = \ln(M^2/T_\mathrm{c}^{3/2}) + C_2 = \ln(M/\rho_\mathrm{c}^{1/2}) + C_3, \qquad (1)$$

where $\mu$ is the mean molecular weight, $H$ atomic mass unit, $k$ Boltsmann constant, and $C_1, C_2$, and $C_3$ additional constants. Equation (1) shows that for the same $T_\mathrm{c}$

(i.e, roughly for the same nuclear burning stage), $s_c$ is higher (and $\rho_c$ is lower) for larger $M$. We need higher entropy to sustain a larger mass against self-gravity.

When nuclear reactions are not active, the star loses entropy by radiation and neutrinos and $s_c$ decreases. Equation (1) shows that both $T_c$ and $\rho_c$ increase as $s_c$ decreases for fixed $M$. In other words, the specific heat in the central region is negative as $c_g \equiv ds_c/d\ln T_c = -1.5k/\mu H$. The increase in $\rho_c$ implies that the star contracts and releases gravitational energy, a part of which goes into internal energy and the rest goes into radiation and neutrino losses.

The sign of the *specfic heat* depends on the equation of state, however. It is *positive* if electrons are degenerate and the pressure does not depend much on temperature. In this case, the loss of entropy hardly induces gravitational contraction, so that $T_c$ decreases as the star loses energy, i.e., $c_g > 0$. Such a change in equation of state occurs for stars with sufficiently low central entopy and thus for a sufficiently small mass, $M$, or small core mass, $M_{core}$, as seen from Eq.(1).

The example of the evolutionary change in the central density and temperature during the gravitational contraction is shown in Figure 1 for neon stars with neon burning artificially suppressed (Nomoto 1984a, 1986a). At the maximum temperature, $c_g$ changes its sign because of electron degeneracy. The maximum temperature is higher for larger $M$. (For evolved cores, the temperature peak appears not in the center but in the outer shell because of neutrino losses. See §3). We see that there exists a minimum mass for the peak temperature to reach the ignition temperature of the nuclear fuel. The minimum masses are 0.08, 0.25, 1.06, and 1.37 $M_\odot$ for hydrogen, helium, carbon, and neon burning, respectively (e.g., Arnett 1978; Nomoto 1981 for a review). Therefore, stars with a core mass, $M_{core}$, smaller than the above minimum mass will not undergo further nuclear burning, but form a white dwarf-like degenerate core. In this sense, the stellar mass, $M$, is a critical parameter to determine the final fate of evolution.

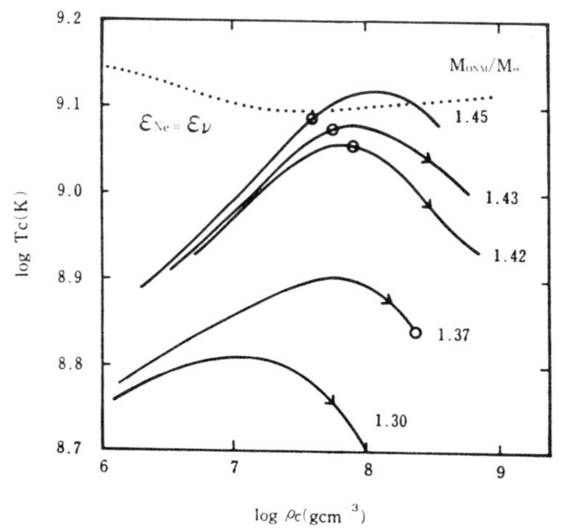


Figure 1: Evolutionary change in the central density and temperature of *neon* stars of mass 1.30 - 1.45 $M_\odot$ during gravitational contraction (Nomoto 1986a). The open circles indicate the stage where the neon ignition temperature (shown by the dotted line) is reached in the outer shell, not in the center. (Further evolution is followed by supressing neon burning.) For stars of mass smaller than 1.37 $M_\odot$, no neon ignition occurs.

## 2.2 Evolution As a Function of Stellar Mass

In Figure 2, the evolution of the central density, $\rho_c$, and temperature, $T_c$, is shown for helium stars (or cores) of mass $M_\alpha$ = 8, 6, 3.3, 3.0, 2.8, and 2.2 $M_\odot$ which correspond to the main-sequence mass of $M_{ms}$ = 25, 19, 13, 12, 11, and 9 $M_\odot$, respectively (Nomoto and Hashimoto 1986). The line for $\psi = 10$ approximately divides the electron degenerate and non-degenerate region, where $\psi$ is the chemical potential of an electron in units of $kT$. These evolutionary paths clearly indicate the effect of electron degeneracy and its dependence of stellar mass. We can classify the stellar evolution by $M$ as follows:

(1) For $M < 0.08\ M_\odot$, the star will become a planet-like black dwarf without igniting hydrogen burning.
(2) For $0.08\ M_\odot < M < 0.45\ M_\odot$, the star will end up as a helium white dwarf, though such a single star has not evolved off the main-sequence in a Hubble time.
(3) For $0.45\ M_\odot < M < 8\ M_\odot$, the star forms a C+O core of mass smaller than 1.06 $M_\odot$, which then become strongly degenerate. Most of them will become a C+O white dwarf by losing their hydrogen-rich envelope, but some of them ($\sim$ 6 - 8 $M_\odot$) could reach a supernova stage by increasing the C+O core mass to the Chandrasekhar mass and ignites a carbon deflagration. The explosion will look like a Type II-L supernova (Doggett and Branch 1985).
(4) For $8\ M_\odot < M < 10\ M_\odot$, an O+Ne+Mg core is smaller than 1.37 $M_\odot$ so that it becomes strongly degenerate. The star will eventually collapse due to electron capture on $^{24}$Mg and $^{20}$Ne as discussed in §4.
(5) For $10\ M_\odot < M < 13\ M_\odot$, neon is ignited in a semi-degenerate O+Ne+Mg core. Whether subsequent evolution leads to a non-degenerate or a degenerate configuration depends on neutronization during oxygen burning and crucial for their final fate. This question is answered in §3.
(6) For $M > 13\ M_\odot$, the star undergo non-degenerate burning to form an iron core.

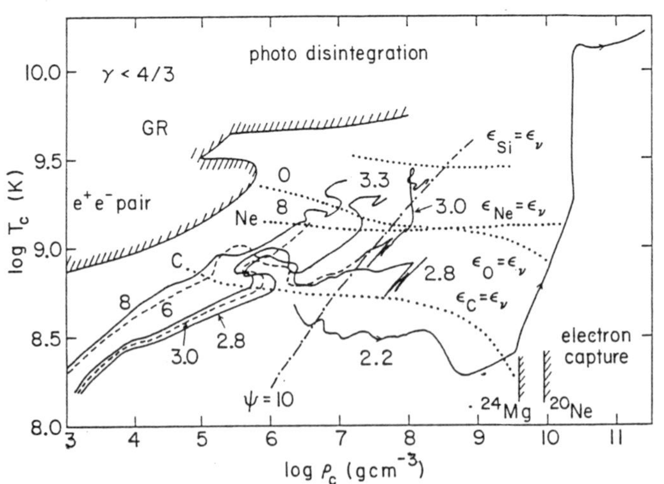


Figure 2: Evolution of the central density, $\rho_c$, and temperature, $T_c$ for helium stars of mass $M_\alpha$ = 8, 6, 3.3, 3.0, 2.8, and 2.2 $M_\odot$. The approximate ignition lines for carbon, neon, oxygen, and silicon, where the nuclear energy generation rate is equal to neutrino energy losses, are shown.

## 3. PRESUPERNOVA EVOLUTION OF 10 - 13 $M_\odot$ STARS

### 3.1 Semi-Degenerate Oxygen-Neon Core and Off-Center Neon Ignition.

The evolution of stars in the mass range of $M_{\rm ms} = 10$ - 13 $M_\odot$ ($M_\alpha \simeq 2.5$ - 3.2 $M_\odot$) is complicated and sensitive to the stellar mass. These stars undergo non-degenerate carbon burning and form an oxygen-neon core whose mass, $M_{\rm ONe}$, is in the range of 1.37 - 1.5 $M_\odot$. The core mass $M_{\rm ONe}$ is large enough to ignite neon, yet the core is semi-degenerate (i.e, $\psi > 10$ in the central region) and the degree of degeneracy depends sensitively on how close $M_{\rm ONe}$ is to the Chandrasekhar mass.

One of the distinct features of a semi-degenerate core is the appearance of a temperature inversion in the central region. This is because neutrino energy losses are faster at higher densities and the degenerate electrons provide the pressure necessary to sustain hydrostatic equilibrium. As a result, neon burning ignites not in the center, but in the outer shell. The location of the neon ignition is $M_{\rm r,ig} = 0.78$ - 0.10 $M_\odot$ for $M_\alpha = 2.6$ - 3.2 $M_\odot$ (Hashimoto and Nomoto 1986; Nomoto 1984a; Habets 1985, 1986). For $M_\alpha = 3.3$ $M_\odot$ ($M_{\rm ms} \sim 13$ $M_\odot$), electron degeneracy is so weak that no temperature inversion appears and neon ignites at the center.

### 3.2 Propagation of Neon-Oxygen Burning Layer

When neon and oxygen ignite off-center, a sharp temperature jump forms. Subsequently the neon/oxygen burning shell propagates inward (Figure 3). An important question is whether or not the burning shell reaches the center. If not, a degenerate oxygen-neon core is left unburned (Barkat et al. 1974; Woosley and Weaver 1986a)

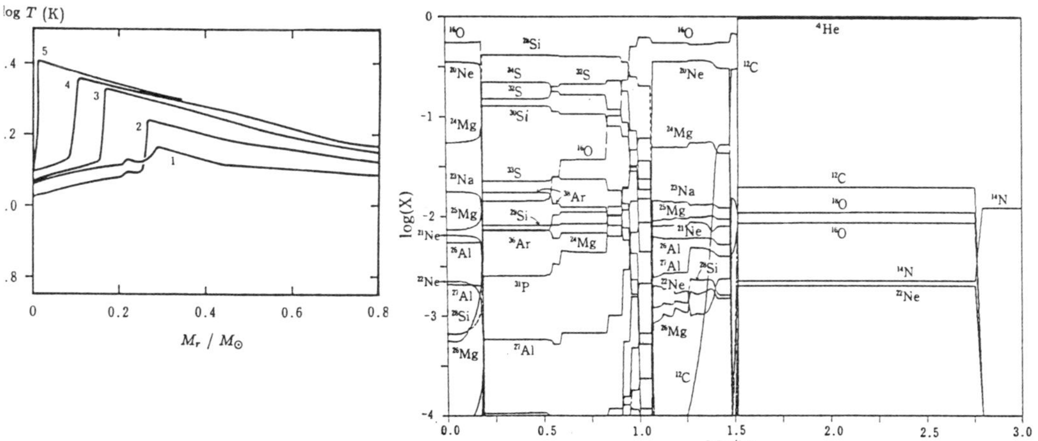


Figure 3 (left): Change in the temperature profile during the propagation of the neon burning front (from stage 1 to 5) for $M_\alpha = 3.0$ $M_\odot$.

Figure 4 (right): Composition of the helium star of $M_\alpha = 3.0$ $M_\odot$ at the shell neon and oxygen burning. The Si-S layer is somewhat neutronized.

and the final fate would be similar to 8 - 10 $M_\odot$ stars. On the other hand, if all neon/oxygen burn, the star will evolve as a non-degenerate massive star and form an iron core.

To answer this question, we have to understand the propagation mode of neon burning front. It is not heat conduction but compressional heating due to gravitational contraction of the oxygen-neon core. If the core mass is larger than a certain critical mass, $c_g < 0$ until $T_c$ reaches the neon ignition temperature. If the core mass is smaller, the temperature decreases ($c_g > 0$) to quench neon burning because of electron degeneracy. For the simple neon star models with the electron mole number $Y_e = 0.5$, this critical mass is 1.44 $M_\odot$ as seen from Figure 1.

For $M_\alpha = 3.0\ M_\odot$ ($M_{ms} \sim 12\ M_\odot$), $M_{ONe}$ is larger than 1.44 $M_\odot$ so that the burning front propagates all the way to the center in 2 yr (Fig. 3). Since the density at the neon burning front does not become so high ($\rho < 10^8$ g cm$^{-3}$) and also neon ignites layer by layer, the released energy in O-Ne flash is too small to induce major dynamical effects.

For slightly smaller mass of $M_\alpha = 2.8\ M_\odot$ ($M_{ms} \sim 11\ M_\odot$), on the other hand, $M_{ONe} = 1.42\ M_\odot$ so that neon burning will be quenched if $Y_e$ were 0.5. However, the density of the oxygen burning front is as high as $10^8$ g cm$^{-3}$ so that $Y_e$ decreases to $\sim 0.48$ due primarily to electron capture on $^{33}$S and $^{35}$Cl (Figure 4). This implies that the core mass exceeds the Chandrasekhar mass, which is now $\sim 1.38$ $M_\odot$, so that $c_g$ remains negative. Therefore, the central temperature will continue to increase up to neon ignition and the burning layer will reach the center. This leads to an important consequnce: the neon burning front reaches densities as high as $\rho > 10^8$ g cm$^{-3}$. Then the neon shell flashes become so explosive that a dynamical event, such as ejection of a helium layer, is expected (Woosley et al. 1980; Hashimoto and Nomoto 1986).

In short, the propagation of the oxygen-neon shell burning is self-sustanined in the sense that neutronization in the burning layer promotes further contraction. Therefore, the quenching of neon burning and the resulting formation of a strongly degenerate oxygen-neon core would be the case only for very limited mass range, if it occurs at all. Most stars of 10 - 13 $M_\odot$ will evolve to form an iron core and, in particular, 10 - 11 $M_\odot$ stars would undergo very exlosive off-center neon flashes.

## 4. PRESUPERNOVA EVOLUTION OF 8 - 10 $M_\odot$ STARS

The evolution of 8 - 10 $M_\odot$ stars ($M_\alpha \simeq 2.0 - 2.5\ M_\odot$) converges to the common path once electrons become strongly degenerate (Nomoto 1984a,b). The representative evolutionary path of ($\rho_c$, $T_c$) for a helium core of $M_\alpha = 2.2\ M_\odot$ is seen in Figure 2. Chemical evolution of the core through the beginning ($M_\alpha = 2.4\ M_\odot$) and the end ($M_\alpha = 2.2\ M_\odot$) of the dredge-up of a helium layer is shown in Figures 4 and 5, respectively.

### 4.1 Formation of Strongly Degenerate O-Ne-Mg Cores

The star in this mass range undergoes non-degenerate carbon burning and forms an O+Ne+Mg core. The advance of the carbon burning shell stops near the helium burning shell at $M_r = 1.28\ M_\odot$ ($M_\alpha = 2.2\ M_\odot$) and $1.34\ M_\odot$ ($M_\alpha = 2.4\ M_\odot$) (Fig. 4 - 5). Hence the oxygen-neon core mass does not exceed the critical mass of $1.37\ M_\odot$ and, therefore, neon is never ignited. After attaining the maximum temperature, which is far below the ignition temperature of neon, $T_c$ decreases due to neutrino emission. Clearly $c_g > 0$, which is an important difference from more massive stars. Afterwards, the core becomes strongly degenerate.

If the core mass stays constant ($\sim 1.3\ M_\odot$), the degenerate core will simply continue to cool like a white dwarf. In a red-giant, the core mass $M_{\mathrm{Heb}}$ (defined as the mass interior to the helium burning shell) grows because hydrogen and helium shell burning processes the material of the hydrogen-rich envelope into carbon-oxygen. (The helium layer has been dredged up by the penetrating surface convection zone as seen in Figure 4). In this sense, the core is essentially the same as accreting white dwarfs. As $M_{\mathrm{Heb}}$ grows, degenerate electrons get more relativistic, i.e., in the equation of state of $P \sim \rho^{(4+\delta)/3}$, $\delta$ changes from 1 to 0. Then $\rho_c$ increases quickly as $\rho_c \sim M_{\mathrm{Heb}}^{2/\delta}$.

### 4.2 Collapse of O+Ne+Mg Core induced by Electron Captures

When $M_{\mathrm{Heb}}$ grows to $1.38\ M_\odot$, $\rho_c$ reaches $4 \times 10^9$ g cm$^{-3}$. Then degenerate electrons provide another effect on stellar evolution, i.e., the electron Fermi energy exceeds the threshold for electron captures $^{24}$Mg (e$^-$, $\nu$) $^{24}$Na (e$^-$, $\nu$) $^{24}$Ne and

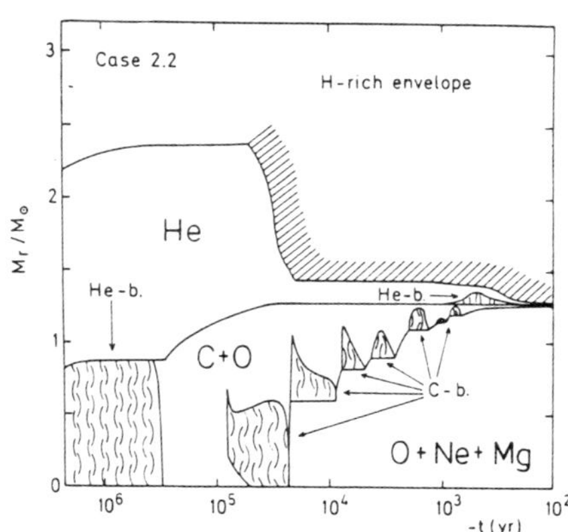


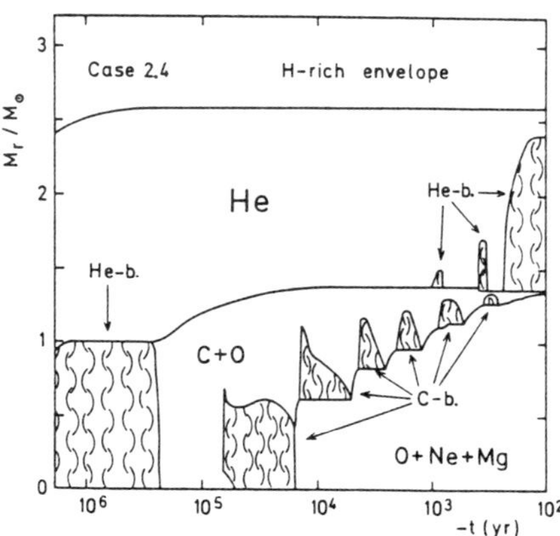


Figure 5 (left): Chemical evolution of the core of $\sim 9\ M_\odot$ star ($M_\alpha = 2.2\ M_\odot$) through the end of the dredge-up of helium layer.

Figure 6 (right): Chemical evolution of the core of $\sim 10\ M_\odot$ star ($M_\alpha = 2.4\ M_\odot$) through the onset of the dredge-up of helium layer.

later for $^{20}$Ne (e$^-$, $\nu$) $^{20}$F (e$^-$, $\nu$) $^{20}$O. Electron captures produce entropy by distorting the Fermi distribution function and emitting $\gamma$-rays (Bisnovatyi-Kogan and Seidov 1970; Sugimoto 1970). The question is whether this heating ignites oxygen burning. We have found that heating due to $^{24}$Na (e$^-$, $\nu$) $^{24}$Ne is slower than the plasmon neutrino cooling but electron capture on $^{20}$Ne and $^{20}$F leads to the ignition of oxygen at $\rho_c \sim 10^{10}$ g cm$^{-3}$ (Miyaji et al. 1980; Nomoto 1984b; Miyaji and Nomoto 1986). However, the oxygen deflagration does not lead to explosion because the decrease in $Y_e$ due to electron capture behind the burning front is fast enough to induce collapse.

The hydrodynamical behavior of collapse is somewhat different from the iron core collapse of more massive stars (see, e.g., Hillebrandt et al. 1984; Burrows and Lattimer 1985; Baron et al. 1986). The O+Ne+Mg core contains nuclear fuel which ignites during infall. Electron captures occur only in the nuclear statistical equilibrium (NSE) layer behind the burning front and, therefore, the region of small $Y_e$ is confined to a central region which grows gradually. The collapse is slower than the collapse of the iron core of a massive star until the burning front has propagated to roughly $\sim 0.8\ M_\odot$ (Hillebrandt et al. 1984). Afterwards the collapse accerelates quickly. $Y_e$ in the NSE region of the O+Ne+Mg core is smaller than $Y_e$ in iron cores, because the entropy at the burning front is higher and thus the proton fraction is larger. These two effects result in a homologous core whose mass is smaller and an outer infalling layer which is less dense than is the case in iron core collpase. Such a structure has two effects on the bounce shock. First, the binding energy of the rebounding core is smaller and, hence, the shock wave is initially weaker (Brown et al. 1982; Lattimer et al. 1985). Secondly, the low density in the outer layers makes the shock propagation easier (Hillebrandt et al. 1984). Which effect dominates depends on the details of collapse hydrodynamics.

## 5. SUPERNOVAE and NEUTRON STARS from LOW MASS PROGENITORS

### 5.1 Neutron Star Mass Inferred from Stellar Structure

Though the hydrodynamic models of collapse and bounce are subject to uncertainties, we can infer some properties of supernovae and neutron stars produced from *low mass* progenitors. Let us compare the density distributions of helium stars of $M_\alpha = 2.4$ and $2.8\ M_\odot$ ($M_{\rm ms} \simeq 10$ and $11\ M_\odot$) with that of $8\ M_\odot$ ($M_{\rm ms} \simeq 25\ M_\odot$). The differences are clearly seen in Figures 7 - 9. Stars forming strongly degenerate or semi-degenerate O+Ne+Mg cores develop a remarkable core-halo structure. The boundary between the *core* and *halo* is located at the helium burning shell. The *core mass* $M_{\rm Heb}$, i.e, mass interior to the helium burning shell, at the precollapse stage is $1.38\ M_\odot$ for $M_\alpha < 2.5\ M_\odot$ and $M_{\rm Heb} = 1.42$, 1.52, and $1.70\ M_\odot$ for $M_\alpha =$ 2.8, 3.0, and $3.2\ M_\odot$, respectively. On the other hand, for non-degenerate massive stars, that feature is much less pronounced.

Suppose that these *low mass stars* give rise to Type II supernova explosions. The mass of a neutron star residue depends on the explosion energy and thus not certain yet. However, we see from the core-halo structure that even a very weak explosion can eject the extended hydrogen-helium layer. Therefore the mass of a neutron star must be less than 1.38 $M_\odot$ (baryon mass) for 8 - 10 $M_\odot$ stars and smaller than 1.4 - 1.7 $M_\odot$ for 10 - 13 $M_\odot$ stars. The baryon mass of 1.4 $M_\odot$ correspond s to the gravitational mass of $\sim$ 1.3 $M_\odot$. Therefore, we can rule out the 8 - 10 $M_\odot$ stars from the progenitors of binary radio pulsars PSR1913+16 because their masses are 1.38 $M_\odot$ and 1.44 $M_\odot$ (e.g., Burrows and Woosley 1986).

## 5.2 Abundance of the Ejecta

What about the elemental abundances of the ejecta? Composition structures near the neon ignition for $M_\alpha = 2.8\ M_\odot$ and near the end of oxygen burning $M_\alpha = 3.0$ and 8 $M_\odot$ are seen in Figures 10, 4, and 11, respectively. If the supernova explosions form a neutron star of mass $M_{\rm NS}$ (baryon mass), they eject a heavy element mantle of mass ($M_{\rm Heb} - M_{\rm NS}$) and a helium envelope. Though $M_{\rm NS}$ depends on the explosion energy, it is likely to be the mass interior to the oxygen burning shell, which is slightly larger than the iron core mass, becasue rather steep density gradient forms there (e.g., Sugimoto and Nomoto 1980; Woosley and Weaver 1986b).

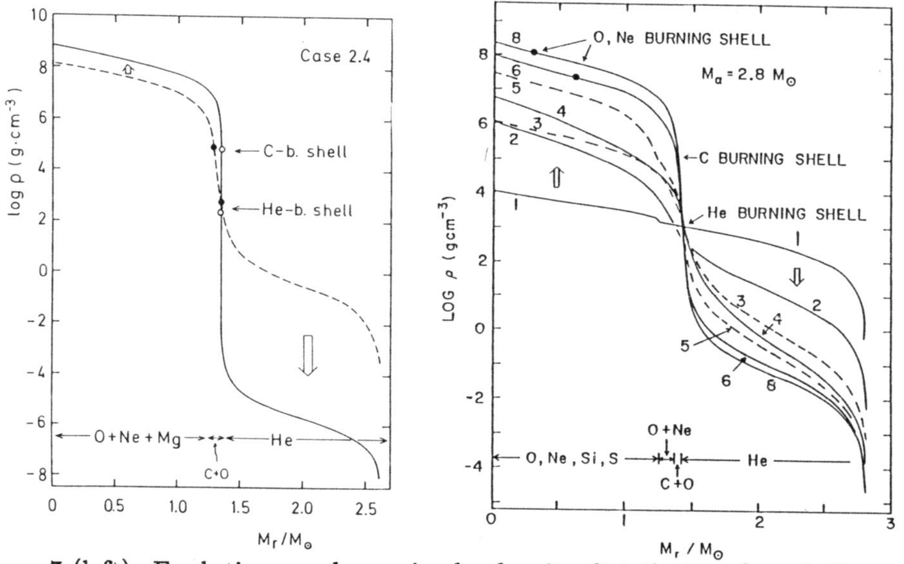


Figure 7 (left): Evolutionary change in the density distribution for a helium core of $M_\alpha = 2.4\ M_\odot$.

Figure 8 (right): Evolutionary change in the density distribution for $M_\alpha = 2.8\ M_\odot$. The stage numbers correspond to 1: exhaustion of helium, 2: ignition of carbon burning, 3: exhaustion of carbon, 4-5: development of a temperature inversion in the O-Ne core, 6: off-center ignition of neon, 7-8: propagation of the neon burning layer.

Then $M_{\rm NS} \simeq 1.35$ - $1.42\ M_\odot$ for $M_\alpha < 4\ M_\odot$ ($M_{\rm ms} < 15\ M_\odot$) (Woosley and Weaver 1986a,b). On the other hand, $M_{\rm Heb}$ and, thus, the ejected mantle mass are sensitive to the stellar mass and smaller for smaller $M_\alpha$. This is an important feature to compare with the elemental abundances of supernova remnants. In particular, there is no significant overabundance of heavy elements in the Crab nebula relateive to the solar values (Davidson and Fesen 1985 and references therein). This would rule out the stars more massive than $12\ M_\odot$ ($M_\alpha > 3\ M_\odot$) from the Crab nebula's progenitor (Arnett 1975; Woosley et al. 1980; Nomoto et al. 1982; Hillebrandt 1983; Nomoto 1985). If $M_{\rm NS} \simeq 1.35\ M_\odot$ in baryon mass for $M_\alpha < 3\ M_\odot$ (Woosley and Weaver 1986a,b), 8 - 10 $M_\odot$ stars would be favored. Moreover, if we take the carbon abundance obtained from the UV observations as a constraint (Davidson

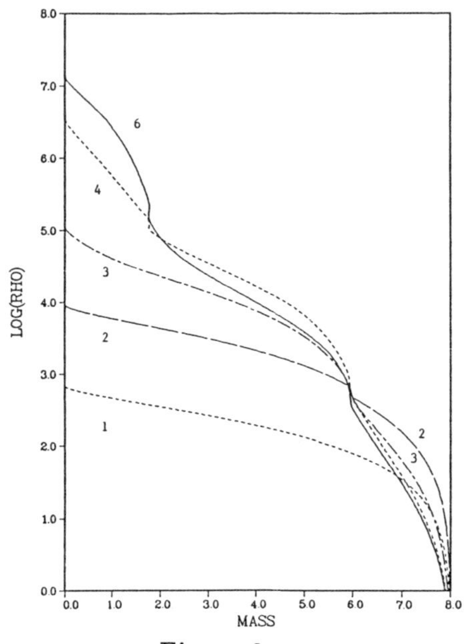


Figure 9

Figure 9: Evolutionary change in the density distribution for the helium star of $M_\alpha = 8\ M_\odot$. Stage numbers refer to: 1) helium burning, 2) helium exhaustion, 3) carbon ignition, 4) neon ignition, 6) oxygen exhaustion.

Figure 10: Compostion of the helium star of $M_\alpha = 2.8\ M_\odot$ near the neon ignition at $M_r = 0.64\ M_\odot$.

Figure 11: Composition of the helium star of $M_\alpha = 8\ M_\odot$ near the end of oxygen burning.

Figure 10

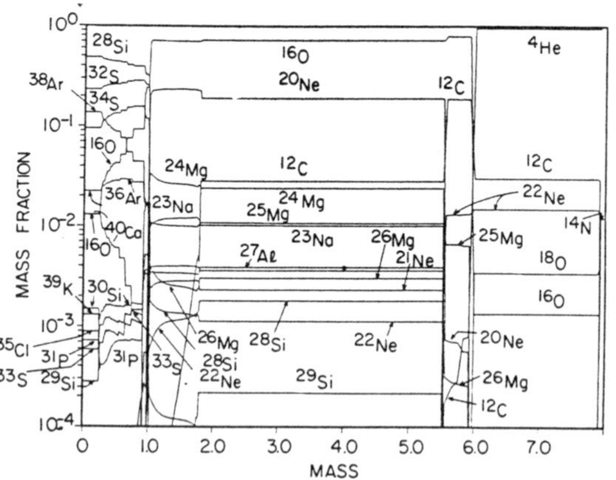


Figure 11

et al. 1982), stars of $M_{ms} \sim 9\ M_\odot$ ($M_\alpha \sim 2.2\ M_\odot$; Fig. 5) would be the likely progenitor because the helium layer of more massive models contain significant amount of carbon as seen in Figures 4, 10, and 11. (Nomoto et al. 1982). On the other hand, if we take little enhancement of nitrogen as a more important constraint, stars of $M_{ms} \sim 10\ M_\odot$ ($M_\alpha \sim 2.4\ M_\odot$; Fig. 6) would be favored (Henry 1986; Nomoto et al. 1982).

### 5.3 Supernova Light Curve and Pulsar Activity

SN 1054 showed a bright optical display with a tail observed for almost 2 yr. What was the energy source of the light curve? The progenitor was expected to be a red-supergiant and thus had a very extended envelope. Therefore the peak light can be accounted for by the shock heating of photoshpere like Type II supernovae (Chevalier 1977). For the tail, possible energy sources are the decay of $^{56}$Ni (which is likely to be the case for some Type II supernovae as well as Type Ia supernovae) and the pulsar activity. From the composition structure, it is clear that the stars smaller than $\sim 15\ M_\odot$ do not produce enough $^{56}$Ni. Therefore progenitor models with mass smaller than $15\ M_\odot$ require the pulsar activity as an energy source for the tail (e.g., Chevalier 1977). This implies that the Crab's neutron star should have strong magnetic field and rapid rotation at its birth. If these requirement cannot be fulfilled, we have to invoke the $^{56}$Ni decay to explain the light curve tail of SN 1054. This would imply that most of heavy elements formed dust.

## 6. THE FATE OF WHITE DWARFS AS A FUNCTION OF MASS AND ACCRETION RATE

Isolated white dwarfs are simply cooling stars that eventually end up as *dark matter*. In binary systems they evolve differently because mass accretion from their companion provides gravitational energy that rejuvenates them. The gravitational energy released at the accretion shock near the stellar surface is radiated away and does not heat the white dwarf interior. However, the compression of the interior by the accreted matter releases additional gravitational energy. Some of this energy goes into thermal energy (compressional heating) and the rest is transported to the surface and radiated away (radiative cooling). Therefore, the interior temperature is determined by the balance between heating and cooling and, thus, strongly depends on the mass accretion rate, $\dot{M}$ (Nomoto 1982, 1984b).

Compression first heats up a layer near the surface because of the small pressure scale height there. Later, heat diffuses inward. The diffusion timescale depends on $\dot{M}$ and is small for larger $\dot{M}$'s because of the large heat flux and steep temperature gradient generated by rapid accretion. For example, the time it takes the heat wave to reache the central region is about $2 \times 10^5$ yr for $\dot{M} \sim 10^{-6}\ M_\odot\ \mathrm{yr}^{-1}$ and $5 \times 10^6$ yr for $\dot{M} \sim 4 \times 10^{-8}\ M_\odot\ \mathrm{yr}^{-1}$. Therefore, if the initial mass of the white dwarf, $M_{CO}$, is smaller than $1.2\ M_\odot$, the entropy in the center increases substantially due to the heat inflow and carbon ignites at relatively low central density ($\rho_c \simeq 3 \times 10^9$ g cm$^{-3}$). On the other hand, if the white dwarf is sufficiently massive and cold at the onset of accretion, the central region is compressed only adiabatically and thus is cold when carbon burning is ignited in the center. In the latter case, the ignition density is as high as $10^{10}$ g cm$^{-3}$ (e.g., Isern et al. 1983).

Accordingly, the ultimate fate of accreting C+O white dwarfs depends on $\dot{M}$ and the inintial mass of the white dwarf $M_{\rm CO}$, as summarized in Figure 12. $\dot{M}$ denotes the growth rate of the C+O white dwarf mass irrespective of the composition of the accreting matter. A similar diagram for the O+Ne+Mg white dwarfs is shown in Figure 13. *Neutron Star* or *NS* in Figure 12 indicates a region of parameter space where neutron star formation by white dwarf collapse is expected. The evolution of the white dwarfs in these three regions is summarized as follows. (The fate of white dwarfs in other regions in Figure 12 is briefly described in §7.)

1) For $\dot{M} > 2.7 \times 10^{-6}$ $M_\odot$ yr$^{-1}$, off-center carbon burning is ignited by rapid compressional heating as indicated in Figure 14 (Nomoto and Iben 1985; Kawai et al. 1986). The C+O white dwarf is peacefully changed into an O+Ne+Mg white dwarf through off-center carbon burning (Saio and Nomoto 1985; Woosley et al. 1986a). The resulting O+Ne+Mg white dwarf will collapse to form a neutron star due to electron capture on $^{24}$Mg and $^{20}$Ne if the Chandrasekhar mass is reached.

2) For $2.7 \times 10^{-6}$ $M_\odot$ yr$^{-1} > \dot{M} > 4 \times 10^{-8}$ $M_\odot$ yr$^{-1}$ and $M_{\rm CO} > 1.2$ $M_\odot$, a central density as high as $10^{10}$ g cm$^{-3}$ is reached by adiabatic compression if the

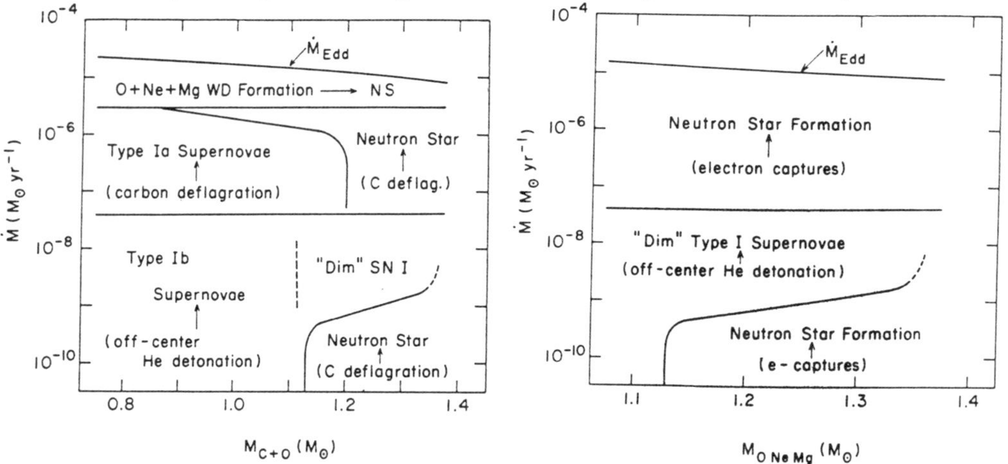


Figure 12 (left): The final fate of accreting C+O white dwarfs expected for their initial mass $M_{\rm CO}$ and accretion rate $\dot{M}$. For two regions indicated by *Neutron Star*, carbon deflagration propagating from $\rho_c \simeq 10^{10}$ g cm$^{-3}$ will induce collapse.

Figure 13 (right): Same as Figure 12 but for O+Ne+Mg white dwarfs. For a wider range of $\dot{M}$ and the initial mass, $M_{\rm ONeMg}$, neutron star formation triggered by electron capture on $^{24}$Mg and $^{20}$Ne is expected.

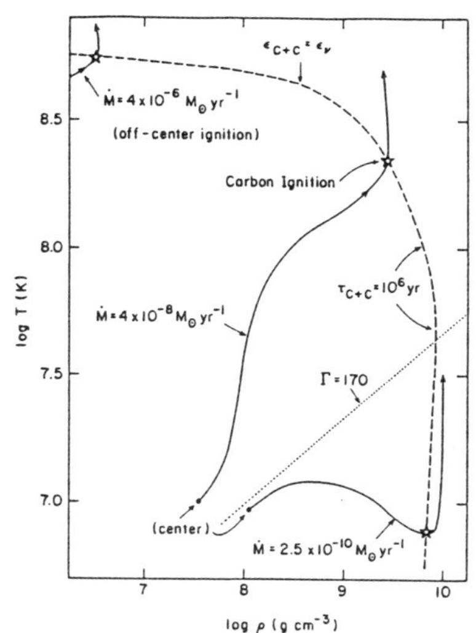


Figure 14: Evolutionary path of ($\rho_c$, $T_c$) at the center of accreting C+O white dwarfs for two cases of $\dot{M}$. For comparison, ignition point of off-center carbon burning is shown for $\dot{M} = 4 \times 10^{-6}\ M_\odot\ yr^{-1}$. The dashed curve is an approximate ignition line where the rate of carbon burning $\epsilon_{C+C}$ is equal to the rate of neutrino losses $\epsilon_\nu$ for $T > 2 \times 10^8$ K and $\tau_{C+C} \equiv c_p\ T/\epsilon_{C+C} = 10^6$ yr for $T \leq 2 \times 10^8$ K ($c_p$ is the specific heat).

white dwarf is sufficiently cold at the onset of accretion. For $\dot{M} > 10^{-6}\ M_\odot\ yr^{-1}$, the lower mass limit is not 1.2 $M_\odot$, but 1.0 $M_\odot$.

3) For $\dot{M} \leq 10^{-9}\ M_\odot\ yr^{-1}$ and $M_{CO} > 1.13\ M_\odot$, the white dwarf is too cold to initiate a helium detonation. Eventually pycnonuclear carbon burning starts in the center when $\rho_c$ reaches $\sim 10^{10}$ g cm$^{-3}$. An evolutionary path of $\rho_c$ - $T_c$ for a model with $\dot{M} = 2.5 \times 10^{-10}\ M_\odot\ yr^{-1}$ and $M_{CO} = 1.16\ M_\odot$ is shown in Figure 14. In this model, the outer layer of 0.24 $M_\odot$ is composed of helium.

## 7. MODELS FOR SUPERNOVAE OF TYPE Ia AND Ib

### 7.1 Type Ia Supernovae

For relatively high accretion rates ($2.7 \times 10^{-6}\ M_\odot\ yr^{-1} > \dot{M} > 4 \times 10^{-8}\ M_\odot\ yr^{-1}$), a carbon deflagration starts in the white dwarf's center at a relatively low central density ($\rho_c \sim 3 \times 10^9$ g cm$^{-3}$) (e.g., Ivanova et al. 1974; Nomoto et al. 1984 and references therein). The *convective* deflagration wave then propagates outward at a subsonic velocity. The density the wave encounters is decreasing due to the expansion of the white dwarf.

The products of explosive nucleosynthesis depend on the temperature and density at the deflagration front and, thus, vary from layer to layer. In the center, iron peak elements are produced. In particular, about 0.6 $M_\odot$ of $^{56}$Ni is synthesized. In the outer layers, intermediate mass elements such as Ca, Ar, S, Si are produced. The white dwarf is disrupted completely and no neutron star residue remains (Nomoto et al. 1984; Thielemann et al. 1986; Woosley et al. 1984).

The carbon deflagration model can account for the light curves, early time specta, and late time spectra of Type Ia supernovae as follows (see Nomoto 1986a; Woosley and Weaver 1986b for reviews and references therein):

(1) The theoretical light curve based on the radioactive decays of $^{56}$Ni and $^{56}$Co into $^{56}$Fe fits the observations well.
(2) The synthetic spectum at maximum light is in excellent agreement with the observed spectum of SN 1981b as seen in Figure 15 (Branch et al. 1985; Wheeler and Harkness 1986).
(3) At late times, the outer layers are transparent and the inner Ni-Co-Fe core is exposed. Synthetic spectra of emission lines of [Fe II] and [Co I] agree quite well with the spectra observed at such phase (Woosley et al. 1984).

### 7.2 Type Ib Supernovae

Recent observations has established the existence of another kind of Type I supernovae, designated Type Ib (SN Ib) (e.g., Wheeler and Harkness 1986 and references therein). The SN Ib spectra are characterized by the lack of the 6125 Å Si feature at maximum light and the appearance of oxygen emission lines at late times (e.g., Gaskel et al. 1986). The currently popular progenitor models for SN Ib are Wolf-Rayet stars (e.g., Wheeler and Levreault 1985; Begelman and Sarazin 1986). However, a large mass of Wolf-Rayet stars may yield a light curve whose decline is too slow to be compatible with SN Ib observations (Wheeler and Levrerault 1985).

Branch and Nomoto (1986) have suggested that the observed spectra are better explained by an accreting white dwarf model. In Figure 16, the maximum-light

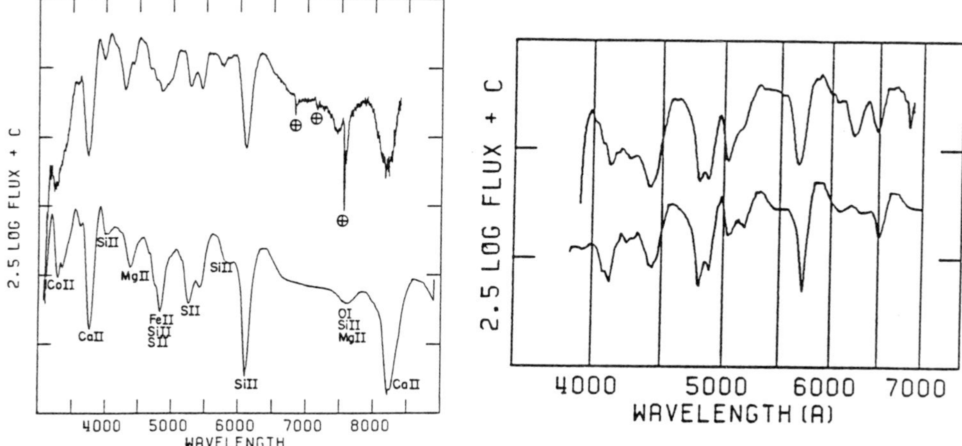


Figure 15 (left): The maximum-light spectrum of SN 1981b (top) is compared to a synthetic spectrum for the carbon deflagration model (Nomoto et al. 1984; Branch et al. 1985). In this model outer layer is assumed to be mixed.

Figure 16 (right): The maximum-light spectrum of the Type Ib SN 1984*l* (upper; Wheeler and Levreault 1985) is compared with a synthetic spectrum (lower; Branch and Nomoto 1986). In the synthetic spectrum the blueshifted absorption component of He I $\lambda$6678 appears near 6500 Å and He I $\lambda$5876 appears near 5850 Å. Other features are produced primarily by Fe II lines.

spectrum of SN 1984$l$ is compared with a synthetic spectrum. Two of the absorption lines in the red are identified as He I lines and other features are well explained as Fe II lines. Here the expansion velocity at the photosphere is 8,000 km $s^{-1}$. In addition, ultraviolet features can fit with a synthetic spectrum of Co II and Fe I lines if the photoshperic velocity is 12,000 km $s^{-1}$ (Branch and Venkatakrishna 1986). The above interpretation suggests that Fe, Co (decaying), and He are in the outer high-velocity layers and oxygen is in the inner layers.

The existence of such high velocity Fe and, especially, Co is difficult to explain with the Wolf-Rayet model. Branch and Nomoto (1986) have speculated that the progenitors of SN Ib are white dwarfs having $\dot{M} < 4 \times 10^{-8}\ M_\odot\ yr^{-1}$. Such a slow accretion induces an off-center helium detonation that will occur at a point rather than all over a spherical shell. The outer helium layer will burn to mostly $^{56}Ni$ with a trace He and the inner C+O core will remain unburned. A part or most of the C+O will be ejected following $^{56}Ni$ layer. Since the ejected mass of $^{56}Ni$ will be as small as 0.1 - 0.3 $M_\odot$, the peak luminosity of this model is lower than that for SN Ia by a factor of 2 - 6. This is consisttent with the observations of SN Ib. If the white dwarf accretes matter with an efficiency of only 0.03 - 0.1 from a wind ($10^{-6}$ - $10^{-7}\ M_\odot\ yr^{-1}$) of a relatively massive (4 - 7 $M_\odot$) red giant companion (Iben and Tutukov 1984), the model would be consistent with relatively young nature of SN Ib. Further, this scenario is consistent with the radio observations of SN Ib in that they can be explained by the interaction of supernova ejecta with the circumstellar shell (Sramek et al. 1984; Chevalier 1984).

### 7.3 Dim Type I Supernovae

If an off-center *single* detonation occurs on a very massive white dwarf ($> \sim 1.1\ M_\odot$), the resulting supernova will be rather *dim*, because the accumulation of only a small amount of helium ($\sim$ 0.01 - 0.1 $M_\odot$) can lead to the helium detonation. In most cases, an unburned C+O core will be left behind as a white dwarf. Such dim supernovae (Branch and Doggett 1985) are more likely to be associated with O+Ne+Mg white dwarfs since their masses are larger than $\sim 1.2\ M_\odot$.

## 8. COLLAPSE INDUCED BY CARBON DEFLAGRATION AT HIGH DENSITY

### 8.1 Conductive Deflagration

As mentioned in §6, there are two scenarios in which a carbon deflagration is initiated in the center when the central density is as high as $10^{10}$ g $cm^{-3}$. At such densities, the carbon deflagration may not lead to an explosion since electron capture is much faster at these high densities than at the lower densities encountered in the models of SN Ia. Moreover, if the central part of the white dwarf is in the solid state, the propagation mode of the burning front could be different. If the solid is strong enough, convection will be suppressed and the burning front will propagate as a *conductive* deflagration wave (Canal et al. 1980; Isern et al. 1983, 1985) though more study is needed on this point. Even for fluid layers, conductive deflagration could dominate convective deflagration in the central region (Woosley and Weaver 1986c). The propagation velocity of a conductive deflagration wave is given approximately by the expression, $v_{\text{def}} \sim \delta/\tau_{\text{n}} \sim (\sigma/c_{\text{v}}\tau_{\text{n}})^{1/2}$, where $\delta$ denotes the width of burning front, $\tau_{\text{n}}$ the nuclear burning timescale, $\sigma$ the conductivity, and $c_{\text{v}}$ the specific heat (Buchler et al. 1980; Woosley and Weaver 1986c). This gives $v_{\text{def}} \sim 100$ km s$^{-1}$ at $\rho \sim 10^{10}$ g cm$^{-3}$, which is about 0.01 $v_{\text{s}}$ (Woosley and Weaver 1986c). Here $v_{\text{s}}$ is the sound speed, equal to 1.0 - 1.3 $\times$ $10^4$ km s$^{-1}$ between $\rho = 10^9$ - $10^{10}$ g cm$^{-3}$.

Whether the white dwarf explodes or collapses depends on whether, behind the deflagration wave, nuclear energy release or electron capture is faster. A white dwarf whose mass is close to the Chandrasekhar mass has an adiabatic index close to 3, so that even a small energy release can cause substantial expansion. However, a slight pressure decrease due to electron capture will easily induce collapse. If $v_{\text{def}}$ is low (high) enough and/or the central density is high (low) enough, a carbon deflagration will lead to collapse (explosion). The outcome is rather sensitive to $v_{\text{def}}$ and the central density.

We have performed numerical simulations of a conductive deflagration started at high central density ($\rho_{\text{c}} \simeq 10^{10}$ g cm$^{-3}$), with $\dot{M} = 2.5 \times 10^{-10}$ $M_{\odot}$ yr$^{-1}$ and

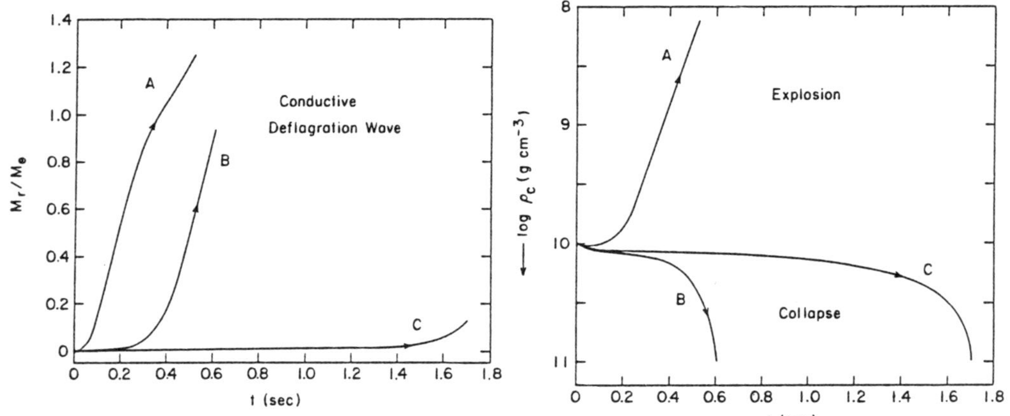


Figure 17 (left): Propagation of the conductive deflagration wave. The location ($M_r$) of the deflagration front is shown as a function of time, $t$, for three cases (A, B, C) of parametrized conductivity.

Figure 18 (right): Change in the central density of the white dwarf associated with the propagation of the conductive deflagration wave. Relatively slow propagation in Cases B and C leads to the increase in $\rho_{\text{c}}$, i.e., collapse of the white dwarf. On the other hand, faster proapagation in Case A induces the explosion of the white dwarf.

$M_{CO} = 1.16\ M_\odot$ as described in §6 and shown in Figure 14 (Nomoto 1986b). Here we have assumed that no chemical separation occurs. Pycnonuclear carbon burning commences in the solid region and turns into thermonuclear runaway. Then, we have parametrized the conductivity to obtain a range of reasonable $v_{def}$. With these, we have explored the range of possible hydrodynamical responses of the white dwarf. In Figure 17, the location of the deflagration front as a function of time, $t$, after initiation is plotted for three cases (A, B, C). In addition, Case D of the slowest propagation has been calculated. The average $v_{def}/v_s$ for Cases A, B, C, and D is about 0.15, 0.1, 0.06, and 0,01, respectively. In Figure 18, the evolution of the central density, $\rho_c$, for Cases A - C is plotted.

## 8.2 Cases of Collapse

As Figure 18 indicates, in Case B, the central density *increases* as the deflagration front propagates outward. When the deflagration front has reached $M_r \simeq 0.9\ M_\odot$ ($t \simeq 0.6$ s), $\rho_c$ is as high as $10^{11}$ g cm$^{-3}$. Clearly, the white dwarf now undergoes quasi-dynamic contraction because electron capture on NSE elements rapidly reduce the electron mole number, $Y_e$, behind the deflagration wave. This effect dominates the opposing effect of nuclear burning. (At $t = 0.16$ s, $Y_e$ has dropped to 0.42 at the center.) Moreover, once the contraction of the white dwarf begins, photodissociation becomes important, further promoting collapse. In Case C, the contraction is much more gradual than in Case B because the deflagration wave is slower. When $\rho_c$ reaches $10^{11}$ g cm$^{-3}$, the burned mass is only 0.13 $M_\odot$ (Figures 17 and 18). In Case D, for which $v_{def}$ ($\sim$ 0.01 $v_s$) is close to the actual conductive deflagration speed, it takes 155 s to reach $\rho_c \simeq 10^{11}$ g cm$^{-3}$. At $t = 155$ s, the mass of the burned region is only 0.03 $M_\odot$ and $Y_e \sim 0.39$.

## 8.3 Case of Explosion

On the other hand, in Case A, $\rho_c$ decreases as the deflagration propagates outward. By the time the front has reached $M_r \simeq 1.2\ M_\odot$, the total energy of the white dwarf is already 1.2 $\times 10^{51}$ ergs and it is clear that it will be completely disrupted. Nuclear energy release dominates electron capture because the front's density, and hence its electron capture rate, decreases as the front propagates outward. When the deflagration wave arrives at the base of the helium layer ($M_r = 1.16\ M_\odot$), it turns into a helium *detonation* because of a low ignition temperature. The explosion starting from high $\rho_c$ should not be frequent since they eject too much neutron-rich iron peak matter into the Galaxy (Nomoto et al. 1984; Woosley and Weaver 1986c).

## 8.4 Convective Deflagration

In some cases of carbon ignition at high densities (e.g., $\dot{M} > 4 \times 10^{-8}\ M_\odot$ yr$^{-1}$), convective deflagration would develop. Even for the solid core, propagation of the deflagration wave is not necessarily due to conduction alone since convection could influence the melting of the solid core (Mochkovitch 1980). To investigate the outcome of convective deflagration, a set of numerical experiments has been performed for $\rho_c \sim 10^{10}$ g cm$^{-3}$ by employing time-dependent mixing length theory with a parameter $\alpha = \ell / H_p$ where $\ell$ is the mixing length and $H_p$ is the pressure scale height (Nomoto et al. 1984).

For $\alpha = 0.7$, the propagation velocity is as slow as $v_{def} < 0.11\ v_s$ at $M_r < 0.6\ M_\odot$ so that the white dwarf collapses as in Case B. On the other hand, for $\alpha = 1.0$, $v_{def} > 0.15\ v_s$ at $M_r > 0.4\ M_\odot$ and the white dwarf explodes completely. Since a value for $\alpha$ of 0.7 is preferred in the low density carbon deflagration model of SN Ia (§7), a plausible choice for $\alpha$ in the present context may be 0.7, not 1.0. (For both low and high central densities, the carbon deflagration with $\alpha = 1.0$ grows into a *detonation* in the outer layer and incinerates almost the entire star to the iron peak. This is incompatible with observations of SN Ia (Nomoto et al. 1984).) Therefore, for plausible choices of the $\alpha$ parameter, a carbon deflagration initiated at high densities will result in white dwarf collapse, not explosion.

## 9. DISCUSSION ON WHITE DWARF COLLAPSE

### 9.1 C+O White Dwarfs

We have shown how the fate of accreting white dwarfs depends on the initial mass, composition, and age of the white dwarf and accretion rate. In particular, we have examined the critical condition for which a carbon deflagration leads to collapse. If a carbon deflagration is initiated in the center of the white dwarf when $\rho_c \simeq 10^{10}$ g cm$^{-3}$ and if the propagation velocity of the deflagration wave is slower than a certain critical speed, $v_{crit}$, the outcome is collapse, not explosion (Nomoto 1986b; also Ivanova et al. 1974, Isern et al. 1985). For $v_{def} > v_{crit}$, complete disruption results (and the ejecta contain too much neutron-rich matter). The value of $v_{crit}$ depends on $\rho_c$ at carbon ignition. For $\rho_c \simeq 1 \times 10^{10}$ g cm$^{-3}$, $v_{crit} \sim 0.15\ v_s$. A lower $\rho_c$ implies a lower $v_{crit}$. In our case of $\rho_c \simeq 1 \times 10^{10}$ g cm$^{-3}$, for both conductive and convective deflagrations $v_{def} < v_{crit}$ and, therefore, collpase will result.

Such a high central density is reached in two regions of the $\dot{M}$ - $M_{CO}$ plane of Figure 12. One is defined by $\dot{M} > 4 \times 10^{-8}\ M_\odot$ yr$^{-1}$ and $M_{CO} > 1.2\ M_\odot$, while the other is defined by $\dot{M} < 10^{-9}\ M_\odot$ yr$^{-1}$ and $M_{CO} > 1.13\ M_\odot$. The frequency of such systems may be small. First, if hydrogen-rich matter accretes at $\dot{M} < 10^{-9}$ $M_\odot$ yr$^{-1}$, nova-like explosions will prevent the white dwarf mass from growing. Secondly, massive C+O white dwarfs ($> 1.2\ M_\odot$) may be rare. The formation of such white dwarfs might be prevented if the precursor star lost its hydrogen-rich envelope by either a stellar wind or Roche-lobe overflow before its degenerate C+O core could grow substantially.

## 9.2 O+Ne+Mg White Dwarfs

As seen in Figure 13, the accretion-induced collapse is the outcome for a wider range of parameter space for O+Ne+Mg white dwarfs. The initial mass of the white dwarf, $M_{\rm ONeMg}$, is larger than $\sim 1.2\ M_\odot$ (Nomoto 1980, 1984a). In many cases, $M_{\rm ONeMg}$ is very close to the Chandrasekhar mass, so that only a small mass increase is enough to trigger collapse. However, an O+Ne+Mg white dwarf formes from an 8 - 10 $M_\odot$ star (Nomoto 1984a). The number of such systems may be significantly smaller than the number of systems containing C+O white dwarfs whose precursors are 1 - 8 $M_\odot$ stars, perhaps, by four order of magnitude (Iben and Tutukov 1984). Even so, the number of low mass X-ray binaries is much smaller than the number of SN I and the statistics may be consistent (Webbink et al. 1983).

## 9.3 Companion Star of the White Dwarfs

Among the possible regions of the parameter space for neutron star formation (Figs. 12 - 13), rather high accretion rates ($\dot{M} > \dot{M}_{\rm det}$) are favored in order to avoid nova explosions and helium detonations (e.g., Nomoto 1982). Possibel binary systems to realize such accretion rate include: (1) Case A mass transfer on thermal timescale from a main-sequence star of $\sim 2\ M_\odot$ ($\dot{M} \sim 10^{-7}\ M_\odot\ {\rm yr}^{-1}$) (Iben and Tutukov 1984), (2) mass transfer on the nuclear timescale from a subgiant of $\sim 0.8\ M_\odot$ ($\dot{M} \sim 10^{-7}$ - $10^{-8}\ M_\odot\ {\rm yr}^{-1}$) (Webbink et al. 1983), and (3) accretion of helium from a helium main-sequence star on a timescale of gravitational wave radiation ($\sim 2$ - $3 \times 10^{-8}\ M_\odot\ {\rm yr}^{-1}$) (Savonije et al. 1986; Iben et al. 1986). For Cases 2 - 3, the $^{14}{\rm N}(e^-, \nu)^{14}{\rm C}(\alpha, \gamma)^{18}{\rm O}$ reaction would be important to lower $\dot{M}_{\rm det}$ from $4 \times 10^{-8}$ to $1 \times 10^{-8}\ M_\odot\ {\rm yr}^{-1}$ (Hashimoto et al. 1986).

## 9.4 Single Millisecond Pulsar from Merging Double White Dwarfs?

As mentioned in §6, if the merging of CO - CO white dwarf pair (Iben and Tutukov 1984; Webbink 1984) leads to a rapid accretion such as $\dot{M} > 2.7 \times 10^{-6}\ M_\odot\ {\rm yr}^{-1}$, the final outcome would be a neutron star. This is also the case with the merging of CO - ONeMg white dwarf pair for all range of accretion rates (Kawai et al. 1986). Since the collapsing white dwarf has gained angular momentum during merging, the resulting neutron star is a rapid rotator. Suppose that the magnetic field of a neutron star originates from a fossil field of the white dwarf and no enhancement of the field occurs after the neutron star formation. Then the neutron star could be weakly magnetized if the magnetic field of the white dwarf was weak. (White dwarfs in cataclysmic variables are classified into two classes, with and without strong magnetic field (King et al. 1984).) The neutron star could continue to rotate rapidly for relatively long time. Forthermore, the precursors of CO or ONeMg white dwarfs are 5 - 10 $M_\odot$ stars so that a location of the resulting neutron star may not be far from the galactic plane. These features could be consistent with the properties of the single millisecond pulsar (Baker et al. 1982). On the other hand, if the magnetic field is enhanced during or after collapse, a normal single pulsar would result but with no supernova remnant around it (see Saio and Nomoto 1985).

### 9.5 Dim or Silent Supernovae?

A hydrodynamical calculation of a white dwarf collapse has not been carried out so that we don't know whether mass is ejected. Even if no mass is ejected, a neutron star will form because the residue's mass (1.4 $M_\odot$ (baryon mass), $\sim$ 1.3 $M_\odot$ (gravitational mass)) is smaller than the maximum mass of a neutron star. Nevertheless, it is important to know whether some mass is ejected by the bounce shock and, if mass is ejected, what its composition is. If some $^{56}$Ni is ejected, the white dwarf collapse can be observed as a *dim* Type I supernova. Otherwise, the collapse would be *silent*, because most of the explosion energy would go into the kinetic energy of expansion. The interior temperatures would be too low to produce a significant optical light curve. If the shock wave is strong enough, some neutron-rich species will be ejected. This might be an important site of some neutron-rich isotopes (Hartman et al. 1985; Takahashi et al. 1986). Mass ejection will affect the binary evolution after the explosion and the results can be compared to the observed neutron star binary systems (e.g., Taam and van den Heuvel 1986).

I would like to thank Drs. Hashimoto, D. Branch, J.C. Wheeler, R. Harkness, and Z. Barkat for informative discussion on massive star evolution and Type I supernovae. It is a pleasure to thank Drs. S.H. Kahana, G.E. Brown, A. Yahil, A. Burrows, J. Cooperstein, and E. Baron for useful discussion and hospitality during my stay in Brookhaven and Stony Brook. This work has been supported in part by the U. S. Department of Energy under Contract No. DE-AC02-76CH00016 and the Japanese Ministry of Education, Science, and Culture through research grant Nos. 59380001 and 60540152.

## REFERENCES

Arnett, W.D. 1975, *Ap. J.*, *195*, 727.

———. 1978, in *Physics and Astrophysics of Neutron Stars and Black Holes*, ed. R. Giaconni and R. Ruffini (Bologna: Soc. Italiana di Fisica), p.356.

Backer, D.C., Kulkarni, S.R., Heiles, C., Davis, M.M., and Goss, W.M. 1982, *Nature*, *300*, 615.

Barkat, Z., Reiss, Y., and Rakavy, G. 1974, *Ap. J. (Letters)*, *193*, L21.

Baron, E., Cooperstein, J., and Kahana, S.H. 1986, private communication.

Begelman, M.C., and Sarazin, C.L. 1986, *Ap. J. (Letters)*, *302*, L59.

Bisnovatyi-Kogan, G.S., and Seidov, A.F. 1970, *Astr. Zh.*, *47*, 139.

Branch, D., and Doggett, J.B. 1985, *A. J.*, *270*, 2218.

Branch, D., Doggett, J.B., Nomoto, K, and Thielemann, F.-K. 1985, *Ap. J.*, *294*, 619.

Branch, D., and Nomoto, K. 1986, *Astr. Ap.*, in press.
Branch, D., and Venkatakrishna, K.L. 1986 *Ap. J. (Letters)*, *306*, L21.
Brown, G.E., Bethe, H.A., and Baym, G. 1982, *Nucl. Phys.*, *A375*, 481.
Buchler, J.R., Colgate, S.A., and Mazurek, T.J. 1980, *J. de Phys. Suppl. 3*, *41*, C2-159.
Burrows, A., and Lattimer, J.M. 1985, *Ap. J. (Letters)*, *299*, L19.
Burrows, A., and Woosley, S.E. 1986, *Ap. J.*, *308*, in press.
Canal, R., Isern, J., and Labay, J. 1980, *Ap. J. (Letters)*, *241*, L33.
Chevalier, R.A. 1977, in *Supernovae*, ed. D.N. Schramm (Dordrecht: Reidel), p.53.
————. 1984, *Ap. J. (Letters)*, *285*, L63.
Davidson, K., and Fesen, R.A. 1985, *Ann. Rev. Astr. Ap.*, *23*, 119.
Davidson, K., et al. 1982, *Ap. J.*, *253*, 696.
Doggett, L.B., and Branch, D. 1985, *A. J.*, *90*, 2303.
Gaskell, C.M., Cappellaro, E., Dinerste in, H., Garnett, D., Harkness, R.P., and Wheeler, J.C. 1986, *Ap. J. (Letters)*, *306*, L77.
Habets, G.M.H.J. 1985, Ph.D. Thesis, University of Amsterdam.
————. 1986, *Astr. Ap.*, submitted.
Hartmann, D., Woosley, S.E., and El Eid, M.F. 1985, *Ap. J.*, *297*, 837.
Hashimoto, M., and Nomoto, K. 1986, in preparation.
Hashimoto, M., Nomoto, K., Arai, K., and Kaminisi, K. 1986, *Ap. J.*, *307*, in press.
Henry, R.B.C. 1986, *Pub. Astr. Soc. Pacific*, in press.
Hillebrandt, W. 1983, *Astr. Ap.*, *110*, L3.
Hillebrandt, W., Nomoto, K., and Wolff, R.G. 1984, *Astr. Ap.*, *133*, 175.
Iben, I. Jr., Nomoto, K., Tornambe, A., and Tutukov, A.V. 1986, *Ap. J.*, submitted.
Iben, I. Jr., and Tutukov, A.V. 1984, *Ap. J. Suppl.*, *54*, 335.
Isern, J., Labay, J., Hernanz, M., and Canal, R. 1983, *Ap. J..*, *273*, 320.
Isern, J., Labay, J., and Canal, R. 1984, *Nature*, *309*, 431.
Ivanova, L.N., Imshennik, V.S., and Chechetkin, V.M. 1974, *Ap. Space Sci.*, *31*, 497.
Kawai, Y., Saio, H., and Nomoto, K. 1986, *Ap. J.*, submitted.
King, A.R., Frank, J., and Ritter, H. 1985, *M.N.R.A.S.*, *213*, 181.
Lattimer, J.M., Burrows, A., and Yahil, A 1985, *Ap. J.*, *288*, 644.
Miyaji, S., Nomoto, K., Yokoi, K., and Sugimoto, D. 1980, *Pub. Astr. Soc. Japan*, *32*, 303
Miyaji, S., and Nomoto, K. 1986, *Ap. J.*, submitted.
Mochkovitch, R. 1980, Thesis, University of Paris.
————. 1983, *Astr. Ap.*, *122*, 212.
Nomoto, K. 1980, in *Type I Supernovae*, ed. J. C. Wheeler (Austin: University of Texas), p. 164.
————. 1981, in *IAU Symposium 93, Fundamental Problems in the Theory of Stellar Evolution*, ed. D. Sugimoto, D.Q. Lamb, and D.N. Schramm (Dordrecht: Reidel), p.295.
————. 1982, *Ap. J.*, *253*, 798.

Nomoto, K. 1984a, *Ap. J.*, *277*, 791.
————. 1984b, in *Stellar Nucleosynthesis*, ed. C. Chiosi and A. Renzini (Dordrecht: Reidel), p. 205 and p. 238.
————. 1985, in *The Crab Nebula and Related Supernova Remnants*, ed. M.C. Kafatos and R.B.C. Henry (Cambridge Univ. Press), p. 97.
————. 1986a, *Ann. N Y Acad. Sci.*, *470*, 294.
————. 1986b, in *Proceedings of VIth Moriond Astrophysics Meeting: Accretion Processes in Astrophysics*, in press.
Nomoto, K., and Hashimoto, M. 1986, *Prog. Part. Nucl. Phys.*, *18*, in press.
Nomoto, K., and Iben, I Jr. 1985, *Ap. J.*, *297*, 531.
Nomoto, K., Miyaji, S., Yokoi, K., and Sugimoto, D. 1979, in *IAU Colloquium 53, White Dwarfs and Variable Degenerate Stars*, ed. H.M. Van Horn and V. Weidmann (Rochester: Univ. of Rochester), p.56.
Nomoto, K., Sparks, W.M., Fesen, R.A., Gull, T.R., Miyaji, S., and Sugimoto, D. 1982, *Nature*, *299*, 803.
Nomoto, K., Thielemann, F.K., and Yokoi, K. 1984, *Ap. J.*, *286*, 644.
Saio, H., and Nomoto, K. 1985, *Astr. Ap.*, *150*, L21.
Savonije, G.J., de Kool, M., and van den Heuvel, E.P.J. 1986, *Astr. Ap.*, *155*, 51.
Sramek, R.A., Panagia, N., and Weiler, K.W. 1984, *Ap. J. (Letters)*, *285*, L59.
Sugimoto, D. 1970, *Ap. J.*, *161*, 1069.
Sugimoto, D., and Nomoto, K. 1980, *Space Sci. Rev.*, *25*, 155.
Taam, R.E., and van den Heuvel, E.P.J. 1986, *Ap. J.*, *305*, 235.
Takahashi, Y., Miyaji, S., Parnell, T.A., Weisskopf, M.C., Hayashi, T. and Nomoto, K. 1986, *Nature*, *326*, 839.
Thielemann, F.-K., Nomoto, K., and Yokoi, K. 1986, *Astr. Ap.*, *158*, 17.
Van den Heuvel, E.P.J. 1984, *J. Ap. Astr.*, *5*, 209.
Webbink, R. 1984, *Ap. J.*, *277*, 355.
Webbink, R.F. Rappaport, S., and Savonije, G.J. 1983, *Ap. J.*, *270*, 678.
Wheeler, J.C., and Harkness, R. 1986, in *Distances to Galaxies and Deviation from the Hubble Flow*, ed. B.M. Madore and R.B. Tully (Dordrecht: Reidel).
Wheeler, J.C., and Levreault, R. 1985, *Ap. J. (Letters)*, *294*, L17.
Woosley, S.E., Axelrod, R.S., and Weaver, T.A. 1984, in *Stellar Nucleosynthesis*, ed. C. Chiosi and A. Renzini, (Dordrecht: Reidel), p.263.
Woosley. S .E., and Weaver, T.A. 1986a, in *Nucleosynthesis and Its Implications for Nuclear and Particle Physics*, ed. J. Audouze and T. van Thuan (Dordrecht: Reidel).
————. 1986b, *Ann. Rev. Astr. Ap.*, in press.
————. 1986c, in *IAU Colloquium 89, Radiation Transport and Hydrodynamics*, ed. D. Mihalas and K.H. Winkler.
Woosley, S.E., Weaver, T.A., and Taam, R.E. 1980, in *Type I Supernovae*, ed. J. C. Wheeler (Austin: University of Texas), p. 96.

## DISCUSSION

Srinivasan: If the pulsar beaming factor is 5, then the analysis of Blaauw suggests that stars with masses as low as 6 $M_{\odot}$ should produce neutron stars. Could you please comment on this.

Nomoto: The smallest mass of stars that produce neutron stars is often called $M_{UP}$, above which a degenerate O+Ne+Mg core forms. $M_{UP}$ depends on the extent of convective overshooting, or convective core size, during H and He burning. $M_{UP} \sim 8\ M_{\odot}$ is obtained from models with negligible overshooting. If the overshooting is significant, $M_{UP}$ could be as low as 5 - 6 $M_{\odot}$ as reported by Italian and Dutch group. However, $M_{UP}$ is quite uncertain because we don't have a good theory of convective overshooting yet.

Bisnovatyi-Kogan: How can you explain theoretically the uniqueness (peculiarity) of the Crab supernova?

Nomoto: The Crab supernova may not be so unique. Increasing number of Crab-like remnants have been observed. The mass range of stars that form O+Ne+Mg cores is rather narrow yet they would produce non-negligible fraction of Type II supernovae because of mass function.

Srinivasan: There are reasons to beleive that the supernova of 1054 was a very low energy explosion. Does this fit to your theoretical picture?

Nomoto: In my scenario, the Crab nebula's progenitor had lost a significant fraction of its hydrogen-rich envelope by wind and became as small as $\sim 4\ M_{\odot}$ when it exploded (Nomoto et al. 1982). In other words, the explosion is supposed to be weak and no high velocity hydrogen-rich material is expected. The progenitor was still a red-supergiant with some circumstellar matter so that even a weak explosion could produce observed optical light.

# ANCIENT GUEST STARS AS HARBINGERS OF NEUTRON STAR FORMATION

Zhen-Ru Wang
Department of Astronomy
Nanjing University
Nanjing
People's Republic of China

## 1. INTRODUTION

It was just after the discovery of neutrons in 1932, Landau suggested the possibility of compact stars composed of neutrons.In 1934 Baade and Zwicky proposed the idea of neutron stars independently and suggested that neutron stars would be formed in supernova explosions.

It is a pity that no any supernova explosion has been observed in our Galaxy since the invention of telescope. We have had no opportunity to catch the formation of neutron stars for these several hundreds of years. It naturally leads to the search for the ancient Guest Stars. "Guest Star" is a term that the ancient Chinese and the people in Far-East used to describe a new star. i.e. a star that suddenly appeared remarkably for some time and then gradually disappeared as if a guest in the sky. For example, the well-known AD 1006, 1054, 1572, and 1604 were all described as Guest Stars by Chinese, Japanese and Korean. In most cases, we might thus expect a Guest Star to be a term of supernova or nova.

There are a lot of records concerning ancient Guest Stars ( AGS ) in Chinese historical books. Two catalogues were compiled by Xi (1955) and Xi and Bo (1965, 1966) that listed 90 probable novae or supernovae observed between 1400 BC and AD 1700 Clark and Stephenson (1977),Ho(1962) and Kanda (1935) collected more or less similar records. Among all the historical records more than 80% are from China. Our following discussion are based on them.

## 2. THE FAMOUS CHINESE ANCIENT GUEST STAR AD 1054 AND THE CRAB

The Crab Nebula is the most fascinating object in our Galaxy (Fig.1) It is a very strong source over all the frequences. A rapidly spinning pulsar lies near its center and emits pulsating radiation with a period of 33 ms from the radio to the $\gamma$-ray frequencies. Many massages about synchrotron radiation, the origin of neutron stars, the sources of cosmic rays, the manufacture of heavy elements, the strange dense matter, the strong magnetic field and gravitation are comming from



*D. J. Helfand and J.-H. Huang (eds.), The Origin and Evolution of Neutron Stars, 305–318.*

the crab. It is no surprising that the Crab Nebula, as a cosmic laboratory, is an interesting topic at a lot of international scientific conferences and gives a good account of itself in the modern astrophysics.

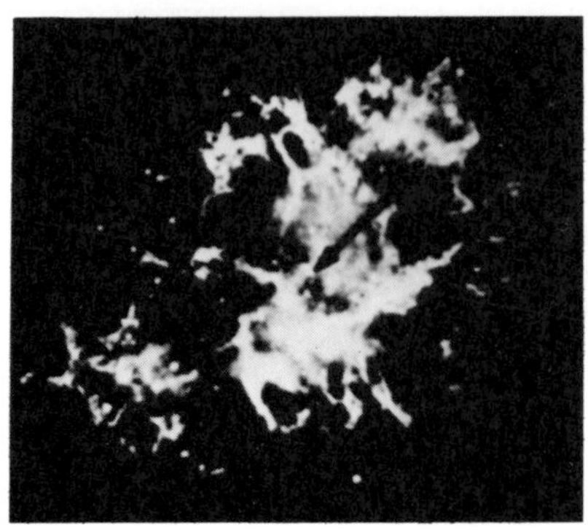

Fig.1. The Crab Nebula. The arrow marks the position of the Crab pulsar

Now it is well-known that the Crab Nebula is the remnant of the supernova explosion that occured in the constellation of Taurus on July 4th,1054 and observed by ancient Chinese astronomers. Some of the historical records of AD 1054 Guest Star are as follows:

1. 至和元年五月己丑客星出天关东南可数寸，岁余消没

《宋史》

"1st year of Zhi-ho reign period, 5th day of Ji-chou,* a Guest Star appeared approximately several inches to the south-east of Tian-guan. After a year and more it gradually vanished"

'Sung-Shi'

*i.e. July 4, 1054

2. 嘉祐元年三月辛未司天监言：'自至和元年五月客星晨出东方，守天關，至是没'

《宋史仁宗本纪》

"1st year of the Jia-hu reign period, 3rd month, day of Xin-wei,* the Director of the Astronomical Bureau reported that since the 5th month of the 1st year of the Zhi-ho reign period a Guest Stay had appeared in the morning in the east guarding Tian-guan and had only now become invisible"

'Sung-Shi-Ren-Zong-Ben-Ji'

*i.e. April 17, 1056

3. 至和元年七月二十二日守将作监致仕杨惟德言："伏覩客星出现，其星上微有光彩，黄色。"

《宋会要》

"1st year of the Zhi-ho reign period, 7th month, 22nd day ···. Yang Weide said:'I humbly observe that a Guest Star has appeared, above the star in question there is a faint glow, yellow in colour.'"

'Sung-Hui-Yao'

"During the 3rd month of the 1st year of the Jia-hu reign period the Director of the Astronomical Bureau said:

4. 嘉祐元年三月，司天监言：
“客星没，客去之兆也。”初，
至和元年五月，晨出東方，守
天關，晝見如太白，芒角四
出，色赤白，凡見二十三日
《宋會要》

5. 天喜二年四月中旬以後，丑
時客星出觜參度，見東方，
孛天關星，大如歲星
《明月记》

'The Guest Star has vanished.'Earlier, during the 1st of the Zhi-ho reign period the Guest Star appeared in the morning at the east, guarding Tian-guan. It was visible in the daytime, like Venus. It had pointed rays on all sides and its colour was reddish-white and visible for 23 days."

'Sung-Hui-Yao'

"2nd year of the Tenki reign period, 4th month, after the middle decade, at the hour Chou (i.e. 1-3am) a Guest star appeared in the degree of Tsue and Shen. It was seen in the east and flared up at Tian-guan. It was as large as Jupiter.'

'Meigetsuki'

Hundreds of years later, a nebula in the constellation of Taurus was discovered by English amateur astronomer Bevis and the expression "Crab Nebula" was used for the first time by Rosse. In 1921 Lundmark first noted that its position is close to the location of the AD 1054 Guest Star and suspected that they were related. Lampland, Duncan and Hubble discovered and detected its expansion. In 1942 Ort and Duyvedark first identified the Crab Nebula as the remnant of the AD 1054 Guest Star. The discovery of the Crab pulsar in 1968 greatly strengthened this identification. The magnetic dipole model of rapidly rotating neutron stars put forward by Gold, Pacini, Goldreich, Ostriker and others successfully accounted for the energy source of the Crab Nebula. The deduced age parameter of the Crab pulsar approximately coincides with the explosion time of the Guest Star AD 1054. Hence, the Crab pulsar is an excellont sample for a neutron star formed in an event of supernova explosion.

Now we know that there are about 425 radio pulsars. The Crab pulsar is the only radio pulsar whose harbinger has been found in the historical records. This is because most of the radio pulsars are too old ($10^4$-$10^7$ yr) compared with the several thousand years of the civilization history of mankind, and are thus not possible to leave historical records.

Are there any other neutron stars except radio pulsars formed in the events of Guest Stars? From the data of hundreds of pulsars and the succesful magnetic dipole model, we know that "pulsar" is only such kind of neutron stars which have a definite range of the rotating period and the magnitude of the magnetic field as well as a certain structure of magnetic field and a limitation on the directions of radiation beam. It is obvious that there should be many neutron stars not detected as pulsars just because of the above physical or geometrical factors beyond the certain range or the limited sensitivity and counts of observation up-to-date. The most apparent example of them are those that exist as the central energy sources of supernova remnants like those in Crab-like SNR and composite SNR (Weiler 1985). In the next three paragraphs. our interests turn to the Guest Stars related to such central sources.

## 3. THE GUEST STAR AD 1181 AND 3C 58

3C 58 was first suggested as a Crab-like SNR (Weiler & Seielstad, 1971)because of its center-filled morphology, flat spectra and strong linear polarization. The compact and extended X-ray emission was first discovered by Becker,Helfand and Szymkowiak(1982) from Einstein observation.

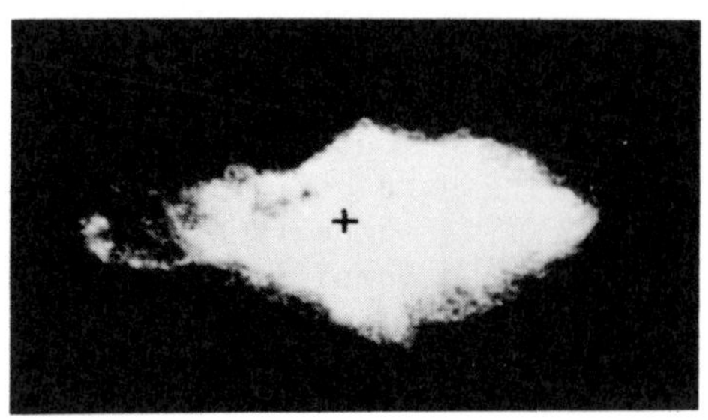

Figure 2. 3C 58 at 2.7 GHz (Green and Gull 1983). The cross marks the position of the X-ray point source.

Stephenson (1971), Wilson and Weiler (1976), Clark and Stephenson (1977) and Liu (1983) have suggested 3C 58 being the remnant of the Guest Star AD 1181. The historical records about it are as follows:

1.(宋)淳熙八年六月己巳客星出奎宿，犯传舍星，至明年正月癸酉，凡一百八十五日始灭

《宋史》

"On the day Ji-si in the 6th month of the 8th year of the Chun-xi reign peroid* a Guest Star appeared in Kui-su and invading Chuan-she until the day Gui-you of the 1st month of the following year**. Altogether 185 days, only then was it extinguished."

'Sung-Shi'

*i.e. August 6, 1181

**i.e. February 6, 1182

2.(金)大定二十一年六月甲戌客星見于华盖，凡一百五十有六日灭

《金史》

"On the day Jia-xu in the 6th month of the 21st year of the Da-ding reign period a Guest Star was seen at the Hua-gai altogether for 156 days, then it was extimguished."

'Jin-Shi'

3.(宋)淳熙八年六月己巳客星出奎宿，犯傳舍.………甲戌，客星守傳舍第五星．九年正月癸酉客星始不見.……

《文献通考》

"On the day Ji-si in the 6th month of the 8th year of the Chun-xi reign period, a Guest Star appeared in Kui-su and invading Chuan-she······ On the day of Jia-xu,the Guest Star guarding the 5th star of Chuan-she. It disappeared on the day Gui-you of the 1st month of the 9th year, altogether it

4. 養和元年六月廿五日庚午客星出北方，近王良，守传舍

《明月记》

5. 治承五年六月二十五日庚午，戌刻客星见艮方，(大如)镇星，色青赤，有芒角，是宽弘三年(1006)出现之后无例云云。

吾妻镜

appeared for 185 days since the day Ji-si of the 6th month last year, then it disappeared."

'Wen Hsien Thung Khao'

"On the day Geng-wu, the 25th day of 6th month of the 1st year of the Yang-he period* a Guest Star appeared at the north near Wang-Liang and Guarding Chuan-she."

'Meigetsuki'

* i.e. August 7, 1181

"At the hour Xu, on the day of Geng-wu, the 25th day of the 6th month of the Zhi-Cheng period* a Guest Star was seen in the north-east. It was like Saturn and its colour was bluish-red and it had rays. There had been no other example since that appearing in the third year of Kuan-Hong (AD 1006)"

'Azuma Kagami'

The appearance of SN 1181 was also shown on the star map of Su-Zhou carved stone in China contributed by Huang Chang in AD 1190 with one more Star (than maps in other era) at the position where 3C 58 locates (Liu 1983).

The compact source in 3C 58 is likely to be a neutron star formed in the event of AD 1181 and was detected as a compact X-ray source by Einstein observation 800 years after the event.

## 4. THE GUEST STAR AD 1408 AND SNR CTB 80

CTB 80 is an unusual SNR. It has a bright core embeded in a very large three-lobed structure in radio (Angerhofer et al 1981). An optical ring of nebulosity is coincident with the central radio core from which some optical filaments extend (Van den Bergh 1980, Blair et al, 1984). a compact X-ray source was discovered by Becker, Helfand and Szymkowiak (1982). Its X-ray structure appears more jet-like than shell-like (Wang and Seward 1984). The compact X-ray source contributes 31% to the total X-ray luminosity. Strom et al (1980) have first suggested that CTB 80 may be the remnant of the Guest Star AD 1408. The ancient records about it are as follows (Li 1978, 1979):

"永乐六年冬十月庚辰，夜中天，辇道东南有星如盏，黄色光润，出而不行，盖周伯、德星也"(《明实录》)

"Reign Ying-le, Year 6, winter, month 10, day Geng-chen*, at night, near the meridian, to the south-east of Niandao (13, $\eta$, $\theta$, Lyr; 4, 17 cyg), there is a star like a lamp, its colour is yellow and its lustre smooth, it shows up and does not move, it is a Zhou-bo, a virtuous star."

'Ming-Shi-Lu'

* i.e. October 24, 1408

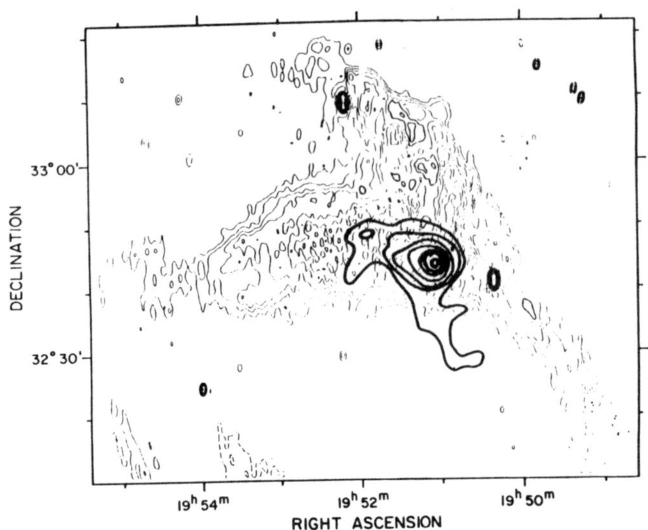


Figure 3. The Einstein IPC image of CTB 80 (Wang and Seward 1984). The thin contours are from 49 cm radio observation of Angerhofer et al. (1981)

Two difficulties have been advanced to this identification. One is about its too large radio size and the other is about its too small ratio of X-ray luminosity to radio luminosity $L_x/L_R$ (Van den Bergh 1980, Dickel et al. 1981 and Becker et al. 1982).Our reanalyses of its Einstein observation (Wang and Seward 1984) showed the ratio of $L_x/L_R \sim 60$ for the whole source and $\sim 150$ for the core-plateau region. This is to the advantage of CTB 80 being a very young SNR. Paying attention to the alignment of the X-ray jet-like structure with the projected magnetic field lines, the size of both radio and X-ray can be explained by the relativistic beam with 0.16c and its synchrotron radiation since the explosion in AD 1408. Strom and Blair (1985) compared its optical image in $H_\alpha$ +[NII] with one taken 28 years before also strongly support CTB 80 to be the remnant of SN 1408. Therefore the compact source in CTB 80 with $L_x = 1.0 \times 10^{34}$ erg $s^{-1}$ (Wang and Seward 1984) is probably a neutron star formed in the event of AD 1408.

## 5. SOME OTHER POSSIBLE IDENTIFICATIONS BETWEEN AGS AND CRAB-LIKE SNRs OR COMPOSITE SNRs

Our discussion is based on the following principal of identification suggested by us ( Wang et al. 1986). For an event of SN explosion, it should be described by the place where it occured and the time when it occured. A reliable identification should be made on the basis of agreement in visual position, distance and age between AGS and SNR. Let us first discuss the visual position. The visual position is exactly known for a known SNR, but is only approximately known for an AGS (e.g.

Lunar Mansion, Yuan or some other Chinese Asterisms). The agreement in visual position means that the visual position of a SNR is just in the region described for an AGS by the ancient records. As to the agreement in distance, we use the well-known relation:

$$M = m + 5 - 5 \times \mathrm{LOG}\ d - A_V \qquad (1)$$

to an Ancient Guest Star at its observation time. Obviously m was its visual magnitude, probably around its maximum luminosity, and could be estimated for an AGS from its historical records. d and $A_V$ should be its distance and absorption correction. It is sure that the ancient people could not know about its distance and absorption. But we can get them from its possible identified candidate of a SNR. Usually, d can be obtained from the neutral hydrogen observation or other method and $A_V$ can be got from the following relation (Gorenstein 1975):

$$A_V = 4.5 \times 10^{-22}\ N_H (\text{ mag. }) \qquad (2)$$

or

$$A_V = 3.0E(B-V) \qquad (3)$$

Where $N_H$ is the column density of neutral hydrogen, E(B-V) is the colour excess, If the value of the absolute magnitude M obtained from eq. (1) is approximately satisfied with the following relation (Trimble 1982)

$$M = -18^m \pm 2^m.5 \qquad (4)$$

the agreement in distance is set up. Finally we discuss the agreement in age-t. t is known for an AGS, and we can get t for a SNR from some theoretical considerations. We demand the age of both SNR and AGS is almost the same for a reliable identification or we demand a reasonable velocity for a SNR from its size and the age of its identified AGS. Using the principal of identification to 90 AGS(Xi and Bo 1965) and >130 SNR(Van den Bergh 1983, Green 1984). We find that the identification in the complete four-dimensions (two dimensions for the visual position and the other two for its distance and age) is a very crucial principal. It gives the identification with a very serious confinement and can exclude effectively some accidental coincidence (Wang et al. 1986). Obviously, the identifications for the Crab, 3C 58 and CTB 80 are satisfied with the principal. Now let us only talk about some other possible identifications related to Crab-like or composite SNRs.

G 332.4-0.4 (RCW 103) -----Guest Star BC 134

RCW103 is a composite SNR with a filament shell (Van den Bergh 1978) and a central compact X-ray source (Tuohy and Garmire 1980) that may be associated with the $\gamma$-ray source CG333+0 (Lamb and Markert 1981). It is considered as a young SNR owing to its small diameter (9 pc) and the high velocity of its optical filament and is estimated to be an order of magnitude younger than the Vela (Tuohy et al.1979), approximately

$1\text{-}2\times10^3$yr old. According to the principal of identification, it should be the remnant of the Guest Star BC 134.

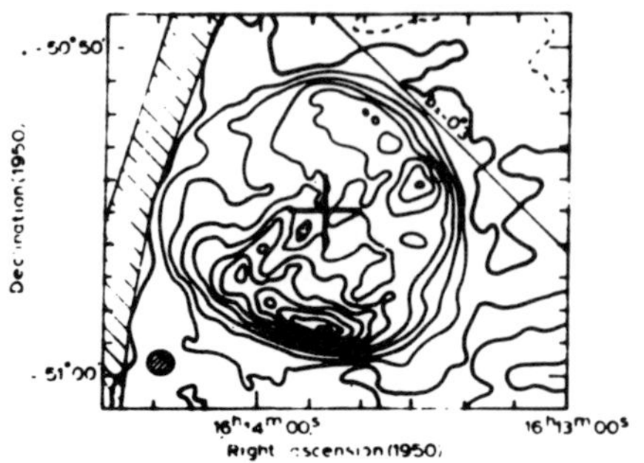


Figre 4. G 332.4-0.4 (RCW 103) at 1.4 GHz (Caswell et al. 1980).The position of the X-ray source is marked with a cross.

The ancient records of the Guest Star BC 134 are as follows:

汉"元光元年六月客見于房"
"During the 6th month of the 1st year of the Yuan-Kuang reign-period (22nd June to 21st July, BC 134), a Guest Star appeared at the Fang"

班固撰，约公元100年，汉书·天文志26卷
Ban Gu, AD 100, Han-Shu, Chapter of Astronomy, Vol.26

徐天麟等撰，约12世纪，西汉会要29卷
Xu Tian-ling, 17th Century, Hsi-Han-Hui-Yao, Vol.29

The Guest Star appeared at the Fang, the sky region of RA between $\pi$ Sco and $\alpha$ Sco. RCW 103 is just in the Lunar Mansion of Fang. The Guest StarBC 134 was discovered by Chinese and Greek astronomer Hipparchus. We estimate its visual magnitued $m=-3$ around its maximum. Taking its distance $d = 3.3$kpc (Caswell et al. 1975), neutral hydrogen column density $N_H=4.8\times10^{21}$ (Tuohy et al. 1979) and absorption correction $A_V$ to be 2.2 from eq.(2). An acceptable value of absolute magnitude for its maximum $M = -17.8$ is obtained. From its diameter and age, a reasonable average expansion velocity $2\times10^3$km.s$^{-1}$ is obtained. So the central compact source sith $L = 10^{34}$ erg.s$^{-1}$(Seward 1985) is probably a neutron stau formed in the explosion of SN BC134.

G21.5-0.9-------Guest Star BC48

G21.5-0.9 is a Crab-like SNR with a small linear size (1.7x2.9pc) and low brightness (Wilson and Weiler 1976). The ratio of $L_X/L_R$=15. It seems to be a young SNR but older than Crab (Becker and Szymkowiak 1981) Its brightest center shown in its Einstein HRI observation implies that a neutron star will be expected by the future X-ray observation with higher resolution and sensitivity. According to the principal of identification. G21.5-0.9 should be the remnant of the Guest Star BC 48. The ancient record about the Guest Star BC 48 is as follows

"汉初元元年四月客星大如瓜，色青白，在南斗第二星东可四尺"
班固和班昭，100年，汉书 天文志

"During the 4th month of the 1st year of Chu Yuan reign period( May 3 to May 31, BC 48), a Guest Star as big as a melon with bluish-white colour was seen about four feet east of the second star of Nan-Tou."
Ban Gu and Ban Zhao, AD 100, Han-Shu, Chapter of Astronomy.

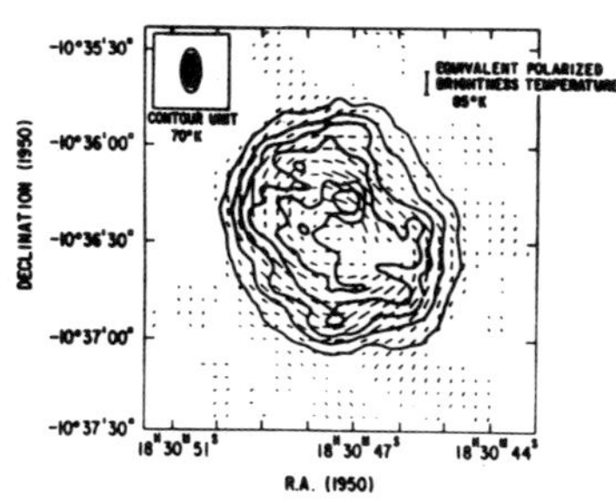


Figure 5. G21.5-0.9 at 5 GHz (Becker & Szymkowiak, 1981)

G21.5-0.9 is just located several degrees north-east to Sgr (the 2nd star of Nan-Tou) and close to the position described in the ancient record. We estimate m=-3 at its maximum. Taking d=4.8 kpc(Caswell et al. 1975), $N_H=10^{22}$(B ecker and Szymkowiak 1981), $A_V=4.5$ from eq. (2), then M=-20.9, also an acceptable value for a supernova at its maximum. From the size (3pc) and its age, an average velocity v=700 km·$S^{-1}$is obtained, slightly lower than the others, perhaps due to its circumferentially oriented magnetic field. The expected neutron Star in G21.5-0.9 is likely to be formed in BC 48.

G74.9+1.2------- New Star BC532

G74.9 +1.2 is a very large Crab-like SNR in both radio and X-ray. It seems to be an older, more evolved remnant and was estimated as $3x10^3$ yr old(Weiler & Panagia 1980). Its X-ray is faint but is clearly brightest at the center(Seward 1985). Hence a compact neutron star will be expected in the future X-ray observation. Our following identification suggests that G74.9+1.2 Should be the remnant of the New Star BC 532. The ancient record about BC 532 is as follows:

"周景王十三年春有星出婺女"
作者不明，公元前295年，竹书纪年

"During the spring of the 13th year of Chau-Ching Wang, a star was seen at the Wu-Nu"
Unknown author, BC 295 Zhu-Shu-Ji-Nian

The New Star BC 532 was in Wu-Nu, a Lunar Mansion at the sky region of R A between $\epsilon$ Aqr and $\beta$ Aqr. Considering the effect of precession for more than $2.5x10^3$ yr (Wu and Liu 1983), the circle of RA at 550 BC is shown as an inclination line in the star map of 1950. G 74.9+

1.2 is just in the Lunar Mansion of Wu-Nu. We estimate m=$1^m$ at its maximum. Taking d=12kpc(Kazes and caswell 1977, Sato 1977), N=$10^{22}$ (Wilson 1980) and $A_V$=4.5 From eq.(2). An expected value M=-18.9 is obtained at its maximum. From the size of G74.9+1.2 and its age, an acceptable v=$5.3x10^3$km.$s^{-1}$ is obtained. The expected neutron star in G 74.9+1.2 is likely to be formed in BC 532.

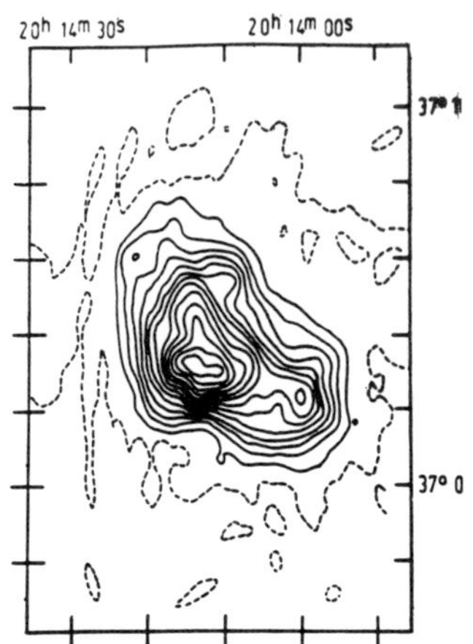


Figure 6. G 74.9+1.2 (CTB 87) at 1.4 GHz (Weiler & Shaver. 1978)

G 29.7=0.3 (Kes 75)----New Star AD 1523

Kes 75 is considered as a composite SNR in radio. X-ray observations showed only the central region ~20" (Seward 1985). A central neutron star will be expected from the future X-ray observation owing to its brightest center. Becker et al. (1983) suggsted an identify crisis for Kes 75 and comsidered it as an extremely young object because of its extremely high ratio of $L_X/L_r$~500, tripe that of the Crab.

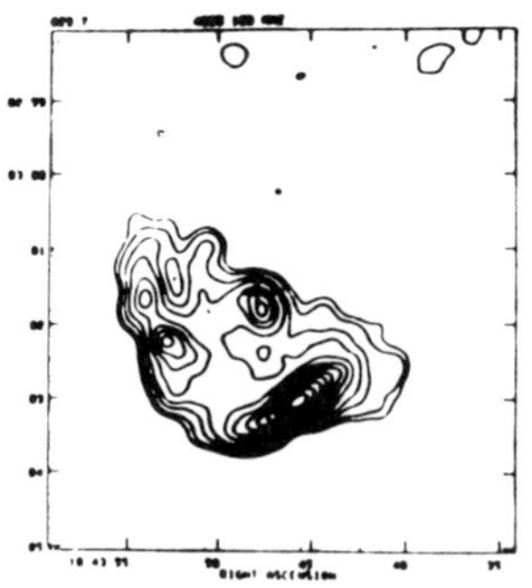

Figure 7. G29.7-0.3 (Kes 75) at 5 GHz (from Van Gorkum,Weiler1985). The position of the X-ray emmision is marked with a cross.

Our following identification shows that Kes 75 is probably the remnant of the New Star AD 1523. The historical records about the New Star AD 1523 are as follows:

明"嘉靖二年六月有星孛于天市"
張廷玉等撰, 1739年, 明史·天文志 27卷
王圻著, 1586年, 续文献通考, 20卷
龙文彬撰, 约17世纪, 明会要, 68卷

"During the 6th month of the 2nd year of the Chia-Ching reign-period (July 13 to August 10,1523) a po star appeared at the Thieh-Shih (Enclosure of Heavenly Market."
Zhang Ting-yu, AD 1739, Ming-Shih, Chapter of Astronomy, Vol.27·Wang Yi. AD 1586, Xu-Wen-Hsien-Thung-khao, Vol 20
Long Wen-bing, 17 AD, Ming-Hui-yao, Vol.68

Kes 75 is located at the position where three constellations Serpens, Aquil and Scutum meet. It is at the Thieh-Shih. One of the three Yuan in the Chinese ancient division of the sky. As it was considered as a po star $m=5^m$ should be reasonable. Taking d=7kpc (Caswellet al. 1975), $N_H=3x10^{22}$ (Becker et al. 1983) and $A_V=13.5$ from eq. (2), $M \simeq -22$ is obtained, slightly brighter than that expected from eq. (4). From its size(5pc) and age 460yr, $v=5x10^3 km.s^{-1}$ is obtained and is reasonable and acceptable for a SNR. The expected neutron star in G29.7-0.3 is likely to be formed in AD 1523.

6. 2 CG 195+4 (GEMINGA)------GUEST STAR AD437

Geminga is a high-energy (>50 Mev) $\gamma$-ray source first discovered by SAS2 satellite in 1973 and then confirmed by COS B in 1975. Bignami et al. (1984) analysed the Einstein data and the new Exosat observation and discovered a period of $\gtrsim$59 sec with a high value of period derivative $\dot{P}$. They also summarized the following available data, two satellite (SAS2 and COS B)>50 Mev $\gamma$-ray observation, two ground-based $>10^{12}$ev r-ray measurement, three sets of Einstein observation and two sets of Exosat observation. They discovered that the very high $\dot{P}$(2x10-9 $s.s^{-1}$) between 1973 and 1979 seems to become higher between 1979 to 1983 ($\dot{p}=4.68x10^{-9}$). Considering the secular change of its period and period derivative, they estimate its age about $1.5x10^3$yr and sugested that it may be related to the Guest Star AD 437 observed by the ancient Chinese. The ancient record of the Star AD 437 is as follows:

魏太延三年正月壬午, 有星晡前会见东北, 在井左右, 色黄, 大如桔
魏收, 572年, 魏书, 105卷

元嘉十四年正月, 有星晡前会见东北维, 在井左右, 黄赤色, 大如桔.
沈约, 500年, 宋书, 26卷

"On a Ren-wu day in the 1st month of the 3rd year of Tai-yan reign period (Feb.2, 437), a star appeared in the afternoon at the north-east near the Tung-Ching (Gem.). It was yellow in colour and as large as an orange"
Wei Shou, AD 572, Wei-Shu, Vol. 105
Shen Yue, AD500, Sung-Shu, Vol. 26

The identification is reasonable in three dimensions (visual position and age). Except that its distance is unknown and it is therefore difficult to complete the four-dimension identification. If we assume that Geminga is in the same distance as the Guest Star AD 437, then Geminga is likely to be a neutron Star formed in AD 437.

## 7. CONCLUSION

In the seventies, only two neutron stars----the Crab pulsar and the one in 3C 58 were known to be born at the events of Guest Stars. Or more exactly to say, it is only the Crab pulsar, because the compact source in 3 C 58 had not been detected and the concept of central energy sources was not very clear at that time. Now, our analyses show that the above 8 neutron stars were formed at the events of Guest Stars in history. If it is true, it will increase the birth rate of neutron stars or the frequency of Typw II SN explosions estimated only from the historical events. This result may also settle the difference between the birth rate of neutron stars and the frequency of Type II SN explosions to a certain degree. If we consider the much shorter lifetime of SNRs than that of pulsars and the incompleteness of samples of both historical enents and the observed SNRs with centeral energy sources, a better settlement could come out. Our above discussion can offer meaningful clues to the origin and evolution of both neutron stars and SNRs from historical facts and gives contemporary value to the ancient records of Guest Stars.

## ACKNOWLEDGEMENT

I wish to thank my collaborators F.D. Seward, P. Gorenstein and M.V. Zombeck in Harvard-Smithsonian Center for Astrophysics and J.Y. Liu in the Institute of History of Natural Science (Beijing) for their help and cooperation in some work before, also thanks to K.W. Weiler for contributing some helpful preprints and S.X. Wu for offering the ancient position of several stars, thanks to the American boy Robert who was 14 years old in 1983 for some calculations. I am particularly indebted to X.F. Hua and W.G. Wu for their rapid and accurate preparations of the manuscript.

## REFERENCES

Angehofer, P.E., Storm, R.G., Velusamy, T. and Kuandu, M.R. 1981 A. & Ap., 94 313

Becker, R.H., Helfand, D.J. and Szymkowiak, A.E., 1982, Ap. J.255, 557

Becker, R.H., Helfand, D.J. and Szymkowiak, A.E., 1983, Ap.J. 268,L91

Becker, R.H. and Szymkowiak, A.E., 1981 Ap. J., 248, L23

Bignami, G.F., Caraveo, P.A. and Paul, J.A., 1984, Nature, 310, 464

Blair, W.P., Kirshner, R.P., Fesen, R.A. and Gull, T.R., 1984, Ap.J. 282, 161

Caswell, J.L. Haynes, R.F., Milne, D.K. and Wellington, K.J., 1980, M.N. R.A.S., 190, 881
Caswell, J.L., Murray, J.D., Roger, R.S., Cole, D.J. and Cooke, D.J., 1975, A. & Ap., 45, 239
Clark, D. H. and Stephenson, F.R., 1977, The Historical Supernovae (Pergamon Oxford).
Dickel, J.R., Angerhofer, P.E., Strom, R.G. and Smith, M.D., 1981, Vistas in Astronomy, 25, 127
Gorenstein, P., 1975, Ap. J., 198, 95
Green, D.A., 1984, M.N.R.A.S. 209, 449
Green, D. A. and Gull, S.F., 1983, in "Supernova Remnants and Their X-ray Emission", Danziger and Gorenstein eds (Dordrecht-Reidel)P.329
Ho, P.Y., 1962, Vistas in Astronomy, 5, 127
Imaeda and Kiang, T., 1980, Journal for the History of Astronomy,11,77
Kanda, 1955, Nihon Temmon Shiryo
Kazes, I. and Caswell, J.L., 1977, A. and Ap., 58, 449
Lamb, R.C. and Markert, T.H. 1981, Ap.J. 244, 94
Li, Q.B., 1979, Chinese Astronomy, 3, 315
Li,, Q.B., 1978, Acta Astr. Sinica, 19, 210
Liu, J.Y., 1983, Studies in the History of Natural Sciences, 2, No. 1
Sato, F., 1977, Pub. Astr. Sco. Japan, 29, 615
Seward, F.D., 1985. Comments Astrophys., 11, No. 1, 15
Stephenson, F.R., 1971, Quart.J. Roy. Astro. Astro. Sco., 12, 10
Strom, R.G., Angerhofer, P.E. and Velusamy, T. 1980, Nature, 284, 38
Strom, R.G. and Blair, W.P., 1985, A. and Ap., 149, 259
Trimble, V., 1982. Rev. Mod. Phys., 54, No. 4
Tuohy, I.R. and Gamire, G., 1980, Ap. J. 239, L 107
Tuohy, I.R., Mason, K.O., Clark, D.H., Cordova, F.A., Charles, P.A., Walter F.M. and Gamire, G.F., 1979, Ap.J.230, L27
Van den Bergh, S., 1983, in "Supernova Remnants and Their X-ray Emission" Danziger and Corenstein eds. (Dordrecht-Reidel), p.597
Van den Bergh, S., 1978, Ap. J. Suppl., 38, 119
Van den Bergh, S., 1980, P.A.S.P. 92, 768
Wang, Z. R. and Seward, F.D., 1984, Ap.J., 285, 607
Wang, Z.R., Liu, J.Y., Gorenstein, P.& Zombeck, M.V., 1986,Highlights Astronomy, Vol.7 (in press)
Weiler, K.W., 1983 The Observatory, 103, No. 1054, 85
Weiler, K.W.,1985, in "The Crab Nebula and Related Supernova Remnants" Kafatos and Henry eds.(Cambridge University Press),P.227 and P.265
Weiler, K.W. and Panagia, N., 1980. A. and Ap., 90, 269
Weiler , K.W. and Seielstad, G. A. 1971, Ap. J. 163, 455
Weiler, K.W. and Shaver, P.A., 1980, A. and Ap., 90, 269
Wilson, A.S., 1980, Ap.J. 241, L 19
Wilson, A.S. and weiler, K.W., 1976, A. and Ap., 53, 89
Wu S. X. and Liu, C, Y., 1983, Publ. of Shannxi Astronomical Observatory 6, No. 2, 55
Xi, Z.Z. and Bo S.R., 1965, Acta Astronomica Sinica, 13, No. 1,1
Xi, Z.Z., 1955, Acta Astronomica, 3, 183
Xi, Z.Z. and Bo S.R., 1966, Science 154, 597

DISCUSSION

KULKARNI: 1. I noticed that you in a few cases you use 21cm absorption to get $A_V$. However at low latitudes the HI column density provides only a lower limit to Av. This can be seen from CTB 87 where you obtain $A_V$= 4.5 but distance is 12 kpc. On the average one gets 1 magnitude per kpc at low latitudes.
2. The association of AD 437 event with Geminga is hard to understand given the absence of a nebula either in radio or optical(see Djorgovski & Kulkarni, 1986, Astron. J. 91, 90)

WANG: 1. It is sure that in a few cases $A_V$ is obtained by 21cm observation. The values of $N_H$ or Av here are obtained by some researchers individually. I think that it is better than the crude estimate of 1 magnitude per kpc. For example, Tycho SNR (G120.1+ 1.4) is also at the low galactic latitude similar to CTB 87 (G74.9 + 1.2), Clark and Stephenson (1977) took its absorption as 0.2 magnitude per kpc for identification, very different from your estimate. In our case of CTB 87, we take the value of $N_H$ from Wilson (1980). He gave a range of value for $N_H$ (1.0---1.6)$\times 10^{22}$, then the value of Av will be 4.5---7.2 magnitude. It is just at the middle value between your estimate and the estimate of Clark and Stephenson. I think that the value of $N_H$ given by Wilson for CTB 87 is probably the best choice for it.
2. The association of AD 437 event with Geminga is based on Geminga probably being a neutron star with a pulsating period and a period deriative reported by Bignami et al. (1984) and the agreement in both position and age between Geminga and the AD 437 event. The absence of a nebula might be caused by some unknown reasons which wait to be explored. Perhaps it was another new type of SN explosion which cause the remnant to disappear in a much shorter time. e.g., the explosion material with low mass and very high velocity or with relativistic velocity in a special structure of magnetic field would disappear soon.

## *IVa. NEUTRON STELLAR EVOLUTION*

### *Quasi Periodic Oscillations as a Clue to Field and Spin Evolution*

*CHAIR: E. P. J. van den Heuvel*

# QUASI-PERIODIC OSCILLATIONS IN LOW-MASS X-RAY BINARIES

M. van der Klis
Space Science Department of ESA
ESTEC, postbus 299
2200 AG Noordwijk
The Netherlands

ABSTRACT. The properties of the rapid, persistent quasi-periodic oscillations (QPO) discovered with EXOSAT in the X-ray flux of at least 7 bright low-mass X-ray binaries are described. Particular attention is given to the various relations observed between QPO frequency and X-ray intensity, the link between QPO and the low-frequency noise in the X-ray intensity and the bimodal properties of in particular Sco X-1, GX 5-1 and Cyg X-2. The merits of the hypothesis that the QPO indicate the presence of a neutron star with a magnetosphere are considered.

## 1. Introduction

Observations with EXOSAT have led to the discovery of rapid quasi-periodic oscillations (QPO) in the X-ray flux of GX 5-1 (van der Klis et al. 1985a,b) and subsequently in at least 6 other low-mass X-ray binaries. The sources are listed in Table 1 in order of decreasing QPO frequency. The first seven sources listed, which show high-frequency (5-50 Hz) low-coherent ($\Delta\nu/\nu \sim 1$) strong (amplitudes typically resulting in a 5% rms flux variation) persistent ($>10^5$ cycles) QPO form the main subject of this review. They are all very luminous ($> 10^{38}$ erg/s) and classified as 'persistently bright' (Bradt and McClintock 1983). The QPO in the rapid burster (1730-33) which were discovered with the Hakucho satellite by Tawara et al. (1982) have slightly different properties: lower frequency range and a persistence of $10^3$ cycles at most (see Stella 1985). The results on Cir X-1 (Tennant 1986), which may be a massive system, and Terzan 2 (Belli et al.1986) are very recent and will not be discussed here.

Five of the seven high-frequency QPO sources are bright galactic bulge sources which notoriously lack diagnostic properties: they rarely show X-ray bursts, never periodic dips or pulsations and have no optical identifications. The QPO, which from their high frequencies probably originate from within approximately $10^2$ km from the central source, are one of the very few available probes into the structure of these systems. The remaining two sources, Sco X-1 and Cyg X-2, are low-mass X-ray binaries which are known to be accreting matter from an evolved



D. J. Helfand and J.-H. Huang (eds.), The Origin and Evolution of Neutron Stars, 321–331.

Table 1 - QPO sources

| Source | $\nu_{QPO}$ (Hz) | Log $L_x$ [1)] | Persistence (log(cycles)) | Remarks [1)] | References |
|---|---|---|---|---|---|
| Cyg X-2 (2142+380) | 5,28-45 | 38.1 | 6 | PB | 1,2,3,4 |
| GX 5-1 (1758-250) | 5?,20-36 | 38.6 [2)] | 6 | PB | 5,6,7,8 |
| 4U 1820-30 (1820-303) | (0.02), 15-35 | 37.9 | (1),6 | PB,Bu,GC | 9,10,11 |
| Sco X-1 (1617-155) | 6-10-20 | 37.9 [3)] | 5 | PB | 12,13,14 15,16 |
| GX 349+2 (1702-363) | 5,11? | 38.5 [2)] | 5 | PB | 17,18,11 |
| GX 3+1 (1744-265) | 8.6 | 38.2 [2)] | 5 | PB,Bu | 19 |
| GX 17+2 (1813-140) | 7.2 | 38.4 [2)] | 5 | PB,Bu? | 20,11 |
| Rap. Burster (1730-333) | 1-5 | 37.0 | 3 | Bu,GC | 21,22,11 |
| Cir X-1 (1516-569) | 1.4 | 38.9 | ? | Bu?,Tr,RV | 23 |
| Terzan 2 (1724-307) | 0.092? | 36.7 | 3 | Bu,GC | 24,25 |

1) Bradt and McClintock (1983)
PB = persistent bright    Tr = transient
Bu = X-ray burst source    RV = rapidly variable
GC = in globular cluster
2) at 10 kpc
3) at 1.2 kpc (White, Peacock and Taylor 1985)
4) ref. 27

References to Table

1. Hasinger et al. 1985a; 2. Hasinger et al. 1986a; 3. Norris and Wood 1985; 4. Hasinger et al. 1985b; 5. Van der Klis et al. 1985a; 6. Van der Klis et al. 1985b; 7. Van der Klis et al. 1985c; 8. Van der Klis, these proc.; 9. Stella et al. 1985c; 10. Stella 1985; 11. Stella et al. 1984; 12. Middleditch and Priedhorsky 1985a; 13. Van der Klis et al. 1985d; 14. Middleditch and Priedhorsky 1985b; 15. Van der Klis et al. 1986a; 16. Priedhorsky et al. 1986; 17. Lewin et al. 1985; 18. Cooke et al. 1985; 19. Lewin et al. 1986; 20. Stella et al. 1985a; 21. Tawara et al. 1982; 22. Stella et al. 1985b; 23. Tennant 1986; 24. Belli et al. 1986; 25. Morini, priv. comm. 1986.

low-mass companion. Except for their position relative to us and the galactic center, they may be very similar to bright galactic bulge sources.

A typical QPO power spectrum is shown in Figure 1. Apart from the part dominated by the broad QPO peak, there is a section that gradually rises towards lower frequencies below the QPO peak. This is due to 'low-frequency noise' (LFN), a common phenomenon of noisy time series. It consists of stochastic variations (here occuring in addition to the QPO) of which the slowest have the largest amplitudes.

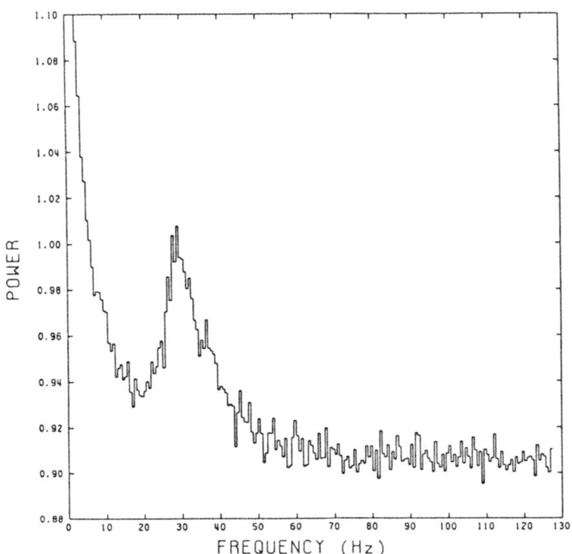


Figure 1. A power spectrum of GX 5-1.

The following description of the observational data of the QPO in these sources concentrates on three main subjects:

1) the relations between X-ray intensity and QPO frequency,
2) the connection between QPO and LFN, and
3) the correlated bimodal X-ray spectral and QPO/LFN behaviour observed in some sources.

This paper will follow a roughly historical line, first discussing the QPO/LFN properties of GX 5-1 and Cyg X-2 in their 'horizontal branch' spectral state, then moving on to the bimodal properties of Sco X-1 and finally comparing this to the bimodal behaviour of GX 5-1 and Cyg X-2. I shall be indicating at some points what is the significance of the discussed observational data for QPO models; for a full review of QPO models see Lewin (these proceedings). We shall be mainly concerned with the three best-studied QPO sources: GX 5-1, Sco X-1 and Cyg X-2.

## 2. GX 5-1 and Cyg X-2 on the 'horizontal branch'

### a) Frequency-intensity relations

Both GX 5-1 and Cyg X-2 show, in their horizontal branch spectral state (see Section 4) a strong positive correlation between X-ray intensity and QPO frequency (van der Klis et al. 1985b, Hasinger et al. 1986a). In both cases the intensity-frequency relation is consistent with a straight line (Figure 2).

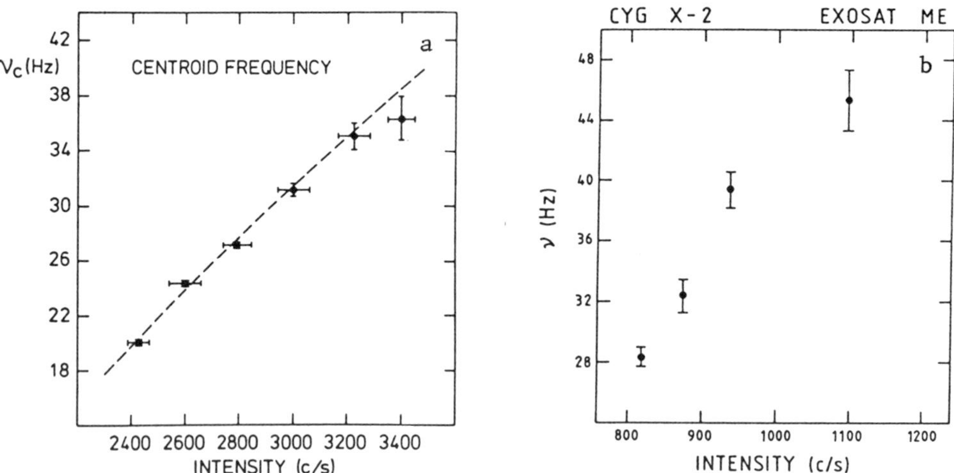


Figure 2. QPO frequency - X-ray intensity relation as observed in GX 5-1 (a, van der Klis et al. 1985b) and Cyg X-2 (b, after Hasinger 1985b) on the horizontal branch.

From the beginning this positive correlation suggested that the QPO frequency might be related to the Keplerian rotation frequency of matter in an accretion disk just outside a magnetosphere. Alpar and Shaham (1985a,b) proposed that the QPO might arise as a beat between the Keplerian rotation of matter in the accretion disk and the spin of a magnetised neutron star. The derived spin frequency would then be of the order of 10 ms and the neutron star surface field $10^9$-$10^{10}$ Gauss.

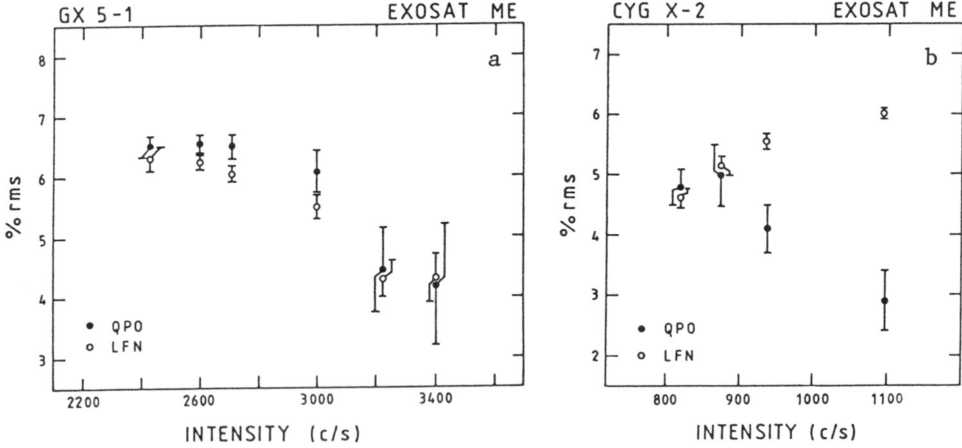


Figure 3. QPO (dots) and LFN (circles) strength as a function of X-ray intensity as observed in GX 5-1 (a, after van der Klis et al. 1985b) and Cyg X-2 (b, after Hasinger et al. 1985b).

b) Low-frequency noise (LFN)

Both sources show LFN of similar strength as the QPO (see Figure 3 for their intensity dependence). This fact prompted the development of a version of the beat-frequency model in which the accreting matter contains random clumps and the
X-ray signal consists of oscillating shots (each shot is the accretion of one clump, the oscillation is due to a magnetic gating mechanism operating at the beat frequency; Lamb et al. 1985). A prediction of this model is that, somewhat dependent on the assumed QPO wave form, the power in the LFN is at least about equal to the power in the QPO.

## 3. Sco X-1

a) Bimodal behaviour

The QPO in Sco X-1 were first discovered in its 'quiescent' state (Middleditch and Priedhorsky 1985a,b). Later QPO were also found in 10-30 min. intensity dips in between flaring episodes (van der Klis et. al. 1985d, 1986a; Priedhorsky et al. 1986). Figure 4a shows both types of QPO in a time vs. frequency diagram where the power in the intensity variations at any given time and frequency is gray-scale coded (darker for higher power). In the active state the QPO centroid frequency varies between 10 and 20 Hz; in quiescence it remains near 6 Hz. A gradual transition between the two states is visible.

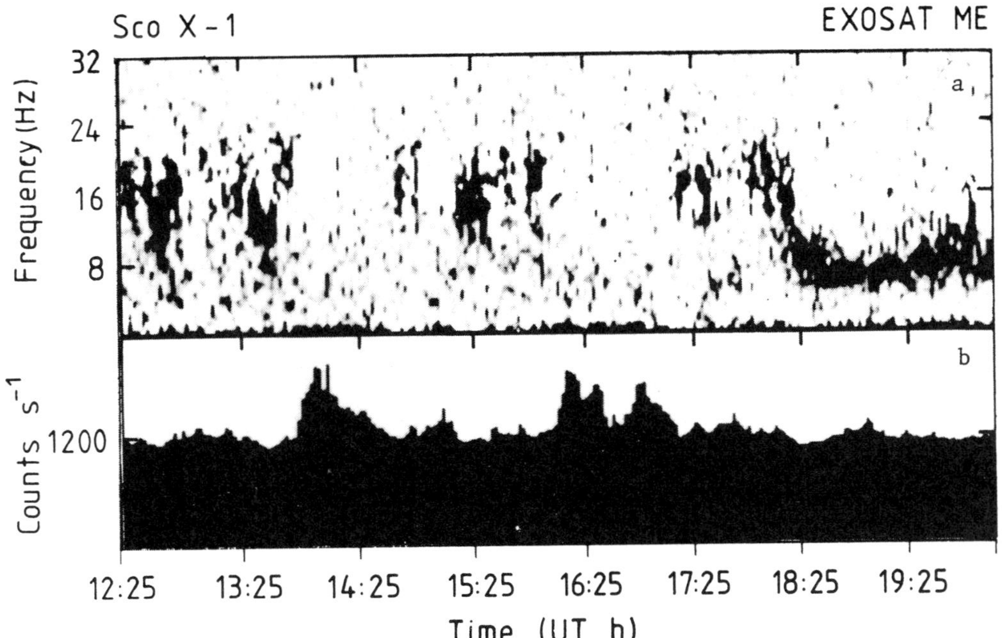


Figure 4. Dynamic power spectrum (a) and X-ray light curve (b) of Sco X-1 (Priedhorsky et al. 1986).

In the dips in the active state a strong positive correlation is observed between frequency and intensity (van der Klis et al. 1985d), in quiescence a weak anti-correlation (Middleditch and Priedhorsky 1985b). QPO which show rapid intensity-uncorrelated variations in frequency have been seen on several occasions ('intermediate' state, van der Klis et al. 1986a; Pollock et al. 1986). The time history of the QPO in the intensity- frequency diagram can in these cases be loop-like (both clock- and counter-clock-wise loops have been seen). Figure 5 shows a sample of the frequency-intensity relations observed in active, 'intermediate' and quiescent states.

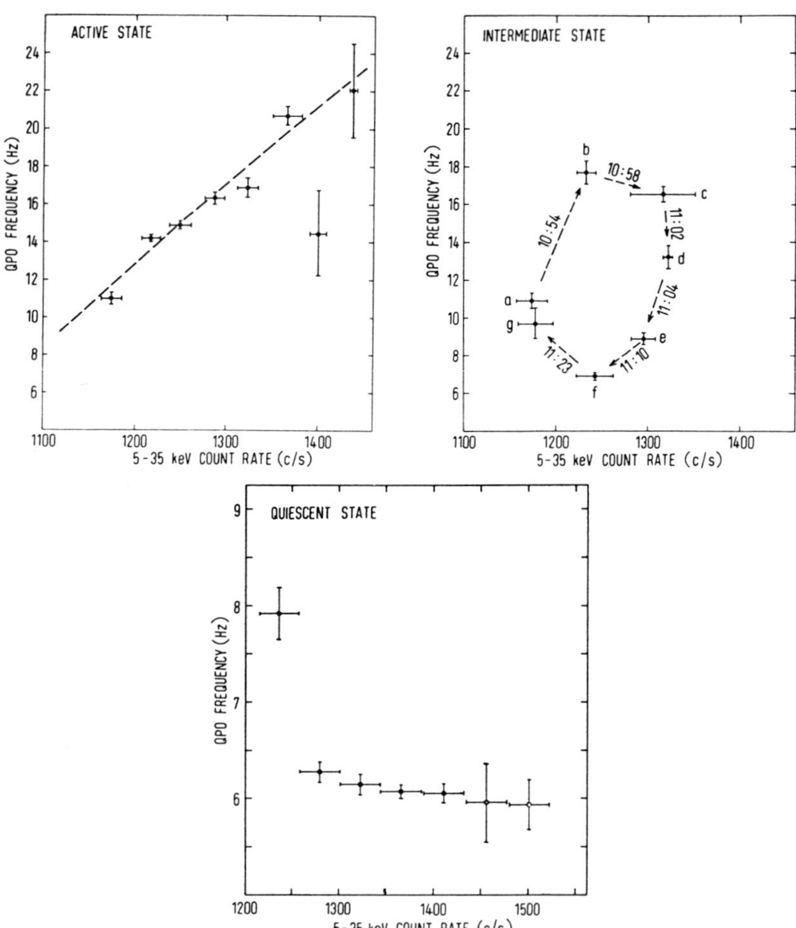


Figure 5. QPO frequency - X-ray intensity relations as observed in Sco X-1 (van der Klis et al. 1986a).

When the source switches from one intensity state to the other, its X-ray spectral characteristics change. There is a gradual change in QPO frequency as the source evolves from one spectral state to the other.

Decomposition of the X-ray spectrum into a blackbody component and an unsaturized comptonization component shows that the flaring is dominated by increases in the blackbody component (White et al. 1985; but see also Mitsuda et al. 1985). Contrary to the case of the relations between QPO frequency and total source intensity (Figure 5), there seems to be one simple relation between frequency and blackbody luminosity (van der Klis et al. 1986a; Figure 6).

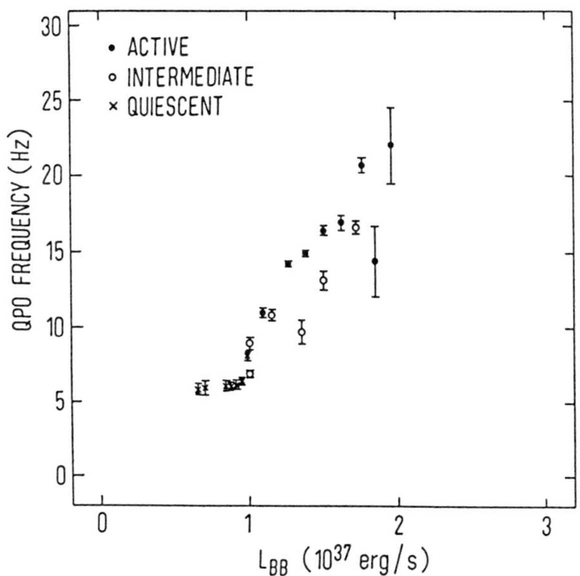


Figure 6. QPO frequency vs. blackbody luminosity in Sco X-1. These are the same data as in Figure 7 (van der Klis et al. 1986a).

The fact that different intensity-frequency relations apply in Sco X-1 makes the magnetospheric hypothesis less compelling.
The basic prediction of magnetospheric QPO models is a correlation between accretion rate and QPO frequency. To explain the complexity of the observed intensity-frequency relations in Sco X-1 within this framework, an additional hypothesis is required about the relation between accretion rate and X-ray intensity (and -spectrum), and this additional hypothesis could be chosen in such a way that other than magnetospheric models can be accommodated.

b) Low-frequency noise

The prediction of the 'random clump accretion beat-frequency model' that QPO and LFN should have approximately the same power is strongly violated by Sco X-1 (van der Klis et al. 1985d; and even more strongly by the rapid burster, Stella 1985). The power in the LFN (down to 1/64 Hz) is usually less than 10% of that of the QPO (van der Klis et al. 1986a). Figure 7 shows how the relative strength of LFN and QPO depend on source intensity: when the source becomes brighter, the QPO disappear but the LFN gets stronger.

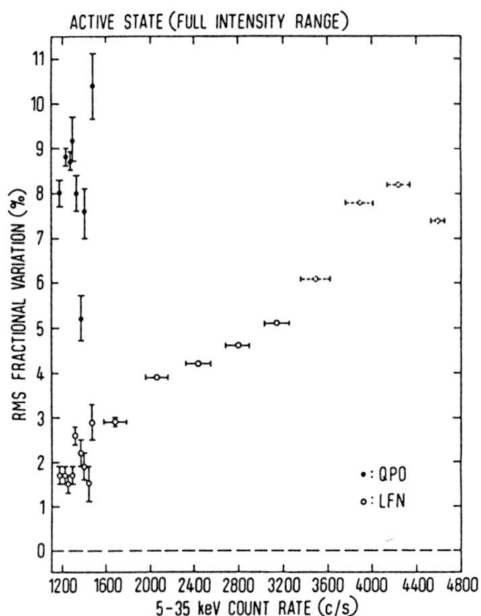


Figure 7. QPO (dots) and LFN (circles) strength as a function of X-ray intensity as observed in Sco X-1 in the active state. The QPO disappears when the source flares (van der Klis et al. 1986a).

To explain the lack of LFN within the beat-frequency model it is necessary to postulate ad-hoc mechanisms to 'line-up' some of the accreting clumps, so that the phase of the QPO can be kept from one oscillating shot to the next (Lamb et al. 1985) or to produce an 'ever-lasting' clump in the Keplerian flow (Shaham 1987). By contrast, in an 'obscuration' model for the QPO (see van der Klis et al. 1986a) it depends on the details of the obscuration geometry (such as inclination) whether or not QPO-related LFN is produced.

## 4. Bimodal behaviour of GX 5-1 and Cyg X-2

Cyg X-2 (Branduardi et al. 1980) and GX 5-1 (Shibazaki and Mitsuda 1983) show both very similar two-branched spectral behaviour. In both cases, the strong 20-50 Hz intensity-dependent QPO and associated LFN described in Section 2 are only seen in the so called 'horizontal branch' (van der Klis et al. 1985d, Hasinger et al. 1985b). In Cyg X-2, 5.6 Hz QPO whose frequency does not depend perceptibly on intensity are seen in parts of the so called 'normal branch' (Hasinger et al. 1986) and there is an indication that ~5 Hz QPO may also occur in GX 5-1 on the normal branch (van der Klis et al. 1986b).
So, also in these sources QPO with different frequency-intensity relations occur in different spectral states.

## 5. Conclusion

Strong, persistent, high-frequency QPO have been discovered with EXOSAT in seven luminous, persistently bright low-mass X-ray binaries. The

detailed properties of these QPO, in particular their dependence on source intensity, are diverse. In the three best-studied sources, GX 5-1, Sco X-1 and Cyg X-2 the known bimodal X-ray spectral behaviour is found to be strictly correlated to bimodal QPO behaviour.
A large variety is observed in the relations between source intensity and QPO frequency. If QPO frequency is simply related to accretion rate, as is the prediction of most proposed QPO models, in particular the magnetospheric ones, then a matching large variety of relations between accretion rate and X-ray intensity will be required - a strong departure from the traditionally assumed strict proportionality.
QPO-related LFN is only detected in part of the QPO sources. This presents problems for oscillating-shot models such as the (magnetospheric) random clump accretion beat-frequency model, and could be more easily accounted for in an obscuration model.

In conclusion, from the point of view of the QPO data, the case for the presence of a magnetic field in the low-mass X-ray binaries is as yet undecided.

Acknowledgements

Stimulating conversations with many participants of the Workshop on 'Accretion onto Compact Objects', April 1986, Tenerife, Spain, the Workshop on 'HE and VHE Behaviour of Accreting X-ray Sources', May 1986, Vulcano, Italy and the 125th IAU Symposium on 'The Origin and Evolution of Neutron Stars', May 1986, Nanjing, China are gratefully acknowledged. I thank Guenther Hasinger and Bill Priedhorsky for permission to reproduce diagrams of their work before publication.

References

- Alpar, M.A. and Shaham, J., 1985a, IAU Circ. **4046**.
- Alpar, M.A. and Shaham, J., 1985b, Nature **316**, 239.
- Belli, B.M., d'Antona, F., Molteni, D. and Morini, M., 1986, IAU Circ. **4174**.
- Bradt, H.V.D. and McClintock, J.E., 1983, Ann. Rev. Astron. Ap. **21**.
- Branduardi, G., Kylafis, N.D., Lamb, D.Q. and Mason, K.O., 1980, Ap.J. **235**, L153.
- Cooke, B.A., Stella, L. and Ponman, T., 1985, IAU Circ. **4116**.
- Forman, W., Jones, C. and Tanaubaum, H., 1976, Ap. J. **208**, 849.
- Gribbin, J.B., Feldman, P.A. and Plagemann, S.H., 1970, Nature **225**, 1123.
- Hasinger, G., Langmeier, A., Sztajno, M. and White, N., 1985a, IAU Circ. **4070**.
- Hasinger, G., Langmeier, A., Sztajno, M., Pietsch, W. and Gottwald, M., 1985b, IAU Circ. **4153**.
- Hasinger, G., Langmeier, A., Sztajno, M., Truemper, J., Lewin, W.H.G. and White, N.E., 1986a, Nature **319**, 469.
- Hasinger, G., Langmeier, A., Sztajno, M. and Pietsch, W., 1986b, in prep.
- Jansen, F.A. et al., 1986, in prep.

- Lamb, F.K., Shibazaki, N., Alpar, M.A. and Shaham, J., 1985, Nature **317**, 681.
- Lewin, W.H.G., van Paradijs, J., Jansen, F., van der Klis, M., Sztajno, M. and Truemper, J.E., 1985, IAU Circ. **4101**.
- Lewin, W.H.G., van Paradijs, J., Hasinger, G., Penninx, W.H., van der Klis, M., Jansen, F., Langmeier, A., Sztajno, M. and Truemper, J., 1986, IAU Circ. **4170**.
- Matsuoka, M., 1985 in Cataclysmic Variables and Low-Mass X-ray Binaries, D.Q. Lamb and J. Patterson (eds.), D. Reidel, p.139.
- Middleditch, J. and Priedhorsky, W., 1985a, IAU Circ. **4060**.
- Middleditch, J. and Priedhorsky, W.C., 1985b, preprint LA-UR-85-3649 to appear in Ap. J.
- Mitsuda, K. et al., 1985, Publ. Astron. Soc. Japan **36**, 741.
- Norris, J.P. and Wood, K.S., 1985, IAU Circ. **4087**.
- Parsignault, D.R. and Grindlay, J.E., 1978, Ap. J. **225**, 970.
- Pollock, A.M.T., Carswell, R.F. and Ponman, T.J., 1986, poster presented at Workshop on Accretion onto Compact Objects, Tenerife, April 1986.
- Ponman, T., 1982, MNRAS **201**, 769.
- Priedhorsky, W., Hasinger, G., Lewin, W.H.G., Middleditch, J., Parmar, A., Stella, L. and White, N., 1986, preprint, to appear in Ap. J. (Lett.).
- Shibazaki, N. and Mitsuda, K., 1983, ISAS RN **234**, 63.
- Shaham, J., 1987, paper presented at IAU Symp. **125**, Nanjing, China, May 1986.
- Stella, L., Kahn, S.M. and Grindlay, J.E., 1984, Ap. J. **282**, 713.
- Stella, L., Parmar, A.N. and White, N.E., 1985a, IAU Circ. **4102**.
- Stella, L., Parmar, A.N., White, N.E., Lewin, W.H.G. and van Paradijs, J., 1985b, IAU Circ. **4110**.
- Stella, L., White, N.E. and Priedhorsky, W., 1985c, IAU Circ. **4117**.
- Stella, L., 1985, EXOSAT prep. **25**.
- Tawara, Y., Hayakawa, S., Kunieda, H., Makino, F. and Nagase, F., 1982, Nature **299**, 38.
- Terrell, N.J., 1972, Ap. J. **174**, L35.
- Tennant, A., talk presented at Workshop on Accretion onto Compact Objects, Tenerife, April 1986.
- Van der Klis, M., Jansen, F., van Paradijs, J., Lewin, W.H.G., Truemper, J. and Sztajno, M., 1985a, IAU Circ. **4043**.
- Van der Klis, M., Jansen, F., van Paradijs, J., Lewin, W.H.G., van den Heuvel, E.P.J., Truemper, J.E. and Sztajno, M., 1985b, Nature **316**, 225.
- Van der Klis, M., Jansen, F., van Paradijs, J., Lewin, W.H.G., Truemper, J. and Sztajno, M., 1985c, IAU Circ. **4140**.
- Van der Klis, M., Jansen, F., White, N., Stella, L. and Peacock, A., 1985d, IAU Circ. **4068**.
- Van der Klis, M., Stella, L., White, N., Jansen, F. and Parmar, A.N., 1986a, preprint, submitted to Ap. J.
- Van der Klis, M. et al. 1986b, in prep.
- White, N.E., Peacock, A. and Taylor, B.G., 1985, Ap. J. **296**, 475.

## DISCUSSION

**F. Frontera:** Have you studied the power behavior vs. frequency of the LFN after subtraction of the Poisson component and after correction for effects of dead time, averaging and samplings?

**M. van der Klis:** Yes. In Sco X-1, the LFN is weak; it is only detected below ~ 1 Hz and it has a power-law shape with index between -1 and -2. In GX5-1 and Cyg X-2 the QPO-related LFN is stronger (about as strong as the QPO) and it is not fit by a power law. An exponential provides a better fit. In GX5-1 we recently found an additional "very low-frequency" component to the LFN. This "VLFN" is much weaker than the QPO, it is present irrespective of whether QPO are present or absent, it is only detected below ~ 0.1 - 1 Hz and it has a power law shape with index ~ 1.7. It is therefore very similar to the LFN in Sco X-1.

**S. Colgate:** The signature of a magnetic field is a cyclotron line and limited luminosity from a polar cap at high-Eddington limit temperature ~ 2 KeV. What is the correlation of either of these with the observed QPO.

**M. van der Klis:** There is no correlation I know of.

**J. Shaham:** The BF model does not necessarily require LFN, only the blob-BF does! Details in my talk.

**M. van der Klis:** The "slightly modulated Keplerian flow" you are invoking to explain QPO without LFN is essentially a big blob which lives forever. You claim that this big blob is created by interaction of the plasma with the magnetic field at the corotation radius, but I do not understand why the big blob is not destroyed on its way down from the corotation radius to the magnetospheric radius, as it will encounter ever higher field strengths. I note that you now need two types of flow ("slightly modulated Keplerian flow" and "chaotic blobs") to explain QPO without and QPO with associated LFN, respectively.

# Observations of Quasi-Periodic Oscillations in Cyg X-2

Günther Hasinger
Max-Planck-Institut für
extraterrestrische Physik
8046 Garching bei München
Federal Republic of Germany

**Abstract.** EXOSAT observations of Cyg X-2 reveal two types of QPO's (quasi-periodic oscillations) which are associated with different spectral behaviour. QPO's in the range 18-55 Hz with a hard spectrum are found in all observations where the source is on the 'horizontal branch' in an hardness-versus-intensity diagram. On the 'vertical branch', however, QPO's with a soft spectrum are found at a stable frequency of 5.6 Hz. A multitude of QPO frequency-intensity relations, found in different 'horizontal branch' observations, coalesce to an almost unique relation when the QPO frequency is correlated with spectral hardness. A cross-correlation analysis of the rapid variability reveals a time-lag of hard photons by 1.5-3.8 ms; the lag being smaller when the QPO frequency is larger. All results are consistent with a model where QPO's represent the Kepler frequency at the edge of the neutron star magnetosphere. In particular the data do not require the assumption of a beat frequency. The derived surface magnetic field strength is of the order of $10^{10}G$.

## 1. Introduction

Quasi-periodic oscillations on timescales of tens of milliseconds have been discovered in the X-ray flux of bright low-mass X-ray binaries (van der Klis et al., 1985a); up to date the class of QPO sources contains about nine objects. For detailed reviews of the observational material (much of which as yes unpublished) see Lewin and van Paradijs (1986) and van der Klis (1987). Several models have been proposed to explain QPO's (see van der Klis, 1987; Lewin, 1986) but there is no general consensus about the production mechanism. In particular it is unclear, whether QPO's are of magnetospheric origin or not. Today it is obvious that the QPO production mechanism is much more complicated than the early results have suggested. The variety of puzzling phenomena increased with every new observation. It is quite conceivable that more than one production mechanism has to be considered.

In this *review* an extended body of QPO observations of Cyg X-2 is discussed. The question, what information QPO's contain about the neutron star magnetic field, is adressed. This paper contains some unpublished material of which a more detailed analysis will be presented elsewhere (Hasinger et al., 1986b).



*D. J. Helfand and J.-H. Huang (eds.), The Origin and Evolution of Neutron Stars, 333–345.*

## 2. Two spectral modes

### 2.1 Hardness -versus-intensity diagrams

Six observations of Cyg X-2 have been carried out with the Medium Energy Detectors (Turner et al., 1981) aboard the European Space Agency's X-ray Observatory EXOSAT, yielding a total net observing time of $3 \cdot 10^5$ s. Figure 1 shows hardness-versus-intensity diagrams for all observations. The data points are samples of 200 s, the different symbols refer to different observations. The abscissa gives total (1-17 keV) source count rates, corrected for deadtime and background, and normalized to 8 observing detectors ($\sim 1500 cm^2$).

The ordinate gives the hardness ratios between two energy channels: (6-17/3-6 keV) for figure 1a and (3-6/1-3 keV) for figure 1b. Although the source was found in several distinct tracks in this diagram, there are only two characteristic modes of spectral behaviour, as was first noted by Branduardi et al. (1979). On the 'horizontal branch' (Shibazaki and Mitsuda, 1984) in figure 1a the 'hard' hardness ratio (6-17/3-6 keV) changes only very little with increasing source intensity. Figure 1b, where the 'soft' hardness ratio (3-6/1-3 keV) is plotted against intensity, demonstrates that the form of the tracks in this diagram is dependent on the choice of energy bands. The tracks of the 'horizontal' branch observations are now actually inclined, i.e. hardness increases with source intensity.

On the 'vertical' branch (which is usually called the 'normal' branch), on the other hand, there is a steep gradient in hardness ratio, which varies dramatically during small intensity changes. This is due to the fact that intensity variations in the high and low bands are anticorrelated (see also Hasinger et al., 1985a), so that decreases at low energies are cancelled by increases at high energies and vice versa. Similar 'vertical' branch behaviour has been reported for Cyg X-2 by Vrtilek et al. (1986).

Transitions between these two modes have been observed several times; they happen on a timescale of ~1000 s.

The differences between figure 1 and 'classical' hardness- intensity diagrams (Branduardi et al., 1979; Shibazaki and Mitsuda, 1983; van der Klis, 1987) are probably only due to the difference in energy bandpass for different measurements: the 1-2 keV range is not accessible for most detectors, nor for heavily absorbed sources.

### 2.2 Spectral decomposition

Analysis of the source spectra in terms of a two-component model (Comptonized thermal and blackbody) indicates, that on the 'horizontal' branch the blackbody component is responsible for most of the intensity variations, while the thermal component stays nearly constant in intensity and temperature (Hasinger et al., 1986a). Consequently the soft hardness ratio (fig. 1b) is a good measure of the

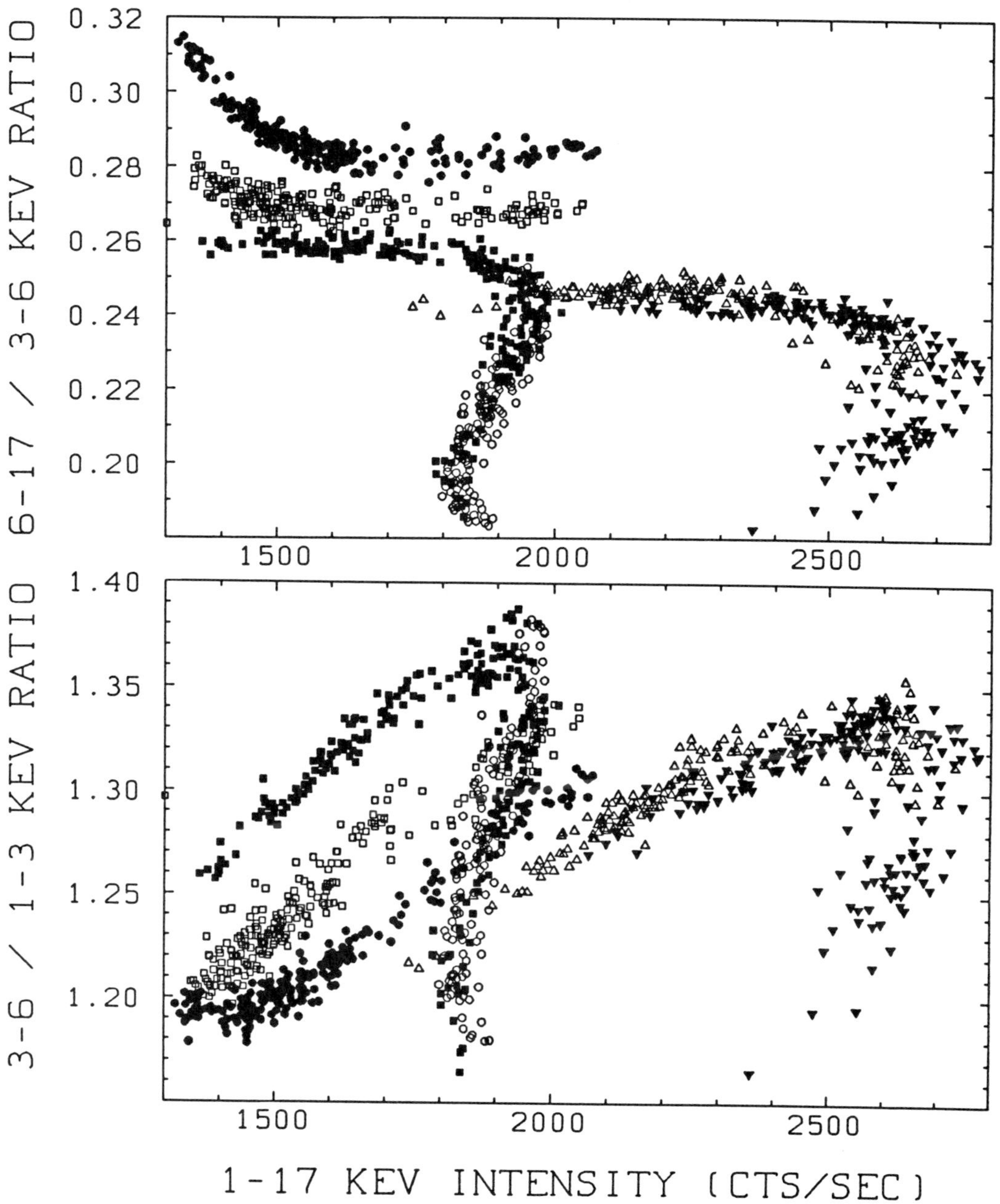


Figure 1: Hardness-versus-intensity diagrams for Cyg X-2. Top panel (fig. 1a): 'hard' hardness ratio; bottom panel (fig. 1b): 'soft' hardness ratio.

blackbody contribution in the 'horizontal' branch.

On the 'vertical' branch, however, preliminary analysis indicates, that the black-body component stays roughly constant around the highest level observed in the 'horizontal' branch (Hasinger et al., 1986b). The large hardness changes are here due to variations in the slope of the 'thermal' component, which tilts around a pivot-point at $\sim 3$ keV. This results roughly in a net conservation of the number of photons in the observed band (1-17 keV), a behaviour very suggestive for Compton scattering through a cloud of variable optical depth.

## 3. QPO analysis

Every observation in figure 1 has been subdivided into several groups of data points at similar intensity and hardness. For these observation subsets the spectral and timing properties of the source were assumed to be constant, so that power spectra could be averaged to perform a 'standard' QPO analysis (van der Klis et al., 1985a; Hasinger et al., 1986a).

### 3.1 'Horizontal' branch

The 'classical', intensity-dependent QPO's, that have been discovered from GX 5-1 (van der Klis et al., 1985) and subsequently from Cyg X-2 (Hasinger et al., 1986a), are observed exclusively during 'horizontal' branch behaviour (see also van der Klis et al., 1985b). Figure 2a shows the resulting power spectra for a typical 'horizontal' branch observation. The average source intensity increases from top to bottom. QPO's and LFN (low-frequency noise) are clearly visible in each panel. The QPO frequency increases from 18 to 45 Hz with increasing source intensity. At low intensities there is a clear indication of additional QPO power at around twice the frequency of the fundamental peak, characteristic of higher harmonics. However, in the second panel (where the addidional peak is most prominent) a fit with two Lorentzian peak profiles yields a frequency ratio of $1.85 \pm 0.03$, smaller than that expected for a second harmonic.

Utilizing an energy-resolved high-time-resolution mode of the EXOSAT on-board computer specifically designed for QPO research (HER7), one can determine the spectral properties of the oscillations. Averaging all data from figure 2a, one finds that in the 'horizontal' branch QPO (and LFN) are much harder than the mean source spectrum: the rms. relative equivalent sinusoidal modulation increases strongly with energy from $\sim 4$ percent at 1 keV to $\sim 10$ percent at 10 keV.

### 3.2 'Vertical' branch

In a preliminary analysis of GX 5-1 no QPO's were found on the 'vertical' branch (van der Klis et al., 1985b). However, for Cyg X-2 QPO's have been found in the 'vertical' branch, though with properties completely different from those in the 'horizontal' branch. Figure 2b shows power spectra averaged for a 'vertical' branch observation. The mean spectral hardness decreases from top to bottom. QPO's at a frequency of 5.6 Hz are observed, but only in the panels 2 and 3, i.e. exclusively at intermediate hardness ratios. Similar QPO's have been observed in two other EXOSAT 'vertical' branch observations (Hasinger et al., 1986b), and by HEAO-1 (Norris and Wood, 1985). In all cases the observed frequency is consistent with

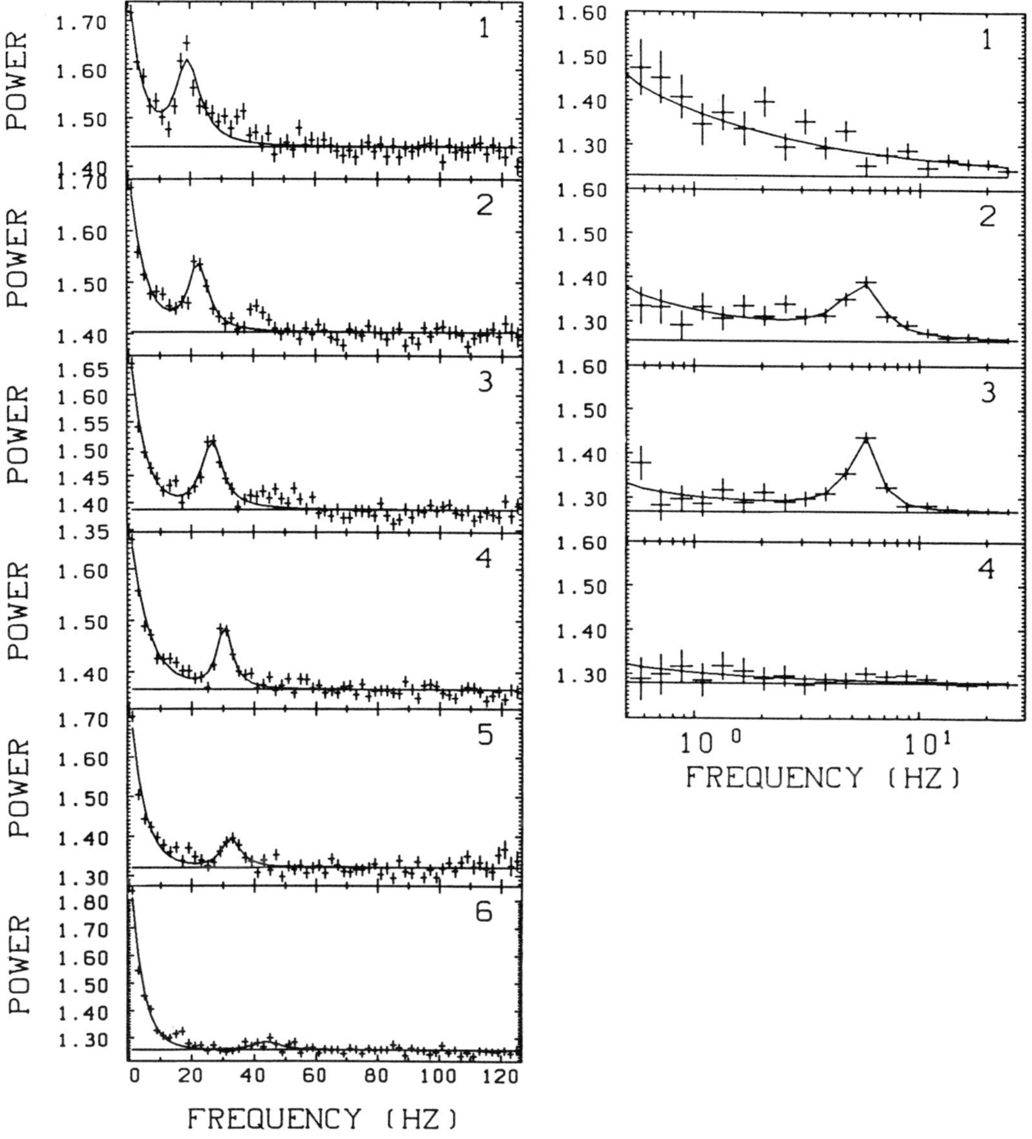


Figure 2: Power spectra of Cyg X-2. 2a (left): A 'horizontal' branch observation; the source intensity increases from 1354 cts/sec in panel 1 to 1990 cts/sec in panel 6. 2b (right): A 'vertical' branch observation; the 'soft' hardness ratio decreases from 1.36 in panel 1 to 1.20 in panel 4.

a value of 5.6 Hz. The remarkable stability of this frequency over timescales of months to years is contrasted by a highly time-variable QPO occurence, which is a consequence of their restriction to a certain band in spectral hardness.

Another characteristic difference compared to the 'horizontal' branch behaviour is the weakness of the low-frequency noise during times when QPO are observed. The LFN decreases steadily with decreasing hardness ratio.

The most distinct difference, however, is the QPO spectrum: on the 'vertical' branch the QPO's are soft: the rms. relative amplitude is $\sim$ 4 percent at 1 keV and drops below 2.3 percent above 4.5 keV.

Recently, an indication for $\sim$ 5 Hz QPO's has also been found in the 'vertical' branch of GX 5-1 (van der Klis, 1987). Taking into account all the differences between 'vertical' and 'horizontal' branch QPO's, it seems very likely that they are produced by different mechanisms or in different geometries. One has therefore to be careful to compare all the properties of QPO's in different sources.

## 4. Which quantity determines the QPO frequency ?

The further discussion will be limited to the 'classical', hard QPO's in the 'horizontal' branch which, due to their strong intensity-dependence enable a comparison to model predictions. The sources GX 5-1, Cyg X-2 and, as will be argued in the following, Sco X-1 belong into this class.

### 4.1 The source spectrum

One of the most striking facts of the early QPO discoveries was the strong intensity dependence of the QPO frequency. In GX 5-1, Cyg X-2 and at times Sco X-1 the frequency $f$ was a steep function of intensity $I$ with a logarithmic slope $\alpha = dlog(f)/dlog(I) = 1.7 - 3$ (van der Klis et al., 1985a,b; Hasinger et al., 1986). The simplicity and uniqueness of this behaviour led to the tentative interpretation of the QPO's as the result of a chaotic process at the boundary of the neutron star magnetosphere (Alpar and Shaham, 1985; van der Klis et al. 1985a).

With the subsequent analysis of more observations, it became clear that a variety of frequency-intensity relationships is possible. Sco X-1 displays a continuous transition from a strongly positive to an erratic and then a slightly negative correlation between frequency and intensity as the source marches through a bimodal hardness-intensity diagram (Priedhorsky et al., 1986; van der Klis, 1987). Cyg X-2, on the other hand, falls on different tracks on the intensity-frequency plane for every other observation (figure 3a), displaying a similar ambiguity as in the hardness-intensity plane (see figure 1). In both sources, Cyg X-2 and Sco X-1 (see also Priedhorsky, 1986) it seems that the total X-ray intensity is an ill-behaving parameter and that a much simpler description of the QPO properties can be found via the source spectrum.

In figure 3b the QPO frequency is correlated to the 'soft' hardness ratio (3-6/1-3 keV) for all horizontal branch observations of Cyg X-2. The individual tracks for different observation coalesce to one, nearly unique relationship in this representation. (This is strictly true for frequencies below 40 Hz, the observation at the highest intensities is slightly offset, possibly because of a hardness variation in the soft component). One therefore has to conclude, that the shape of the X-ray spectrum is a much better indicator of the QPO frequency than the total X-ray intensity. In a two-component spectral fit the soft hardness ratio is a measure of the

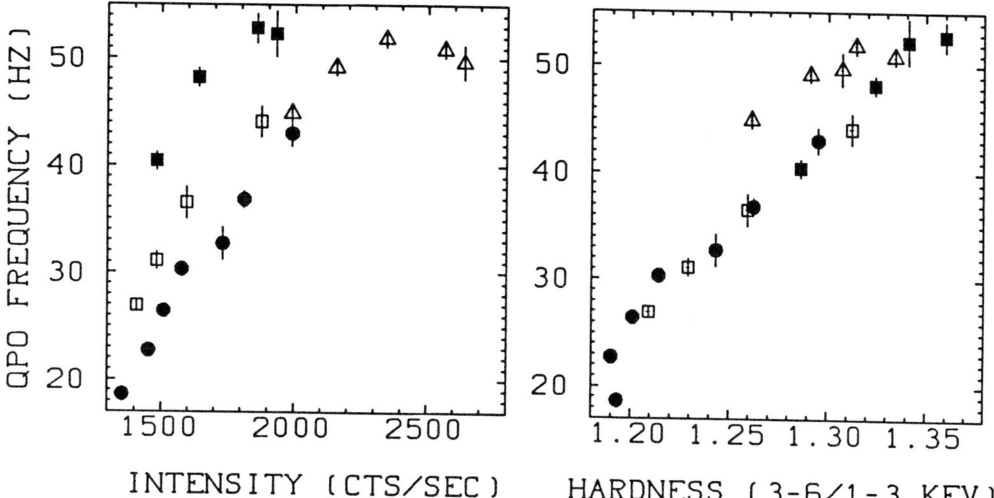


**Figure 3: Correlation of the QPO frequenies from 'horizontal' branch observations of Cyg X-2. Left panel (fig. 3a): frequency versus intensity; right panel (fig. 3a): frequency versus 'soft' hardness ratio.**

blackbody contribution (see above), and preliminary spectral fits show indeed, that the QPO frequency is better related with the blackbody flux than with the total X-ray flux (Hasinger, 1985). Due to the relatively large variations of the blackbody component the QPO frequency is a rather flat function of the blackbody flux $F_{bb}$: the logarithmic derivative $\alpha_{bb} = dlog(f)/dlog(F_{bb})$ is only ~0.5 for Cyg X-2 (Hasinger et al., 1986a). This is consistent with the value $\alpha = 3/7$ expected for the Kepler frequency at the boundary of the neutron star magnetosphere (Bath, 1973).

Also for Sco X-1 the complicated bimodal spectral and QPO behaviour reduces to a rather simple dependence of QPO frequency on blackbody flux (van der Klis, 1986). Together with the fact that all QPO's from Sco X-1 have a hard spectrum (Priedhorsky et al., 1986) this indicates that there is no principal difference to the QPO's in GX 5-1 and Cyg X-2 on one hand and Sco X-1 on the other hand.

The decomposition of low-mass X-ray binary spectra into two simple components is the subject of some controversy; the true source spectra are certainly more complicated. The limited bandwidth of X-ray observations, the rather coarse spectral resolution, the interstellar absorption and other systematic effects lead to ambiguities in the parameter estimation. In addition the decomposition is not unique, and different schools of thought prefer different representations. In particular there is no consensus regarding the 'thermal' or 'soft' component, which is modelled as a 'multitemperature blackbody' by the TENMA team (Mitsuda et al., 1984) and as 'unsaturated Comptonization' by EXOSAT observers (White et al., 1986). Consequently the derived blackbody temperatures and fluxes are very different in the two representations.

The strong correlation of an independent parameter, namely the QPO frequency, to the blackbody flux is supporting the 'unsaturated Comptonization' model, and may indicate that the blackbody flux is a real physical quantity and not just the figment of a mathematical parametrization.

## 4.2 The size of the X-ray emitting region

The ambiguities in total intensity and source spectra outlined above, make it desirable to find an independent physical quantity to describe the QPO frequency behaviour. Until now QPO research was mainly based on power spectral analysis, which maintains the frequency information but throws away all phase information involved in the quasi-periodic process. With the use of the energy-resolved high-time-resolution mode of the EXOSAT on-board computer (HER7), a cross-correlation analysis between the light curves in different energy bands can be performed.

The model of a Comptonized spectrum used for the spectral decomposition (see above) makes an important prediction: soft photons are upscattered in a cloud of hot electrons, thus the rapid fluctuations at high energies should lag behind those at low energies. The time-lag is an indirect measure of the size of the scattering cloud, and is expected to be on the order of several milliseconds (Shapiro et al., 1976; Priedhorsky et al., 1979; Page, 1985).

For Cyg X-2 subsamples of 'horizontal' branch observations at roughly the same QPO frequency have been selected (typically 12-16 s). The light curve in the high-energy band (4.5-17 keV) was cross-correlated with the one in the low-energy band (1-4.5 keV). The cross-correlation function (CCF) was derived by averaging the results from many of these subsets. Figure 4 displays a CCF, obtained by averaging ~38400 s of data with a time resolution of 3.91ms. Error bars are estimated from the variance of 2401 subsamples. The CCF is asymmetric, indicating that the average properties of the rapid fluctuations are different in the two energy bands. In addition the CCF has sinusoidal ondulations which are due to the quasi-periodic modulation. The most important finding is, however, a time lag of hard versus soft photons. This is clearly visible for the data points at ±3.91ms, which should have roughly the same height for a zero lag.

The solid line represents a fit to the data of the form:

$$CCF = Ae^{-\frac{t-\Delta}{\tau_1}}(1 + Bcos(\omega(t-\Delta))) + C \qquad \text{for } t > \Delta \text{ and}$$
$$CCF = Ae^{-\frac{\Delta-t}{\tau_2}}(1 + Bcos(\omega(\Delta-t))) + C \qquad \text{for } t < \Delta,$$

i.e. a delayed, asymmetrically decaying exponential function modulated by a delayed, exponentially decaying cosine. $A$, $B$ and $C$ are normalization constants, $\tau_1$ and $\tau_2$ are the exponential decay times for positive and negative lags, respectively, $\omega$ is the QPO frequency and $\Delta$ is the time delay. The significance of the parameter $\Delta$ has been estimated with an F-test (Bevington, 1969): the same fit as above was applied to the data, but $\Delta$ was forced to zero. Comparing the reduction in $\chi^2$ the significance of the additional parameter was estimated.

Since a large amount of data had to be averaged for every single CCF and since the HER7 mode was not available for earlier EXOSAT observations, only three independent CCFs could be derived for Cyg X-2. Table 1 summarizes the results of these

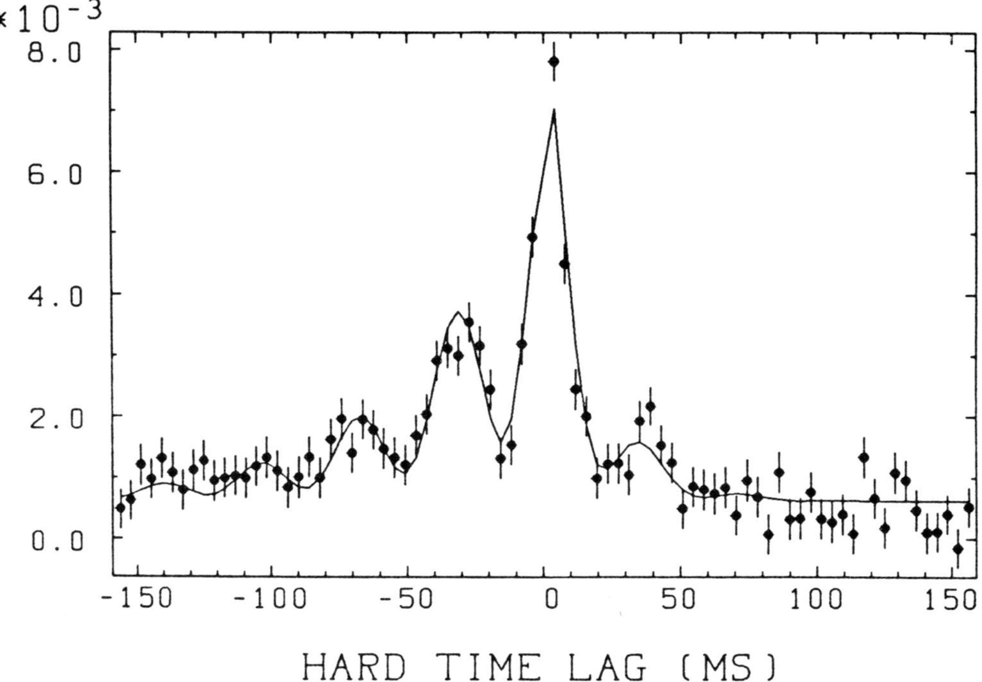


Figure 4: Sample cross-correlation function for Cyg X-2. Data points are seperated by 3.91 ms. The error bars are standard deviations derived by the variance of the 2401 samples averaged.

time-lag measurements at different QPO frequencies. Column 1 gives the mean QPO frequency and column 2 the measured time lag. Errors are standard deviations for one parameter of interest (Avni, 1976). Column 3 gives the time-resolution for the relevant measurement and columns 4, 5 and 6 give the F-value for the time lag, the probability of this value occurring by chance and the derived significance of the lag value.

| Frequency (Hz) | Time lag (ms) | Bin (ms) | F-value | Prob. | Sign. $\sigma$ |
|---|---|---|---|---|---|
| 20.7±0.3 | 3.8±0.3 | 3.91 | 62.2 | $4.8 \cdot 10^{-9}$ | $> 6.6$ |
| 28.4±0.3 | 2.7±0.3 | 3.91 | 34.3 | $1.1 \cdot 10^{-7}$ | 5.2 |
| 50.4±1.9 | 1.4±0.3 | 2.92 | 13.5 | $8.6 \cdot 10^{-4}$ | 3.3 |

Table 1: Results of the cross-correlation analysis for Cyg X-2.

## 5. Consequences for QPO models

The measured time delay is in the right sense, i.e. hard fluctuations lag, in agreement with the Comptonization model. The detailed CCF's provide a variety of constraints on the quasi-periodic process. The asymmetry of the profiles gives direct evidence that the average lifetime of the fluctuations is different at high and low energies. The fact that the sidelobes on the CCF also seem to be delayed indicates that the QPO's are formed inside the Compton scattering cloud. The time delay $\Delta$ gives a measure of the size $r_s$ of the scatterer: $\Delta = \frac{r_s \cdot \tau_c}{c}$; $\tau_c$ being the Compton optical depth and $c$ the velocity of light. With an optical depth of $\tau_c = 5$ (which is consistent with the Comptonized spectral fits), we find $r_s \approx 170$ km for a time lag of 3 ms. At this radius the Kepler frequency of material orbiting the neutron star is $\sim 20$ Hz, i.e. consistent with the observed QPO frequency.

The most important finding is, however, that the time-lag gets smaller for larger QPO frequency, as displayed in figure 5. This is the first independent indication that a change in QPO frequency is really associated with a considerable change in the system size, placing severe constraints on QPO models. The measured data points are consistent with a power law of a slope -2/3 (solid line in figure 5), which is the expected dependence of the time delay on the QPO frequency under the assumption that the QPO frequency is identical to the Kepler frequency at $r_s$ , and that $\tau_c$ stays roughly constant. The slope implied by the data is even somewhat steeper.

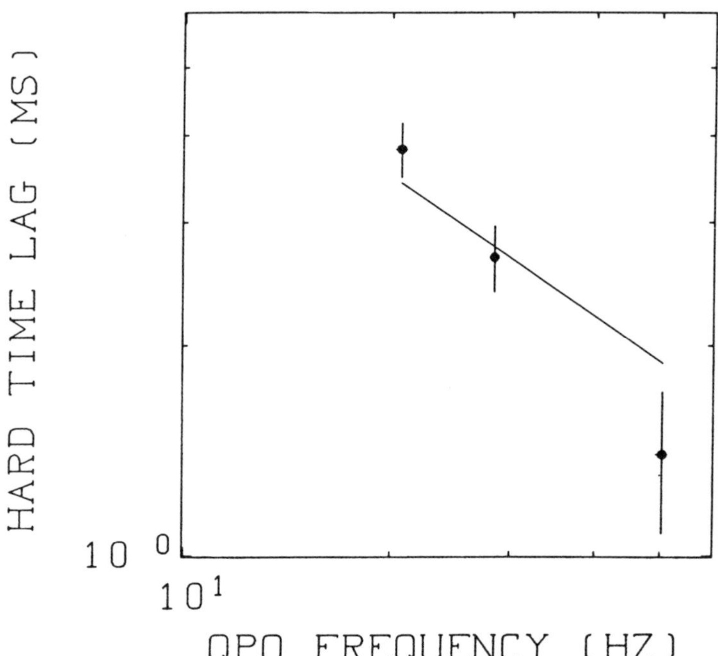


Figure 5: Time delay as a function of QPO frequency for Cyg X-2. The solid line is the dependence expected for the Kepler frequency.

Although the exact mechanism that produces QPO's is far from understood, the new observational results of Cyg X-2 strongly favour the view that in the 'horizontal branch' we are dealing with the Kepler frequency at the magnetospheric boundary. The dependence of QPO frequency on the blackbody luminosity and independently the relation of the time delay on QPO frequency are both perfectly consistent with this conception. The assumption of addidional parameters, as e.g. in the beat-frequency model (Alpar and Shaham, 1985; Lamb et al., 1985), seem unnecessary. Trying to estimate the surface magnetic field of the neutron star in the framework of the Bath model (van der Klis et al., 1985a; Hasinger et al., 1986a) there is a remaining ambiguity as to which value for the mass accretion rate $\dot{M}$ should be taken. The ill-behaved total luminosity is certainly not a good measure for $\dot{M}$. The blackbody luminosity, on the other hand, seems to be proportional to the mass accretion rate, but the absolute value of $\dot{M}$ remains unknown. Thus only a rough estimate of the magnetic field ($B \approx 10^{10} G$) can be derived.

## Acknowledgements

I thank my collaborators in this project: Andreas Langmeier, Drs. Mirek Sztajno, Wolfgang Pietsch, Manfred Gottwald and Bill Priedhorsky. I acknowledge clarifying discussions with Drs. Jan van Paradijs and Frank Verbunt, and I thank the organizers of the 125th IAU symposium in Nanjing, China for an enchanting meeting.

## References

Avni, Y. 1976, *Astrophys. J.* **210**, 642.

Alpar, M.A. and Shaham, J. 1985, *Nature* **316**, 239.

Bath, G.T. 1973, *Nature Phys. Sci* **246**, 84.

Bevington, P.R. 1969, in: *Data Reduction and Error Analysis for the Physical Sciences* (New York: McGraw Hill).

Branduardi, G., Kylafis, N.D., Lamb, D.Q. and Mason, K.O. 1980, *Astrophys. J. (Letters)* **235**, L153.

Hasinger, G. 1985, talk at the *'Discussion Group on EXOSAT Results'*, Cambridge, England, November 1985.

Hasinger, G., Langmeier, A., Pietsch, W. and Sztajno, M. 1985a *Space Sci. Rev.* **40**, 233.

Hasinger, G., Langmeier, A., Sztajno, M., Pietsch, W. and Gottwald, M. 1985b, *IAU Circ. No.* 4153.

Hasinger, G., Langmeier, A., Sztajno, M., Trümper, J., Lewin, W.H.G. and White, N.E. 1986a, *Nature* **319**, 469.

Hasinger, G., Langmeier, A., Sztajno, M., Pietsch, W., Gottwald, M. and Priedhorsky, W. 1986b (in preparation)

Lamb, F.K., Shibazaki, N., Shaham, J. and Alpar, M.A. 1985, *Nature* **317**, 681.

Lewin, W.H.G. 1987, this volume.

Lewin, W.H.G. and van Paradijs, J. 1986, *Comments on Astrophysics* (in press).

Mitsuda K. et al. 1984, *Publ. astr. Soc. Jap.* **36**, 741.

Norris, J.P. and Wood, K.S. 1985, *IAU Circ. No.* 4087.

Page, C.G. 1985, *Space Sci. Rev.* **40**, 387.

Priedhorsky, W., Garmire, G.P., Rothschild, R., Boldt, E., Serlemitsos, P. and Holt, S. 1979, *Astrophys. J.* **233**, 350.

Priedhorsky, W. 1986, *Astrophys. J. (Letters)* (in press).

Priedhorsky, W., Hasinger, G., Lewin, W.H.G., Middleditch, J., Parmar, A. Stella, L. and White, N.E. 1986, *Astrophys. J. (Letters)* (in press).

Shapiro, S.L., Lightman, A.P. and Eardley, D.M. 1976, *Astrophys. J.* **204**, 187.

Shibazaki, N. and Mitsuda, K. 1984, in: *High Energy Transients in Astrophysics* , ed. S.E.Woosley, p.49, American Institute of Physics, New York.

Turner, M.J.L., Smith, A. and Zimmermann, H.U. 1981, *Space Sci. Rev.* **30**, 513.

van der Klis, M., Jansen, F., van Paradijs, J., Lewin, W.H.G., Trümper, J., van den Heuvel, E.P.J. and Sztajno M. 1985a, *Nature* **316**, 225.

van der Klis, M., Jansen, F., van Paradijs, J., Lewin, W.H.G., Trümper, J. and Sztajno, M. 1985b, *IAU Circ. No.* 4140.

van der Klis, M. 1987 this volume.

van der Klis, M., Jansen, F., White, N.E. and Stella, L. 1985, *IAU Circ. No.* 4068.

Vrtilek, S.D., Kahn, S.M., Grindlay, J.E., Helfand, D.J. and Seward, F.D. 1985, *Astrophys. J.* (in press).

White, N.E., Peacock, A., Hasinger, G., Mason, K.O., Manzo, G., Taylor, B.G., Branduardi-Raymont, G. 1986, *Mon. Not. R. astr. Soc.* **218**, 487.

## DISCUSSION

**D. Eichler:** What is the electron temperature of your Comptonizing ball?

**G. Hasinger:** 3 to 4 KeV.

**M. van der Klis:** Did you also observe a delay in the 5 ms QPO?

**G. Hasinger:** No, the statistics are not yet good enough. The intrinsic time structure in the vertical branch is too long (20 ms) to find an expected 10 ms delay.

**W. Sieber:** How does the LFN fit into your shot-noise model?

**G. Hasinger:** Since I get a good fit to the Autocorrelation Function, and the ACF contains the same information as the Power Spectrum, the LFN in the power spectrum is reproduced by the simulations.

**N. White:** As you point out, the hardness ratio versus intensity plot you get is very sensitive to how you choose your energy bands. This makes comparison between different sources very difficult and perhaps misleading. I would strongly urge that these plots be used only as guides and that more emphasis be given to using the spectral deconvolution when comparing different sources.

**G. Hasinger:** I admit that it is hard to compare the Hardness ratio vs. Intensity diagram for different sources 1) because of selection of energy bands and effective bandwidth of the detectors used and 2) because of the effect of interstellar absorption. But spectral fits are model dependent, and as long as we do not agree on using the same spectral model, the results are even less comparable.

**N. White:** With regard to the screening of millisecond pulsations, John Middleditch pointed out at the Taos meeting last summer (1985) that millisecond pulsations will still survive at an amplitude of a few percent even if the pulsar is in an optically thick cloud.

**G. Hasinger:** The simulation of J. Middleditch and B. Priedhorsky, to my knowledge, used an infinitely thin slab of finite Compton optical depth, thus taking into account the degradation in intrinsic beaming, but not taking into account the light travel time effects.

# THE BEAT FREQUENCY MODEL FOR QPOs

Jacob Shaham
Physics Department and Columbia Astrophysics Laboratory
Columbia University
New York, New York 10027

## 1. INTRODUCTORY REMARKS

We have to date reports of Quasi-Periodic-Oscillation (QPO) observations in some twelve X-ray source, of which at least seven are low mass X-ray binaries (van der Klis 1987). They constitute a formidable zoo of phenomenae with so much variety that they, at times, do not at all even seem amenable to a single model. Some of the other QPO talks in these proceedings will try and present observations in the context of various models. My task is to talk about the Beat Frequency Model which, it seems to me, is by far the prime model for at least some of the QPOs.

The Beat-Frequency-Model (BFM) was constructed in order to account for what is, perhaps, the simplest of the QPOs, GX5-1. As you know, by some good fortune (or bad?) GX5-1 happened to have also been discovered first. I would therefore like to open my talk by first describing the GX5-1 story.

To remind everyone of the observations: van der Klis et al. (1985a,b) reported the discovery of quasi-periods between about 25 and 50 msec in GX5-1 with a Q value ($\equiv f/\Delta f$, where f is the corresponding frequency) of at most 4. The quasi periods were correlated with the overall source's 1-18 keV count rate I, such that $\Lambda \equiv d\log f/d\log I$ was about 2.5, while the size of the modulation seemed to decrease steadily with count rate - from a value of 8% for $I = 2.4\times10^3$ cps to zero at $I = 3.1\times10^3$ cps. In addition, low frequency noise (LFN) was observed in GX5-1. Its intensity correlated well with the size of the QPOs modulation.

A periodicity - albeit a quasi-periodicity - so short, has triggered in Ali Alpar and me (and, presumably, in many others) the following immediate reaction: In our previous scenario for the formation of millisecond pulsars (Alpar et al. 1982) we have argued, that they were weak-field neutron stars spun up by accretion from a companion in a galactic-bulge-type binary system. Millisecond periods were therefore expected in these binaries while they were still X-ray-active, and yet none has yet been observed.



D. J. Helfand and J.-H. Huang (eds.), The Origin and Evolution of Neutron Stars, 347–361.

The GX5-1 observation was the first such kind of period reported. Yet, The large variability of the period length ruled out stellar rotation because of the needed large variability in angular momentum, so that another - external - location for the origin of the oscillations had be be considered. Alpar and I concluded it had to be the magnetospheric boundary: We recalled that accepted models for the region just outside the surface of an accreting neutron star (see, e.g. Ghosh and Lamb 1979) talk about a magnetic-field dominated regime of plasma motion (for surface fields $B_s \gtrsim 10^6$ G) which turns over into a gravity-dominated one (a "Keplerian disk") outside of some boundary region at a radius $R_o$. $R_o$ can be roughly determined by equating the ram pressure of the Keplerian plasma, $\rho v_K^2$, to the magnetic pressure $B^2/8\pi$; it is given, more or less, by

$$R_0 \sim 10^6 B_8^{4/7} R_6^{12/7} \dot{M}_{17}^{-2/7} (M/M_\odot)^{-1/7} \text{ cm} \tag{1}$$

where in (1), $B_8$ is the surface magnetic dipolar field in units of $10^8$G, $R_6$ is the stellar radius $R_S$ in units of $10^6$ cm, M is the neutron stellar mass, $M_\odot$ is the solar mass and $\dot{M}_{17}$ is the total mass accretion rate in units of $10^{17}$ g/sec. The Keplerian angular frequency at $R_0$, $\Omega_0$, is

$$\Omega_0 \equiv \left(\frac{GM}{R_0^3}\right)^{1/2} \sim 10^4 B_8^{-6/7} R_6^{-18/7} \dot{M}_{17}^{3/7} (M/M_\odot)^{5/7} \text{ rad/sec} \tag{2}$$

$\Omega_0$ is therefore a prime candidate for variability because of its dependence on $\dot{M}$ hence on the count rate I; however, it only varies weakly with $\dot{M}$ to account for a $\Lambda$ of 2.5. So, Alpar and I (1985) postulated that the QPO frequency, $\Omega_{QPO} \equiv f/2\pi$, is to be identified with $\Omega_B$, the beat frequency,

$$\Omega_B = \Omega_0 - \Omega_S \equiv A\dot{M}^{3/7} - \Omega_S \tag{3}$$

where A is some constant and $\Omega_S$ is the stellar spin. This, provided the stellar rotation axis is perpendicular to the disk. We shall make this assumption throughout this talk, and only modify it towards the end (p. 12).

While we are at it, let us recall that, even though it was hard to see how this might be the source of the GX5-1 period, there are two additional points of interest outside of a rotating, accreting neutron star:

The corotation radius $R_c$ is that radius for which the Keplerian angular velocity equals the stellar spin angular frequency $\Omega_S$:

$$\left(\frac{GM}{R_c}\right)^{3/2} = \Omega_S$$

hence

$$R_c \sim 1.5\times10^6\ (M/M_\odot)^{1/3}\ (P_S/1\ \text{msec})^{2/3}\ \text{cm} \qquad (4)$$

where $P_S$ ($\equiv 2\pi/\Omega_S$) is the neutron stellar period. In a rotating star with a magnetic field structure attached, $R_c$ plays a very important role: If any field lines are threading the accretion disk inside of $R_c$, they develop an azimuthal component $B_\phi$ downstream which tends to slow down the plasma - hence spin-up the star; the reverse holds outside of $R_c$. Thus, spin-up and spin-down torques act simultaneously on the star; for some value of $R_c$ their resultant should vanish - that determines the equilibrium period $P_{eq}$ of the accreting star. It is roughly given by (Alpar _et al._ 1982)

$$P_{eq} \sim (0.33\ \text{msec})B_8^{6/7}(M/M_\odot)^{-5/7}(\dot{M}/10^{-8}M_\odot\text{yr}^{-1})^{-3/7}R_6^{15/7} \qquad (5)$$

Secondly, the velocity-of-light radius $R_L$ is the distance for which corotation with the star would mean moving at the speed of light,

$$R_L \sim 5\times10^6\ \text{cm}\ (P_S/1\ \text{msec}) \qquad (6)$$

For millisecond pulsars $R_0$, $R_c$ and $R_L$ close in on each other, still with $R_0 < R_L$; and, as long as the neutron stars are still spinning up, then, on the average, $R_0 < R_c < R_L$. If the mass accretion rate fluctuates, $R_0$ fluctuates too, but $R_c$ and $R_L$ do not. Now, in (3), A is some constant which depends on $B_S$, the surface magnetic dipolar field. Comparing (3) with the correlation data of $\Omega_{QPO}$ and the count rate I yielded $2\pi/\Omega_S \sim 7$ msec, $B_S \sim 5\times10^9$ G which were amazingly close to what one may have expected from the above millisecond pulsar model. Also, (3) gives

$$d\log f/d\log I = (3/7)\left(1-\Omega_S/(\Omega_S+\Omega_{QPO})\right) \simeq 2.7\ \text{average},$$

in close agreement with the GX5-1 observations.

What was the physical mechanism, giving rise to the beat frequency (3), going to be be? We had originally felt that a variety of physical phenomenae at $R_0$ would cause a modulation of flow onto the star at the beat frequency (BF) and, to be specific, suggested one class of phenomenae - a 'magnetic gate' process: Assume that plasma coming down to $R_0$ is magnetized (for example, after having been temporarily magnetized by the stationary stellar field at $R_c$ or by some random magnetization process). Then, it will cross $R_0$ preferentially depending on the stellar field at $R_0$ (e.g., where the stellar field is opposite in direction to the plasma magnetization so that field reconnections could occur); that will give a modulation of the crossing rate at $\Omega_B$.

Naturally, any BF process can only come about if the plasma does not flow into $R_0$ hydrodynamically. Instead, every region in the plamsa must carry its own magnetization along and remain pretty localized until crossing $R_0$. Fred Lamb has suggested to us that perhaps matter actually comes down in blobs - evidence for the blobby nature of accretion flow onto neutron stars in some compact X-ray sources has been around for a while. As we shall see below, blobs do make a fine physical scenario for realizing the $\Omega_{QPO}$; however, as we shall also see below, they are not the only one.

It is also important to realize, that the magnetic axes of the neutron stars in QPO sources may well be at an angle to their rotation axes, perhaps even perpendicular to the latter. Therefore, the magnetospheric boundary may not be a circle in the accretion plane at all, but only some closed curve with mirror symmetry through the center, even if the <u>average</u> curve for large scale phenomenae <u>is</u> circular. This, of course, can provide for an $\Omega_B$ modulation without the incoming plasma being previously magnetized, and even in the hydrodynamic flow limit there may be an $\Omega_B$ component. These things are very hard to discuss in any great detail because of their great mathematical complexity, and we are (nevertheless) presently involved in having a crack at them. In the meantime, however, I shall assume that some kind of function of stellar and material field strengths and directions determines an $\Omega_B$ modulation such that any group of particles at $R_0$ will produce a luminosity signal modulated by the $\Omega_B$ clock.

## 2. THE BFM

There are, essentially, four basic assumptions that go into the BFM:

(1) That the QPO phenomenon, being essentially universal (from having now observed several sources), may actually be independent of the observer's relative geometry and can therefore depend, at most, on internal stellar geometry;

(2) that some special radius exists outside the star, that is the source of the oscillations, possibly $R_0$;

(3) that $\Omega_{QPO}$ is related to the Keplerian frequency $\Omega_0$ - and with the given variety of observed $\Omega_{QPO}$ values possibly related to $\Omega_B \equiv \Omega_0 - \Omega_S$; and

(4) that any observed intensity-correlated variability of $\Omega_{QPO}$ is related to variability in mass accretion rate $\dot{M}$ , even

if intensity changes do not necessarily relate to changes in $\dot{M}$ in any simple way (Priedhorsky 1986).

Given the above, we now describe the QPO modulation as follows:

Let us view everything in a coordinate system corotating with the matter at $R_0$ and have a density delta function disturbance $\delta(\phi-\phi_i, t-t_i)$ appear on the $R_0$ circle at azimuth $\phi_i$ and time $t_i$. As that disturbance goes down into the neutron star, let it produce a luminosity signal $\ell(\Omega_B t_i-\phi_i,\ t-t_i)$ which depends on the starting position relative to the $\Omega_B \equiv \Omega_0 - \Omega_s$ clock. Note, that an extra phase parameter should come into $\ell$ due to, say, a random direction of matter field $\Omega_B t_i-\phi_i \rightarrow \Omega_B t_i-\phi_i-\alpha_i$; but we shall, for simplicity, absorb $\alpha_i$ into $\phi_i$. Assume now that the luminosity arising from a superposition of many delta-function density disturbances at $R_0$ is the sum of the individual signals - i.e., no non-linear effects on the way down to the star. Then, if the linear density profile at $R_0$ is

$$g(\phi,t) = \int d\phi_i dt_i\ g(\phi_i,t_i)\ \delta(\phi-\phi_i,t-t_i) \tag{7}$$

we have

$$L(t) \equiv \int d\phi_i dt_i\ g(\phi_i,t_i)\ \ell(\Omega_B t_i-\phi_i,\ t-t_i) \tag{8}$$

Let us go over to Fourier components,

$$g(\phi,t) = \sum_{\nu} \int d\omega\ e^{i\nu\phi} e^{i\omega t}\ \tilde{\tilde{g}}(\nu,\omega) \tag{9}$$

$$\ell(\psi,\tau) = \sum_{\bar{\nu}} \int d\bar{\omega}\ e^{i\bar{\nu}\psi}\ e^{i\bar{\omega}\tau}\ \tilde{\tilde{f}}(\bar{\nu},\bar{\omega}) \tag{10}$$

Hence

$$L(\omega) = \sum_{\nu} \tilde{\tilde{g}}(\nu,\ \omega - \nu\Omega_B)\ \tilde{\tilde{f}}(\nu,\omega) \tag{11}$$

The simplest BFM (Lamb et al. 1985) utilizes a simple form for $\ell(\psi,\tau)$,

$$\ell(\psi,\tau) = \theta(\tau)\ e^{-\gamma\tau}\ \left[1 + \beta\cos(\Omega_B\tau + \psi)\right] \tag{12}$$

where $\theta(\tau)$ is the step function: 1 for $\tau > 0$, 0 for $\tau < 0$. (8) and (12) describe a rather subtle process. One firstly needs, of course, the linearity, namely that various regimes on the $R_0$ circle do not influence each other's dynamics as they go down the magnetospheric boundary. Secondlay (but also related to the first point), one requires that the specific $R_0$ bunch of particles retains it coherence during the process; otherwise, if the coherence lifetime is shorter than $\gamma^{-1}$, we may need to replace (12) by

$$\ell(\psi,\tau) = e^{-\gamma\tau}\left[1 + \beta e^{-\Gamma\tau}\cos(\Omega_B\tau + \psi)\right]$$

and, for $\Gamma^{-1} << \gamma^{-1}$, the BF essentially disappears from the spectrum. We ask for the process to be, in that sense, not a topologically simple or a hydrodynamical flow process between $R_0$ and the luminosity-generating location. We do not know yet whether the inner disk dynamics really allows for these physical assumptions to be valid-disk dynamics is still too hard to handle down to such detail. We shall, however, assume that they are valid. Work on testing that is in progress, and it seems that an embedded magnetic field has much to do in securing a small $\Gamma$.

Before we proceed, we ought to comment on the possible values for $\beta$, since some recent data analysis suggests that $\beta > 1$ while a negative $\ell$, which could result from such $\beta$, is certainly nonphysical.

We note, that $\ell(\psi,\tau)$ could, in principle, also include higher $\Omega_B$ harmonics, as long as their magnitude is consistent with the data. It is well known, that cosine or sine Fourier transforms $\tilde{f}(\nu)$ of positive functions $f(\tau)$ have the following property:

$$\left|\tilde{f}(\nu)\right| \equiv \frac{1}{\pi}\left|\int f(t)\begin{matrix}\cos\ \nu t\\ \sin\ \nu t\end{matrix} dt\right| \leqslant \frac{1}{\pi}\int \left|f(t)\right|\left|\begin{matrix}\cos\ \nu t\\ \sin\ \nu t\end{matrix}\right| dt \leqslant$$

$$\leqslant \frac{1}{\pi}\int f(\tau)\ \alpha t \equiv 2\tilde{f}(0)$$

Hence, in (12), the only strict requirement on the coefficient of $\cos\ (\Omega_B t + \psi)$ is really $|\beta| \leqslant 2$, provided of course, higher harmonics are present in $\ell(\psi,\tau)$ or else $\ell$ will indeed not be always positive. For example, notice that

$$0 \leqslant \sin^{2k}(\theta/2) = 2^{-2k}\left[\binom{2k}{k} + 2\sum_{p=0}^{k-1}(-)^{k-p}\binom{2k}{p}\cos\left[(k-p)\theta\right]\right]$$

$$\equiv 2^{-2k}\binom{2k}{k}\left\{1+2\sum_{p=0}^{k-1}(-)^{k-p}\frac{\binom{2k}{p}}{\binom{2k}{k}}\cos\left[(k-p)\theta\right]\right\}$$

$$[\theta \equiv \Omega_B t + \psi]$$

and the effective β value (i.e., the coefficient of cos θ) here is

$$\beta = \binom{2k}{k-1}\binom{2k}{k}^{-1} \equiv \frac{2k}{k+1} \; (\to 2 \text{ for large } k)$$

So, β can be above 1 provided the observational upper bound on the relative power at higher harmonics is not below the corresponding value of $\binom{2k}{p}^2 \binom{2k}{k}^{-2}$. If β = 1.35 is found (Elsner et al. 1986), then we can take k=2, and expect $\ell(\psi,\tau) = e^{-\gamma\tau}(1 - \frac{4}{3}\cos\theta + \frac{1}{3}\cos 2\theta)$ hence a relative power of 2nd to 1st harmonic 1.4%!

After this detour - back to the calculation of L(ω). Given the expression for ℓ, we find

$$\tilde{\tilde{\ell}}(0,\omega) = (\gamma + i\omega)^{-1}$$

$$\tilde{\tilde{\ell}}(\pm 1,\omega) = \frac{1}{2}\beta\,[\gamma + i(\omega \mp \Omega_B)]^{-1} \tag{13}$$

hence

$$L(\omega) = \tilde{\tilde{g}}(0,\omega)\,\frac{1}{\gamma + i\omega}$$

$$+ \frac{1}{2}\beta\tilde{\tilde{g}}\,(1,\omega-\Omega_B)\,\frac{1}{\gamma + i(\omega-\Omega_B)}$$

$$+ \frac{1}{2}\beta\tilde{\tilde{g}}\,(-1,\omega+\Omega_B)\,\frac{1}{\gamma + i(\omega+\Omega_B)} \tag{14}$$

(14) is the fundamental equation of the BFM. The fundamental conclusion we draw from it is that to observe the QPO line, g must have power at $e^{\pm i\phi}$; to observe the LFN, g must have a ϕ-independent (i.e. isotropic) component. These are necessary, not sufficient requirements.

Anyone who can imagine processes involving various contributions from an isotropic and/or a cosϕ term in the density distribution on the $R_0$ circle can explain any corresponding ratio of QPO to LFN intensities. I have heard it said that the BFM cannot cope with a finite-QPO no -LFN situation. This, however, is not the case, as the following example would show:

Consider

$$g(\phi,t) = A + B(t)\cos(\phi + \alpha) \tag{15}$$

where $\alpha$ is a fixed phase. What (15) describes may be a toy model which works as follows: Imagine that at $R_c$, plasma does not only get partly magnetized with azimuthal dependence but also acquires - again, due to the magnetic field - an azimuthal density variation by magnetostriction. On the way down to $R_0$ that density variation, like the magnetization, will be partially preserved; that will lead to (15). (15) may actually be a good approximation for any non-noisy accretion flow at the boundary region, outside of $R_0$. We should keep in mind that the non-steady-state apsect of the BFM is reflected in (12) and is not necessarily needed for (8).

(15) could also describe the following: Remember that we have absorbed the magnetic field angle into $\phi$ (p. 5). Suppose the azimuth angle is, in fact, random, but that the probability for a chunk of matter to begin entry into the magnetosphere depends on its relative magnetization angle. A situation like that may arise when the process is sufficiently non-linear. This will bring about a probability function of type (15), which can therefore be treated as an effective $R_0$ density.

We now have

$$\tilde{\tilde{g}}(0,\omega) \propto A\delta(\omega)$$

$$\tilde{\tilde{g}}(\pm 1,\omega) \propto \frac{1}{2}\,\tilde{B}\,(\omega)^{\pm i\alpha}$$

hence

$$L(\omega) \propto c\delta(\omega) + \frac{1}{2}\,\tilde{B}(\omega-\Omega_B)\;\beta\;\frac{e^{i\alpha}}{\gamma+i(\omega-\Omega_B)} + \text{c.c.} \tag{16}$$

where c is some constant. The $c\delta(\omega)$ represents a dc term in the spectrum, not LFN; hence only the QPO lines exist, shaped according to $\tilde{B}(\omega)$ (or rather $|\tilde{B}(\omega)|^2$). It is easy to see how the dc term appears: If the density distribution at $R_0$ is isotropic, interference from all $\phi$ regimes will destroy the $\Omega_B$ modulation. If, furthermore, the flow has no temporal structure, no frequency except for the zero one will appear. Also note, that B should only have power around $\omega = 0$, but can be a random process with the $\cos\phi$ term turning randomly on and off. If (15) represents the magnetic field angle, then in (16) $\tilde{B}(\omega-\Omega_B)$ should be replaced by $\tilde{B}(\omega)$, so that B should have power around $\Omega_B$.

Naturally, if A were time dependent, LFN would arise, shaped by $\tilde{A}(\omega)$. But, in general, different LFN and QPO lineshapes can be easily understood via different time dependence of A(t) and B(t).

The easiest way of envisaging power at $e^{\pm i\phi}$ is, of course, in a process which has power at <u>every</u> $\phi$ harmonic (including $\nu = 0$). A blobby flow can accomplish that (Lamb *et al.* 1985), with

$$g(\phi,t) = \sum_j a_j \delta(\phi-\phi_j)\ \delta(t-t_j) \tag{17}$$

where $a_j$, $\phi_j$ and $t_j$ are all random variables. Since independent evidence for the blobby nature of flow in some X-ray sources does exist, this is a natural process to examine. Then

$$\tilde{\tilde{g}}(\nu,\omega) \propto \sum_j a_j e^{-i\omega t_j} e^{-i\nu\phi_j}$$

and the corresponding LFN and QPO powers are

$$\text{LFN} \to \frac{1}{\gamma^2+\omega^2} \sum_{p,q} \bar{a}_p a_q\ e^{i\omega(t_p-t_q)} \equiv \frac{1}{\gamma^2+\omega^2} F(\omega) \tag{18}$$

$$\text{QPO} \to \frac{(\frac{1}{4})\ \beta^2}{\gamma^2+(\omega-\Omega_B)^2} \sum_{p,q} \bar{a}_p a_q\ e^{i(\phi_p-\phi_q)}\ e^{i(\omega-\Omega_B)(t_p-t_q)}$$

$$\equiv \frac{(\frac{1}{4})\ \beta^2}{\gamma^2+(\omega-\Omega_B)^2}\ G(\omega-\Omega_B) \tag{19}$$

For uncorrelated $a_j$, $\phi_j$ and $t_j$,

$$F(\omega) = G(\omega-\Omega_B) \equiv \Sigma|a_p|^2$$

hence the LFN and QPO powers are always correlated here, much as they are found to be in GX5-1 (van der Klis et al. 1985) with power ratio of (1/2) $\beta^2$; as we pointed out earlier, under some circumstances that ratio may be close to 2. Processes in which some cross correlations between $a_j$, $\phi_j$ and $t_j$ can be imagined, which therefore have various degrees of coherence, can change the LFN/QPO power.

## 3. PROBLEMS

The preceding discussion represents, probably, all the distance in mathematics that one could dare to cover before more detailed physical computations will be at hand. Let me, therefore, address myself to some of the obvious problems the BFM may have - which may all be attributed to our ignorance, not necessarily to the inadequacy of the BFM.

The most important problems are:

(i) Underlying all is the assumption that $\Omega_S$ exists, i.e. that the neutron star rotates at angular frequency $\Omega_S$ which, in the case of GX5-1, say, is of order 10 msec. Yet, $\Omega_S$ is yet to be detected in the X-ray luminosity of any of the LMXBs, let alone the QPO ones. This, while the very assumption of the existence of $R_0 > R_S$ means that if the magnetic field is dipolar one should see an X-ray modulation across the stellar surface. Some ways of understanding that have been discussed in Lamb et al. (1985). Effects like the low efficiency of channeling of plasma onto polar caps due to the relative closeness of $R_0$ and $R_S$, gravitational lensing (but see Chen and Shaham 1987), optical scattering in the disk corona, preferential viewing along the direction perpendicular to the disk (hence the magnetic field?) because the disk is optically thick - all of these may combine to quench the $\Omega_S$ modulation substantially, but possibly not below some of the observational upper bounds. A promising possibility is, however, the following:

Eichler and Wang show in these proceedings, that as the magnetic field in neutron stars decays, higher multipoles become relatively stronger on the surface. The resultant multi-peaked surface field will cause a very different X-ray pulsation pattern than one finds in simple dipole fields: With sufficiently large "cap" areas, already an octupolar field component may smear the $\Omega_S$ pulsations substantially; higher mutlipoles will produce low level modulation at much higher - and at a wide range - of frequencies. We note, that this complex surface field need not be relevant to $R_0$ (hence the accretion process) or $R_L$ (hence the pulsar action after the X-ray source became a pulsar): Even for a 1.5 msec pulsar, $(R_L/R_S)^2 \sim 50$, which is roughly by how much the dipole will be amplified over any other multipole at $R_L$ over their surface ratio. The amplification at $R_0$ will be [see (5)] ~50 for a 10 msec neutron star.

(ii) The simple f vs. I relation in GX5-1 becomes inverted and even bimodal or indefinite in ScoX-1 (Priedhorsky et al. 1986). How to interpret that? We note, that the BFM predicts an f vs. $\dot{M}$ correlation, not directly an f vs. I. I itself is really not defined until one specifies the frequency range, $\Delta\nu_i$. Now, since torques (magnetic? material?) on the disk vary with $\Omega_S/\Omega_0 \equiv (1 + \Omega_B/\Omega)^{-1}$, and since these fine-tune the amount of total enery available (accretion energy onto the neutron stellar surface ± $\text{spin}^{\text{down}}_{\text{up}}$ energy of the neutron star), a marked difference may develop between dlog $\Omega_0$/dlog $\dot{M}$ and the various dlog $\Omega_0$/dlogI($\Delta\nu_i$), as was first pointed out by Priedhorsky (1986; see also Shaham and Tavani 1987). In fact, Priedhorsky writes

$$I(\text{hard X-rays}) = \dot{M}\left[\frac{GM}{R_S}\left\{1 - \frac{1}{2}\left(\frac{R_S}{R_c}\right)^3\right\} - f(\dot{M}, \Omega_S)\right]$$

$$I(\text{soft X-rays}) = \dot{M}\, f\,(\dot{M}, \Omega_S) - N\Omega_S$$

where N is the total torque on the star and f is some function. Priedhorsky suggests

$$f = \frac{GM}{2R_0} + \Omega_S(R_0^2\,\Omega_0 - R_S^2\Omega_S)$$

while Shaham and Tavani (1986) investigated the possibility

$$f = \frac{GM}{2R_0}$$

We find, that for a variety of choices of f, dlog I(hard)/dlog$\dot{M}$ indeed changes sign for some value of $\dot{M}$. Furthermore, an extended region with dlogI(hard)/dlog$\dot{M}$ >> 1 can exist, where f is essentially constant with I. A well defined I(hard) vs. $\dot{M}$ curve can be constructed to correspond to the full extent of the $\Omega_{QPO}$ vs. I ScoX-1 observations. Thus, even ScoX-1 can be interpreted on the basis of the BFM.

There is certainly a lot more that we do not know about the magnetospheric boundary, and hence about the detailed physics of the BFM. One may want to see the observations in order to either probe the magnetospheric boundary with the aid of the BFM or come up with a clearly better model for QPOs. A group of us is actively following the first option, mostly because the BFM looks, to date, more promising than other suggested models (see W.H.G. Lewin 1987).

One important physical method of probing an unknown region is to have it subject to some controlled external perturbation. Within the BFM, one important perturbation of the magnetospheric boundary can be neutron stellar free precession (as in Her X-1, see Trumper et al. 1985). Free precession will periodically change the opening angle of the cone, on the surface of which any particular magnetic field line segment rotates, so that the actual beat frequency may change (for details see Shaham 1986).

To fix our ideas, assume that the magnetic gate operates via $(\vec{B}\cdot\vec{r})^n$ where $\vec{r}$ is the radius vector of material orbiting at $R_0$, $\vec{B}$ is the local field at $R_0$ and n is some integer. Then $\vec{B}\cdot\vec{r} \propto \vec{\mu}\cdot\vec{r}$ where $\vec{\mu}$ is the stellar dipole. An extreme case is when the stellar rotation axis is in the disk plane with $\vec{\mu}$ perpendicular to it. Then $\vec{\mu}$ crosses the disk twice each $2\pi/\Omega_S$ period, and for $n = 1$ this gate gives both $\Omega_0 \pm \Omega_S$ freqeuncies at equal powers. Another extreme case for $n = 1$ is with $\vec{\mu}$ aligned with the rotation axis which is, again, in the plane of the disk. Then only $\Omega_0$ appears, i.e. the BF is replaced by the magnetospheric boundary Keplerian frequency. In the most general orientation for $n = 1$, one obtains the $\Omega_0 \pm \Omega_S$ and $\Omega_0$ frequencies at different powers, generally modulated by the free precession frequency. For larger n values more frequenices show up, among them $\Omega_S$, $2\Omega_S$ and $2\Omega_0$, and I find it interesting that in CygX-2 there is some evidence for $\Omega_0$ showing up (Hasinger 1987). In CygX-2 these things also correlate with hardness ratio (Hasinger 1986). The relation between the above mentioned cone angle and the hardness ratio is to date a complete unknown, but it seems quite worthwhile to look into free precession of QPOs with the Cyg X-2 observations in mind.

## 4. CONCLUSION

The QPO phenomenon has grown rapidly from a clean, simple one when only GX5-1 was known to a complex, puzzling one as other sources were discovered. Does it mean that there are many ways to produce QPOs - or that the geometries and environments change from source to source while the fundamental process remains the same? Scientific debate is always the hottest when very little is known. So, one should keep looking to present and future observations for guidance and also try to get a better theoretical handle on the magnetospheric boundary. Hopefully, the next IAU symposium on neutron stars will bring with it cooler temperatures and better answers.

This work was supported under NAGW-567 and NAG8-497. This is contribution number 320 of the Columbia Astrophysics Laboratory.

## 4. REFERENCES

Alpar, M.A. and Shaham, J. 1985, Nature, **316**, 239.
Alpar, M.A., Cheng, A.F., Ruderman, M.A., and Shaham, J. 1982, Nature, **300**, 728.
Chen, K.Y and Shaham, J. 1987, this volume.
Ghosh, P. and Lamb, F.K. 1979, Ap.J., **234**, 296.
Hasinger, G. 1987, this volume.
Lamb, F.K., Shibazaki, N., Alpar, M.A., and Shaham, J. 1985, Nature, **317**, 681.
Lewin, W.H.G. 1987, this volume.
Priedhorsky, W. 1986, Ap.J. (Letters), **306**, L97.
Priedhorsky, W., Hasinger, G., Lewin, W.H.G., Middleditch, J., Parmar, A., Stella, L., and White, N. 1986, Ap.J. (Lett.), **306**, L91.
Shaham, J. and Tavani, M. 1987, this volume.
Shaham, J. 1986, Ap.J. (in press).
Trumper, J., Kahabka, P., Ogelman, H., Pietsch, W. and Voges, W. 1985, preprint.
van der Klis, M. 1987, this volume.
van der Klis, M., Jansen, F., Van Paradijs, J., Lewin, W.H.G., Trumper, J. and Sztajno, M. 1985a, IAU circular 4043.
van der Klis, M., Jansen, F., Van Paradijs, J., Lewin, W.H.G., Trumper, J. and Sztajno, M. 1985b, Nature, **316**, 225.

## DISCUSSION

**R. Narayan:** It seems to me that a natural way to obtain blobs is through an instability in the accretion disk, in which case a modulated density might be more likely than random blobs. However, the azimuthal wave-number m of the instability could be greater than 1. Some of the formulae you showed would then change. How does this affect your estimates of magnetic fields, neutron star period and accretion radius?

**J. Shaham:** In the mathematical picture that I presented, only m=1 will contribute ($e^{\pm i\phi}$). Higher m values will only contribute when $m(\Omega_B\tau+\psi)$ is present in eq. (12). Then, of course, the real $\Omega_B = \Omega_{QPO}/m$. That will reduce $\Omega_s$ by m, amplify B by $m^{7/6}$ and multiply $R_0$ by $m^{2/3}$.

**J. Arons:** Comment 1: Every speaker attributes the "natural" frequency for QPO to the phenomenon with which he/she is most familiar. You call the frequencies of orbital mechanics/stellar rotation the most natural. Narayan, who has done lovely work on disk instability, suggests in a preceding question the instability of a disk is most "natural." In the interests of continuity, I suggest a third "natural" means of forming QPO. As I showed you on Tuesday, polar cap accretion is prone to form bubbling and boiling motions whose onset time scale is ~ msec; while the ultimate fate of these motions is still unknown, we think they lead to fluctuations in the flow on the order of 10s to 100s of msec. Therefore, in the interests of "natural" theories, let me suggest QPO reflect "convection" at the stellar surface in the accretion flow, rather than an orbital or a disk phenomenon. Comment 2: On a (slightly) more serious note, any magnetospheric model, blob frequency or not, requires a *small* magnetosphere, consistent with the commonly held thoughts about "decay" of magnetic field in old neutron stars. The only decay model in the literature that makes any sense is that of Flowers and Ruderman, recently given a first quantitative form by Eichler and Wang (this meeting). Here, the MHD motions of the core, constrained to crustal ohmic rates by the penetration of field lines through the crust, cause the dipole field to *disappear*, in favor of higher order multipoles (perhaps to stop at field complexity ~ octopole). Thus, one gets a small magnetosphere (needed for spin up) with stronger surface fields, and a complex pattern of accretion flow onto the surface rather than two simple polar caps (good for having a *broad* QPO line, since now many emission sites on the rotating star contribute). *But*, a field of this structure doesn't give $f_{QPO} \propto (\text{luminosity})^{3/7}$. For example, a quadrupole field gives a magnetopause radius $R_m \propto L^{-2/11}$ instead of the dipolar $R_m \propto L^{-2/7}$ (Arons & Lea, 1980, *Ap. J*), which in turn gives $f_{QPO} \propto L^{3/11}$; dominance by higher order structure makes the exponent still smaller. If the observers can show that the power really is 3/7 (modulo the bulk of the data which doesn't fit well anyway), I would regard that as evidence for the stars having been born with weak fields, with the currents in the star sufficiently deep down to give a persistent, dipole component which dominates even as close as several stellar radii. I doubt this is right, because we don't see pulses, and prefer the complex field model, but this is clearly a question which needs observational solution. Question: Do you refer to flux transfer events in your reference to geophysical evidence for mass entry when fields are opposed?

**J. Shaham:** Yes. Some observational evidence exists to suggest that flux entering the Earth's atmosphere is maximum at locations where fields oppose.

**F. Verbunt:** Does the beat frequency model also apply to pulsars with a strong magnetic field? If so, what frequencies would one expect?

**J. Shaham:** The question I have in my mind is what role does the region between $R_c$ and $R_0$ play in preserving the coherence of the matter flow and what role does it play in the transition region at $R_0$. In principle, large B's increase the scale of everything and less coherence may ensue. If one can still keep some of it, one might expect frequencies to scale as $B^{6/7}$ (Eq. 5).

**M. van der Klis:** How do you preserve the pattern of the "slightly modulated Keplerian flow" from the corotation radius down to the magnetospheric radius in the presence of the neutron star's magnetic field, the strength of which increases when you go down?

**J. Shaham:** When $R_c$ and $R_0$ are sufficiently close, the stellar field will not increase much. But the question is a relevant one even then, as I mentioned in my talk, because flow must be essentially non-interactive, and at present the theoretical situation is not clear even when blobs exist. As mentioned before, an embedded field will help.

**W. Brinkmann:** So far only power spectra were used in the data analysis. Generally you lose information in going from Fourier transforms to power spectra. Do you think there is any useful information in the amplitude-phase relations?

**J. Shaham:** There is importance in the phases of the power spectra. One way of utilizing them is described by Gunther Hasinger in his talk, namely, with cross correlation functions yielding information on time delays. Also, the degree of coherence of processes (blobbiness?) may be revealed.

# Quasi-Periodic Oscillations

## The Rapid Burster (MXB 1730-335)

## Models vs. Observations - a Brief Review

Walter H. G. Lewin
Massachusetts Institue of Technology
Cambridge, MA 02139, USA.

**Abstract.** The salient features of quasi-periodic oscillations (QPO) observed in type 2 bursts and in the persistent emission from the Rapid Burster are discussed. In addition, a brief review is given of the models that have recently been proposed to explain high-frequency QPO observed in several bright low-mass X-ray binaries. We do not yet know the mechanism(s) of the QPO, not even whether they are magnetospheric in origin. However, some of the proposed ideas could well be relevant to the various rather complex aspects of the QPO. It is likely that more than one mechanism is at work.

## 1. The Rapid Burster (MXB 1730-335)

### 1.1. Introduction

The Rapid Burster is a recurrent transient (probably a low mass X-ray binary, LMXB) located in a globular cluster. When the source is active, the accretion onto the neutron star can occur spasmodically resulting in type 2 bursts, or the accretion can be continuous, resulting in a persistent (though variable) X-ray flux. Type 2 bursts and persistent emission have been simultaneously observed. The type 2 bursts can last a few sec up to ~10 min. When they last longer than ~15 sec, in general, they have more or less "flat tops" (saturated flux) which can differ in their peak fluxes by factors up to 4. Therefore the saturated type 2 burst fluxes probably do not reflect an Eddington luminosity (except, perhaps for the highest bursts). For type 2 bursts, the waiting time to burst #n+1 is approximately linearly proportional to the integrated energy in burst #n; thus type 2 bursts behave like a relaxation oscillator. Type 1 bursts (presumably due to thermonuclear flashes on the surface of a neutron star) have also been observed from the Rapid Burster.

For references, see e.g., a review by Lewin and Joss 1983; Lewin et al. 1976; Lewin 1977; Ulmer et al. 1977; Hoffman, Marshall and Lewin 1978; White et al. 1978; Marshall et al. 1979; Van Paradijs, Cominsky and Lewin 1979; Basinska et al. 1980; Inoue et al. 1980; Tawara et al. 1982; Tanaka 1983; Pollard et al. 1983; Kunieda et al. 1984; Makino 1984; Lewin 1985; Tawara and De Yu Wang 1985; Lewin 1986.

The Rapid Burster is unique. There are other sources such as e.g., Cyg X-1, GX 301-2, (Hoffman, Marshall and Lewin 1978), and Cir X-1 (Tennant 1986) which show hiccups in their accretion, and, following Hoffman, Marshall and Lewin (1978), we call those type 2 bursts. However, none show the relaxation oscillator behavior as observed in the Rapid Burster. **The instability operating in the Rapid Burster is therefore probably different from those that produce type 2 bursts in other sources.**



*D. J. Helfand and J.-H. Huang (eds.), The Origin and Evolution of Neutron Stars, 363–374.*

Many long type 2 bursts (with duration of a few times $10^2$ sec) were detected by Tawara et al. (1982), using Hakucho. In 2 of 63 of these bursts, they discovered quasi-periodic oscillations (QPO) of ~2 Hz. The strength of the QPO (rms variation) was about 30%; in those bursts in which QPO were not observed, the upper limits were ~10%.

On August 28, 1985, EXOSAT observations were scheduled of 4U/MXB 1728-34 within ~0.5° of the Rapid Burster. Since the Rapid Burster was burst-active, the observations of 1728-34 were terminated, and the viewing direction of EXOSAT was changed for maximum possible exposure of the Rapid Burster, with 1728-34 just outside the field of view. Persistent emission and very long type 2 bursts (~3-12 min) were observed (no type 1 bursts were detected from the Rapid Burster). There was an interesting anti-correlation between burst duration and mean peak burst flux; the higher the burst flux, the shorter was its duration (I do not recall that this has previously been reported).

The observations were made in collaboration with Luigi Stella, Arvind Parmar, Nick White, and Jan van Paradijs (Stella et al. 1985). Luigi is not present at this meeting, and he has asked me to present some of our results. [In this paper, I will only describe some of the salient features; I will not present the figures that I did show at the meeting. A paper, presently in preparation, will contain a complete description and figures (Stella et al. 1986; see also the EXOSAT Calendar September 1986; Stella 1985,1986)]. I will first describe the QPO characteristics observed in the type 2 bursts, and then those observed in the persistent emission between these bursts.

### 1.2. QPO in the Type 2 Bursts

QPO were observed in many (not all) type 2 bursts. The **mean** centroid QPO frequency, $\nu$, in the bursts ranges from ~2-5 Hz; it is strongly anti-correlated with the **mean** peak burst flux, $I$, [$\nu \propto I^{(-0.9\pm0.1)}$]. During the type 2 bursts, the peak burst flux can vary (up to ~20%), and the centroid QPO frequency can also change. This change can be correlated, or anti-correlated, or not correlated at all with the variable peak burst flux.

**1.2.1. Strength of QPO and LFN.** The strength of the QPO is the highest (rms variation ~20%) when the **mean** peak burst flux, $I$, is relatively low (400 cts/sec, 1-15 keV); it decreases when $I$ increases. For values of $I$ between ~1000 and ~1200 cts/sec, the strength varies between ~3% and ~10%. Tawara et al. (1982) observed in 2 of 63 type 2 bursts a strength of ~30%.

The strength of the low-frequency noise (LFN) is relatively low (up to a few %). As an example, **when the QPO strength during a 12-min type 2 burst was ~21%, the strength of the LFN in the range 0.0039 Hz to 0.3 Hz was only ~1.4%.**

### 1.3. QPO in the Persistent Emission

In between the type 2 bursts, a relatively strong (~200 cts/sec; for comparison with type 2 bursts, see above), and variable "persistent" flux is observed. As previously observed with the SAS-3 Observatory, this persistent flux declines substantially just before burst onset, and just after a burst has occurred (Van Paradijs, Cominsky and Lewin, 1979; Marshall et al. 1979). QPO is, in general, detected in the persistent emission when the burst intervals are relatively short (~600-1000 sec); it is not observed when the burst intervals are in excess of ~1500 sec. The QPO centroid frequency between bursts typically changes from ~4 Hz after a burst, when the persistent emission has just come out of its few-minute decline, to ~2.5 Hz, just before the next decline which is then followed within ~1 min by another type 2 burst.

There are times, particularly when the persistent emission shows flare-like events, that the QPO centroid frequency is anti-correlated with the persistent flux; but no general relation exists. There is, however, a clear correlation between the spectral hardness of the persistent emission (~4-10 keV/~1-4 keV) and the QPO frequency; the higher the hardness ratio, the higher is the frequency.

**1.3.1. QPO Strength.** The strength of the QPO in the persistent emission can be as high as ~30%. This is higher than what we observed in the type 2 bursts but comparable to what Tawara et al. (1982) observed in 2 of 63 type 2 bursts (see above).

**1.3.2. Fundamental and First Harmonic.** During a particular ~4-min interval, starting ~6 min before the onset of a type 2 burst, and ~12 minute after the end of the previous burst, two QPO frequencies of ~0.44 Hz and ~0.88 Hz were simultaneously observed. Their strength (rms variation) was ~24% and ~14%, respectively. It is likely that they represent the fundamental and first harmonic of a quasi-periodic variability in the persistent flux.

### 1.4. Different Origins of QPO?

The complexity in QPO characteristics in the Rapid Burster is unlike that in any other source; the mechanism responsible for its QPO may be different. It is also possible that the simultaneously observed ~0.44 Hz and ~0.88 Hz QPO have a different origin than the ~2-5 Hz QPO observed in both the persistent emission and in the type 2 bursts.

## 2. Models vs Observations - a Brief Review.

### 2.1. QPO in Cataclysmic Variables

Quasi-Periodic Oscillations are commonly observed in the optical flux (in a few cases also in X rays) of dwarf novae in outburst (dwarf novae are accreting white dwarfs; for reviews see e.g., Robinson and Nather 1978; Patterson 1981; Cordova and Mason, 1983). The timescale of these oscillations ranges from ~10-$10^3$ sec (frequency ~1 mHz to 0.1 Hz) and the coherence ranges from a few oscillations up to ~$10^5$ oscillations. QPO with very different coherence can be observed simultaneously at two frequencies. The strength of the optical oscillations is typically less than 1%. Many models have been proposed to explain these oscillations. Not one alone can explain the whole range of complex phenomena; it is almost certain that more than one mechanism is at work.

It is not surprising that the recent models on high-frequency QPO observed in the bright LMXB reflect some ideas put forward earlier for optical QPO in the Cataclysmic Variables. In scaling the geometry of an accreting white dwarf to that of an accreting neutron star, it is not too difficult to imagine a frequency increase by a factor ~$10^{2-3}$.

### 2.2. Long-Period QPO in LMXB

Before I discuss some of the current models for the high-frequency QPO in the bright LMXB, I want to mention that long-period QPO of ~10-$10^3$ sec have also been observed in several X-ray binaries. The QPO spectrum can be softer as wel as harder than the mean source spectrum. The fraction of the modulations in the flux (thus the strength of the QPO) can be enormous (typically ~50%). I list here only those cases that I am aware of in sequence of increasing frequency from ~1 mHz to ~2 Hz (1626-67 Joss, Avni and Rappaport 1978; Li et al. 1980; Cyg X-3 Van der Klis and Jansen 1985; GX 349+2 Matsuoka 1985; Her X-1 Voges et al. 1985; 1820-30 Stella, Kahn and Grindlay 1984;

GX 339-4 Maejima et al. 1984). Perhaps the 1.5-h oscillations observed by Langmeier et al. (1985) in GX 17+2 are also quasi-periodic; this would then extend the range of long-period QPO in LMXB up to periods of ~$5 \times 10^3$ sec. I suspect that these long-period QPO have a different origin than the short-period QPO. However, it is unclear as yet where the dividing line is (perhaps somewhere between 0.01 and 1 Hz).

### 2.3. QPO Models

The models that I will now discuss have been proposed to explain some of the recent observations of high-frequency QPO in LMXB (Alpar and Shaham 1985a,b; Berman and Stollman 1985; Lamb et al. 1985; Lamb 1986; Hameury, King and Lasota 1986; Boyle, Fabian and Guilbert 1986; Morfill and Truemper 1986a,b). It is interesting to look at the evolution of some of them in historical perspective. The excitement, and activity started with the discovery of the intensity dependent QPO in GX 5-1 in early 1985 (Van der Klis et al. 1985a,b). The centroid frequency of the QPO, $\nu$, was linearly related to the observed source intensity, $I$, ($\nu \simeq 1.9 \times 10^{-2} I - 25$ Hz) over the observed range of $I$ from about 2400-3400 cts/sec (1-18 keV). No one paid much attention to this linear relation then, and not now, and even though we mention the linearity in our paper, we do not give it quantitatively (Van der Klis et al. 1985b). Our data can also be fit by a power-law dependence ($\nu \simeq 6.9 \times 10^{-6} I^{1.9}$ Hz). The observed exponent (which I will designate $\alpha$) led Alpar and Shaham to the idea that QPO could be the result of a beat phenomenon, as earlier suggested by Warner (1983) for optical QPO in cataclysmic variables. Another striking relation was present between the strength of the QPO and that of the low-frequency noise (LFN); the two went "hand in hand" (Van der Klis et al. 1985b).

**2.3.1. The Bath Model.** Before I expand on the beat frequency idea, I want to remind you of a model suggested 13 years ago by Geoffrey Bath to explain optical QPO in cataclysmic variables (Bath 1973). He suggested that the QPO frequency was that of the Kepler frequency of matter orbiting a magnetized white dwarf at the magnetopause (inner edge of the accretion disk). This idea can be adopted to magnetized accreting neutron stars in an effort to explain the observed high-frequency QPO in X rays; I will refer to this hereafter as "the Bath model". The radius of the inner edge of the accretion disk, r, depends on the mass and radius of the neutron star, on the magnetic dipole field strength at the surface of the neutron star, and on the mass transfer rate $\dot{M}$ at that radius (see Lamb, Pethick and Pines 1973). Combining the relation between r and $\dot{M}$ with Kepler's law, and assuming that the Kepler frequency, $\nu_K$, equals the QPO frequency, $\nu$, (Bath model) one can easily find that

$$\nu \propto \dot{M}^{3/7} \quad (1).$$

If we now make the assumption that $\dot{M}$ depends linearly on the observed broad-band X-ray intensity, $I$, we find that the observed QPO frequency, $\nu$, should be proportional to $I$ to the power 3/7, thus:

$$\nu \propto I^{3/7} \quad (2).$$

This, however, was not observed; $\alpha$ was ~2, and not 3/7 (Van der Klis et al. 1985a,b).

**2.3.2. The Beat Frequency Model .** In the beat frequency model (Alpar and Shaham 1985a,b), the observed QPO frequency is the difference between the Kepler frequency, $\nu_K$, and the rotation frequency, $\nu_n$, of the neutron star

$$\nu = \nu_K - \nu_n \quad (3).$$

If again the assumption is made that the mass transfer rate, $\dot{M}$, at the magnetopause depends linearly on the observed broad-band X-ray intensity, one finds that

$$\alpha = 3\nu_K / 7\nu \quad (4).$$

Combining eqs. 3 and 4 leads to:

$$\nu_n = \nu(7\alpha/3 - 1) \quad (5).$$

If, as an example, we take an approximate average value of 30 Hz for the observed QPO frequency, $\nu$, in GX 5-1, then, with $\alpha \approx 2$ (as observed), $\nu_K \approx 140$ Hz (eq. 4), and $\nu_n \approx 110$ Hz (eq. 5). Assuming a mass for the neutron star of 1.4 $M_\odot$, with Kepler's law we can then also find the radius of the inner edge of the accretion disk (here $\approx$55 km). If, in addition, one assumes a radius for the neutron star (e.g., 10 km), and one estimates the mass transfer rate at the magnetopause (this can be done from an estimate of the distance to the source, and from the observed broad-band X-ray intensity), one also finds the magnetic dipole field strength, $B$, at the surface of the neutron star, using the equations given by Lamb, Pethick and Pines (1973). For GX 5-1, one finds $B \approx 6 \times 10^9$ G.

As we just saw, in the case of GX 5-1, the beat frequency model (Alpar and Shaham 1985a,b), leads to the neutron star rotation period, and to its magnetic dipole field strength. These results were very pleasing as they fitted well into an existing evolutionary scenario in which the neutron stars in the bright LMXB are spun up by accretion and form the progenitors of msec binary radio pulsars (see e.g., Smarr and Blandford 1976; Van den Heuvel 1981; Radhakrishnan 1981; Srinivasan and Van den Heuvel 1982; Alpar et al. 1982; Radhakrishnan and Srinivasan 1982; Webbink, Rappaport and Savonije 1983; Taam 1983; Joss and Rappaport 1983; Paczynski 1983; Savonije 1983).

**Problems?** If the beat frequency idea is correct, it is puzzling that we do not observe in our GX 5-1 data ~110 Hz **coherent** X-ray pulsations (however, see Lamb et al. 1985, and Lamb 1986). The absence of coherent pulsations may not be the only problem. A few months after QPO were found in GX 5-1, they were also discovered in Sco X-1 (Middleditch and Priedhorsky 1985, 1986; see also Priedhorsky et al. 1986; Van der Klis et al. 1986). Here the observed QPO frequency was **anti-correlated** with the observed source intensity ($\nu \propto I^{-0.6}$). Thus, $\alpha$ was negative. As long as we make the above assumption that $\dot{M}$ is linearly proportional to the observed X-ray intensity, $\alpha$ can not be negative. This can easily be seen as follows. Any increase in $I$ will be associated with an increase in $\dot{M}$. This leads to a smaller radius of the magnetopause, and thus to an increase in the Kepler frequency. If the Kepler frequency goes up, the QPO frequency will also go up (see eq. 3).

**Introduction of the $\beta$ parameter.** Lamb et al. (1985) do not assume that the observed X-ray intensity is linearly proportional to $\dot{M}$, but rather

$$I \propto \dot{M}^{\beta} \quad (6).$$

Here $\beta$ can have all values between $-\infty$ and $+\infty$, as can easily be seen. Suppose there is no change in $\dot{M}$, but there is, e.g., a minute decrease in the observed X-ray intensity as the result of a small increase in absorption along the line of sight; $\beta$ would then be $-\infty$. If the absorption were to decrease, $\beta$ would become $+\infty$; $\beta$ can have all values in between. For "n-fold symmetry" (Lamb et al. 1985; see also below) eq. 3 becomes

$$\nu = n(\nu_K - \nu_n) \quad [n=1,2,3,......] \quad (7).$$

With the introduction of $\beta$, we find for the beat frequency model

$$\alpha = 3n\nu_K/7\beta\nu \quad (8).$$

For n=1, and $\nu_K = \nu$ (Bath model), we find

$$\alpha\beta = 3/7 \quad (9).$$

Combining eqs. 7 and 8, the beat frequency model leads to

$$n\nu_n = \nu(7\alpha\beta/3 - 1] \quad (10).$$

The rotation frequency of the neutron star plays no role in the Bath model. The latter is mathematically equivalent to the beat frequency model with $n\nu_n=0$, and $\alpha\beta=3/7$ (eq. 10).

**With the introduction of the β parameter (with β≠1) the beat frequency model can obtain any value for the rotation frequency of the neutron star.** This can best be illustrated with an example. Priedhorsky et al. (1986) found in Sco X-1 a "6-Hz QPO branch" which was anti-correlated with the source intensity ($\alpha\simeq-0.6$), and a "15-20 Hz branch" which was correlated with the source intensity ($\alpha\simeq+3$) (see also Van der Klis et al. 1986). Their results are shown in the figure below. With the observed values for $\nu$ and $\alpha$ (using eq. 10), we can now find values for β to obtain any neutron star rotation rate. For instance for a rotation frequency $n\nu_n=1000$ Hz, β would have to be ~+9 in the 15-20 Hz branch, and ~−120 in the 6-Hz branch. For a rotation rate ($n\nu_n$) of 100 Hz, the corresponding values for β would have to be ~+1, and ~−13, respectively. The rotation rate can also be zero (this is mathematically equivalent to the Bath model). Then, the values for β would have to be ~+0.14, and ~−0.7, respectively (here $\alpha\beta=3/7$). Near the apex of the "curve" (see figure below), $\alpha=\pm\infty$, (observational fact). If at the same time $\beta=0$ (i.e., the mass transfer rate changes, but no simultaneous source intensity change is observed), the product $\alpha\beta$ becomes undetermined, and this is consistent with any neutron star rotation frequency.

In this context it is interesting to mention that Priedhorsky et al. 1986 (see also Priedhorsky 1986) suggest that in "tracing" the curve of the figure below in a clockwise direction, there is a continuous monotic increase in the mass transfer rate which, however, is not reflected in a continuous increase in the **observed** broad-band X-ray intensity. With increasing mass transfer rate, the **observed** intensity at first decreases (6-Hz branch), and then increases (15-20 Hz branch). If this suggestion is correct, it would mean that near the apex, where the two branches meet, $\beta=0$.

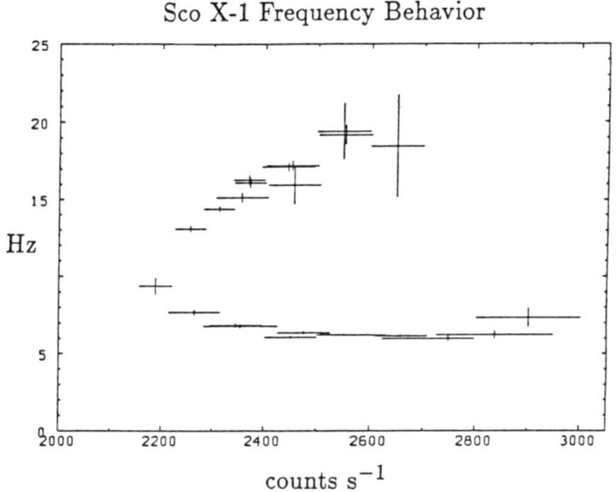


Centroid QPO frequency, $\nu$ (in Hz), vs. the 8-20 keV X-ray intensity, $I$ (in cts/sec), observed for Sco X-1 with EXOSAT. In the upper QPO branch (15-20 Hz) the frequency and source intensity are correlated; $\nu\propto I^{+3}$. In the lower branch (~6 Hz), the QPO frequency is anti-correlated with $I$, here $\nu\propto I^{-0.6}$. This figure is from Priedhorsky et al. 1986.

In my opinion, the main reason for the early rejection of the Bath model in favor of the beat frequency model is no longer valid. With the introduction of β (≠1), which seems entirely reasonable, the product $\alpha\beta$ counts, and no longer $\alpha$ alone. In the Bath model $\alpha\beta=3/7$, and any observed value of $\alpha$ is allowed (thus also $\alpha\simeq2$ as observed for

GX 5-1, as long as $\beta \approx 0.21$). In the absence of knowledge of $\beta$ ($\beta$ can vary on a timescale of minutes), we are at a loss (however, see Priedhorsky 1986). **Obviously, this does not mean that the beat frequency model is incorrect.** It is too early to decide (see also Hasinger, 1987, and the note added at the end of my text).

A specific beat frequency scenario has been proposed by Lamb, Shibazaki, Alpar and Shaham (1985) (see also Van der Klis et al. 1985b; Berman and Stollman 1985; Lamb 1986). They suggest that the beat phenomenon is the result of blob formation at the inner edge of the accretion disk. The matter in the blobs is gradually stripped off by interaction with the magnetic field which is anchored into a spinning neutron star. The matter will then reach the surface of the neutron star in a quasi-periodic fashion with a frequency equal to the difference between the (variable) Kepler frequency of the blobs and the (fixed) rotation frequency of the neutron star (see eq. 3) [or n times that for n-fold symmetry (see eq. 7)]. While a blob is "milked" the X-ray flux oscillates (producing QPO), and the mean X-ray flux is temporarily increased. The latter gives rise to low-frequency noise (LFN). The model can be represented by oscillating "shots". The oscillatory part of the shots determines the QPO characteristics; the envelopes of the shots determine the characteristics of the LFN. The blobs are produced at random times, and with a random equatorial azimuthal distribution. For a sufficiently large number of blobs, the "crests" of the oscillations, produced by one blob, will always coincide (or nearly so) with the "valleys" of those produced by others, and both the QPO and the associated LFN will disappear.

In this scenario, LFN is **"a logical consequence"** of the QPO, and one would expect to observe strong LFN whenever strong QPO are observed. This was the case in GX 5-1 where the QPO and the LFN strength went "hand in hand". However, in later observations of e.g., Sco X-1 and the Rapid Burster this was not the case (sect. 1.2; Stella 1985,1986; Stella et al. 1986; Van der Klis et al. 1986). Fred Lamb kindly pointed out to me (see also Lamb 1986, and Shaham 1987) that one can adjust the mathematics of the shots so that the blobs oscillate more or less in unison (crests from one blob support crests from other blobs). This could give strong QPO in the near abscence of LFN. It remains to be seen, however, whether this mathematical adjustment is physically meaningful.

**2.3.3. Other Models.** I will now discuss some other models. They will not get as much attention as the model proposed by Alpar and Shaham (1985a,b). This does not mean that I favor their model. It merely reflects that it has been around longer and evolved (Lamb et al. 1985; Lamb 1986) as the observations revealed more complex behavior and richness in the QPO, and LFN. I thought it was worthwhile to discuss some aspects of this evolution.

**Morfill & Truemper** (1986a) suggested a magnetospheric beat frequency model in a heavily obscured binary. The supersonic stirring of the disk by an inclined magnetic field causes shock waves which interact with disk inhomogeneities (plasmoids) in Keplerian rotation. The plasmoids are hotter than their environtment. This results in a quasi-periodic signal at the beat frequency (eq. 3). Unlike in the beat frequency model discussed above, here the QPO signal does not come from the surface of the neutron star but from the plasmoids in the magnetopause (self-luminous plasmoid model). Thus the available energy in the QPO is dictated by the radius of the magnetopause. If the latter were e.g. ten times that of the neutron star, the maximum available energy (to produce QPO) would be ~10% of the total X-ray flux (for larger radii it would be lower, for smaller radii larger). The QPO mechanism could not be 100% efficient (see Lamb 1986). "Self luminous" models like this one, can therefore perhaps explain QPO with a modest strength, but it is hard to see how they could explain the high percentages observed in e.g., GX 5-1 (up to ~6%), Sco X-1 (up to ~8% at high energy X rays), and the Rapid Burster (up to ~30% ). [For references see above].

In a later version, Morfill and Truemper (1986b) point out that one may expect the formation of large vortices in the disk flow (near the magnetopause), and that vortex shedding (perhaps associated with a characteristic Strouhal frequency) is likely to occur. They show that under certain conditions, and with a constant Strouhal number of ~1.3, the QPO frequency could be about twice the Keplerian frequency (i.e., ~$2\nu_K$). They also mention that other phenomena such as frequency halving or doubling may add further complexity. They point out that the observed QPO frequencies could perhaps also be due to the single Keplerian frequency of the plasmoids and/or to a beat frequency.

These are all versions of "self-luminous" models, and my comments regarding the limited available energy for the QPO should also be valid here. The authors, however, allow for the possibilty that the QPO are the result of shadowing of the central source by hot cocoon gas, in which case the limitation in available energy is not a problem (see sect. 2.4).

In their closing comments, Morfill and Truemper (1986b) also suggest that the interaction of the plasmoids with the shocks (the latter are revolving with the neutron star rotation frequency) may bring accretion disk material into co-rotation, and thus making accretion along the field lines onto the polar caps feasible. If that happens, the QPO emission comes from the neutron star surface with a beat frequency (see eqs. 3 and 7); this scenario has similarities with that proposed by Lamb et al. (1985) (see also Lamb 1986).

**Hameury, King and Lasota** (1985) proposed a scenario which does not require the presence of a magnetopause. The QPO are the result of hot spots rotating in the boundary layer where the accreting matter settles onto the surface of a "slowly" rotating (with a period of order ~1 sec) neutron star. They suggest that transient magnetic fields are generated due to turbulent dynamo action, and that the local hot spots are heated by conduction due to these magnetic fields. QPO are then the result of repeated occultations of the hot spots when they rotate out of our line of sight to the "back side" of the neutron star. The low-frequency noise results from the lifetime of the hot spots, and from those spots that are never occulted due to their high lattitude. The authors predict that the QPO spectrum is that of a blackbody with a radius less than that of the neutron star. There is some growing evidence that the QPO spectra are indeed blackbodies for Cyg X-2 (Hasinger, 1987), and Sco X-1 (Van der Klis et al. 1986). A blackbody spectrum, however, is not unique to this model.

The "available" frequencies in this model range all the way from ~1 Hz (at the bottom of the boundary layer, rotating with the neutron star frequency) to ~$10^3$ Hz (at the very top of the boundary layer rotating with the Keplerian frequency). If this scenario were operating, it is puzzling why the observed QPO frequencies vary only in such a limited range (e.g., in GX 5-1 by a factor of ~2), and why the width of the peaks in the power spectra of most QPO sources are so narrow (typically ~10% to ~30%). Since here the LFN is a logical consequence of the QPO, it may also be difficult to explain those cases of observed near absence of LFN in the presence of strong QPO (see sect. 1.2, and the end of sect. 2.3.2). It is also unclear whether the hot spots could contain such a high fraction of the available gravitational potential energy to explain the observed high strength of QPO (typically ~5% rms variation, but up to ~30% in the Rapid Burster, see sects. 1.2 and 1.3).

A particular hot spot would have to appear and disappear (occultation) at least three or four times to produce the observed QPO. The QPO with a fundamental frequency of ~0.44 Hz in the Rapid Burster (sect. 1.3.2) would thus require a lifetime of the hot spots of at least ~10 sec. According to Jean Paul Lasota (private communication), the lifetime of a hot spot is probably less than a few seconds. Thus, if this is so, the very low-frequency QPO could not be explained with this "sun spot" scenario. It seems quite possible that more than one mechanism is responsible for the various "kinds" and different characteristics of the QPO (see also sect. 2.2, and Hasinger 1987), and it would be somewhat naive to expect that one model could explain them all.

**Boyle, Fabian and Guilbert** (1986) proposed that the QPO are produced by a hot (>$10^7$ °K) accretion disk corona. X rays produced at the central source (the neutron star) scatter off oscillating disturbances in this corona near the inner disk. The oscillations are in a direction perpendicular to the disk plane (disk oscillations); they have a frequency approximately equal to the local Keplerian frequency. Thus, QPO frequencies of ~100 Hz and ~1 Hz, would correspond to an effective scatter radius of ~$8\times10^4$ m (~8 stellar radii) and ~$1.6\times10^6$ m (160 stellar radii), respectively. No magnetic field, and no neutron star rotation are required.

The X rays scatter off the $\tau\simeq1$ surface. Since the observed peaks (QPO) in the power spectra have widths of typically ~10 to ~20%, a very restricted range of radii of this scatter surface is required. The spread in the radii of the effectively contributing part of this surface (as seen from Earth) should be no larger than ~30%. Thus this scatter surface must be very steep and well localized. It is not sufficient that the disk oscillations cause this surface to move closer to, or farther away from, the neutron star by ~30%, as this would not modulate the scattered X rays sufficiently to produce the observed oscillations. The scatter surface should more or less "come and go" (with the frequency of the disk oscillations) in order to obtain a strong modulation in the scattered X rays. It should act as a "wall" that rises and falls at the required frequency. The authors predict that those systems seen at low inclinations are favored (no QPO should be seen from eclipsed systems).

The fraction of X rays that scatter off the accretion disk coronae in 1820-37 and 2129+47 is no more than ~10% (Keith Mason and Nick White, private communication). If this number is typical for LMXB, it is very hard to see how the disk oscillations could produce the high strength of QPO (typically ~5% rms variation, but up to ~30% in the Rapid Burster, sects. 1.2 and 1.3). If the modulation efficiency (of the scattered X rays) were as high as ~10%, one would expect to observe a modulation in the total X-ray signal (scattered and non-scattered) of only ~1%. If the central source were obscured, the situation would be different, and the observed percentage of modulation would be much higher since then only the scattered X rays are seen. This, however, is not the case for most (if not all) sources in which QPO have been detected (the sources are very bright).

The authors appreciate the problems and suggest that non-linear effects might help to localize the $\tau\simeq1$ surface. Such non-linear effects are unexplored. The physics is not understood, and so far it appears to be only an interesting mathematical exercise until more theoretical work is done.

### 2.4. Occultation Models

Some of the ideas as put forward by Morfill and Truemper (1986a,b) may work if the plasmoids are not self luminous (see above) but occult the central source. Similarly, the disk oscillations (Boyle, Fabian and Guilbert 1986) could produce a high degree of modulation if the neutron star is obscured by the oscillating disturbances. Van der Klis et al. (1986) (see also Van der Klis 1987) show that certain occultation models can explain the absence of LFN in the presence of strong QPO. Occultation models do not suffer from a lack of available energy for the QPO. However, they probably require that the systems that exhibit QPO are seen at a rather large inclination, and it is puzzling why that would be the case for so many bright low-mass X-ray binaries.

## 3. Concluding Remarks

**As of today (May 29, 1986) we do not yet know what causes the QPO, not even whether they are magnetospheric in origin** (see note added below). It appears that none of the proposed scenarios can alone explain the richness and complexity in the high-frequency quasi-periodic X-ray oscillations now observed in about eight bright low-mass X-Ray binaries (above references; Stella 1985; Lewin and Van Paradijs 1986; Van der Klis 1986; and references therein). However, some of the proposed ideas could well be relevant to some aspects of the QPO. In any case it is likely that more than one mechanism is at work.

In the light of this ignorance, I would like to finish with a humorous quote from Harlow Shapley which was brought to my attention by Ed Chupp (1984).

> A hypothesis or theory is clear, decisive and positive but it is believed by no one but the man who created it. Experimental findings, on the other hand, are messy, inexact things which are believed by everyone except the man who did the work.
>
> Harlow Shapley

### Acknowledgments

During the past 18 months I have had numerous dicussions with many colleagues and friends about QPO. Their contributions have become an essential and integral part of my thinking. I particularly want to thank Jan van Paradijs, Michiel van der Klis, Guenther Hasinger, Luigi Stella, Ed van den Heuvel and Fred Lamb. I am grateful for a generous award from the Alexander von Humboldt Stiftung, and for a John Simon Guggenheim fellowship. My work was also supported by NASA grant NAG8-571.

### Note added

Guenther Hasinger (1987) has shown at this meeting that in the case of Cyg X-2 the rapid variabilty in high-energy photons lags that of the low-energy photons by several msec; the time lag is smaller when the QPO frequency is larger. He suggests that Comptonization is responsible for the time lag. For a delay of ~3 msec, and a Compton optical depth of ~5, the size of the scattering cloud would be ~170 km. At that radius the Kepler frequency of matter orbiting the neutron star is ~20 Hz (as observed in Cyg X-2). When the Compton scattering cloud moves closer to the neutron star the QPO frequency would become larger, and the time lag smaller. Guenther's new, and very interesting findings are in general support of the Bath model which explains the QPO frequencies in terms of the Keplerian frequencies of orbiting matter at the magnetopause; this model provides no information on the rotation frequency of the neutron star (see text).

# References


Alpar, M.A., Cheng, A.F. et al., Nature, **300**, 729, 1982.

Alpar, M.A. and Shaham, J., IAU Circ. No. 4046, 1985a.

Alpar, M.A. and Shaham, J., Nature, **316**, 239, 1985b.

Basinska, E.M., Lewin, W.H.G., Cominsky, L., et al., Ap. J., **241**, 787, 1980.

Bath, G.J., Nature Phys. Sci. **246**, 84, 1973.

Berman, N. and Stollman, G., Astron. Astrop. **154**, L23, 1985.

Boyle, C.B., Fabian, A.C. and Guilbert, P.W., Nature, **319**, 648, 1986.

Chupp, E.L., Ann. Rev. Astron. Astrophys., **22**, 359, 1984.

Hameury, J.M., King, A.R. and Lasota, J.P., Nature, **317**, 597, 1985.

Hasinger, G., Talk presented at the IAU Symp. No. 125 on "The Origin and Evolution of Neutron Stars", Nanjing, China, May 1987, this volume.

Hoffman, J., Marshall, H., and Lewin, W.H.G., Nature, **271**, 630, 1978.

Inoue H., Koyama, K., Makishima, K., Matsuoka, M., Murakami, T., et al., Nature, **283**, 358, 1980.

Joss, P.C., Avni, Y. and Rappaport R., Ap. J., **221**, 645, 1978.

Joss, P.C. and Rappaport, S.A., Nature, **304**, 419, 1983.

Kunieda, H., Tawara, Y., Hayakawa, S. and Nagase, F., Publ. Astr. Soc. Japan, **36**, 807, 1984.

Lamb, F.K., in: "The Evolution of Galactic X-Ray Binaries", eds. J. Truemper, W.H.G. Lewin and W. Brinkmann, NATO ASI Series C: Math. & Phys. Sci., **167**, 151, 1986 (Reidel, Dordrecht).

Lamb, F.K., Pethick, C.J. and Pines, D., Ap. J., **184**, 271, 1973.

Lamb, F.K., Shibazaki, N., Shaham, J. and Alpar, M.A., Nature, **317**, 681, 1985.

Langmeier, A. et al., in: "The Evolution of Galactic X-Ray Binaries", eds. J. Truemper, W.H.G. Lewin and W. Brinkmann, NATO ASI Series C: Math. & Phys. Sci., **167**, 253, 1986 (Reidel, Dordrecht).

Lewin, W.H.G., Ann. N.Y. Acad. Sci., **302**, 210, 1977.

Lewin, W.H.G., Proceedings of the Japan-US Seminar on "Galactic and Extragalactic Compact X-Ray Sources", eds. Y. Tanaka, and W.H.G. Lewin, ISAS, Tokyo, 1985.

Lewin, W.H.G., in: "Cosmic Radiation in Contemporary Astrophysics", ed. M.M. Shapiro, NATO ASI Series (Reidel Dordrecht) 1986.

Lewin, W.H.G., Doty, J., Clark, G.W., Rappaport, S.A., et al., Ap. J. Lett. **207**, L95, 1976.

Lewin, W.H.G. and Joss, P.C., in: "Accretion Driven Stellar X-Ray Sources", eds. W.H.G. Lewin and E.P.J. v.d. Heuvel (Cambridge University Press) page 41, 1983.

Lewin, W.H.G. and Van Paradijs J., Comments Astrophys., **11**, No. **3**, 127, 1986.

Li, F.K., Joss, P.C., McClintock, J.E., et al., Ap. J., **240**, 628, 1980.

Maejima, Y., Makishima, K., Matsuoka, M., Ogawara, Y., Oda, M., et al., Ap J. **285**, 714, 1984.

Makino, F., IAU Circ. No. 3957, 1984.

Marshall, H.L., Ulmer, M.P., Hoffman, J.A., Doty, J., and Lewin, W.H.G., Ap. J. **227**, 555, 1979.

Matsuoka, M., in: "Cataclysmic Variables and Low-Mass X-Ray Binaries", eds. D.Q. Lamb, and J. Patterson (Reidel, Dordrecht), page 139, 1985.

Middleditch, J. and Priedhorsky, J., IAU Circ. No. 4060, 1985.

Middleditch, J. and Priedhorsky, J., Ap. J., **306**, 230, 1986.

Morfill, G.E. and Truemper, J., in: "The Evolution of Galactic X-Ray Binaries", eds. J. Truemper, W.H.G. Lewin and W. Brinkmann, NATO ASI Series C: Math. & Phys. Sci., **167**, 173, 1986a (Reidel, Dordrecht).

Morfill, G.E. and Truemper, J., submitted to Nature, 1986b.

Paczynski, B., Nature, **304**, 421, 1983.

Patterson, J., Ap. J. Suppl., **45**, 517, 1981.

Pollard, G., White, N., Barr, P. and Stella, L., IAU Circ. No. 3854, 1983.

Priedhorsky, W., Ap. J. (Lett), **306**, L97, 1986.

Priedhorsky, W., Hasinger, G., Lewin, W.H.G., Middleditch, J., Parmar, A., Stella, L. and White, N.E., Ap. J. (Lett), **306**, L91, 1986.

Radhakrishnan, V., IAU Asian-Pacific Regional Meeting, Bandung, ed. B. Hydyat, 1981.

Radhakrishnan, V. and Srinivasan, G.M., Current Sci., **51**, 1096, 1982.

Robinson, E.L. and Nather, R.E., Ap J. Suppl. **39**, 461, 1979.

Savonije, G.J., Nature, **304**, 422, 1983.

Shaham, J., this volume, 1987.

Smarr, L.L. and Blandford, R.D., Ap. J., **207**, 574, 1976.

Srinivasan, G. and Van den Heuvel, E.P.J., Astron. Astrop., **108**, 143, 1982.

Stella, L., Proceedings of the "Mediterranean School on Plasma Astrophysics", Colymbari, Crete, Greece, Sept-Oct 1985.

Stella, L., Talk presented at the ESA conference on "The Physics of Accretion onto Compact Objects", Tenerife, Spain, April, 1986.

Stella, L., Kahn, S.M. and Grindlay, J.E., Ap. J., **282**, 713, 1984.

Stella, L., Haberl, F., Parmar, A., White, N., Lewin, W., and Van Paradijs, J., to be submitted to Ap. J. 1986.

Stella L., Parmar, A., White, N.E., Lewin, W.H.G., and Van Paradijs, J., IAU Circ. No. 4110, 1985.

Taam, R.E., Ap. J., **270**, 694, 1983.

Tanaka, Y., IAU Circ. No. 3852, 1983.

Tawara, Y., Hayakawa, S., Kunieda, H., Makino, F., and Nagase, F., Nature, **299**, 38, 1982.

Tawara, Y. and De Yu Wang, Proceedings of the Japan-US Seminar on "Galactic and Extragalactic Compact X-Ray Sources", eds. Y. Tanaka, and W.H.G. Lewin, ISAS, Tokyo, 1985.

Tennant, A., talk presented at the ESA meeting on "The Physics of Accretion onto Compact Objects", Tenerife, Spain, April, 1986.

Ulmer, M.P., Lewin, W.H.G., Hoffman, J.A., et al., Ap. J. (Lett) **214**, L11, 1977.

Van den Heuvel, E.P.J., IAU Symp No. 95, eds. W. Sieber and R. Wielebinsky, page 379, 1981 (Reidel).

Van der Klis, M., This volume, and Talk presented at the ESA conference on "The Physics of Accretion onto Compact Objects", Tenerife, Spain, April, 1986.

Van der Klis, M., and Jansen, F.A., Nature, **313**, 768, 1985.

Van der Klis, M., Jansen, F., Van Paradijs, J., Lewin, W.H.G., Truemper, J. and Sztajno, M., IAU Circ. No. 4043, 1985a.

Van der Klis, M., Jansen, F., Van Paradijs, J., Lewin, W.H.G., Truemper, J., Van den Heuvel, E.P.J. and Sztajno, M., Nature, **316**, 225, 1985b.

Van der Klis, M., Stella, L., White, N., Jansen, F. and Parmar, A.N., submitted to Ap. J., 1986.

Van Paradijs, J., Cominsky, L. and Lewin, W.H.G., MNRAS, **189**, 387, 1979.

Warner, B., in: "Cataclysmic Variables and Related Objects", eds. M. Livio and G. Shaviv, page 155, 1983 (Reidel, Dordrecht).

Webbink, R.F., Rappaport, S.A. and Savonije, G.J., Ap. J., **270**, 678, 1983.

White N.E., Mason, K.O., Carpenter, G.F., and Skinner, G.K., MNRAS, **184**, 1P, 1978.

# ON THE LONG TERM STABILITY OF NEUTRON STAR MAGNETIC FIELDS

David Eichler[1,2] and Zhengzhi Wang[1]
1. University of Maryland, College Park, Md.
2. Ben Gurion University, Beer Sheva, Israel

It is known that fluid object that it magnetized with a purely poloidal field is unstable to perturbations that overturn the field lines in one hemisphere relative to those in the other. We have shown that this instability occurs even when the entire object is encased by a solid crust, as conjectured by Flowers and Ruderman (FR) nearly a decade ago. We have derived the exact dispersion relation that displays this instability for an infinite, cylindrical geometry, which will be published elsewhere. The timescale for the instability is roughly the ohmic dissipation timescale of the crust.

We suggest that accreting neutron stars in the galactic bulge, which mostly display no periodicity, are distinguished from those in the disk, which mostly do, in that they are old enough for the instability to have occurred in them. Because the instability is likely to destroy almost all of the magnetic dipole moment of the neutron star, the older population of accreting neutron stars is not expected to funnel the accretion flow onto polar caps, and the periodic component of emission sources. Similarly, the death of pulsars after $\sim 10^7$ yr., which is implied by pulsar demographics, could also be accounted for by this instability, as originally suggested by FR.

Invoking field line overturning, rather than decay of a purely crustal field, allows old neutron stars to have magnetospheres, which have been invoked to explain rapid bursting and quasi-periodic oscillations, and which are implied by cyclotron features in gamma ray bursts. The recent dating of the millisecond pulsar via its white dwarf companion establishes that old neutron stars can have surface fields of order $10^9$ gauss, which is enough to funnel Eddington accretion into polar caps if it is in purely dipolar form.



D. J. Helfand and J.-H. Huang (eds.), The Origin and Evolution of Neutron Stars, 375.

ON THE EVOLUTION OF MAGNETIC INCLINATION WITH AGE

L.S. Li
Department of physics, Northeast Normal University
Changchun, China

ABSTRACT. In this paper, the author studied the evolution of magnetic inclination with age for several pulsars. According to the results obtained by the author ( 1981 ), the formula for this evolution is given below

$$\sin\alpha = \sin\alpha_o \exp[-(1-e^{-\xi t})\frac{\dot{P}_o}{\xi P_o}\cot^2\alpha_o]$$

The evolutionary curves of magnetic inclination, calculated from this formula, for three pulsars — PSR 0525+21, PSR 0329+54 and PSR 1133+16 — are presented in Fig. 1.

It can be seen that the magnetic inclination for these three pulsars are all gradually decreasing with time. The older the pulsar is, the smaller the magnetic inclination will be.

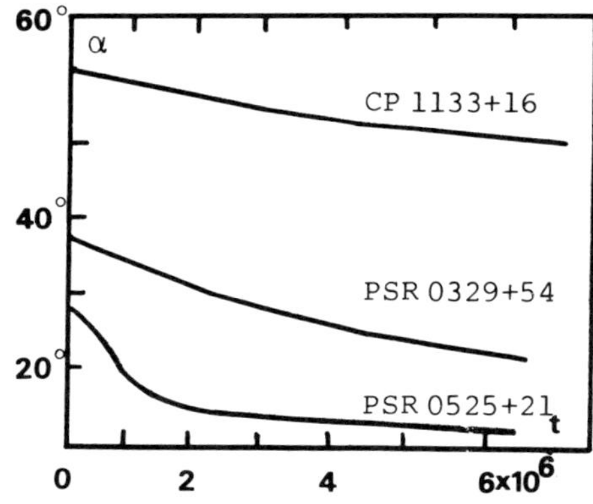


Fig. 1. The evolutionary curves of magnetic inclination for several pulsars

Reference

Li, L.S.: 1981, J.Northeast Normal Univ., No.1, 37



*D. J. Helfand and J.-H. Huang (eds.), The Origin and Evolution of Neutron Stars, 376.*

# TIMESCALE FOR THE DECAY OF MAGNETIC FIELDS OF PULSARS

S. Krishnamohan
Radio Astronomy Centre
Tata Institute of Fundamental Research
P.O. Box 1234
Bangalore 560012
India

ABSTRACT. It is generally agreed that the magnetic fields of pulsars decay with the time. The previous estimates for the decay timescale range from 2Myr to < 10Myr based on the evolutionary tracks in the $(P,\dot{P})$ diagram and the kinematics of pulsars. We make a new estimate, by using the 'measured fields' themselves, to be about 20Myr and show that the z-distribution of pulsars is consistent with the new estimate if the known dependence of velocities of pulsars on the magnetic moment is taken into account.

If the magnetic moment decays exponentially the initial and the present magnetic moments M and m are related by $M^2 = m^2(1+\tau/\tau_B)$, where $\tau$ is the characteristic time and $\tau_B$ is the field decay time. It is unlikely that all the pulsars decay with the same $\tau_B$. When a single $\tau_B$ is assumed for the whole pulsar population two types of errors occur : i) deviation of the assumed $\tau_B$ from the mean value for the population and ii) deviation of $\tau_B$ of the individual pulsars from the mean of the population. Our aim is to find out the correct mean value of $\tau_B$ for the population by minimising the error due to the first cause. The error due to the second cause cannot be reduced anyway by assuming a single $\tau_B$. The r.m.s. error in the initial fields calculated by assuming a single $\tau_B$ is given by,

$$\sigma_M = \frac{1}{2} \frac{1}{\tau_B^2} < \frac{m_j^2 \tau_j^2}{1+\tau_j/\tau_B} > \sigma_{\tau_B} ,$$

where the suffix j indicates the jth pulsar. The value of $\tau_B$ which minimises $\sigma_{\tau_B}$ is the correct mean value.

$\sigma_{\tau_B}$ is computed by using all the 293 pulsars for which magnetic field strengths are listed in the pulsar table by Manchester & Taylor (1981). The results are presented in Figure 1. $\sigma_{\tau_B}$ decreases rapidly as $\tau_B$ is increased from 1Myr to ~20Myr and thereafter it goes into a shallow minimum at 33 Myr, suggesting that $\tau_B$ is about 20Myr.

Now, one may ask whether such a large value for $\tau_B$ is reasonable ? In particular, is it consistent with the observed z-distribution of pulsars ? To check this we computed the expected z-distribution as



D. J. Helfand and J.-H. Huang (eds.), The Origin and Evolution of Neutron Stars, 377–378.

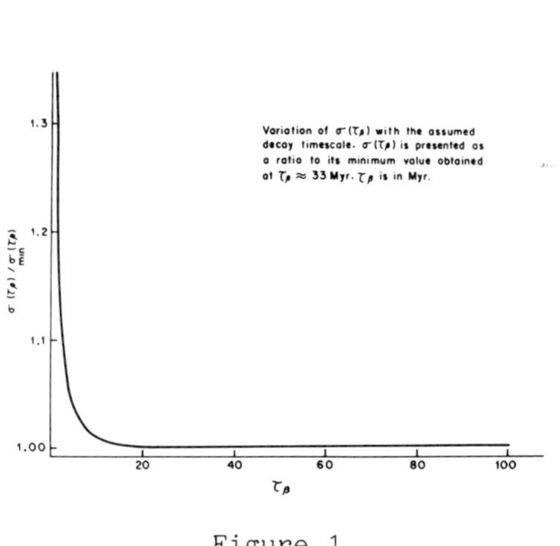


Figure 1

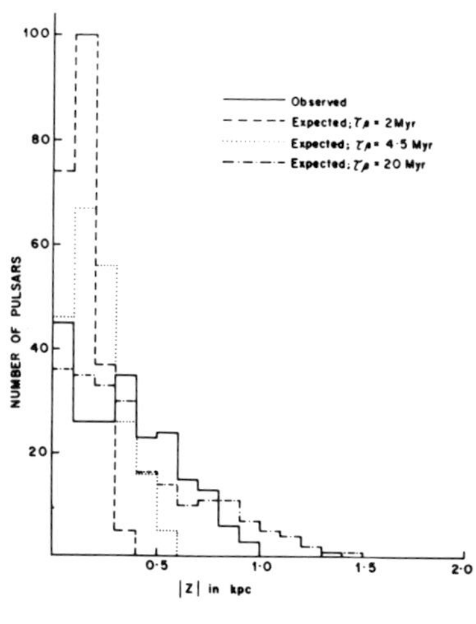


Figure 2

follows. Anderson & Lyne (1983) have pointed out a correlation between the measured transverse velocities of 26 pulsars and their magnetic moments. There is a spread of nearly a factor of 30 in velocities such that higher the magnetic moment, higher the velocity. The correlation has been confirmed by using a sample of 59 pulsars for which the speeds have been obtained through interstellar scintillation measurements (Cordes, 1987). This coupled with the fact that pulsars with larger magnetic moments have smaller ages helps in containing pulsars to smaller z values than are possible otherwise. To demonstrate this, we have restricted ourselves to the second Molonglo sample (Manchester et.al.,1978) to minimise selection effects. For each pulsar, an age is calculated, from the observed characteristic age, by assuming a $\tau_B$. We get a z height for the pulsar by multiplying the age with a mean z-velocity appropriate for its magnetic moment. The z-velocities are computed by using the relationship between the transverse velocities and the magnetic moments observed by Anderson & Lyne. The expected z-distributions obtained by using different $\tau_B$ are shown in Figure 2. A comparison with the observed z-distribution, shows that a $\tau_B$ of 20Myr is consistent with the observations. While comparing, one should remember that if proper allowance is given for projection of velocities the calculated distributions would become more concentrated towards low z values.

REFERENCES

Anderson, B. and Lyne, A.G., 1983. *Nature*, **303**, 597.
Cordes, J.M., 1987. *This proceedings*.
Manchester, R.N., Lyne, A.G., Taylor, J.H., Durdin, M.I., Large, M.I. and Little, A.G., 1978. *Mon. Not. R. astr. Soc.*, **185**, 409.
Manchester, R.N. and Taylor, J.H., 1981. *Astron. J.*, **86**, 1953.

# TORTION INFLUENCE ON THE MAGNETIC FIELD OF ROTATING NEUTRON STARS

C.N.Zhang, J.G.Shi, F.P.Chen
Department of Physics
Dalian Institute of Technology
Dalian, China

ABSTRACT. Using the axisymmetric tortion tensor given by J.Nitsh in $U_4$ gravitational gauge theory. Applying G.S.Yang and L.F.Luo's magnetic field model of neutron stars, i.e. magnetic field is originated from long-arranged neutron spin. Precession of spin coupled with tortion has been studied,indicating an influence on the magnetic field. The results show that the initial magnetic filed B made of long-arranged spin magnetic moment is at an angle of $\alpha$ with rotating axis, if $|\frac{\pi}{2} - \alpha| > 0.1$ radians, B makes a so small angle to the rotating axis at time $10^8$ sec. that no pulses could be radiated from the neutron stars. Only when $|\frac{\pi}{2} - \alpha| < 0.1$ radians, the pulse radiation could be done. Based on the assumption that the orientation of the initial magnetic fields is isotropic,we can get then that only about ten percent of the rotating neutron stars radiate pulses.



*D. J. Helfand and J.-H. Huang (eds.), The Origin and Evolution of Neutron Stars, 379.*

# IVb. NEUTRON STELLAR EVOLUTION

## Evolution in Binaries

CHAIR: A. Blaauw

# BINARY PULSARS: OBSERVATIONS AND IMPLICATIONS

J. H. Taylor
Joseph Henry Laboratories and Physics Department
Princeton University
Princeton, NJ 08544 USA

ABSTRACT. The Galaxy contains a large number of neutron stars in gravitationally bound binary systems. Among the most fruitful of these to study have been the binary radio pulsars, of which seven are now known. Unlike the "accretion-powered" neutron stars located in mass-exchanging X-ray binary systems, the "rotation-powered" binary radio pulsars are found in dynamically simple, clean systems in which both stellar components have already completed their nuclear evolution, thereby shedding their atmospheres and most of their mass. In such circumstances the orbital parameters of the system and the rotational parameters of the pulsar can be determined with high precision from analysis of pulse timing data. These measurements constrain the component masses and yield an estimate of the pulsar's magnetic dipole moment, which turns out to be an essential parameter in understanding the evolution of the systems. In this paper I review the known facts concerning binary pulsars, and then briefly discuss some implications for our understanding of the place of neutron stars in stellar evolution.

Let me first describe the distribution of binary pulsar systems in the Galaxy. The galactic coordinates of all 430 presently known pulsars are illustrated in Figure 1, with a rough indication of dispersion measure (and hence relative distance) encoded in the sizes of the circles. Locations of the seven binary pulsars and the millisecond pulsar PSR 1937+21, which I believe is closely related to the binaries, are explicitly identified. Five of the eight named objects lie at very small galactic latitudes $|b| < 4°$, and the remaining three have small dispersion measures and thus relatively small distances from the Sun. Based on distance estimates determined from the model of Lyne, Manchester & Taylor (1985), the mean distance of the group from the galactic plane is $< |z| > = 165$ pc, about half the corresponding scale height of the pulsar population as a whole. Their smaller $z$-distances imply that pulsars in binary systems have smaller peculiar velocities than single pulsars, a conclusion sure to be of significance in understanding their evolution.



*D. J. Helfand and J.-H. Huang (eds.), The Origin and Evolution of Neutron Stars, 383–392.*

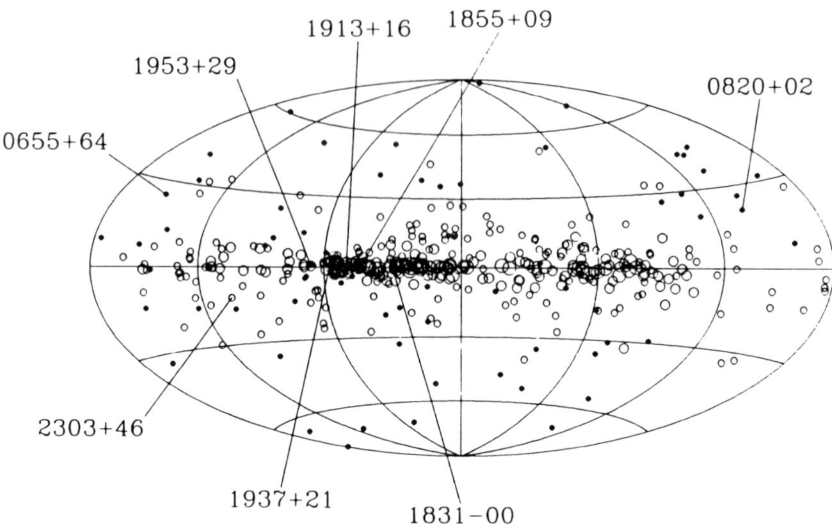


Fig. 1. Distribution of 430 pulsars in galactic coordinates. Sizes of circles correspond to dispersion measures in the ranges $< 30$, 30 to 100, 100 to 300, and $> 300\ \mathrm{cm}^{-3}$ pc. Binary and millisecond pulsars are identified by name.

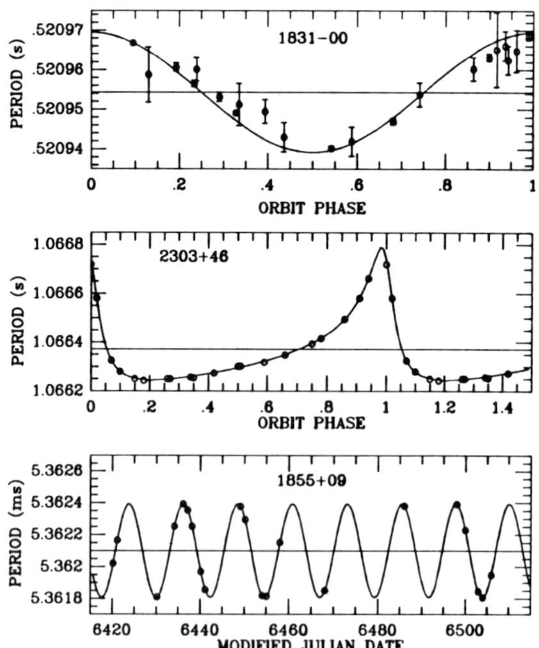


Fig. 2. Velocity curves of the three most recently discovered binary pulsars (Dewey *et al.* 1986, Stokes *et al.* 1985, Segelstein *et al.* 1986).

Table I. Parameters of binary and millisecond pulsars.

| PSR | P (ms) | $\log \dot{P}$ | $\log B$ (G) | $\lvert z\rvert$ (pc) | $a$ ($R_\odot$) | $P_b$ (days) | $e$ | $f(m_1,m_2)$ ($M_\odot$) | Likely $m_2$ ($M_\odot$) |
|---|---|---|---|---|---|---|---|---|---|
| 1937+21 | 1.6 | −19.0 | 8.6 | 20 | — | — | — | — | — |
| 1855+09 | 5.4 | −19.7 | 8.5 | 20 | 23 | 12.33 | 0.00002 | 0.0052 | 0.2 - 0.4 |
| 1953+29 | 6.1 | −19.5 | 8.6 | 20 | 100 | 117.35 | 0.0003 | 0.0027 | 0.2 - 0.4 |
| 0655+64 | 195.6 | −18.2 | 10.0 | 120 | 5 | 1.03 | $<0.00005$ | 0.0712 | 0.7 - 0.8 |
| 1913+16 | 59.0 | −17.1 | 10.3 | 190 | 2.8 | 0.32 | 0.6171 | 0.1322 | 1.38 |
| 1831−00 | 520.9 | $<-17.0$ | $<10.9$ | 190 | 6 | 1.81 | $<0.005$ | 0.00012 | 0.06 - 0.13 |
| 0820+02 | 864.9 | −16.0 | 11.0 | 280 | 500 | 1232.40 | 0.0119 | 0.0030 | 0.2 - 0.4 |
| 2303+46 | 1066.4 | −15.4 | 11.8 | 480 | 29 | 12.34 | 0.6584 | 0.2463 | 1.2 - 1.8 |

Timing data for binary pulsars yield the same kind of Doppler information as that available for single-line spectroscopic binary stars. Three examples of binary pulsar velocity curves are shown in Figure 2. In practice, analysis of the orbits is carried out using absolute pulse arrival times rather than inferred velocities, since the available precision is much higher with this approach (Taylor *et al.* 1976). Good orbital solutions are now available for all seven binary pulsars, and Table I summarizes some of the astrophysically interesting parameters.

As can be seen in the table, orbital periods of the seven binary pulsars range from less than eight hours to more than three years. Five of the orbits are essentially circular, with eccentricities $e \lesssim 0.01$. Four of these also have very small mass functions

$$f_1(m_1, m_2) = \frac{(m_2 \sin i)^3}{(m_1 + m_2)^2} \quad , \tag{1}$$

implying companion star masses no more than a few tenths of a solar mass. (Here $m_1$ and $m_2$ are the masses of the pulsar and companion, respectively, and $i$ is the inclination between the plane of the orbit and the plane of the sky.) The remaining two systems, in contrast, have orbits with $e > 0.6$ and much larger mass functions. Table I also lists the approximate separations between the pulsars and their companion stars, computed (except for PSR 1913+16) under the assumptions $m_1 = 1.4\ M_\odot$ and $\cos i = 0.5$. The estimated separations range from 2.8 to 500 solar radii.

Masses of the component stars are of particular importance, but unfortunately are not uniquely determined by the mass function. However, Equation (1) provides a useful constraining relation between $m_1$, $m_2$, and $i$, as illustrated for each of the seven systems in Figures 3 through 5. Figure 3 shows curves of $m_2$ vs. $m_1$

for assumed values of $\cos i = 0.0$, 0.5, and 0.8. With no other information on $i$ available, one can only conclude that there is a 50% chance that the correct masses lie between the lower two curves in each sub-figure, and an 80% chance that they lie between the top and bottom curves. Under the reasonable assumption that the pulsar mass is close to the Chandrasekhar mass, 1.4 $M_{\odot}$, it is clear that the companions of PSRs 0820+02, 1831-00, 1855+09, and 1953+29 are all low-mass objects, probably in the range $\sim 0.06$ to 0.4 $M_{\odot}$.

For the remaining three binary pulsars some additional information on the masses is available. Lyne (1984) has used interstellar scintillation observations to estimate the transverse velocity of PSR 0655+64 as a function of orbital phase. Combined with timing results for the radial velocity, his data yield two possible solutions for the orbital inclination. These values constrain the component masses as illustrated in Figure 4. For a pulsar mass of 1.4 $M_{\odot}$, the companion of PSR 0655+64 must have a mass very close to either 0.66 $M_{\odot}$ or 0.78 $M_{\odot}$.

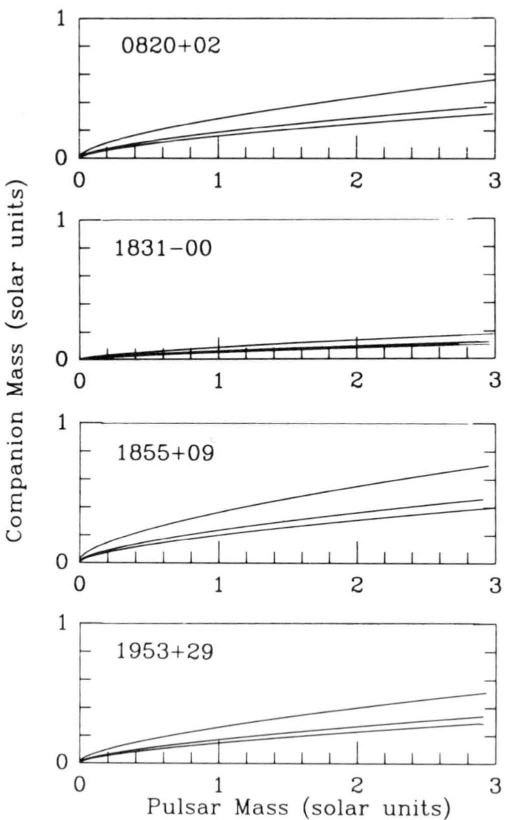


Fig. 3. Constraints on the masses of PSRs 0820+02, 1831-00, 1855+09, and 1953+29 and their companion stars. In each panel the three curves correspond to assumed values $\cos i = 0.0$, 0.5, and 0.8 (from bottom to top).

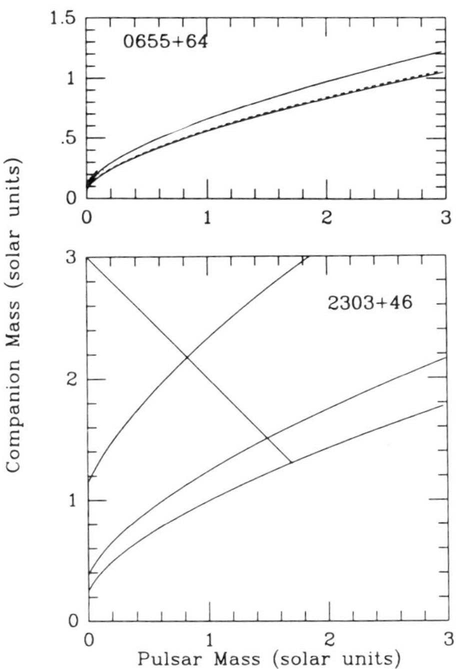


Fig. 4. Constraints on the masses of PSRs 0655+64 and 2303+46 and their companion stars. For 0655+64 the masses must lie on or close to one of the two heavy curves, which correspond to the values cos $i = 0.10$ and cos $i = 0.47$ consistent with the scintillation measurements of Lyne (1984). The lighter curves indicate ranges of uncertainty. For 2303+46, as in Fig. 3, the three curves correspond to assumed values cos $i = 0.0$, 0.5, and 0.8. The observed rate of periastron advance constrains the masses of this system to lie on or near the sloping straight line (see text).

My colleagues and I have recently succeeded in measuring the general relativistic periastron advance of the orbit of PSR 2303+46, one of the two binary pulsars with high orbital eccentricity and large mass function. For this system, as for PSR 1913+16, there can be little doubt that it consists of two neutron stars. The rate of periastron advance, currently estimated as $\dot{\omega} = 0.0109 \pm 0.0023$ (Taylor *et al.* 1986), implies a total system mass $m_1 + m_2 = 3.0 \pm 0.9\ M_{\odot}$ — a very plausible result for the sum of two neutron stars. The sloping straight line in Figure 4 illustrates for PSR 2303+46 the additional constraint imposed by $m_1 + m_2 = 3.0\ M_{\odot}$. Much better precision should be obtainable for $\dot{\omega}$ within a year or so.

PSR 1913+16, the best studied of all binary pulsars, has the ideal combination of short orbital period, short pulsar period, and large eccentricity which combine to make possible the detection of higher-order secular and periodic relativistic effects, and therefore to specify the component masses quite accurately. In addition to the rate of precession of the orbit, it is possible to determine the variable part of second-order Doppler shift and gravitational redshift, $\gamma$; the rate of decay of the orbit due to gravitational radiation, $\dot{P}_b$; and the orbital inclination, $\sin i$. The most recent analysis (Taylor 1986, Taylor & Weisberg 1986; see also Weisberg & Taylor 1984) yields for the masses $m_1 = 1.451 \pm 0.007$ and $m_2 = 1.378 \pm 0.007$ (see Figure 5). These results are extremely important in that they provide a firm experimental foundation on which to build models of neutron stars and their formation.

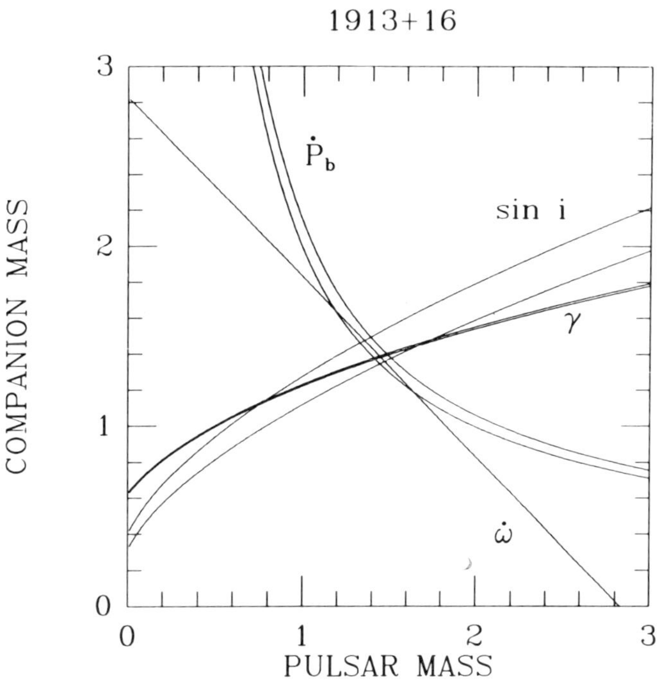


Fig. 5. Constraints on the masses of PSR 1913+16 and its companion. Families of curves bracket the range of masses consistent with each of the relativistic parameters $\dot{\omega}$, $\gamma$, $\dot{P}_b$, and sin $i$. (The uncertainties of $\dot{\omega}$ and $\gamma$ are too small to show.)

Let me now turn attention from orbital and dynamical matters to the characteristics of the binary pulsars themselves. Figure 6, an up-to-date version of the pulsar $P, \dot{P}$ diagram, shows that pulsars found in binary systems have unusually small spin-down rates $\dot{P}$, and that most of them have unusually small periods $P$ as well. There is some evidence that their luminosities are somewhat smaller (by a factor of about three) than those of single pulsars, though the large dispersion of pulsar luminosities makes it difficult to state this with confidence. A better statement concerning luminosities may be that pulsars located close to the "death line" in Figure 6 (Ruderman & Sutherland 1975) tend to have smaller radio luminosities (Taylor & Stinebring 1986; see also Prószyński & Przybycień 1984). All of the binary pulsars either have unusually short periods or are found not far from the death line.

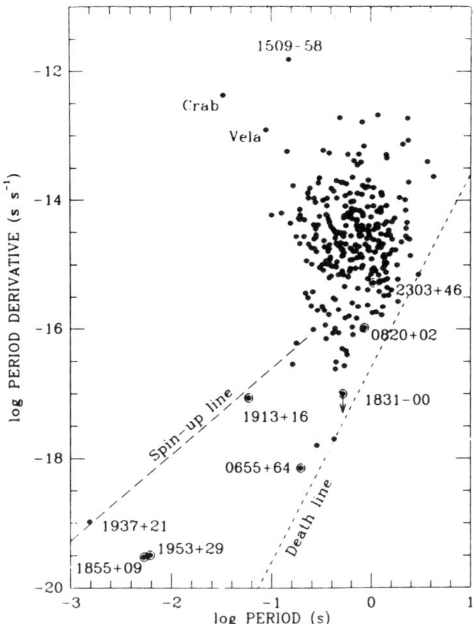


Fig. 6. Periods and period derivatives of 349 pulsars.

Pulsars with small values of $\dot{P}$ generally show little if any timing noise (Cordes and Downs 1985). Although in some cases the span of observations is not yet very long, no evidence for *any* timing noise has yet turned up in our timing observations of the binary and millisecond pulsars. PSRs 1855+09 and 1937+21, in particular, appear to be extraordinarily stable clocks (Davis *et al.* 1985, Segelstein *et al.* 1986). This fact is consistent with the understanding that as recycled objects, binary pulsars are likely to be very old. It is probable that most, if not all of them exist in the "active" region of Figure 6 (to the left of the death-line) solely because they acquired extra angular momentum while accreting matter from their evolving companions.

Other characteristics of the binary radio pulsars appear to be indistinguishable from those of single pulsars. Pulse shapes and pulse widths are normal; average polarization characteristics, when they have been measured (e.g. Stinebring *et al.* 1984, Segelstein *et al.* 1986), appear to be unextraordinary.

Although fewer than 2% of the 430 presently known pulsars are in binary systems, the fraction increases to 13% (4 of 30) in the period range $P < 200$ ms, and to 60% (3 of 5) of those with $P < 60$ ms. Six percent of the 50 pulsars with the smallest dispersion measures, and therefore presumably closest to the sun, are binaries. Until very recently all major pulsar surveys were relatively insensitive for periods $\lesssim 200$ ms (Dewey *et al.* 1984), and only one high-sensitivity survey has been made looking explicitly for low-luminosity pulsars in the solar vicinity (Dewey *et al.* 1985). It therefore seems likely that as more extensive surveys for fast and nearby pulsars are undertaken, many more binaries will be found. Experience gained in the Princeton/Arecibo survey (Segelstein *et al.* 1986; Stokes *et al.* 1986) suggests that at least $\sim 10\%$ of the pulsars detected in a survey with good sensitivity down to $P \lesssim 5$ ms will be millisecond pulsars, and probably most of these will be binaries. Future fulfillment of these expectations will be a great boon for understanding of the physics of neutron stars and their evolution, and perhaps for gravitation physics as well.

## REFERENCES

Cordes, J.M., and Downs, G.S. 1985, *Astrophys. J. Supp. Ser.*, **59**, 343.

Davis, M.M., Taylor, J.H., Weisberg, J.M., and Backer, D.C. 1985, *Nature*, **315**, 547.

Dewey, R.J., Stokes, G.H., Segelstein, D.J., Taylor, J.H., and Weisberg, J.M. 1984, in *Millisecond Pulsars*, ed. S.P. Reynolds and D.R. Stinebring, (National Radio Astronomy Observatory: Green Bank, WV), p. 234.

Dewey, R.J., Taylor, J.H., Weisberg, J.M., and Stokes, G.H. 1985, *Ap. J. (Letters)*, **294**, L25.

Dewey, R.J., Maguire, C.M., Rawley, L.A., Stokes, G.H., and Taylor, J.H. 1986, *Nature*, in press.

Lyne, A.G. 1984, *Nature*, **310**, 310.

Lyne, A.G., Manchester, R.N., and Taylor, J.H. 1985, *M.N.R.A.S.*, **213**, 613.

Prószyński, M., and Przybycień, D. 1984, in *Millisecond Pulsars*, ed. S.P. Reynolds and D.R. Stinebring, (National Radio Astronomy Observatory: Green Bank, WV), p. 151.

Rawley, L.A., Taylor, J.H., and Davis, M.M. 1986, *Nature*, **319**, 383.

Ruderman, M.A., and Sutherland, P.G. 1975, *Astrophys. J.*, **196**, 51.

Segelstein, D.J., Rawley, L.A., Stinebring, D.R., Fruchter, A.S., and Taylor, J.H. 1986, *Nature*, in press.

Stinebring, D.R., Boriakoff, V., Cordes, J.M., Deich, W., and Wolszczan, A. 1984, in *Millisecond Pulsars*, ed. S.P. Reynolds and D.R. Stinebring, (National Radio Astronomy Observatory: Green Bank, WV), p. 32.

Stokes, G.H., Segelstein, D.J., Taylor, J.H., and Dewey, R.J. 1986, *Astrophys. J.*, in press.

Stokes, G.H., Taylor, J.H., Weisberg, J.M., and Dewey, R.J. 1985, *Nature*, **317**, 787.

Taylor, J.H. 1986, *11th Intl. Conf. on General Relativity and Gravitation*, Stockholm, July 1986.

Taylor, J.H., Dewey, R.J., Weisberg, J.M., Maguire, C.M. 1986, unpublished results.

Taylor, J.H., Hulse, R.A., Fowler, L.A., Gullahorn, G.E., and Rankin, J.M. 1976, *Astrophys. J. (Letters)*, **206**, L53.

Taylor, J.H., and Stinebring, D.R. 1986, *Ann. Rev. Astron. Astrophys.*, in press.

Taylor, J.H., and Weisberg, J.M. 1986, in preparation.

Weisberg, J.M., and Taylor, J.H. 1984, *Phys. Rev. Letters*, **52**, 1348.

## DISCUSSION

**J. Arons:** On Monday you remarked that you thought the millisecond pulsars are a different population. Could you elaborate, since they are part of the binaries?

**J. Taylor:** I believe that a bi-modal period distribution probably exists because most spin-ups will occur after a time long enough for the magnetic fields to have decayed to less than $10^9$ gauss, so the equilibrium periods will be very short.

**J. Arons:** Then how does 1913 fit in, since it is an example of a pulsar in a binary with a period between 10 and 100 msec, exactly where you are suggesting there will be a gap?

**J. Taylor:** Pulsars like 1913+16 are the cases in which the original mass ratio was close to unity, so much less time elapsed between the times when the two stars evolved off the main sequence.

**J.H. Huang:** You put two theoretical lines on logP vs. log$\dot{P}$ plot; i.e., a death line for isolated pulsars and a spin-up line for binaries. Does it mean you need a special evolution from isolated pulsars to binaries when you put it together?

**J. Taylor:** No. The death line applies for any kind of pulsar, but pulsars which cross the death line while still in a binary system may get a second life after being spun up.

**J. Dickel:** Any thoughts or comments yet on the implications of the fact that the two binaries in which both components are neutron stars have only 1 pulsar each?

**J. Taylor:** There are many reasons why the companion neutron stars may not be observable as pulsars; probably the most likely are beaming effects and the rather short pulsar lifetime.

**R. Narayan:** If 2% of observed pulsars are in binary systems, and those are on the average harder to see, both because they are under-luminous and because many of them have short periods, then would you agree that a substantial fraction of the active pulsars in the Galaxy are in binaries? In fact, could one even say that a substantial fraction are millisecond binary pulsars?

**J. Taylor:** I think that 10% is a lower limit on this substantial fraction, and it could, indeed, be considerably higher. Of course the lifetime of the millisecond pulsars is much longer so their birthrate could still be much smaller than that of ordinary pulsars.

**L. Woltjer:** You have 7 binaries with velocity amplitude larger than 8 km/sec. How many are single to the same level of accuracy?

**J. Taylor:** At least 340 of the known pulsars - essentially all of those with measured period derivatives, and which are not already known to be binaries - are certainly not binaries down to limits around one meter per second in velocity.

**L. Woltjer:** Do you have proper motions for any of the binaries?

**J. Taylor:** Only upper limits for proper motions are presently available for any of the binaries.

**J. Shaham:** Do you think that the $P\dot{P}$ vs. $V_t$ relation for pulsars holds all the way down to the millisecond pulsars. If so, what does it imply regarding their expected abundance in the galaxy?

**J. Taylor:** Unless our ideas concerning the origin of millisecond pulsars are far wrong, they will not have large peculiar velocities. Within present uncertainties the abundance estimates will not be seriously affected; the z-distribution of millisecond pulsars would be made smaller by the low velocities, but larger by the long lifetimes.

# MILLISECOND PULSAR FORMATION AND EVOLUTION

E.P.J. van den Heuvel
Astronomical Institute
University of Amsterdam
Roetersstraat 15
1018 WB Amsterdam, The Netherlands

ABSTRACT. The evolutionary history of binary radio pulsars, including the two millisecond binary pulsars, is reviewed. There are two groups of binary pulsars, the PSR 1913+16-group, which descended from massive X-ray binaries, and the PSR 1953+29-group, which descended from fairly wide low-mass X-ray binaries. The neutron stars in the second group probably formed by the accretion-induced collapse of a massive white dwarf. The companion stars in both groups of systems are expected to be dead stars, i.e. white dwarfs or neutron stars.

The large total number of millisecond binary pulsars in the galaxy ($\sim 10^4$), indicates that magnetic fields of neutron stars do not decay below a value of order $10^9$ G. Possible explanations for this phenomenon are discussed.

Coalescence with a close degenerate companion provides a viable model for the formation of the single millisecond pulsar.

## 1. INTRODUCTION

Three millisecond pulsars are known, two of which are in binary systems. With the other binary radio pulsars, they have in common (see table 1):

- an unusually rapid rotation: three out of the five most rapidly spinning pulsars known are in binaries, while in total only seven out of the about 500 known pulsars are in binaries.
- an unusually weak surface dipole magnetic field; figure 1 depicts this clearly: only 2 out of the 7 binary pulsars have magnetic field strengths in the same range as that of the bulk of the single pulsars.

The two binary millisecond pulsars thus appear to form the short-period portion of the pulse period distribution of the general binary pulsar population. Their evolutionary history is therefore expected to have been similar to that of the other binary pulsars.

The single millisecond pulsar PSR 1937+21 is a different case but, as we will argue in the last section, also its origin is likely to be tied with the evolution of binary systems.

For detailed reviews of the evolutionary history of binary and



*D. J. Helfand and J.-H. Huang (eds.), The Origin and Evolution of Neutron Stars, 393–406.*

millisecond radio pulsars and X-ray binaries we refer to Van den Heuvel (1984, 1986a, b), Van den Heuvel and Habets (1985) and Taam and Van den Heuvel (1986). Here we only summarize the most important recent developments.

TABLE 1. Observed and derived parameters of binary and millisecond pulsars. The maximum ages correspond to spindown at constant B from the spin-up line up to the present spin period. The times till turn off are the times required for reaching the deathline by spindown at constant B. For PSR 1913+16 the latter time is much longer than the coalescence time of the system ($\sim 3.10^8$ yrs). PSR 1831-00 is assumed to be situated above the deathline (see figure 1). (Observed parameters from Dewey et al. 1986).

| PSR | P (ms) | log $\dot{P}$ | log B (G) | $P_b$ (days) | e | $f(m_1,m_2)$ ($M_\odot$) | Likely $m_2$ ($M_\odot$) | Max. Age ($10^9$yr) | Time till turn-off ($10^9$yr) |
|---|---|---|---|---|---|---|---|---|---|
| 1937+21 | 1.6 | -19.0 | 8.65 | --- | --- | --- | --- | 0.1 | $>t_{Hubble}$ |
| 1953+29 | 6.1 | -19.5 | 8.65 | 117.35 | 0.0003 | 0.0027 | 0.2 - 0.4 | 2.9 | $>t_{Hubble}$ |
| 1855+09 | 5.4 | -19.4 | 8.7 | 12.33 | 0.00002 | 0.0052 | 0.2 - 0.4 | 2.1 | $>t_{Hubble}$ |
| 0655+64 | 195.6 | -18.2 | 10.0 | 1.03 | <0.00005 | 0.0712 | 0.7 - 1.3 | 3.9 | 5.3 |
| 1913+16 | 59.0 | -17.1 | 10.3 | 0.32 | 0.6171 | 0.1322 | 1.4 | 0.097 | 4.2 |
| 1831-00 | 520.9 | <-17.0 | <10.9 | 1.81 | <0.005 | 0.00012 | 0.06- 0.13 | 0.7-2.1 | <0.3 |
| 0820+02 | 864.9 | -16.0 | 11.5 | 1232.40 | 0.0119 | 0.0030 | 0.2 - 0.4 | 0.12 | 0.2 |
| 2303+46 | 1066.4 | -15.4 | 11.8 | 12.34 | 0.6584 | 0.2463 | 1.2 - 2.5 | 0.033 | 0.08 |

## 2. THE BINARY "RECYCLING" MODEL FOR THE ORIGIN OF BINARY RADIO PULSARS.

### 2.1. Period and magnetic field evolution of newborn neutron stars

Statistical investigations of radio pulsars show that the quantity $(P\dot{P})$ decreases in the course of time on a timescale of order $5.10^6$ yrs (cf. Lyne, this volume). This quantity is related to the surface dipole magnetic field of the neutron star by the relation (cf. Flowers and Ruderman 1977)

$$(P.\dot{P}) = K\ (B_\perp^2 + \ \xi\ B_{/\!/}^2) \qquad (1)$$

where $B_\perp$ is the component of the magnetic field perpendicular to the rotation axis, $B_{/\!/}$ is the component parallel to the rotation axis, and K is a constant which contains the moment of inertia of the neutron star. The quantity $\xi$ is related to the charge flow in the magnetosphere of the neutron star. Its expected value depends on the adopted model for the emission of the radio pulses. Ruderman and Sutherland (1976) estimate $\xi$ to be close to unity, but there are also models with a much lower value of $\xi$ (cf. Arons 1981).

We will, for the sake of argument, adopt with Ruderman and Sutherland that $\xi \simeq 1$. In that case $(P.\dot{P})$ is a measure of the absolute value of the magnetic field strength. The magnetic field strengths in figure 1 were derived in this way, by adopting a standard moment of inertia of neutron stars of $10^{45}$ g.cm$^2$ (for details: see Radhakrishnan 1982, 1986; van den Heuvel 1984).

The decrease of $(P.\dot{P})$ in the course of time implies, in terms of this model, that the surface dipole magnetic field strength decays on a timescale of $10^6$-$10^7$ yrs.

The dashed tracks in figure 1 indicate how a pulsar will evolve through the B-P diagram when its field decays on a timescale of $2.10^6$ yrs. When the pulsar passes a "deathline" the emission of pulsed radiation ceases and the pulsar enters the "graveyard", which is filled with extinct pulsars. The deathline drawn in figure 1 is that given by the Ruderman-Sutherland model.

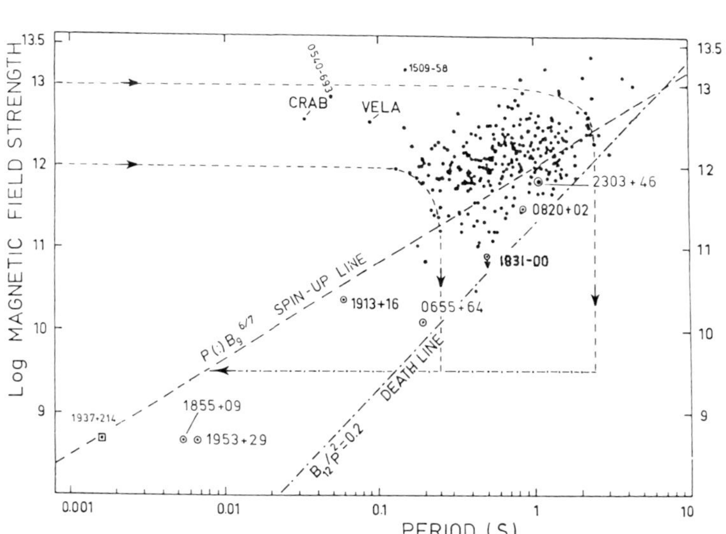


Figure 1. Surface dipole magnetic field strength $B_s$ (in Gauss) vs. pulse period for radio pulsars (sources of the observational data: see Van den Heuvel 1984, and table 1). The binary and millisecond pulsars are indicated. The meanings of the various lines as well as the derivation of the $B_s$ values are explained in the text. The dashed lines are evolutionary tracks of pulsars born with magnetic fields of $10^{12}$ and $10^{13}$ G, respectively, for a field decay timescale of $2.10^6$ yr. (Diagram based on figures 12 and 13 of Radhakrishnan 1982).

## 2.2. Resurrection from the graveyard

For single pulsars there is no way back from the graveyard. However, when a neutron star is in a binary system and its companion begins to transfer matter to it, its rotation will be accelerated due to the angular momentum carried with the transferred matter. This is clear from the gradual spin-up observed for all neutron stars in X-ray binaries which are fed by disk-accretion (cf. Rappaport and Joss 1983).

When it is being spun up the neutron star may already be quite old such that its magnetic field may have partially decayed. The dash-dotted horizontal line in figure 1 depicts schematically how an old neutron star in a binary may be spun-up to the region of the "living" radio pulsars once its companion evolves away from the main-sequence and begins to transfer matter to it (cf. Bisnovaty-Kogan and Komberg 1974; Smarr and Blandford 1976).

When the gas density around the neutron star has become very low, such that the radio signal will not be dispersed beyond recognition, the neutron star may become observable as a radio pulsar. This requires that also its companion must have terminated its life, and has become a compact object, either a neutron star, a white dwarf or a black hole.

### 2.3. The spin-up line

The shortest possible spin period that can be reached by accretion is the equilibrium spin period $P_{eq}$ (for the definition: see e.g. Henrichs 1983), corresponding to the maximum possible accretion rate $\dot{M}_{edd} \simeq 1.5 \times 10^{-8}\ M_\odot$/yr. $P_{eq}$ is given by (see Van den Heuvel 1977):

$$P_{eq} = (1.9\ \text{ms})\ (B_9)^{6/7}\ M^{-5/7}\ (\dot{M}/\dot{M}_{Edd})^{-3/7}\ R_6^{\ 18/7} \qquad (2)$$

where $B_9$ is the surface dipole field strength in units of $10^9$ G, $\dot{M}$ is the accretion rate, and M and R are the mass and radius of the neutron star, in units of 1.4 $M_\odot$ and $10^6$ cm, respectively. Adopting the latter two quantities to be unity, one observes that $P_{eq}$ is only a function of $B_9$ and $\dot{M}$. Hence, the shortest attainable value of $P_{eq}$, which occurs for $\dot{M} = \dot{M}_{Edd}$, is only a function of $B_9$:

$$P_{eq,min} = (1.9\ \text{ms})\ B_9^{\ 6/7} \qquad (3)$$

This relation defines the "spin-up" line, which is indicated in figure 1. All "recycled" pulsars are expected to be found in the wedge-shaped region between this line and the deathline. As figure 1 shows, all the 7 binary radio pulsars, as well as the single millisecond pulsar PSR1937+21 are indeed located in this region. This strongly suggests that they are recycled old pulsars.

### 2.4. The Evolutionary Status of the Companions of the binary radio pulsars

The fact that these pulsars are recycled at the same time, according to the above, implies that their companions themselves must now also be "dead" stars.

This expectation is confirmed in the cases of PSR 0655+64 and PSR 0820+02 by their optical identification (Kulkarni 1986a,b) which shows that the companions are located below the main sequence, in the HR-diagram, i.e.: are white dwarfs.

### 2.5. The production of millisecond pulsars

Figure 1 shows that in order to reach P < 10 ms, B should at the time of the spin up be smaller than $6.10^9$ G. At larger field strengths, millisecond periods cannot be reached. A further constraint is that enough mass should be transferred, since otherwise not enough angular momentum can be transferred to achieve spin-up to a short period. For example, to reach P = 1.55 ms at $B_s = 5.10^8$ G requires an amount of accretion of at least 0.12 $M_\odot$.

# 3. THE TWO TYPES OF BINARY RADIO PULSARS AND THEIR ORIGIN

## 3.1. Introduction

The 7 binary radio pulsars can be divided into two distinct classes, related to the two broad classes of binary X-ray sources, as follows (see table 1):

a. The "PSR 1913+16 class".
   These are the three systems PSR 1913+16, PSR 2303+46 and PSR 0655+64, which have (i) relatively short orbital periods, (ii) relatively high companion masses [$\sim 1.4\ M_\odot$ in PSR 1913+16 and PSR 2303+46, and $\sim 1\ M_\odot$ in PSR 0655+64], and (iii) very eccentric orbits in two out of three cases.

b. The "PSR 1953+29 class".
   These are the four other systems, which have:
   (i) relatively long orbital periods, of order of weeks to years (the only exception is PSR 1831+00), (ii) very low mass functions, indicating companion masses of $\sim 0.2 - 0.4\ M_\odot$. (iii) circular orbits.

The existence of these two classes can be understood on the basis of an evolutionary history with "spiral-in" for the first group, and "spiral-out" for the second group (van den Heuvel and Taam 1984), relating them to the Massive X-ray binaries (MXBs) and the Low-mass X-ray binaries (LMXBs), respectively.

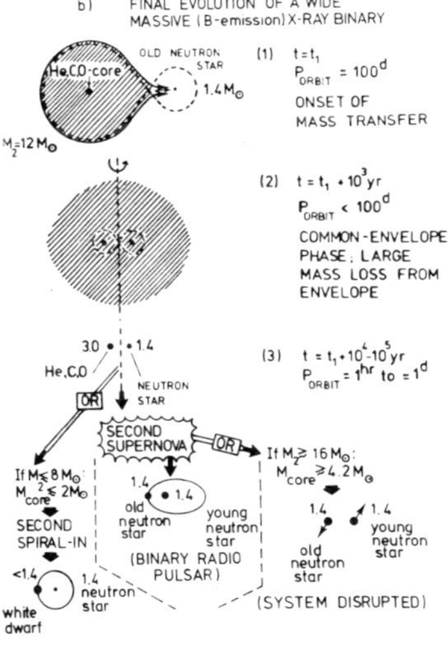


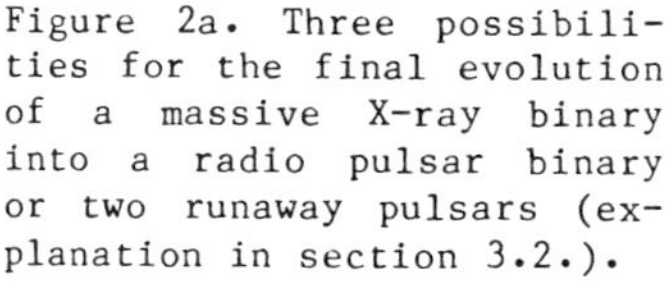

Figure 2a. Three possibilities for the final evolution of a massive X-ray binary into a radio pulsar binary or two runaway pulsars (explanation in section 3.2.).

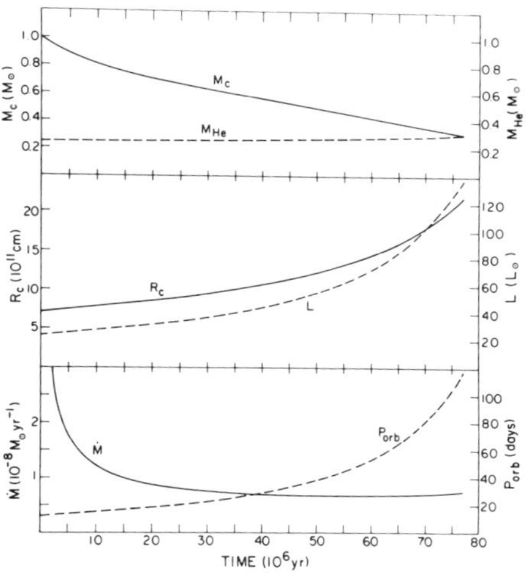


Figure 2b. Evolution of a wide low-mass X-ray binary into a radio pulsar binary of the PSR 1953+29-type, as calculated by Joss and Rappaport (1983). The system started out with a $1\ M_\odot$ lower giant-branch star with a $1.3\ M_\odot$ neutron star companion in a 12.5 day orbit. Shown are the evolution of the companion mass $M_c$, its helium core mass $M_{He}$, its radius $R_c$, luminosity L, mass transfer rate $\dot{M}$ and orbital period. Explanation in the text.

Figures 2a and b depict the evolution in these two cases. Basically, the difference between the evolution of these two types of systems is due to the fact that, as a consequence of conservation of orbital angular momentum, transfer of mass from a more massive companion to the neutron star causes the orbit to shrink, whereas transfer of mass from a less massive companion causes the orbit to widen (see, e.g. Paczynski 1971). We will assume here that in both cases, the mass transfer is driven by the envelope expansion caused by the interior nuclear evolution of the star. We will now consider the two cases depicted in figure 2 in more detail.

### 3.2. Origin of the "PSR 1913+16 group"

If the companion to a neutron star has a mass considerably larger than that of the neutron star, like in the MXBs, one expects that once the companion overflows its Roche lobe and begins to transfer matter to the neutron star, the neutron star will soon be entirely engulfed by the envelope of its companion. It will then rapidly spiral down into this envelope, as a consequence of the large frictional drag excerted on its orbital motion (see Taam et al. 1978; Bodenheimer and Taam 1985; Meyer and Meyer-Hofmeister 1979; Van den Heuvel 1986b). The expected outcome of the spiral-in is a very close binary system, consisting of the heavy-element core (He, and/or C, O, etc.) of the companion together with the neutron star. Due to the frictional drag the orbit of this narrow system will be perfectly circular. For the further evolution of this narrow system there are three possibilities, depending on the core mass, as depicted in figure 2a (see also Habets 1985, 1986; Van den Heuvel and Habets 1985; Van den Heuvel 1986b):

(i) If $M_{core} < 2\text{-}2.5\ M_{\odot}$ ($M_{initial} < 8\text{-}10\ M_{\odot}$) a narrow system results consisting of a C-O white dwarf and the old "recycled" neutron star, in a circular orbit. PSR 0655+64 presumably formed in this way.

(ii) For $2.5\ M_{\odot} < M_{core} < 4.2\ M_{\odot}$ ($M_{initial}$ in the range 10-15 $M_{\odot}$) the heavy-element core explodes as a supernova.
If there are no asymmetries in the explosion the system will not be disrupted and will consist of two neutron stars - one of them "recycled" - in an eccentric orbit. PSR 1913+16 and PSR 2303+46 are thought to have formed in this way (Flannery and van den Heuvel 1975; Smarr and Blandford 1976).

(iii) $M_{core} > 4.2\ M_{\odot}$ ($M_{initial} > 15\ M_{\odot}$).
These systems are not expected to undergo further mass transfer before the core explodes as a SN. Again, adopting a circular orbit, a neutron star mass of 1.4 $M_{\odot}$ (for both stars) and no asymmetries, the system will be disrupted, since more than half of the system mass is ejected (Blaauw 1961). The result is: two runaway pulsars, one young and one old ("recycled").

### 3.3. Origin of the "PSR 1953+29-group" of binary pulsars

Figure 3 depicts the orbital dimensions and periods of 16 LMXBs with detected orbital motion. In the short-period LMXBs ($P_{orb} \leqslant 12^h$) the mass transfer can be driven by gravitational radiation (GR) losses and magnetic braking (cf. Rappaport et al. 1983). On the other hand, in the wider systems such as Sco X-1, Cyg X-2, 2S0921-63 and GX 1+4 the mass transfer must be driven by the interior nuclear evolution of the companion star (Webbink et al. 1983; Taam 1983). This mechanism, moreover, is the only one that can explain the high X-ray luminosities of $\sim 10^{38}$ ergs/s of the about one dozen of strong bulge X-ray sources (Webbink et al. 1983). The mass-transfer rates in this case appear to be, throughout the evolution, given by simple formulae of the type (for X = 0.70, Z = 0.02):

$$\dot{M} = - 6 \times 10^{-10} \; P_o \; (d) \; M_\odot/yr \qquad (4)$$

where $P_o$ is the initial orbital period in days (this formula holds for $P_o \geqslant 2^d$). Fig. 2b depicts the evolution of a system with $P_o = 12^d.5$ as calculated by Joss and Rappaport (1983). The figure shows that at the end of the evolution only the helium core of the companion is left, with mass $M_c = 0.31\ M_\odot$, in an orbit with $P = 117^d$, i.e. exactly reproducing the system parameters of PSR 1953+29. This model explains in a natural way two most important characteristics of the PSR 1953+29-group of systems:

(i) the low mass of the companion star, which is always the remnant helium core of 0.2 - 0.4 $M_\odot$ of the solar-type companion star.

(ii) the circularity of the orbits, since during the long time interval that the low-mass giant filled its Roche-lobe, the tidal interaction in the systems was very strong.

It thus appears that the PSR 1953+29-group of systems are the natural descendants of the wide LMXBs such as Cyg X-2, 2S0921-63, Sco X-1, GX 1+4 and the bright bulge sources.

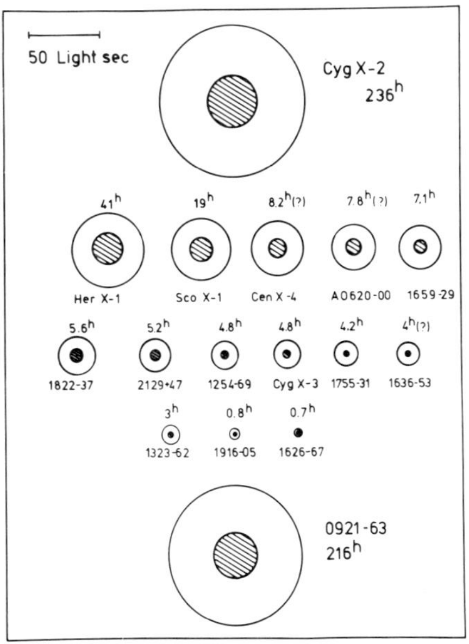


Figure 3. Orbital dimensions and periods of 16 low-mass X-ray binaries (adapted from McClintock and Rappaport 1984).

3.4. The formation mechanism of the neutron stars in the two types of binary pulsar systems

In the PSR 1913+16-group the one or two neutron stars present in the system originated by core collapse of a massive star at the end of its nuclear evolution.

On the other hand, for the 1953+29-group and their progenitors, the wide LMXBs, such a scenario seems unlikely for several reasons, notably because in the core collapse of a massive star a considerable amount of mass is being ejected which most probably will have unbound a system in which the companion is a low-mass star (Helfand et al. 1983). A variety of arguments indicate that these neutron stars formed recently, most probably during the accretion process itself, by the accretion-induced collapse of a white dwarf (Taam and Van den Heuvel 1986), abbreviated here as AIC.
The systems with evolved low-mass companions are indeed ideally suited for the occurrence of AIC, since AIC requires a mass-transfer rate $\geqslant 10^{-8.5}$ ($10^{-9}$) $M_{\odot}$/yr. This is because at these accretion rates the accreted hydrogen burns steadily or in weak flashes such that little or no mass is being ejected and the white dwarfs can grow considerably in mass. On the other hand, for $\dot{M} < 10^{-9}$ $M_{\odot}$/yr, the accreted matter burns in strong nova-like flashes (Nomoto, this volume), and most or all of the accreted matter is ejected.
Just in the relatively wide systems with evolved low-mass companions a steady high mass transfer rate is maintained (see above and figure 2b), making these systems very suitable for AIC to occur. On the other hand, the short period nova-like binaries in general have much lower transfer rates ($\leqslant 10^{-10}$ - $10^{-9}$ $M_{\odot}$/yr) and are therefore less well suited for the AIC model. For a more detailed discussion for the conditions of occurrence of AIC we refer to Taam and Van den Heuvel (1986).

3.5. Relation between msec pulsar binaries and QPO sources.

The discovery of QPO s in the two wide LMXBs Cyg X-2 and Sco X-1 as well as in a number of bright galactic bulge X-ray sources provides supporting evidence for the wide LMXB-nature of the bright bulge sources (see section 3.3.). The most straightforward interpretation of the QPO-phenomenon is provided by the Alpar-Shaham (1985a, b) beat-frequency model (for details: see Shaham, this volume; and Lewin, this volume). This interpretation yields typical rotation periods and magnetic field strengths of the neutron stars in these systems of $\leqslant$ 10 ms and $10^{9}$ - $10^{10}$ gauss, respectively. These values are very similar to those found in the millisecond binary radio pulsars. Since also the system dimensions of Cyg X-2 and Sco X-1 are similar to those of the progenitors of the PSR 1953+29-group of radio pulsar binaries, the QPO-sources seem ideal progenitors for the 1953+29-group of radio pulsar binaries.

### 3.6. The formation rate of binary millisecond pulsars and the magnetic-field-strength evolution of neutron stars

The total galactic number of bright wide LMXBs is at most some 30 to 50. Assuming a mass transfer time of $\sim 10^8$ yrs (see section 3.3) and assuming that half the mass transfer life-time is spent in the pre-collapse phase, with a white dwarf companion, the galactic formation rate of these systems is $(3 - 5) \times 10^{-7}$ yr$^{-1}$. In a steady state this is also the formation rate of the systems of the 1953+29-group. This low formation rate provides very strong evidence for the existence of a "bottom" value for the magnetic fields of neutron stars of $\sim 5.10^8$ G, below which the fields do not decay (cf. Bhattacharya and Srinivasan 1986; Van den Heuvel et al. 1986).

The reasoning is as follows: if one assumes that neutron-star magnetic fields decay on an $\sim 10^7$ yr timescale, as indicated by radio pulsar observations (cf. Lyne, this volume) then systems of the PSR 1953+29-group will be observable as radio pulsars for only $\sim 2.10^7$ yrs. This is because in order to allow spin up to $\sim 5$ msec, the magnetic field strength of a neutron star must already be $\leqslant 3.10^9$ G. From this value, the field will decay to $\leqslant 5.10^8$ G within about two foldings, i.e. $\sim 2.10^7$ yrs. Thus, if at $B \leqslant 3.10^9$ G the decay timescale of the magnetic fields of neutron stars is still $\sim 10^7$ yrs, one would expect the total number of observable binary millisecond pulsars in the galaxy to be of order $\leqslant 6 - 10$. However, among the $\sim 500$ known pulsars there are already 2 such systems known. As the distances of these systems are of the same order as those of the bulk of the radio pulsars (namely 0.3 and 3 kpc) these two systems give a measure for the actual percentage of these systems among the radio pulsars, of order 0.4%, uncorrected for selection effects. With an estimated total galactic number of active pulsars of $(2 - 5) \times 10^5$, the total number of active binary millisecond pulsars becomes $\geqslant 800$ to 2000. Stokes et al. (1986) taking observational selection effects properly into account estimate that the actual galactic number of binary millisecond pulsars is even higher, of order $\sim 10^4$.

The total number of remnants of LMXBs produced in the entire lifetime of the galaxy ($\sim 1.5 \times 10^{10}$ yr) is, for a constant formation rate equal to the present one, of a similar order, i.e. $\sim (4.5 - 7.5) \times 10^3$. The only way to explain why these numbers are of the same order is that magnetic fields of neutron stars do not decay any more below a value of $\sim (0.5 - 1) \times 10^9$ G. If the magnetic field strengths remain fixed around such a value, the spindown timescale (by magnetic dipole radiation, etc.) is longer than a Hubble time, and all binary millisecond pulsars ever produced in the galaxy will still presently be observable.

This theoretical argument confirms and provides independent support for the earlier findings by Kulkarni (1986a, b) indicating that the cooling age of the white dwarf optical companion of PSR 0655+64 is $\geqslant (0.5) \times 10^9$ yrs. Since the field strength of PSR 0655+46 is still $\sim 10^{10}$ G it seems that in some neutron stars the field decay may already stop at this value.

### 3.7. Possible interpretations for the existence of a "bottom field strength" in neutron stars

Two interpretations seem possible, namely:

A. In terms of the Flowers and Ruderman (1977) model. In this model the magnetic field of a neutron star consists of two parts: (i) the strong observable outside field of $\sim 10^{12-13}$ G in young pulsars, which is anchored in the crust, and supported by crustal electric currents. These currents resulted from flux-freezing and dynamo action shortly before and during the collapse, and (ii) a field seated in the superconducting interior.

The crustal part of the field will, in this model, decay on a relatively short timescale $\sim 10^7$ yrs, due to ohmic dissipation of the crustal electric currents.

On the other hand, the field in the superconducting interior will never decay.

In terms of this model, the final outside field of $\sim 5.10^8$ G is connected with superconducting interior and will remain forever.

B. Accretion-induced decay, as suggested by Taam and Van den Heuvel (1986), as an alternative interpretation of the observations. In this semi-empirical "model", magnetic fields of non-accreting neutron stars may not decay at all (cf. Kundt 1981), but accretion of matter would gradually decrease the strength of the exterior dipole field.

All our knowledge about magnetic fields of neutron stars in X-ray binaries and radio pulsar binaries is consistent with such an interpretation (Taam and Van den Heuvel 1986). Those neutron stars that have the weakest fields (the binary millisecond pulsars) are indeed just the ones that underwent the largest amount of accretion, and vice versa. (Of course, this inverse correlation can also be seen as an inverse correlation between magnetic field strenth and time, i.e.: as just another manifestation of the spontaneous decay of neutron star magnetic fields in general. This is, because the amount of matter accreted is, roughly, proportional to time, as the accretion rate is, - within an order of magnitude - equal to the Eddington limit in most systems).

A possible theoretical model in which the field strength can be reduced by accretion has been suggested by Blondin and Freese (1986).

## 4. THE FORMATION OF THE SINGLE MILLISECOND PULSAR PSR 1937+21

The location of this pulsar in the B vs. P diagram (figure 1) close to the spin-up line, suggests strongly that it was also recycled in a binary system (Radhakrishnan and Srinivasan 1982; Alpar et al. 1982). In this case the companion must have been lost. This can have happened in two ways, either (i) by a supernova explosion of the companion which caused the system to become unbound, or (ii) by coalescence with the companion, in a close system in which the orbital angular momentum was lost, for example, by GR-losses.
The first possibility seems ruled out (see Henrichs and van den Heuvel 1983; Van den Heuvel 1986b). Thus, the only binary models which seem to

remain are the coalescence models, which we will now briefly discuss.

Three types of coalescence models have been proposed which all seem - in principle - feasible, namely:

(i) coalescence with a second neutron star (Henrichs and van den Heuvel 1983);

(ii) coalescence with a massive white dwarf ($\geqslant 0.65\ M_\odot$) (Van den Heuvel and Bonsema 1984; Bonsema and van den Heuvel 1985).

(iii) coalescence with a low-mass degenerate dwarf, either hydrogen- or helium rich (Alpar et. al. 1982; Ruderman and Shaham 1983, 1985; Fabian et al. 1983).

For each of these models examples of progenitor systems are known, viz.:

(i) PSR 1913+16, consists of two neutron stars, which will certainly coalesce in the future, since the observed timescale of orbital decay by GR losses is $3.10^8$ yrs (Weisberg and Taylor 1984).

(ii) PSR 0655+64 consists of a neutron star and a massive white dwarf. If such a system were born with a slightly shorter orbital period, $< 12^h$ (which is certainly expected in some cases of spiral-in as depicted in figure 2a), then coalescence due to GR losses will occur within a Hubble time.

(iii) Systems with a red dwarf companion are the LMXBs with short orbital periods, less than $\sim 8^h$ (see figure 3).

The advantages and disadvantages of each of these three models are extensively discussed elsewhere (Van den Heuvel 1986b) to which we refer for details. The third model is possible only if large angular momentum losses from the system are invoked, to a rather unlikely degree (see Taam and Wade 1985). Furthermore, at the time of coalescence the mass-transfer rate in these systems has become so low that the neutron star will have been spun down to a long period, of the order of $\sim 0.1$ second. During the coalescence the amount of mass accreted ($\leqslant 0.005\ M_\odot$) is insufficient to spin the neutron star back up to a millisecond period (Jeffrey 1985). For these reasons the other two coalescence models seem preferrable for the origin of PSR 1937+21. The progenitor systems in both these models were massive binary systems (see figure 2b), contrary to the progenitors of the two binary millisecond pulsars.

REFERENCES


Alpar, M.A., Cheng, A.F., Ruderman, M.A. and Shaham, J., 1982, Nature **300**, 728.

Alpar, M.A. and Shaham, J., 1985a, IAU Circ. No.**4046**.

Alpar, M.A. and Shaham, J., 1985b, Nature **316**, 239.

Arons, J., 1981, IAU Symp. No.**95**, 69.

Blaauw, A., 1961, Bull. Astron. Inst. Netherlands **15**, 265.

Bhattacharya, D. and Srinivasan, G., 1986, Current Science **55**, 327.

Bisnovaty-Kogan, G.S. and Komberg, B.V., 1974, Astron. Zh. **51**, 373 (Soviet Astron. **18**, 217).

Blondin, J.M. and Freese, K., 1986, Nature (in the press).

Bodenheimer, P. and Taam, R.E., 1984, Astrophys. J. **280**, 771.

Bonsema, P.F.J. and van den Heuvel, E.P.J., 1985, Astr. Astrophys. **146**, L3.

Fabian, A.C., Pringle, J.E., Verbunt, F., Wade, R., 1983, Nature **301**, 222.

Flowers, E. and Ruderman, M.A., 1977, Astrophys. J. **215**, 302.

Habets, G.M.H.J., 1985, Ph.D. Thesis, University of Amsterdam.
Habets, G.M.H.J., 1986, Astr. Astrophys. (in the press).
Helfland, D.J., Ruderman, M.A. and Shaham, J., 1983, Nature **304**, 423.
Henrichs, H.F., 1983, in: Accretion-Driven Stellar X-ray Sources, W.H.G. Lewin and E.P.J. van den Heuvel (eds.), Cambridge Univ. Press, 393.
Henrichs, H.F. and van den Heuvel, E.P.J., 1983, Nature **303**, 213.
Jeffrey, L.C., 1985, Nature **319**, 384.
Joss, P.C. and Rappaport, S.A., 1983, Nature **304**, 419.
Kulkarni, S.R., 1986a, IAU Circ. No.**4160.**
Kulkarni, S.R., 1986b, Astrophys. J. (in the press).
Kundt, W., 1981, Astr. Astrophys **98**, 207.
McClintock, J.E. and Rappaport, S.A. Proc. Seventh North American Workshop on Cataclysmic Binaries and Low-Mass X-ray Binaries Center for Astrophys.). In the press.
Meyer, F. and Meyer-Hofmeister, E., 1979, Astr. Astrophys. **78**, 167.
Paczynski, B., 1971, Ann. Rev. Astr. Astrophys. **9**, 183.
Radhakrishnan, V., 1982, Contemp. Phys. **23**, 207.
Radhakrishnan, V., 1986, in: Highlights of Astronomy **7**, Reidel. Pub. Comp. Dordrecht.
Radhakrishnan, V. and Srinivasan, G., 1982, Curr. Science **51**, 1096.
Rappaport, S.A. and Joss, P.C., 1983, in: Accretion-Driven Stellar X-ray Sources, Eds. W.H.G. Lewin and E.P.J. van den Heuvel, Cambridge Univ. Press, 1.
Rappaport, S.A., Verbunt, F. and Joss, P.C., 1983, Astrophys. J. **275**, 713.
Ruderman, M.A. and Shaham, J., 1983, Nature **304**, 425.
Ruderman, M.A. and Shaham, J., 1985, Astrophys. J. **289**, 244.
Ruderman, M.A. and Sutherland, P.G., 1976, Astrophys. J. **196**, 51.
Smarr, L.L. and Blandford, R.D., 1976, Astrophys. J. **207**, 574.
Stokes, G.H., Segelstein, D.J., Taylor, J.H. and Dewey, R.J., 1986, Astrophys. J. (in the press).
Taam, R.E., 1983, Astrophys. J. **270**, 694.
Taam, R.E., Bodenheimer, P., Ostriker, J.P., 1978 Astrophys. J. **222**, 269.
Taam, R.E., and van den Heuvel, E.P.J., 1986, Astrophys. J. **305**, 235.
Taam, R.E. and Wade, R.A., 1985, Astrophys. J. **293**, 504.
van den Heuvel, E.P.J., 1984, J. Astrophys. Astr. **5**, 209.
van den Heuvel, E.P.J., 1986a, in: The Evolution of Galactic X-ray Binaries, eds. J. Truemper et al., Reidel Pub. Comp. Dordrecht, 107.
van den Heuvel, E.P.J., 1986b, in: High-Energy Phenomena Around Collapsed Stars, Ed. F. Pacini, Reidel Pub. Comp, Dordrecht (in the press).
van den Heuvel, E.P.J. and Bonsema, P.F.J., 1984, Astr. Astrophys. **139**, L16.
van den Heuvel, E.P.J. and Habets, G.M.H.J., 1985, in: Supernovae, Their Progenitors and Remnants, Eds. G. Srinivasan and V. Radhakrishnan, Indian Acad. Sc., Bangalore, 129.
van den Heuvel, E.P.J. and Taam, R.E., 1984, Nature **309**, 235.
van den Heuvel, E.P.J., van Paradijs, J.A. and Taam, R.E., 1986, Nature (in the press).
Webbink, R.F., Rappaport, S.A. and Savonije, G.J., 1983, Astrophys. J. **270**, 678.
Weisberg, J.M. and Taylor, J.H., 1984, Phys. Rev. Letters **52**, 1348.

## DISCUSSION

**D. Arnett:** Have you considered the effects on the companion of pushing a white dwarf over the Chandrasekhar limit?

**E. van den Heuvel:** Taam and I have looked into this problem. If the ejected amount of mass is not more than a few tenths of a solar mass, we find, using the results of Fryxell and Arnett, that the effects on a red dwarf companion are negligible, although the system gets a runaway velocity due to the impact. In the case of a giant companion, where the envelope is less strongly bound, the effects might be stronger.

**G. Hasinger:** If you do not believe in field decay, do you still need the accretion-induced collapse of a white dwarf?

**E. van den Heuvel:** If one does not believe in field decay, my reasoning that PSR 0820+02 must be young will no longer hold and accretion-induced collapse (AIC) is no longer required to explain its strong field. There are, however, additional reasons which make AIC likely for these systems, namely that the mass loss associated with the core collapse of a massive star would have unbound a system with a low-mass companion, as was pointed out by Helfand *et al.* (1983).

**J.H. Huang:** It may be worthwhile to point out that the neutron star formed by accretion-induced collapse in a binary will, after the collapse, continue to accrete. So, it may grow to a mass of up to 2 solar masses (by accreting for $10^8$ yrs in a low-mass X-ray binary). I could imagine that such neutron stars have a completely different interior structure than neutron stars formed in a type II supernova. This might also be a reason for finding two types of pulsars.

**N. White:** Could you push the neutron star over its maximum mass into a black hole?

**E. van den Heuvel:** In principle this is a viable possibility in the low-mass systems, where the neutron star may accrete as much as a solar mass. If the upper mass limit for neutron stars is somewhere near 2.0 to 2.2 solar masses, as preferred equations of state seem to indicate, this might indeed happen.

**J. Shaham:** Let's go back for a moment to the "0.005 $M_\odot$" scenario you mentioned earlier. What happens to a system of a neutron star and a 0.4 $M_\odot$ main sequence star, say, that is driven by gravitational radiation, as it evolves? According to the Jeffery paper, it spends *most of its life* at rates of $10^{-12}$ $M_\odot$/year or *less*. However - one does not see many systems like that among the LMXBs. That may indicate, that either some process prevents $\dot{M}$ from getting that small (onset of large mass and angular momentum loss from the system, for example, which speeds up the evolution) or that the starting binary does not exist. What is your view on that?

**E. van den Heuvel:** This is an important point, to which I do not have a ready answer. We indeed do not know many systems with X-ray luminosities between $10^{33}$ and $10^{35}$ ergs $s^{-1}$, although according to the models you mentioned, many such sources would be expected to exist. A possible answer might be that once the mass tranfer rate drops below $10^{-10}$ $M_{\odot}$/yr, the sources become transients instead of steady sources. We do indeed observe quite a number of soft X-ray transients, and their total number may be very large. Verbunt has in fact, in his talk here, suggested that the many weak sources seen in globular clusters are such transients in their quiescent state.

**R. Manchester:** Would you expect the process of coalescence in the formation of PSR1937+21 to cause the period to depart from the equilibrium spin-up value?

**E. van den Heuvel:** In the red-dwarf case this is indeed expected, as was shown by Jeffrey (Nature **319**, 384-386, 1986): she expects that the neutron star then settles at the equilibrium period corresponding to the final mass transfer rate of $\lesssim 10^{-12}$ $M_{\odot}$/yr, which is of the order of one second. Also in the case of coalescence with a massive white dwarf it is not clear whether it will settle near the spin-up line. If super-Eddington accretion would occur - since the mass transfer is highly super-Eddington - it might end up to the left of the spin-up line. The fact that it is just located on this line may, however, well indicate that super-Eddington accretion cannot occur over long intervals of time. In the case of coalescence of two neutron stars the end-product has a rotation period of 1.5 msec which, by coincidence, just happens to fall on the spin-up line for the present field strength of the millisecond pulsar.

**S. Kulkarni:** The model you describe for low-mass radio binary pulsars will fail when $P_0$, the initial orbital period is less 10 days since the accretion rate on the white dwarf will then be less that $10^{-9}$ $M_{\odot}$/yr and hence does not result in an accretion induced neutron star. The existence of LMXBs with present periods less than 10 days (ex. 4U1626-67 and 1E2259+586) and the low z-distribution of low-mass binary pulsars suggest that, in fact, the radio binary pulsars come from Population I (or intermediate population) stars. Could you please comment on this radical proposal?

**E. van den Heuvel:** The formula for the evolution-driven accretion rate which I used yields $\dot{M} < 10^{-9}$ $M_{\odot}$/yr only for $P \lesssim 2$ days. However, for just these short periods the formula (which is an approximation, and moreover, also depends on the heavy-element abundance) no longer yields the correct mass-transfer rate, since for these short periods ($\lesssim 4^d$) the effects of angular momentum loss by magnetic braking also become important. In fact, the effects of magnetic braking increase towards the shorter periods, causing the mass-transfer rate to remain $\gtrsim 10^{-9}$ $M_{\odot}$/yr in the period range of $\sim 0^d.5$ to $\gtrsim 2^d$. So even for LMXBs and binary pulsars with orbital periods in this range, the neutron stars may still have formed by accretion-induced collapse. For the shorter-period systems, such as 4U1626-67 this may still be the case, provided that the companion was a hydrogen-poor star (cf. Taam and van den Heuvel 1986; Savonye, de Kool and van den Heuvel 1986).

# SECONDARY COMPONENTS OF BINARY PULSARS & MAGNETIC FIELD DECAY IN NEUTRON STARS

Shrinivas R. Kulkarni
Department of Astronomy, 105-24
Caltech, Pasadena
CA 91125, U. S. A.

**ABSTRACT.** We report the discovery of white dwarf secondaries in 0655+64 and 0820+02 systems. In the 2303+46 system, we do *not* find any optical counterpart suggesting that the companion is another neutron star. The existence of a *cool* and therefore *old* white dwarf in the 0655+64 system implies that the surface magnetic field of neutron stars stops decaying beyond some value(s) of field strength.

a)**0655+64:** Coincident with the timing position (Damashek *et al.* 1982) we find a $m_R \sim 22^m$, cool ($T \sim 8000$ K) white dwarf. The cooling age of the white dwarf ($\sim 2 \times 10^9$ y) is comparable to the spin-down age, $\frac{P}{2\dot{P}}$. In any evolutionary scenario, the white dwarf is formed *after* the primary. This suggests that in weak field pulsars such as PSR 0655+64 ($B_s \sim 10^{10}$ G) the magnetic field does not decay as rapidly as in the isolated, strong field pulsars (see Kulkarni (1986) for more details).

b)**0820+02:** A faint ($m_R \sim 22.8^m$), blue (V−R $\sim -0.2^m$) star is coincident with the position given by Manchester *et al.* 1983. A preliminary U−B determination by Liebert and Spinrad (*pers. comm.*) suggests the candidate is a *hot* white dwarf. Unlike the secondary in 0655+64, the white dwarf cooling age is much smaller than $\frac{P}{2\dot{P}} \sim 10^8$y (see Kulkarni (1986) for further details). Consistent with this young age the field strength of the pulsar is $\sim 3.3 \times 10^{11}$ G – not too different from that of a typical single pulsar.

c)**2303+46:** The large eccentricity of this system immediately suggests that the secondary is another neutron star (Stokes *et al.* 1985) – consistent with the lack of optical detection ($m_R \gtrsim 23^m$). These limits will be improved during the coming Fall season.

d)**1855+09:** In collaboration with S. Djorgovski we tentatively report a faint ($m_V \sim 23^m$) star which is within $3\sigma$ of the best timing position (Segelstein *et al.* 1986). Pending further optical data all we can say at this point is that the candidate (if not a chance background source) is consistent with the low-mass, old white dwarf.

*References*

Damashek *et al.* (1982), *Ap. J.* **253**, L57.

Kulkarni, S. R. (1986), *Ap. J. July issue.*

Manchester, R. N. *et al.* (1983), *Ap. J.* **268**, 832.

Segelstein, D. J. *et al.* (1986), *Nature, in press.*



*D. J. Helfand and J.-H. Huang (eds.), The Origin and Evolution of Neutron Stars, 407.*

# MONTE CARLO SIMULATIONS OF RADIO PULSARS AND THEIR PROGENITORS

Rachel J. Dewey and James M. Cordes
Center for Radiophysics and Space Research
Cornell University
Ithaca NY 14853 USA

ABSTRACT. The formation of neutron stars in binary systems is often used to explain the nature of specific radio pulsars and characteristics of the pulsar population as a whole. We have investigated the extent to which such scenarios provide a self-consistent description of the pulsar population. Using a computer simulation, we modeled the evolution of the main sequence stellar population and compared the predicted neutron star population to the observed radio pulsar population, focusing our attention on the pulsar velocity distribution and the incidence of binary pulsars. These characteristics relate very directly to the binary nature of pulsar progenitors, and are not strongly dependent on models of pulsar magentic field and luminosity evolution.

The need to reproduce both the high velocities typical of pulsars and the low incidence of binary pulsars strongly constrains the formation of pulsars in binary systems. Unless one assumes that virtually all pulsars originate in close binary systems, the observed velocity distribution cannot result from the disruption of binary systems by symmetric supernova explosions; some additional acceleration process (e.g. asymmetric supernova mass ejection or asymmetries in pulsar radiation) must act during or soon after a pulsar's formation. It is possible to reproduce the velocity distribution by assuming that all pulsars are born in binary systems with initial orbital periods less than about 30 years. However, the predicted incidence of binaries is then too large by more than an order of magnitude, unless one also assumes that the process of mass transfer from the primary to the secondary is almost always non-conservative, or that the minimum mass necessary for a stripped helium core to explode as a supernova is larger (over $4\,M_\odot$) than currently believed. Further analyses of the radio pulsar population, the X-ray binary population and the abundances of elements ejected in supernovae should help determine which of these alternatives is most reasonble. Additional studies of the main sequence stellar population, accounting more accurately for evolutionary and observational selection effects, will reduce the uncertainties in modeling the formation of the neutron star population.

It has also been suggested that the observed correlation between pulsar velocities and magnetic moments (see Cordes, these Proceedings) is induced by the differing evolutionary paths by which stars in binary systems form radio pulsars. Our simulation does not reproduce this correlation, and we do not find any paths likely to produce low velocity, low magnetic field neutron stars not in binary systems.

We are submitting a full description of our model and results to *The Astrophysical Journal*.



*D. J. Helfand and J.-H. Huang (eds.), The Origin and Evolution of Neutron Stars, 408.*

# CONSTRAINTS TO POSSIBLE PROGENITOR SYSTEMS OF PSR 1831-00

E.H.P. Pylyser
Astronomical Institute "Anton Pannekoek"
Roetersstraat 15
1018 WB Amsterdam
The Netherlands

Abstract. Numerical computations were carried out to simulate the evolution of the recently discovered radio pulsar binary PSR 1831-00 with a very low mass function ($P_{orb}$ = 1.81 days, Dewey et al. 1986). These show that the minimum value of the initial mass of the progenitor of the mass-losing secondary is about 1 $M_{\odot}$. All computed systems with a final period equal to the observed one, and an initial progenitor mass of 1 $M_{\odot}$ have an age of the order of the Hubble time ($\sim 1.3 \times 10^{10}$ years). For a 1.5 $M_{\odot}$ progenitor of the mass-losing secondary we find that the final (= presently observed) orbital period sets strong constraints to the possible initial system configuration: the binary period should have been such that the 1.5 $M_{\odot}$ star was just at the onset of the over-all contraction phase ($P_{orb}$ = 0.74 days, $X_c$ = 0.05) when it began to fill its Roche-lobe, and the initial mass of the compact mass-accreting star must have been < 0.85 $M_{\odot}$. (This star was then presumably a white dwarf, which later collapsed due to the accretion). For example, the final period and remnant secondary mass for a system with an initial configuration of 1.5 + 0.8 $M_{\odot}$ are 1.83 days and 0.20 $M_{\odot}$, respectively.

If one would start out from a more evolved initial state of the 1.5 $M_{\odot}$ secondary than the end of the overall-contraction phase, the maximum initial mass of the compact object will become drastically smaller and the final remnant mass of the secondary becomes larger than 0.20 $M_{\odot}$ in all cases.

Calculations for a 2.0 $M_{\odot}$ progenitor of the secondary suggest that if such a progenitor can produce the present system, the initial mass of the accreting primary was at maximum 1.30 $M_{\odot}$.

All the calculations were performed with the inclusion of magnetic braking in so far as the extent of the convective envelope allowed this (see Pylyser & Savonije 1986, in preparation). The effects of the explosion effects due to the accretion induced collapse of the white dwarf were not taken into account.

Reference

Dewey, R.J., Maguire, C.M., Rawley, L.A. Stokes, G.H., Taylor, S.H., 1986, preprint.

This work is supported by the Netherlands Organisation for the Advancement of Pure Reseaerch ZWO, contract no. 782-371-017.



*D. J. Helfand and J.-H. Huang (eds.), The Origin and Evolution of Neutron Stars, 409.*

# GEODETIC PRECESSION IN BINARY PULSARS

M. Bailes
Mount Stromlo and Siding Spring Observatories
Canberra ACT 2606 Australia

ABSTRACT. We have calculated the probability of observing geodetic precession in the binary pulsar PSR1913+16 for several different progenitor systems. Such an observation would support the asymmetric kick hypothesis for the origin of pular velocities. The results are shown to be dependent on the assumed progenitor but not to a strong degree. It is concluded that the probability of an observation to date is less than 20 percent in contrast to past predictions. In 15 years we expect this figure to be near 60 percent. We conclude that the null result to date cannot be taken as evidence against the asymmetric kick hypothesis for the origin of pulsar velocities. We develop our model and apply it to binary pulsars in general. We conclude that the likelihood of observing geodetic precession in the near future is low.

A complete version of this paper is soon to be submitted to a scientific journal.



*D. J. Helfand and J.-H. Huang (eds.), The Origin and Evolution of Neutron Stars, 410.*

# V. NEUTRON STAR PHYSICS

*CHAIR: T. Weaver*

NEUTRON STAR INTERIORS

Charles Alcock
Department of Physics
Massachusetts Institute of Technology, Cambridge, MA 02139 U.S.A.

ABSTRACT. The current status of the theory of neutron star interiors is reviewed. The various phases of matter that might exist are discussed and their relation to other areas of astrophysics briefly noted. Observational distinction between the models is not simply obtained; analysis of pulsar post-glitch relation, of neutron star precession and neutron star cooling are promising in this regard.

## 1. INTRODUCTION

This meeting has demonstrated how much our understanding of the properties of neutron stars has grown since their identification in 1968. It is no longer reasonable to describe neutron stars as unusual objects. However, in spite of the great theoretical progress in the past eighteen years, it is still not possible to describe the interior of a neutron star with certainty; indeed, it may turn out that there are no neutrons at all in these objects, and that the title of this meeting is a misnomer! Nevertheless, in spite of the significant conceptual differences between the various models for neutron star interiors, the gross properties of the objects (of mass $1.4\,M_{\odot}$) are well-described by all the models. Observable differences between the models are subtle.

There have been a number of excellent review articles on neutron star interiors since IAU Symposium No. 53 met in 1972. Prominent among these are the two reviews of Baym and Pethick (1975, 1979) and the textbook of Shapiro and Teukolsky (1983). More ambitious recent work on the attempts to use neutron stars as "hadron physics laboratories" was recently reviewed by Pines (1985). This brief review cannot compete in depth with these precedessers; instead, a slightly different slant will be used to illuminate the range of possibilities for descriptions of neutron star interiors.

## 2. STATES OF MATTER

The interior conditions in neutron stars are conventionally described by the physics of "cold, dense matter". This is entirely appropriate, even when finite temperature effects are taken into account (as in the cooling of neutron stars)



D. J. Helfand and J.-H. Huang (eds.), The Origin and Evolution of Neutron Stars, 413–424.

because the total heat content of neutron stars (more than a few seconds after formation) is small compared to their energy density.

It was remarked in the introduction that there is considerable uncertainty regarding the state of the interior of a neutron star. It is informative to explore the physics involved using a phase diagram. Phase diagrams customarily show the material of phase of some substance at all loci in a plane projection of an $(n+1)$-dimensional space where the independent variables are temperature $(T)$, pressure $(P)$, and the $(n-1)$ independent abundances of the $n$ distinct components; the equilibrium phase is obtained by minimizing the Gibbs potential. In principle, our job is easier since only two independent variables, $T$ and the baryon number density $(n_B)$ need be specified; the equilibrium phase may be found by minimizing the free energy. (Minimizing the Gibbs potential at each $(T,P)$ is formally equivalent, but less illuminating in this context).

Of course, most phase diagrams are obtained empirically, usually by materials scientists or geophysicists. Much of the $(T,n_B)$ plane that is of interest to us is inaccessible experimentally, and the diagram is filled out theoretically. There are many sources of uncertainty in this process, which stem from the inherent difficulties in carrying out any strong interaction calculation. The two phase diagrams that are presented here are representative of the uncertainties involved, and in addition, both are consistent with all known facts (*i.e.* nuclear masses, scattering data).

## 3. THE STANDARD MODEL

In Fig. 1, a phase diagram for the hadrons that will be called the "standard model" is presented. For completeness, the plane includes both "hot" and "cold" regions; specifically, both the temperatures and the chemical potentials range from $\sim$ 1 KeV to $\sim$ 1 GeV. Except for the possibility discussed in §4, all of the features in Fig. 1 are very likely to be correct. The precise locations of the boundaries are not at all certain, especially in the hot, dense region; however, the existence of these boundaries and the phases that are illustrated is reasonably well-established.

At temperatures below $3\times10^9\,K$ and densities below $8\times10^6$ g cm$^3$, the equilibrium phase is ionized $^{56}$Fe. At higher temperatures thermodynamic equilibrium shifts in favor of lighter nuclei (shaded region), and for temperatures in the range $10^{10}\,K$ to $10^{12}\,K$ we find protons, neutrons, and electrons, with most of the energy in photons and neutrinos. These composition changes are gradual; no phase transitions occur in this region.

Somewhat near $T=2\times10^{12}\,K$ a dramatic phase transition between nucleonic matter and free quarks occurs. The existence of this phase transition is more or less assured; this statement depends on the existence of quarks as the fundamental constituents of nucleons and the property of asymptotic freedom (*i.e.* that at high energies the quark-quark interactions are not strong.) There is some evidence that this is a first order phase transition. There is good reason to believe it occurs near $T\sim2\times10^{12}\,K$, but a precise calculation of the transition temperature is not

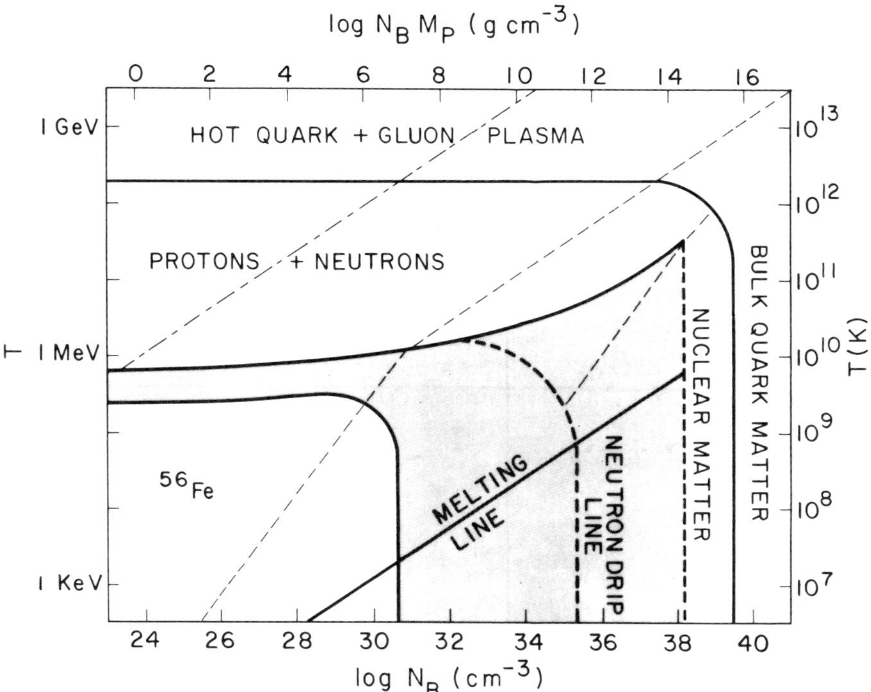


Fig. 1: Phase diagram for the hadrons in the standard model. The regions in which the hadrons exist as $^{56}$Fe, as protons and neutrons, or as free quarks, are marked. In the shaded area, intermediate nuclei are found: at $n_B \lesssim 10^{30}$ cm$^{-3}$ these are nuclei lighter than $^{56}$Fe, while in the higher density shaded region heavier nuclei are found. First order phase transitions separate the quark phase from the nucleonic phase and the solid ionic lattice from the liquid ionic phase (melting line). The neutron drip line and nuclear density line are marked. The long dashed line marks the onset of electron degeneracy. The shorter dashed line marks the onset of neutron degeneracy. The universe evolved along the dot-dashed line.

possible at present. This is not the place to review these issues carefully; access to the literature may be made through Hogan (1983) and Witten (1984).

The universe evolved through the hot, low density portion of Fig. 1; however, for kinetic reasons the universe went out of thermodynamic equilibrium as it cooled below $10^{10}$ $K$, which is why the universe today is not made of lumps of iron.

Now consider the phase plane for $T \gtrsim 3 \times 10^9$ K and for densities increasing up to $4 \times 10^{11}$ g cm$^{-3}$. When the electron degeneracy line is crossed, the material may be described as "cold". Several interesting things occur in this region, which represents the "outer crust" of a neutron star. This discussion is based on that of Baym, Pethick and Sutherland (1971). First, when the melting line is crossed, the ions arrange themselves into a solid lattice; this is a first order phase transition, but very little latent heat is evolved. Second, the increasing electron potential induces electron captures, altering the equilibrium nuclide from $^{56}Fe$ at low densities through eleven steps up to $^{118}Kr$. Each change in equilibrium nuclide is a first order phase transition. It is interesting to note that the equilibrium nuclide at a given pressure is determined, in part, by the lattice contribution to the Gibbs potential.

Continuing at low temperatures but to higher densities, one encounters at $4.1 \times 10^{11}$ g cm$^{-3}$ the phenomenon of "neutron drip". Free neutrons appear in the system. As the density increases the abundance of neutrons increases, and the equilibrium nuclide shifts toward superheavy, very neutron rich species which are arranged in a solid lattice. The equation of state for the region, which represents the "inner crust" of a neutron star, has been discussed by Baym, Bethe and Pethick (1971), Negele and Vautherin (1972) and Negele (1974), among others. Recent work on the superfluid behavior of the neutrons in this region has been reviewed by Pines and Alpar (1985) and by Pines (1985).

At nuclear density ($2.8 \times 10^{14}$ g cm$^{-3}$) and above, the identity of individual nuclei is clearly lost. This material is an interacting gas comprising neutrons with a smaller number of protons and electrons; at densities above a few times nuclear $\Sigma^-$ and then $\Lambda$ hyperons appear, but do not significantly affect the equation of state.

While there is general agreement on the constituent particles in this density regime (but see the discussion in §4 for some disagreement!), there is a wide variety of proposed equations of state for this region. This variety reflects the inherent difficulties of nuclear many-body theory, in that a good model for the interaction potential does not exist, and further that the computational techniques needed to construct an equation of state are not fully developed. There is insufficient space here to appraise the literature critically, and instead representative results will be presented for (a) the mean field (MF) theory calculation of Pandharipande and Smith (1975a); (b) the tensor interaction (TI) model of Pandharipande and Smith (1975b); and (c) the pure neutron gas with Reid potential (R) (Pandharipande 1971). MF and TI are "stiff" models and R is a "soft" model.

At the high density end of the nuclear matter region the electron chemical potential may rise above the effective mass for the negative pion, resulting in the appearance of pions and the possibility of a Bose condensation of pions. This condensation adds massive particles into states of low momentum, thus softening the equation of state. Additionally, the pions greatly enhance the cooling rate due to neutrino emission. Pandharipande (1971) has computed a model equation of state (P) using a Reid potential and a low enough pion effective mass that the condensation appears at a density of $6 \times 10^{14}$ g cm$^{-3}$.

Since it is clear that neutrons and protons are comprised of quarks, and further that the nucleons have finite sizes, it is obvious that at some density the identity of individual nucleons is lost and that a bulk quark phase results. The existence of a phase transition of this kind is not in doubt, however, the density at which this transition occurs is not known. If the transition occurs at a density of $\sim 10^{15}$ g cm$^{-3}$, neutron stars would have a core of quark matter: such stars are known as "quark stars" and have been discussed by, among others, Chapline and Nauenberg (1977), and Fechner and Joss (1978). The uncertainties involved in a discussion of quark matter are discussed in §4.

It is worth remarking on the different natures of the uncertainties regarding the existence of quark matter and of pion condensates. There is no doubt that above *some* density, quark matter exists; there is considerable doubt about whether it exists in neutron stars. On the other hand, while a pion condensate will certainly form in a gas of nucleons above some density, it is possible that quark matter forms out of nucleons before that density is reached; there is, therefore, real doubt about whether pion condensates exist under any conditions.

The high temperature, high density region of Fig. 1 is the least certain portion of the phase plane; the various boundaries were joined together in a plausible fashion. This region of the plane appears to have little interest to astrophysics, except perhaps to model neutron star collisions. It is this region that will be explored in relativistic heavy ion collisions.

## 4. STRANGE MATTER

A radically new possibility regarding the equilibrium state of matter has been raised by Witten (1984). Witten proposed that the true ground state of the hadrons is "strange matter", not $^{56}$Fe. Strange matter is a bulk quark phase that consists of roughly equal numbers of up, down and strange quarks, plus a smaller number of electrons (to guarantee charge neutrality) which is conjectured to have a lower energy per baryon than ordinary nuclei. This means that strange matter is absolutely stable at zero pressure.

A phase diagram that is consistent with this picture is shown in Fig. 2. At densities $\gtrsim 4 \times 10^{14}$ g cm$^{-3}$ and temperatures $\lesssim 10^{11}$ K, there is bulk strange matter, which in many respects is similar to the quark matter phase discussed above. What is different here is that the pressure of strange matter vanishes at $\rho \approx 4 \times 10^{14}$ g cm$^{-3}$. If the mean baryon density of the universe falls below this value, the equilibrium consists of a gas of lumps of strange matter (strangelets), each of which have internal densities of $4 \times 10^{14}$ g cm$^{-3}$. These lumps may have baryon numbers between $\sim 100$ and $2.5 \times 10^{57}$; the upper limit is set by gravitational collapse. The pressure of this phase is entirely due to the photons and neutrinos.

There is a region in the phase plane where neutrons and protons are the thermodynamic equilibrium. This occurs for the same reason that $^{56}$Fe is not favored in this region: strange matter is a minimum of the energy, but not of the free energy.

As in the standard model discussed in §3, a first order phase transition between nucleons and free quarks occurs at $T \sim 2 \times 10^{12}$ K. Note that the universe

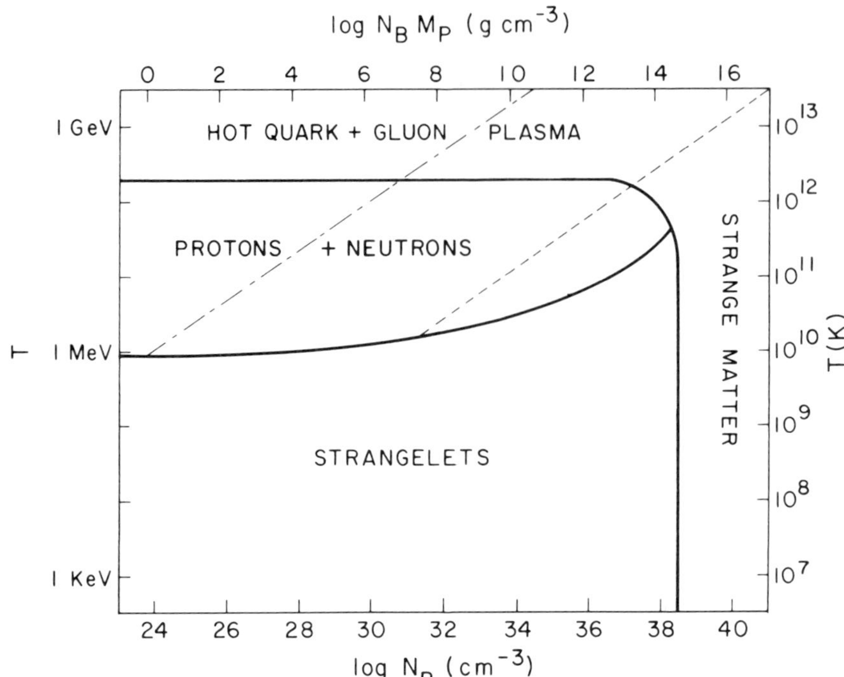


Fig. 2: Non-standard phase diagram for the hadrons. The hadrons exist either as a gas of strangelets, as neutrons and protons, or as free quarks. The neutrons and protons are separated from the quark phase by a first order phase transition. The dashed line marks the onset of electron degeneracy; note that there are almost no free electrons in the gas of strangelets. The universe evolved along the dot-dashed line.

evolves through the "neutron and proton" region; it has been shown that any strangelets which form during the phase transition (a possibility proposed by Witten (1984) and disputed by Applegate and Hogan (1985)) evaporate completely into protons and neutrons, which means that there is very little cosmic strange matter today (Alcock and Farhi 1985).

It is puzzling, given the long history of discussions of quark matter, that the strange matter hypothesis was made only recently. Various forms of quark matter have been discussed by, among others, Ivanenko and Kurdgelaidze (1969), Itoh (1970), Collins and Perry (1975), Freedman and McLerran (1978), Baluni (1978), and Chin and Kerman (1978). The possibility that a form of quark matter with a significant fraction of strange quarks might be absolutely stable was raised by Witten (1984). A detailed study by Farhi and Jaffe (1984) showed that, with the uncertainties inherent in a strong interaction calculation, the existence of strange matter is reasonable. This means that no decision can be made at this point about which of Figs. 1 and 2 represents the equilibrium states of the hadrons. It then behooves us to examine the consequence of this phase for neutron stars; this has been done by Haensal, Zdunik and Schaeffer (1985), Baym *et al.* (1985) and by Alcock, Farhi and Olinto (1986).

Since strange matter is absolutely stable, a star may be made entirely of strange matter. Such an object (a "strange star") has an exposed quark matter surfaces which, since it is held together by the strong force, is not subject to the Eddington limit. However, this surface is probably highly reflective in the X-ray (Alcock, Farhi and Olinto (1986), and very high luminosities will only occur in very hot, transient events. Furthermore, the star does not necessarily have an exposed quark matter surface, since a layer of material identical to the outer crust of a neutron star can be supported electrostatically above the quark surface. There is no possibility of an inner crust in a strange star because the strange matter absorbs all the free neutrons; this may have important consequences for the interpretation of observations, as will be discussed in §6.

## 5. GLOBAL PROPERTIES OF NEUTRON STARS

It is well-known how to construct models of compact stars (see *e.g.* Shapiro and Teukolsky 1983). The vast amount of work in this area is summarized in Fig. 3 which shows mass-radius relations for models constructed with the various equations of state for nuclear matter that were discussed in §3, and for the strange matter equation of state (S) that was derived by Alcock, Farhi and Olinto (1986).

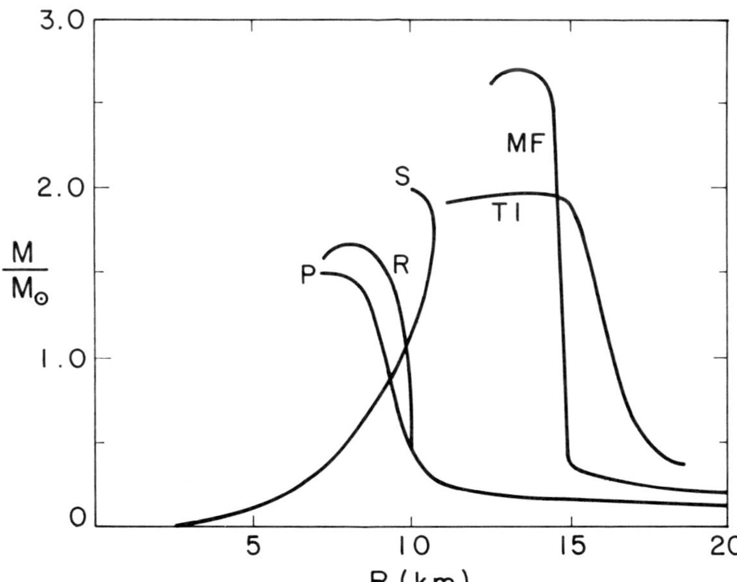


Fig. 3: Mass vs. radius for four neutron star equations of state and for one strange star equation of state.

Most of this figure (*i.e.* the curves for the four nuclear matter equations of state) is familiar. The strange star mass-radius relation is very different. For low masses, the stars have uniform density and $M \propto R^3$; the minimum mass is $\sim 10^{-55} M_\odot$. As the mass increases, gravity becomes important and there is a maximum radius of $\sim 11$ km, followed by a maximum mass $\sim 2 M_\odot$.

However different the strange star mass radius relation is overall, in the mass range that matters ($\sim 1.4\ M_\odot$) it is not greatly different from a neutron star with a Reid potential equation of state. In fact, any of the mass-radius relations shown in Fig. 3 can be reconciled with the observational data on neutron stars; this reflects our ignorance of neutron star radii.

## 6. DISTINGUISHING THE MODELS

A decision between the models for neutron star interiors may be made on the basis of improved theory. Alternatively, a decision may be reached by analysis of observations of neutron stars, an approach advocated by Pines (1985). The observational distinctions may arise from analysis of pulsar glitches, from evidence of neutron star precession, or from analysis of thermal X-rays from cooling neutron stars.

There is a theory for pulsar post-glitch relaxation that involves the massive uncoupling of neutron superfluid vortices in the inner crust during a glitch, followed by a gradual re-coupling of the vortices to the charged component in the star, and thus to the magnetic field. This theory, which has been reviewed by Pines and Alpar (1985) and by Pines (1985), is attractive in that a natural explanation of the long post-glitch relaxation time is possible.

Since all the action in the neutron superfluid model for post-glitch relaxation occurs in the inner crust, it is useful to see how this portion of the star depends on the equation of state in the interior. The radial fractions of core, inner crust and outer crust for three models of $1.4\,M_\odot$ stars are shown in Fig. 4. Note that the stiff (TI) model has a substantial inner crust, the soft (R) model much less, and the strange star (S) has *none*. Clearly, the neutron superfluid model does not apply to the strange star picture; if no model for glitches is developed in that picture, the success of the neutron superfluid models could be interpreted as evidence against the strange matter hypothesis. In this fashion, which unfortunately is model dependent, observations of neutron stars may shed light on fundamental physics.

Another model dependent statement regarding the inner crust has been made recently. Trumper *et al.* (1986) interpret that *EXOSAT* data on the 35-day cycle in Hercules X-1 as evidence for large amplitude ($\sim 25°$) free precession of the neutron star. This model requires a rather thick inner crust, which leads Pines (1985) to argue that the equations of state must be comparatively stiff.

The final remark in this section concerns the cooling of neutron stars. There is evidence for a contradiction between X-ray observations of young supernova rem-

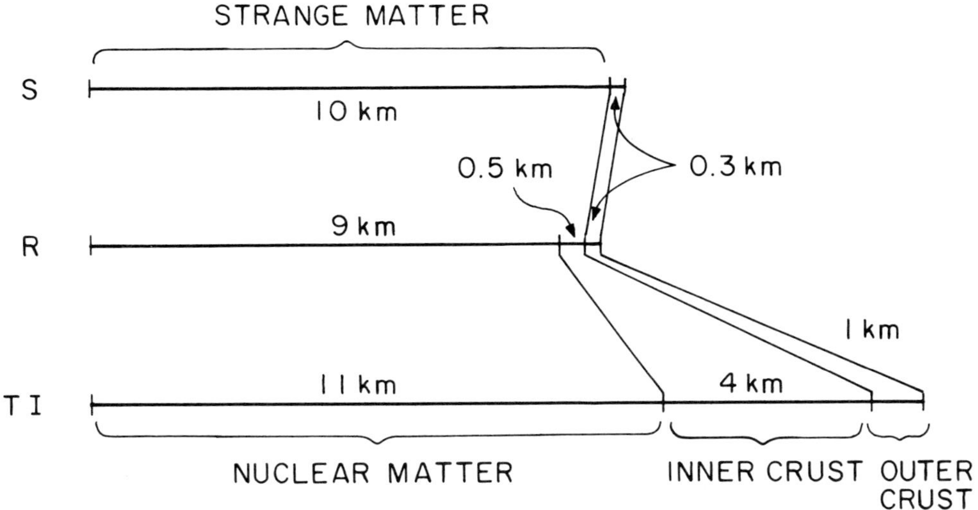


Fig. 4: Schematic showing the relative proportions of the core, the inner crust, and the outer crust for three models of a 1.4 $M_\odot$ object.

nants and the cooling of neutron stars in the standard models (see *e.g.* Glen and Sutherland 1980; Van Riper and Lamb 1981; Yakovlev and Urpin 1981; Nomoto and Tsuruta 1981). Stars with pion condensates cool much faster (Maxwell *et al.* 1977) and so do stars containing quark matter (see *e.g.* Iwamoto 1980; Duncan, Wasserman and Shapiro 1983; Alcock, Farhi and Olinto 1986). Should unambigious evidence for "cool, young" neutron stars come available, this could naturally be interpreted as evidence for pion condensates or quark matter.

In summary, it is possible in principle to draw observational distinctions between the various models of neutron stars. However, such distinctions are often highly model dependent, and we may be testing the imagination of astrophysicists as much as the hadron physics.

## 7. CONCLUSIONS

A "safe" conclusion is that neutron star interiors remains a fascinating area of research. It is not possible to conclude firmly what are the microscopic constituents of a neutron star; there may even be no neutrons in a neutron star.

The best prospects for deciding between the competing models appears to be in the analysis of pulsar glitches, and the study of the dynamics of precessing neutron stars may also prove fruitful; in both of these cases the conclusions will inevitably be model dependent. X-ray observations of young neutron stars are potentially a less model dependent tool.

## REFERENCES


Alcock, C., and Farhi, E. 1985, *Phys. Rev. D*, **32**, 1273.

Alcock, C., Farhi, E., and Olinto A. 1986, *to appear in Ap. J.*

Applegate, J., and Hogan, C. 1985 *Phys. Rev. D.*, **31**, 3037.

Baluni, V. 1978, *Phys. Lett.*, **72B**, 381.

Baym, G., Bethe, H. A., and Pethick, C. J. 1971, *Nucl. Phys.*, **A175**, 225.

Baym, G., Jaffe, R. L., Kolb, E. W., McLerran, L., and Walker, T. P. 1985, *Phys. Lett.*, **160B**, 181.

Baym, G., and Pethick, C. J. 1975, *Ann. Rev. Nucl. Sci.*, **25**, 27.

Baym, G., and Pethick, C. J. 1979, *Ann. Rev. Astr. Ap.*, **17**, 415.

Baym, G., Pethick, C. J., and Sutherland, P. G. 1971, *Ap. J.*, **170**, 299.

Chapline, G., and Nauenberg, N. 1977, *Phys. Rev. D*, **16**, 450.

Chin, S., and Kerman, A. 1978, *Phys. Rev. Lett.*, **43**, 1292.

Collins, J., and Perry, M. J. 1975, *Phys. Rev. Lett.*, **34**, 1353.

Duncan, R. C., Shapiro, S. L., and Wasserman, I. 1983, *Ap. J.*, **267**, 358.

Farhi, E., and Jaffe, R. L. 1985, *Phys. Rev. D.*, **32**, 2452.

Freedman, B., and McLerran, L. 1978, *Phys. Rev. D*, **17**, 1109.

Glen, G., and Sutherland, P. 1980, *Ap. J.*, **239**, 671.

Haensel, P., Zdunik, J. L., and Schaeffer, R. 1985, *preprint.*

Hogan, C. J. 1983, *Phys. Rev. Lett.*, **51**, 1488.

Itoh, N. 1970, *Prog. Theor. Phys.*, **44**, 291.

Ivanenko, D., and Kurdgelaidze, D. F. 1969, *Lett. Nuovo Cimento*, **2**, 13.

Iwamoto, N. 1980, *Phys. Rev. Lett.*, **44**, 1637.

Maxwell, O., Brown, G. E., Campbell, D. K., Dashen, R. F., and Manassah, J. T. 1977, *Ap. J.*, **216**, 77.

Negele, J. W. 1974, in *Physics of Dense Matter*, ed., C. J. Hansen (Dordrecht:Reidel), 1.

Negele, J. W., and Vautherin, D. 1972, *Phys. Rev. C.*, **5**, 1472.

Nomoto, K., and Tsuruta, S. 1981, *Ap. J. (Letters)*, **250**, L19.

Pandharipande, V. R. 1971, *Nucl. Phys. A*, **178**, 123.

Pandharipande, V. R., and Smith, R. A. 1975a, *Nucl. Phys. A.*, **237**, 507.

Pandharipande, V. R., and Smith, R. A. 1975b, *Phys. Lett.*, **59B**, 15.

Pines, D. 1985, Lectures presented at NATO Institute "High Energy Phenomena and Collapsed Stars".

Pines, D., and Alpar, M. A. 1985, *Nature*, **316**, 27.

Shapiro, S. L., and Teukolsky, S. A. 1983, "Black Holes, White Dwarfs, and Neutron Stars" (Wiley: New York).

Trümper, J., Kahabka, P. Ögelman, H., Pietsch, W., and Voges, W. 1986, *Ap. J. (Letters)*, **300**, L63.

Van Riper, K. A., and Lamb, D. Q. , *Ap. J. (Letters)*, **244**, L13.

Witten, E. 1984, *Phys. Rev. D*, **30**, 272.

Yakovlev, D. G., and Urpin, V. A., *Soviet Astr. Letters*, **7**, 88.

## DISCUSSION

**F. Frontera:** Can the period noise observed in X-ray pulsars be interpreted as due to the interval structure of "Strange Stars"?

**C. Alcock:** Probably not, since the strange matter behaves like a Newtonian fluid; however, we have not looked at this issue carefully. Of course, the period noise may be related to the accretion process.

**S. Colgate:** Wang and Eichler point out that a dipole magnetic field's decay is dependent upon the crustal conductivity. The interior fluid allows multipoles to form. Would not a strange star with no crust allow the rapid decay of the presumed dipole field?

**C. Alcock:** My guess is that the answer is yes, but I have not yet fully understood the calculation of Wang and Eicher. I should point out that the maximum crust allowed on a strange star is modest compared to that of a neutron star, so that in any event the Wang and Eichler mechanism should proceed rapidly.

**J. Shaham:** Is there enough crust in "strange stars" to still accomodate Bisnovatyi-Kogan's $\gamma$-ray burst model?

**C. Alcock:** I don't think so, since much of the physics in that model occurs deep inside the inner crust.

**M. Bailes:** Is strangeness conserved and what do you make a strange quark out of?

**C. Alcock:** Strangeness is not conserved in the weak interaction. Strange quarks may be made in a number of reactions, the most important being the reaction $u + d \rightarrow u + s$.

**A. Burrows:** Please describe the processes involved in the propagation of the strange front to the surface and give an estimate of its velocity.

**C. Alcock:** A strangeness front absorbs neutrons, liberating u and d quarks behind the front. The equilibration with s quarks is achieved by both weak interactions and diffusion of s quarks from behind the front. As a result, the front moves about one strong interaction length per weak interaction time. We have not calculated this yet, but the velocity may be as slow as 1 mm/s.

**V. Trimble:** Can you predict the decay time for a magnetic field anchored in a quark star rather than a neutron star?

**C. Alcock:** We have not done a decent calculation yet. The result may be very similar to that for a neutron star. The quark fluid may become a superconductor, a possibility that is still being investigated.

**J. Taylor:** Is there a simple-minded explanation for what seems to me to be an astonishing fact: that your strange stars have the same masses and radii as the neutron stars that we all know and love?

**C. Alcock:** This is because the natural density scale for strange matter ($4\times10^{14}$ g cm$^{-3}$) is close to the mean density for a neutron star. The neutron star achieves this mean density only very close to its maximum mass, which is why the mass-radius relations are so similar here.

**N. Itoh:** In your talk you mentioned the increase of the electron chemical potential as the main cause of the appearance of the pion condensate. I believe the pion-nucleon interaction is essential in bringing about the pion condensate.

**C. Alcock:** Maybe I was unclear on this point. The true condition for the appearance of pions is that the electron chemical potential be larger than the lowest energy state for the pions. The pion-nucleon interaction is crucial in lowering the energy of this state. I had intended to convey this by referrring to the "effective mass" for the pion.

**G. Bisnovatyi-Kogan:** The pion-condensation is connected with the strong interaction between nucleons and not related to the fermi energy of electrons. At densities close to nuclear the potential hole is so deep, that $\pi^\circ$ and $\pi^+\pi^-$ pair birth is energetically favored. This process is similar to the birth of $e^+e^-$ pairs in a strong electrical field.

**C. Alcock:** This is the same as Itoh's question, which I have answered. The pion chemical potential is equal to the electron chemical potential, which is crucial in this issue.

# TWO TYPES OF PULSARS

J.H. Huang
Astrophysics Institute
Nanking University
Nanking
China

ABSTRACT. To sort out the whole sample of pulsars with measured P and $\dot{P}$ into two types has much something to do with the origin and evolution of neutron stars. Under the configuration of two types of pulsars with different spindown mechanism, we have discussed a variety of their properties, including their radio emission mechanism, space velocities, interior structures and evolutionary modes. The fact that different type of pulsars does have quite different properties indicates that the processes to create neutron stars may have two distinct types, say, Type II supernova explosion and the collapse of accreting white dwarfs. The evolutionary mode for our Type I pulsars provides such a key link between binary pulsars and X-ray binary pulsars that we may propose a self-consistent scenario for binary pulsars, X-ray binary pulsars, fast pulsars as well as Type I pulsars.

## 1. INTRODUCTION

The idea of ' Two Types of pulsars ' is not a new one proposed recently. Till now, at least four different kind of arguments have appeared.

The first one came out in 1977, when Helfand and Tademaru suggested that there may have two types of pulsars with different space velocities, scale heights and the initial field strengths, Class A and Class B ( Helfand and Tademaru, 1977 ), based on their analyses of pulsars' proper motion and $|Z|$ distribution. They also postulated that Class A sources may originate in tight binaries where their impulse acceleration at birth is insufficient to remove them from the system, while Class B sources may arise from single stars or loosely bound binaries and are accelerated to high velocities by their asymmetric radiation pressure. Obviously, this is a very suggestive idea, especially for the origin of neutron stars. Thereafter, Radhakrishnan ( 1984 ) further advanced some arguments to show that pulsar velocities are determined by their binary histories and not governed in any way by their magnetic fields.

Also in 1977, Manchester and Lyne ( 1977 ) proposed that there may exist two types of pulsars with different beaming mechanism in order to explain the behavior of interpulse in some pulsars, one forming a coni-



D. J. Helfand and J.-H. Huang (eds.), The Origin and Evolution of Neutron Stars, 425–437.

cal beam directed outward along the open field lines and the other forming a beam directed approximately perpendicular to the magnetic axis.

In the IAU Symposium held in Bonn in 1981, this idea got ahead. As pointed out by Radhakrishnan ( 1981 ) in his concluding review, that the very young and very old pulsars may function in a different way. Since then, this concept has been expounded further mainly by Soviet astronomers, Malov and his collaborators. They suggested ( Malov, 1985 ) that the radio emission from short period pulsars could be described by the light cylinder model and that from long period pulsars conforms to the polar cap model, mainly based on the distinct relation between pulse width and period.

The problem here is that we need to clarify the relation between the pulse width and the angular width of the radiation cone as pointed out by Prószyński ( 1979 ).

In 1982, Alpar and his collaborators ( 1982 ) considered that the millisecond pulsars and some pulsars with short periods, long spindown ages as well as a few binary pulsars may constitute a new type of pulsars with accretion spun-up histories.

There were several talks given in this conference ( see, e.g. van den Heuvel ) about the millisecond pulsar formation and evolution. They all suggest an evolutionary scenario for binary pulsars that the spin-up by accretion moves the pulsars back from the ' graveyard ' into the region of ' living ' pulsars along the horizontal tracks in the $B_S$ vs P diagram ( see, e.g. van den Heuvel, 1984, Fig. 2 ). Surely, this evolutionary stage should have something to do with the X-ray binary pulsars. However, some very large periods for X-ray binary pulsars ( > 1000. second ) are incompatible with the evolutionary tracks from the standard magnetic decay model in which the tracks in the diagram of log $B_S$ vs log P become nearly vertical in the end, caused by the magnetic field decay when their age exceeds the decay time scale. It means that the pulsars stayed in the ' graveyard ' will evolve furhter but almost with the same periods as they had at the time of entering the ' graveyard ', according to the magnetic decay model. And these periods would be less than 1000. sec.

We will show briefly in section 3 that this problem could be overcome, all the binary pulsars, X-ray binary pulsars and millisecond pulsars would find their right places in our proposed evolutionary scenario for Type I pulsars.

In 1983, my collaborators and I proposed that there may exist two types of pulsars with different spindown mechanism ( Huang et al, 1983 ). Now we find that this idea may be very useful for studying the origin of neutron stars. The emphasis of my talk will be put on it.

## 2. TWO TYPES OF PULSARS WITH DIFFERENT SPIN DOWN MECHANISM

### 2.1. New Spin Down Mechanism ( $\dot{P} \propto P^2$ )

As we have shown ( Huang et al, 1982 ) that there is a net magnetic moment for neutron pairs coupled by the $^3P_2$ wave interaction. Thus, there should exist an electromagnetic interaction between $^3P_2$ superfluid neutrons and the internal magnetic field, leading to dipole radiation. The $^3P_2$ superfluid neutrons should lose energy by this radiation, and the total power of it , $W_m$ , is proportional to the total number of superfluid vortexes, which is proportional to the angular velocity of pulsar, $\Omega$ , or

$$W_m = C\ P^{-1}$$

here C is a coefficient relating to the interior structure of neutron stars, P the period of pulsars. Because of the heavy absorption in the interior of neutron stars, not a bit of such emission could come out of the surface of neutron stars. And it is easy to see that the resulting period derivative is proportional to P squared

$$\dot{P} = B\ P^2$$

So, generally the spindown rate of pulsars should have a form as follows

$$\dot{P} = A\ e^{-2t/\tau_D}\ P^{-1} + B\ P^2$$

$\tau_D$ here is the decay time scale. The first term on the right-hand side is just one accepted generally, and the coefficient B is constant with time, which implies that the internal magnetic field is constant.

The statistical analysis shows that the relation between lg $\dot{P}$ and lg P changes from an inverse proportionality to a direct one as pulsars evolve ( Peng et al, 1982 ). For $P < 0.5$ sec and $P > 1.25$ sec, this relation is significantly different which is just what was suggested from our model. And the two characteristic periods, 0.5 sec and 1.25 sec, here have their physical implication as we'll show below.

2.2. Two Types of Pulsars with Different Interior Structures

The distribution of pulsar periods is characterized by the presence of two maxima at about 0.5 sec and 1.25 sec, see Fig.1. The selection effect caused by the instrumental bias against the detection of short period pulsars may affect the distribution only at short period end ( Lyne et al, 1985 ), but will not change the positions of the two maxima.

Fig.2 shows a distribution of periods for pulsars with distances $D \leqq 1.5$ kpc. The features in this distribution are similar to that for the whole sample of pulsars except

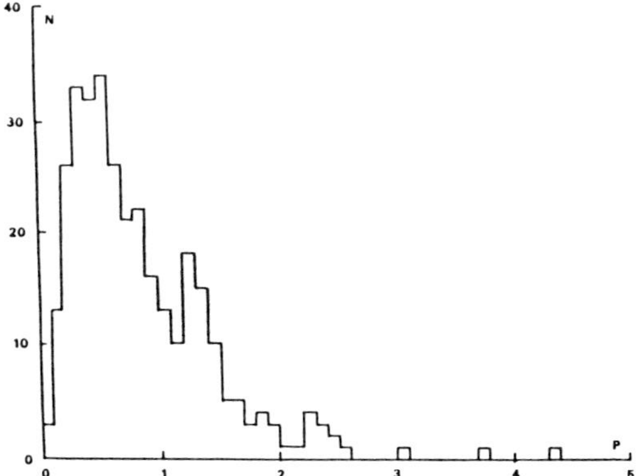


Fig. 1. Histogram of pulsar periods, showing the two maxima at 0.5 sec and 1.25 sec

that the second maximum near 1.25 sec is more striking.

This difference simply shows that the pulsars round 1.25 sec are mainly nearby ones with low luminosity. This conjecture can be confirmed from the luminosity distribution. Fig.3a illustrates the luminosity distribution for the whole sample of pulsars, and that for pulsars within 1.5 kpc of the sun is indicated in Fig.3b.

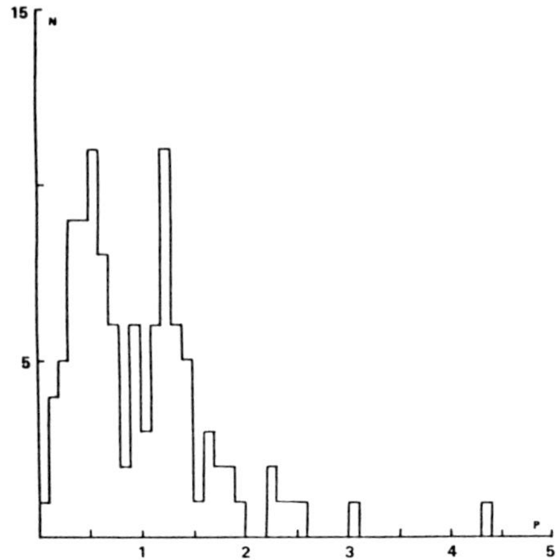


Fig. 2. Same as Fig. 1, but the histogram is for pulsars with D ≦ 1.5kpc

As you can see that the low luminosity pulsars dominate in Fig.3b. The broken line histogram in this diagram shows the distribution for pulsars with P > 1.2 sec and < 1.4 sec, which again indicates that pulsars round 1.25 sec are mainly that with low luminosity.

The theoretical distributions to fit these features are shown in Fig. 4. The dotted line illustrates the theoretical one derived from $\dot{P} = A \exp(-2t/\tau_D) P^{-1}$ model with the value of $\tau_D \sim 10^7$ yr. And the solid line represents the one from the $\dot{P} = A P^{-1} + B P^2$ model with the value of $B = 5A$, no matter whether magnetic decay exists or not.

From this, we infer that there may have two types of pulsars. Type I pulsars could be depicted with the model of $\dot{P} = A P^{-1} + B P^2$ and $B = 5A$; Type II pulsars can be represented with the model of $\dot{P} = A \exp(-2t/\tau_D) P^{-1}$ and $\tau_D \sim 10^7$ yr.

Such a difference in spindown mechanism between Type I and II pulsars suggests a rather large difference in the $^3P_2$ superfluid region between

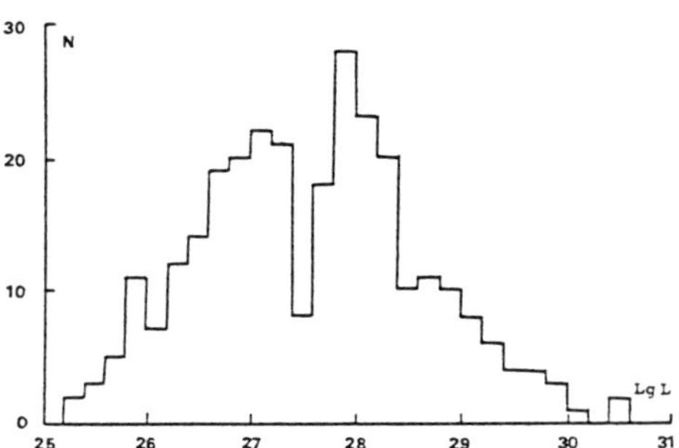


Fig. 3a. Histogram of pulsar luminosity

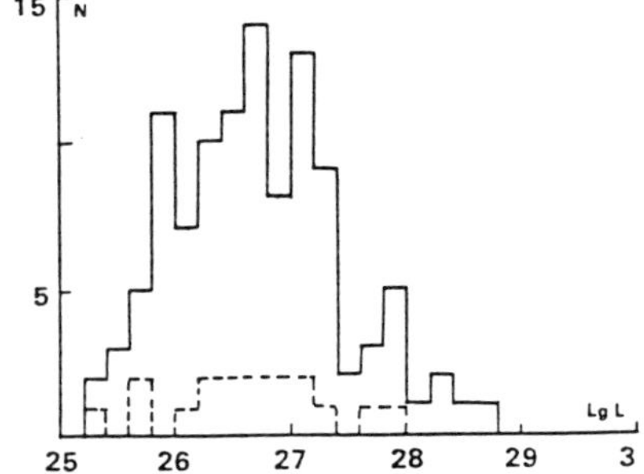


Fig. 3b. Same as Fig. 3b, but the histogram now refers to pulsars with D ≦ 1.5kpc, broken line one for 1.2<P<1.4

these types. That is, the $^3P_2$ superfluid region for Type I pulsars is so large that the effects of new spindown mechanism ( $\dot{P} \propto P^2$ ) on pulsar evolution are significant. On the contrary, the Type II pulsars have very small $^3P_2$ superfluid region, so only one mechanism, $\dot{P} \propto P^{-1}$, is effective on this type of pulsars. When we say, then, that there may exist two types of pulsars with different spindown mechanism, one should realize that it means with different interior structure indeed.

Now the question is how to classify these these two types of pulsars. The criteria to sort out the whole sample of pulsars with measured P and $\dot{P}$ into two types are as follows ( Huang et al, 1985 ).

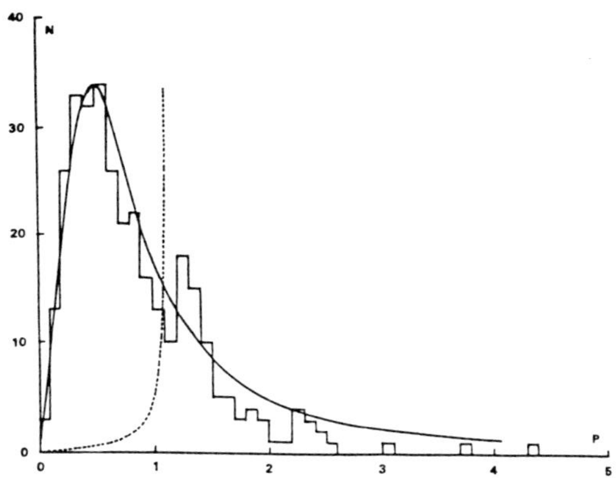


Fig. 4. Histogram of pulsar periods, compared with the prediction of a $\dot{P} \propto P^{-1}$ model ( dotted line ) and of our model ( solid line )

The first one is a relation that Type II pulsars should obey:

$$\lg P = 2 \lg P_\infty - \lg (P + T_D \dot{P})$$

here $P_\infty$ is the maximum period for Type II pulsars. Another one is that the surface magnetic field of two typical pulsars at their birth should follow a relation

$$B_{II} \approx 29 B_I$$

based on the assumptions that the two typical pulsars of each type are different in their interior structure as shown above and that the magnetic fields of the progenitors for these two typical pulsars are the same.

The total number of the resulting sample of Type I pulsars is 130 and that of Type II is 164. The period distribution for Type II pulsars with distances $D \leqq 1.5$ kpc is really typical of the $\dot{P} \propto P^{-1}$ spindown mechanism, represented by the dashed line in Fig. 5. The peak round 0.6 sec in this figure could be produced by the uncertainties in the derivation of pulsars' distances or it just reflects the uncertainties in our Type II sample. In view of the crudeness of our method to distiguish between Type I and II pulsars, however, this histogram still convinces us that our sample of Type II pulsars is basically reliable.

## 3. BASIC PROPERTIES FOR TWO TYPES OF PULSARS

### 3.1. Evolutionary Modes for Two Types of Pulsars

Type II pulsars are those typical of $\dot{P} \sim P^{-1}$ spindown mechanism with mag-

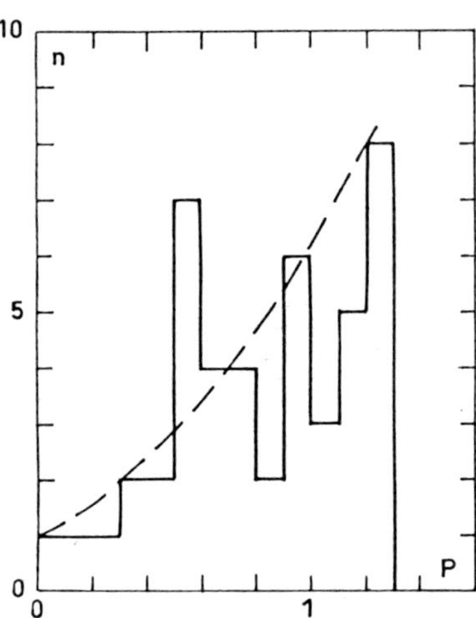


Fig. 5. Period distribution for Type II pulsars with D≦1.5 kpc

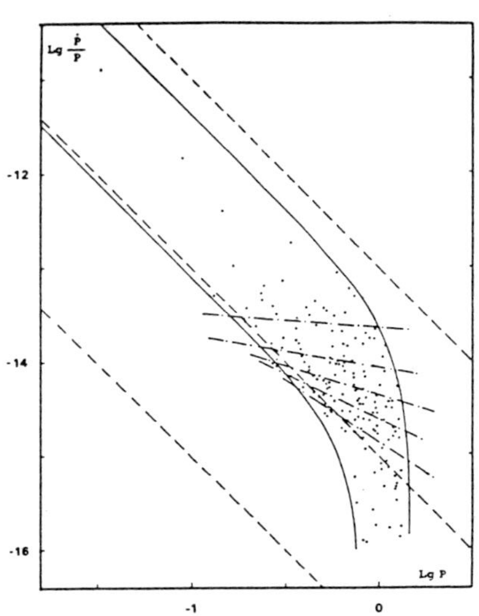


Fig. 6. Distribution for Type II pulsars, showing the magnetic decay

---

netic decay, suggesting an usual evolutionary mode as can be seen in Fig. 6. The initial magnetic fields for this type of pulsars are distributed over a narrow range, $9.4\times10^{11}$ - $6.6\times10^{12}$ Gauss, consistent with those predicted theoretically by Flowers and Ruderman ( 1977 ).

We can also illustrate the evolutionary tracks in another way ( Deng et al, 1987 ), i.e. in a diagram of lg t vs lg P. The true ages for Type II pulsars are given below

$$t = \frac{\tau_D}{2} \ln \left( \frac{2}{\tau_D} \frac{P}{2\dot{P}} + 1 \right)$$

From our considrations ( Huang et al, 1987a ), the value of decay time scale, , is about $10^7$ yr, one obtained by Lyne et al ( 1985 ) is about $9\times10^6$ yr, and that by Krishnamohan ( 1987 ) is about $2\times10^7$ yr. Taking $10^7$ yr, we can get evolutionary tracks for Type II pulsars, represented by dashed lines in Fig. 7. It shows a maximum period of about 1.3 sec.

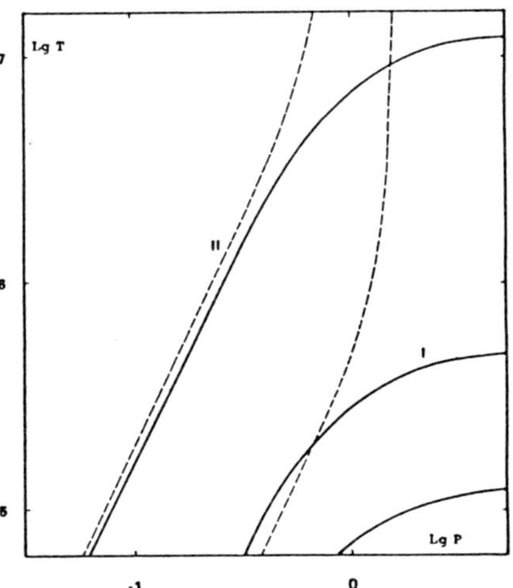


Fig. 7. Evolutionary modes for Type I pulsars ( solid lines ), and Type II pulsars ( dashed lines )

However, The evolutionary mode for Type I pulsars is totally different from that of Type II. The true ages for Type I pulsars are determined with the following formula

$$t_{II} = \frac{(1+5P^3)}{3\ 5^{2/3}P\dot{P}}\left[\frac{1}{2}\ln\left(\frac{5^{2/3}P^2 - 5^{1/3}P + 1}{(5^{1/3}P+1)^2}\right.\right.$$
$$\left.\left. + 3^{\frac{1}{2}}\left(tg^{-1}\frac{2\ 5^{1/3}P+1}{3^{\frac{1}{2}}} + \frac{\pi}{6}\right)\right]$$

As you can see from Fig. 7 that the solid lines, the tracks for Type I, go another way, indicating a rapidly growing of periods for Type I pulsars after P > 1 sec. In other words, Type I pulsars will soon die after P is larger than some typical value.

This evolutionary mode does supply a key link between "binary pulsars" and X-ray binary pulsars. The "binary pulsars" here refer to those systems composed of a pulsar with a normal non-degenerate star as a companion, but they are not expected to be observable as a radio pulsar, as even a tenuous wind or corona will disperse the pulsed signal beyond detectability as was pointed out by van den Heuvel (1984). A scenario for millisecond pulsar formation and evolution (Xia et al, 1986) we proposed is shown in a schematic diagram, Fig 8. The "binary pulsars" (Type I) evolve along the track I first, then they will enter the "graveyard" along the evolutionary tracks for Type I pulsars when their periods are large enough. Before their companions fill their Roche lobe to start the accretion phase, these binaries would evolve further along the track II. The positions to start accreting is dependent on the evolutionary rate of their companions. At these point the "binary pulsars" would become observable X-ray binary pulsars with spun-up by accretion and then jump to the lower part of this figure because of their negative period derivatives.

The dots in the lower part of this figure are those for observed X-ray binary pulsars, that would evolve along the track III till their periods approach to $P_{eq}$. The accreted matter would be swung out by the centrifugal forces at that time, and the X-ray binary pulsars would become observable radio ones with very short periods. The advantage of this scenario is that the observed X-ray binary pulsars get their right place because of our evolutionary mode for Type I pulsars. We will discuss this scenario in detail elsewhere.

## 3.2. Radio Emission Mechanism for Two Types of pulsars

One of the most important parameters for pulsars is the radio luminosity. Gunn and Ostriker (1970) assume a model in which the radio lumi-

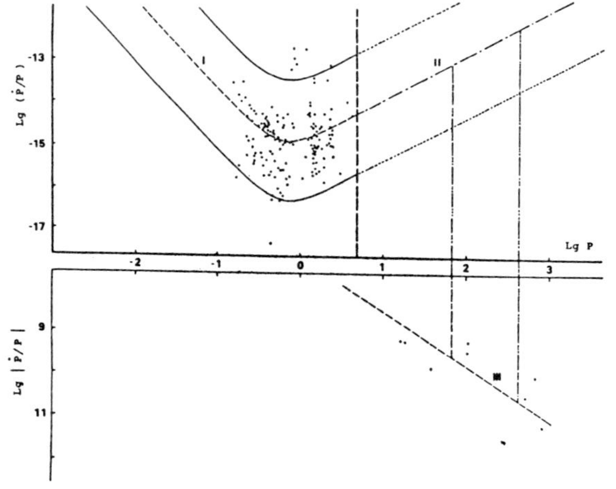


Fig. 8. Schematic diagram for millisecond pulsar formation

nosity is proportional to $B^2$, which was accepted by Lyne et al ( 1985 ). Proszynski and Przybycien ( 1985 ), however, tried to search the relation between the radio luminosity and $P, \dot{P}$ through non-parameter statistics which suggests that the pulsar radio luminosity various roughly as the cube root of the total loss of rotational energy. Pineault ( 1987 ) also tried to explore such a relation using statistical analysis, resulting in a relation of $L \propto B/P$.But the large dispersions over luminosities, periods and period derivatives make these results uncertain. Besides, a method like this has no abilities to clarify the dispute on the radio emission mechanism.

We are trying to do the same work but through a new approach, i.e., to fit in observational true age distributions. Considering the fact that the overwhelming majority of radio pulsars in the Galaxy can be detected with current searching techniques only if they lie within a kiloparsec or two of the sun, we can express the theoretical age distribution for pulsars within 5 kpc of the sun as follows

$$\frac{dN}{dt} \propto \frac{L(t)}{\left|\frac{dL}{dt}\right|} R(L)$$

here $L(t)$ is the radio luminosity of pulsars, $\frac{dL}{dt}$ the luminosity derivative, and R the birth rate.

Assuming that the radio luminosity of pulsars is related to P and $\dot{P}$ with the following formula

$$L = \alpha P^a \dot{P}^b (1+5P^3)^c$$

here $\alpha$ is coefficient, a, b, and c are constants to be determined. The factor $(1+5P^3)$ was taken because the surface magnetic fields for Type I pulsars are given below

$$B_{s,I} = \left( \frac{3c^3 I}{8\pi^2 R^6} \frac{P\dot{P}}{1+5P^3} \right)^{\frac{1}{2}}$$

Then the constants a, b, and c can be determined from the fitness as shown in Fig. 9. for Type II and Fig. 10 for Type I pulsars. The histograms in these two figures are observational true age distributions.

For Type II pulsars, we get $a=-1, b=1$, and $c=0$, suggesting that $L \propto B_s^2/P^2$, or that the radio emission mechanism we should consider for Type II pulsars is the polar cap model. One by-product from this fitness is that the decay time scale, $\tau_D$, must be larger than $5 \times 10^6$ yr, somewhere round $10^7$ yr.

For Type I pulsars, we obtain two sets of values: $a=-6, b=2, c=-1$; or $a=1, b=1, c=-1$, meaning that $L \propto B_L^2 \dot{P}/P$ or $L \propto B_s^2$. This result is consistent with that from multi-element regression analysis, but there may still have two possibilities. One is that we should consider both the polar cap model and the light cylinder model for Type I pulsars. The other is that the approach we use here may not be sensitive enough to distinguish the two mechanisms. We will discuss this problem elsewhere.

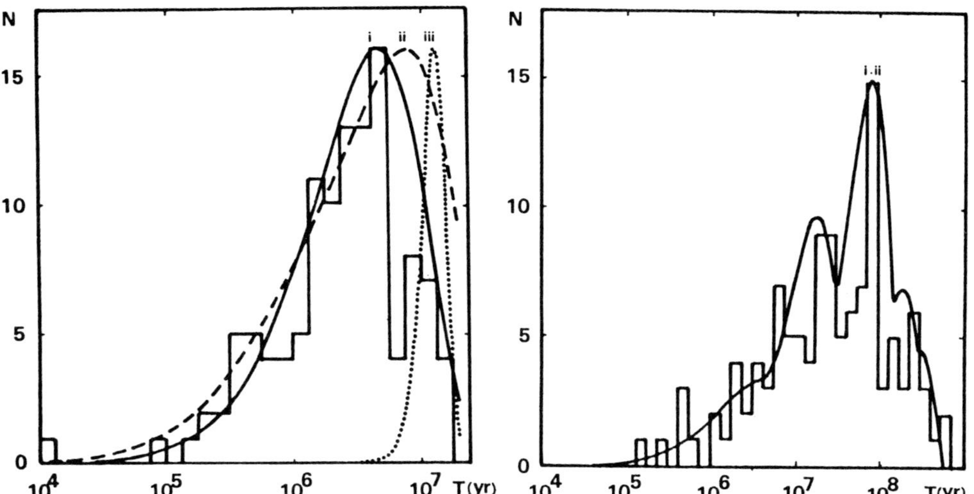


Fig. 9. Histogram of Type II pulsar ages,compared with the theoretical distributions for i) a=-1,b=1,c=0;ii) a=1,b=1,c=0;iii)a=-5,b=1, c=0

Fig. 10. Histogram of Type I pulsar ages, compared with the theoretical distributions for i) a= -6, b= 2, c = -1; ii) a = 1, b = 1, c = - 1

---

## 3.3. Kinematic Properties for Two Types of Pulsars

Following the way that Helfand and Tademaru took ( 1977 ) in their analysis, we find that the kinematic properties for our two types of pulsars just follow what was suggested by them. The average space velocities for Type I pulsars are about 40 - 50 km/sec, which is quite small as compared with that for Type II pulsars, 160 km/sec. ( see the figures in Huang et al, 1987b )

Here we should point out that in analysing the space velocities for Type I pulsars, the data used are only those for ages of less than about $10^7$ yr, while the ages for Type I pulsars are distributed over a range of $10^5$ - $10^9$ yr, because the predicted oscillation period available now is about $7\times10^7$ yr (Arnaud and Rothenlfug, 1981 ).

## 3.4. Interior Structures for Two Types of Pulsars

In a poster paper presented in this conference, we have shown two diagrams with very interesting features ( see Fig. 1a and 2a in Huang et al, 1987c ). Fig. 1a in that paper illustrates a plot of lg $\frac{\dot{P}}{P}$ vs lg $\frac{P\dot{P}}{1+5P^3}$ for Type I pulsars, here $\dot{P}/P$ is a relative energy loss rate, $P\dot{P}/(1+5P^3)$ a quantity proportional to the surface magnetic fields for Type I.

The characteristic of this distribution is that it is dispersed over a very narrow band with a sharp cut-off at its right-hand side.

Fig. 2a ( Huang et al, 1987c ) is a diagram showing the relation between lg $P^2\dot{P}$ and lg $P\dot{P}$ for Type II pulsars, here $P^2\dot{P}$ is a quantity proportional to the magnetogyro ratio, and $P\dot{P}$ the one proportional to the surface magnetic fields for Type I. This distribution is characterized by its triangular dispersion with a cut-off at its left-hand side.

The important parameters entering the relations between the surface magnetic field and $P\dot{P}$ or $P\dot{P}/(1+5P^3)$ are radius, R, and moment of inertia, I, which means that the characteristic of these two distributions should have much something to do with these two parameters. Trying to get some information from these distributions may set certain constraints on the interior structures of neutron stars. A way to do this is the Monte-Carlo simulation.

Fig. 1b and 2b in our poster paper (Huang et al, 1987c) are two plots from Monte-Carlo simulation, both of which fit with their corresponding observational distributions quite well. The premise to get these fitnesses is that the coefficients in these two simulations must differ by an order of magnitude

$$\langle A_{II} \rangle \approx 15 \langle A_I \rangle$$

or

$$\langle B_{II}^2 \left(\frac{R^6}{I}\right)_{II} \rangle \approx 15 \langle B_I^2 \left(\frac{R^6}{I}\right)_I \rangle$$

If we use $A_I$ for Type II simulation or vice versa, what we have will be totally different from the observational ones. To be typical pulsars of each type, it would be very hard to imagine that the interior structures are the same, but the surface magnetic fields are different, or vice versa.

But we have shown that (Huang et al, 1985) if the interior structures for typical pulsars of each type are different, soft EOS for Type II and stiff EOS for Type I, then their surface magnetic fields will be different, the resulting coefficient A for each type will be different too and it just satisfies the need for the Monte-Carlo simulation.

## 4. CONCLUDING REMARKS

In conclusion, the properties of two types of pulsars proposed by us can be summarized in Table 1.

TABLE 1 Basic Properties of Two Types of pulsars

| Type | I | II |
|---|---|---|
| $\rho_c$ | low | high |
| R | large | small |
| $R/I_{45}^{1/6}$ | 1.3 - 1.4 | .65 - 1.05 |
| EOS | stiff | soft |

| | | |
|---|---|---|
| $\dot{P}$ | $A P^{-1} + B P^2$<br>$B = 5A$ | $A \exp(-2t/\tau_D) P^{-1}$<br>$10^7$ |
| $B_s$ | weak | strong |
| L | $B_L^2 \dot{P}/P$<br>$B_s^2$ | $B_s^2/P^2$ |
| Emission Mechanism | Light Cylinder Model<br>Polar Cap Model | Polar Cap Model |
| Evolution | dying of steep increase in P | dying of magnetic field decay |
| Origin | Collapse of accreting WD | SN II |

Now the question is what causes the differences for these two types of pulsars? The very reasonable answer might be the origin of neutron stars. Here we would like to say that Type I pulsars might be formed through the collapse of accreting WD, and the explosion of SN II might create Type II pulsars. But in order to give you more conceivable picture, there are still a lot of work to be done.

References

Alpar,M.A.,Cheng,A.F.,Ruderman,M.A., and Shaham,J.: 1982, Nature,300,728
Arnaud,M., and Rothenflug,R.: 1981, Astron.Astrophys., 103,263
Deng,Z.G., Huang,J.H., and Xia,X.Y.: 1987, this volume
Gunn,J.H., and Ostriker,J.P.: 1970, Astrophys.J.,160,979
Helfand,D.J., and Tademaru,E.: 1977, Astrophys.J.,216,842
Huang,J.H., Lingenfelter,R.E., Peng,Q.H., and Huang,K.L.: 1982, Astron. Astrophys.,113,9
Huang,J.H., Huang,K.L., and Peng,Q.H.: 1983, Astron.Astrophys,117,205
Huang,J.H., Huang,K.L., and Peng,Q.H.: 1985, Astron.Astrophys.,148,391
Huang,J.H., Deng,Z.G., and Xia,X.Y.: 1987a,this volume,'Radio Emission Mechanism for Two Types of Pulsars'
Huang,J.H., Deng,Z.G., and Xia,X.Y.: 1987b,this volume,'Kinematic Properties for Two Types of Pulsars'
Huang,J.H., Deng,Z.G., and Xia,X.Y.: 1987c,this volume,'Interior Structures for Two Types of Pulsars'
Krishnamohan,S.: 1987, this volume
Lyne,A.G., Manchester,R.N., and Taylor,J.H.: 1985, Mon.Not.R.Astron.Soc., 213,613
Malov,I.F.: 1985, Sov.Astron.,29,144
Manchester,R.N., and Lyne,A.G.: 1977, Mon.Not.R.Astron.Soc.,181,761
Peng,Q.H.,Huang,K.L., and Huang,J.H.: 1982, Astron.Astrophys.,107,258
Pineault,S.: 1987, this volume

Prószyński.M.:1979, Astron.Astrophys.,79,8
Prószyñaki,M., and Przybycien,D.: 1984, STScI preprint,No.22
Radhakrishnan,V.: 1981,in Pulsars, ed. W.Sieber,R.Wielebinski,Reidel, Dordrecht,p.449
Radhakrishnan,V.: 1984, in Birth and Evolution of Netron Stars: issues raised by millisecond pulsars, ed. S.P.Reynolds,D.R.Stinebring, p.130
van den Heuvel,E.: 1984, J.Astron.Astrophys., 5,209
van den Heuvel,E.: 1987, this volume
Xia,X.Y.,Deng,Z.G., and Huang,J.H.: 1986,in preparation

## DISCUSSION

**S. Kulkarni:** If the age of Type I pulsars is as large as $10^9$ y, then type I pulsars should form a relaxed system. Is the z-distribution of such pulsars consistent with this expectation? Are some of these pulsars returning back to the galactic plane?

**J.H. Huang:** Yes - indeed. The $|z|$-distribution of Type I pulsars really shows a pattern characteristic of a relaxed system. As you can see from the figure given above, a typical star has oscillated several times when the true age is larger than a few times $10^7$ yr, which is consistent with the predicted oscillation period of about $7\times10^7$ yr. But the data for pulsars' proper motions are not sufficient for judging whether our determined ages for Type I pulsars are correct or not. We have to wait and see.

**J. Shaham:** Could you please elaborate some more on the role of $^3P_2$ neutron superfluidity in changing the slow-down pattern of pulsars?

**J.H. Huang:** In the interior of neutron stars, the $^3P_2$ superfluid neutrons should lose energy through interaction with the internal magnetic field. Then they will move out of the vortex axes without any change in their quantum number. In order to stabilize the superfluid neutron vortices, we assume that the Ekman pumping process may function, i.e., the normal neutrons in the vortex axes will move out and become superfluid ones, then the normal neutrons at the normal neutron layer will flow into the vortex axes. This is a way to transfer the rotational energy of neutron stars into superfluid vortices because there is an interaction between the magnetic moments of normal neutrons and electrons in the inner crust, and the relaxation time scale for this interaction is only seconds. This is how we imagine $^3P_2$ superfluid neutrons slow down the pulsars.

**C. Alcock:** While it is not possible to decide whether the equation of state is stiff or soft, it is difficult to believe that both types of EOS co-exist in nature. The reason for this is that at neutron star densities all reactions proceed to completion, and only one state is allowed.

**J.H. Huang:** If two different processes to create neutron stars exist, say, Type II SN explosion and the collapse of accreting WD, then the rather violent explosion of Type II SN may generate neutron stars with quite small radii and high central densities, i.e., with soft EOS while the rather mild collapse of the accreting WD may create neutron stars with large radii and low central densities, i.e., with stiff EOS. Of course, I can't prove it now, but it would be very hard to believe that different physical processes to create neutron stars lead to only one type of EOS.

**G. Srinivasan:** As we have heard, PSR 0655 may be very old, implying field saturation around $2\times10^{10}$ Gauss. The three millisecond pulsars, on the other hand, have a magnetic field of $5\times10^{8}$ Gauss. This suggests that the asymptotic field strength of neutron stars formed in accretion induced collapse of white dwarfs may be more than an order of magnitude smaller than that of neutron stars formed by direct core collapse. This, again, may be pointing to differences in structure and equation of state of these two types (or classes) of neutron stars.

# NEUTRON STAR COOLING: CRITICICAL TEST OF DENSE MATTER PHYSICS

Naoki Itoh
Department of Physics
Sophia University
7-1, Kioi-cho, Chiyoda-ku, Tokyo 102
Japan

ABSTRACT

Recent developments in the standard theory of neutron star cooling is critically reviewed. Emphasis is placed on the recent developments in the calculations of thermal conductivity and neutrino energy loss rates.

## 1. INTRODUCTION

Comparison of the neutron star cooling theory with the results of the X-ray obsevations provides us with an ideal test of dense matter physics. In this paper I will critically review the neutron star cooling and report on the state-of-the-art calculations.

The Einstein Observatory was launched in 1979 and offered the first opportunity for possible detection of thermal radiation directly coming from neutron star surfaces. Most of the analysis is now completed(Tuohy and Garmire 1980; Pye et al. 1981; Heefand 1981; Harnden and Seward 1984; Harnden et al. 1985).

On the theoretical side progress has been made in recent years concerning the calculations of the thermal conductivity of dense matter(Yakovlev and Urpin 1980; Itoh et al. 1983; Mitake, Ichimaru, and Itoh 1984; Itoh et al. 1984c) and neutrino energy loss rates involving electrons(Itoh and Kohyama 1983; Itoh et al. 1984d; Itoh et al. 1984a; Itoh et al. 1984b; Munakata, Kohyama, and Itoh 1985, 1986). In additon to these standard cooling mechanisms it is possible that neutron stars cool off rapidly due to the efficient neutrino emission in the presence of the pion condensate or the quark matter(Tsuruta 1985) or due to the emission of axions (Iwamoto 1984; Raffelt 1985; Kohyama, Nakagawa, and Itoh 1986).

Nomoto and Tsuruta(1986) have recently carried out the standard neutron star cooling calculation incorporating the recent theoretical developments and compared with the results of the X-ray observations. In this paper I will focus on the



*D. J. Helfand and J.-H. Huang (eds.), The Origin and Evolution of Neutron Stars, 439–446.*

recent developments in the standard(non-exotic) cooling model. Our strategy is as follows: We first work on the standard cooling theory and compare the results with the X-ray observations. If there exist irreconcilable discrepancies, then we introduce exotic cooling mechanisms.

The present paper is organized as follows: Comparison of the cooling calculations with the X-ray obsevations is made in §2. Recent developments in the calculations of the thermal conductivity of dense matter are reviewed in §3. Recent developments in the calculations of the neutrino energy loss rates are reviewed in §4. Concluding remarks are given in §5.

## 2. COMPARISON OF COOLING CALCULATIONS WITH OBSERVATIONS

In Figure 1 I show Nomoto and Tsuruta's(1986) result on the comparison of the cooling calculations with the observations. According to Nomoto and Tsuruta it is likely that no neutron stars are left in Tycho and SN1006 as Type I supernova explosions may very well have disrupted the stars completely. On the other hand Seward(1987) has pointed out in this Symposium the possibility that a neutron star is hidden in Cas A. In that case Cas A and Vela pulsar are much cooler than the standard cooling theory predicts whereas the other sources are not inconsistent with the standard cooling theory. However, one should note that no definitive obsevation of the neutron star surface temperature has been so far made because of the complete lack of the spectral obsevation.

It is possible that exotic fast cooling mechanisms such as pion condensate, quark matter, and /or axion cooling are responsible for the fast cooling of Cas A and Vela pulsar.

## 3. THERMAL CONDUCTIVITY OF DENSE MATTER

Thermal conductivity of the dense matter plays an essential rele in the cooling of neutron stars. There has been recent progress in the calculation of the thermal conductivity relevant to neutron stars. Flowers and Itoh(1976, 1979) presented detailed results of the extensive calculations of the transport properties of dense matter. More recently Yakovlev and Urpin(1980) improved upon the Flowers-Itoh conductivity, and presented more accurate results. Their paper marked the beginning of the precision calculations of the elementary processes occurring inside neutron stars. Stimulated by their work Itoh and his collaborators embarked on further accurate calculations of the electrical and thermal conductivities of the dense matter and published the results recently(Itoh et al. 1983; Mitake, Ichimaru, and Itoh 1984; Itoh et al. 1984c).

In this section I will briefly summarize the calculation of the thermal conductivity of dense matter in the liquid phase. At high densities the polarizability of the degenerate electrons can be neglected as a first approximation, and

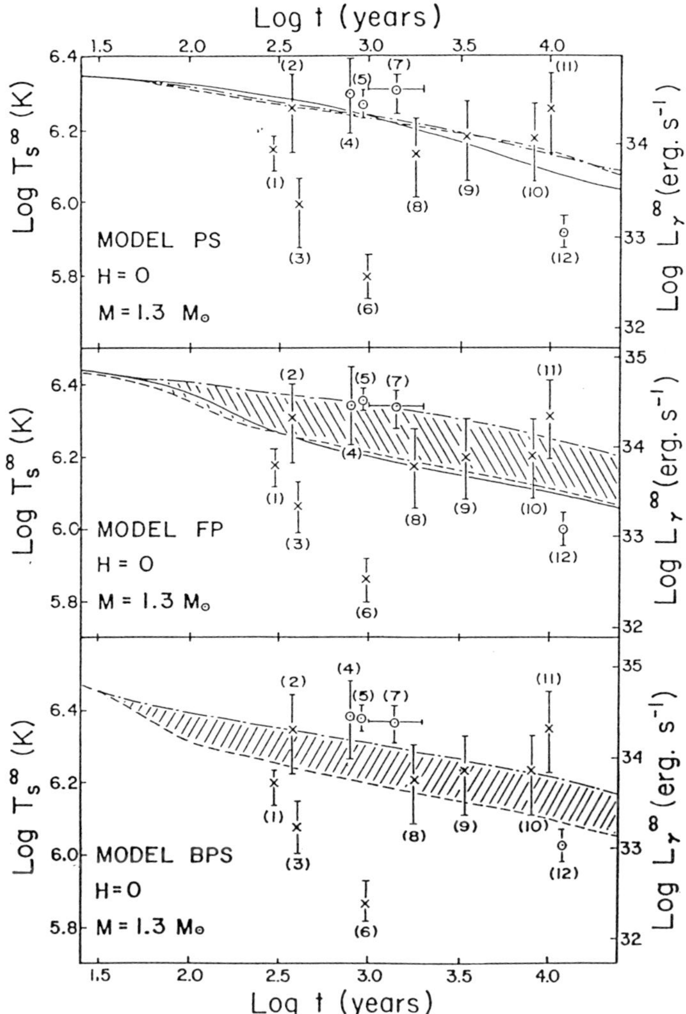


FIG.1. Comparison between the temperature upper limits set by the Einstein observations and the latest theoretical models of neutron star "standard" cooling obtained with the "exact" method. Surface temperatures(left) and photon luminosities(right), both to be observed at infinity, are shown as a function of age, for models PS(top), FP(middle), and BPS(bottom), with $M_A$ = 1.4$M_\odot$ and H = 0. In the dashed curves all standard neutrino emissivities are included. Dot-dashed and solid curves show the maximum and the "best guess" effect of superfluidity, respectively. The numbers refer to the ollowing sources: (1)Cas A, (2)Kepler, (3)Tycho, (4)3C58, (5)Crab, (6)SN1006, (7)RCW 103, (8)RCW86,(9)W28, (10)G350.0-18, (11)G22.7-0.2, and (12)Vela. The circles and crosses refer, respectively, to the temperature upper limits for SNRs with and without detected point sources. The error bars indicate the uncertainty in these upper limits due mainly to the interstellar absorption.

the dense matter can be described as a classical one-component plasma(OCP) embedded in the negative background of electrons. Crystallization of OCP is given by the condition (Slattery, Doolen, and DeWitt 1982)

$$\Gamma \equiv \frac{Z^2e^2}{ak_BT} = 2.275\times10^{-1}\frac{Z^2}{T_8}\left(\frac{\rho_6}{A}\right)^{1/3} \geqq 178, \tag{1}$$

$$a \equiv [3/(4\pi n_i)]^{1/3}, \tag{2}$$

where Ze is the ionic charge, $T_8$ is the temperature measured in units of $10^8$ K, $\rho_6$ is the mass density in units of $10^6$ gcm$^{-3}$, and $n_i$ is the number density of ions. Therefore the liquid state of OCP corresponds to the condition $\Gamma<178$. The ionic correlation in the liquid phase crucially decides the thermal conductivity. In order to incorporate the ionic correlation effect in the calculation of the thermal conductivity, one uses the liquid structure factor of OCP. In Figures 2 and 3 I show the results of the calculation. The ordinate shows a quantity which is proportional to thermal resistivity. Yakovlev and Urpin(1980) completely neglected the effects of electron screening. This accounts for the most part of the discrepancy between their calculation and that of Itoh et al.(1983). In the recent calculations of Itoh and his collaborators accuracy of about 10% is aimed at.

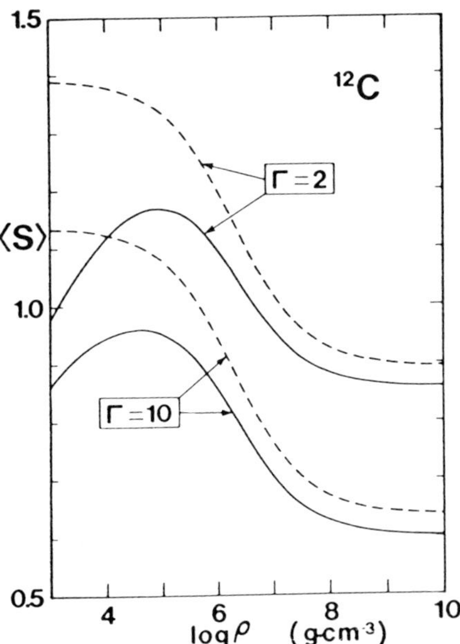


FIG.2. Comparison of Yakovlev and Urpin's resistivity(dashed curves) with the present results(solid curves) for the $^{12}$C matter.

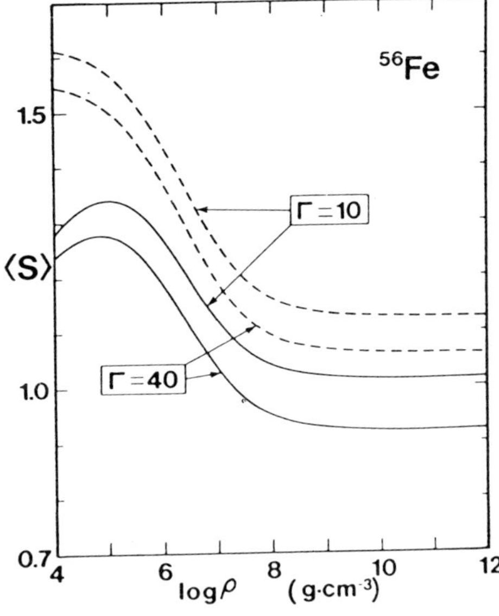


FIG.3. Comparison of Yakovlev and Urpin's resistivity(dashed curves) with present results(solid curves) for the $^{56}$Fe matter.

## 4. NEUTRINO ENERGY LOSS RATES

Neutrino emission processes are efficient cooling mechanisms for neutron stars. There have been important developments in recent years in the calculations of the neutrino energy loss processes involving electrons. There are four major neutrino emission porocesses involving electrons. They are pair, photo-, plasma, and bremsstrahlung neutrino processes. Concerning the former three processes Beaudet, Petrosian, and Salpeter(1967) presented the detailed results of the calculation based on the Feynman-Gell-Mann(1958) theory. More recently Dicus(1972) recalculated these three processes using the Weinberg-Salam theory(Weinberg 1967; Salam 1968). However his numerical calculation was carried out for special densities and temperatures, and hence his results were of little use for stellar evolution computations. Therefore Beaudet, Petrosian, and Salpeter's results were still widely used in stellar evolution computations until recently.

In order to remedy this unsatisfactory situation Munakata, Kohyama, and Itoh (1985, 1986) carried out the calculation of the neutrino energy loss rates due to pair, photo-, and plasma neutrino processes using the Weinberg-Salam theory, and presented detailed results for a wide range of densities and temperatures. In Figure 4 I show an example of the results of the calculation. The results of Munakata, Kohyama, and Itoh's(1985, 1986) calculation are summarized in the following way. Let n be the number of neutrino species other than electron neutrino whose mass is negligible compared with kT. Let $Q_{MIK}$ and $Q_{BPS}$ denote the neutrino energy loss rates calculated by Munakata, Kohyama, and Itoh(1985,1986) and by Beaudet, Petrosian, and Salpeter(1967), respectivly. Then for n = 0 one has $0.35 \lesssim Q_{MKI}/Q_{BPS} \lesssim 0.87$ corresponding to various densities and temperatures; for n = 1 one has $0.56 \lesssim Q_{MKI}/Q_{BPS} \lesssim 0.88$; for n = 2 one has $0.77 \lesssim Q_{MKI}/Q_{BPS} \lesssim 0.88$.

Concerning bremsstrahlung neutrino process Festa and Rudreman(1969) calculated the neutrino energy loss rate using the Feynman-Gell-Mann theory. More recently Dicus, Kolb, Schramm, and Tubbs(1976) calculated this process using the Weinberg -Salam theory. However they did not take into account the ionic correlation effect correctly. Therefore from the point of view of quantitative applications their result was not reliable.

Itoh and his collaborators(Itoh and Kohyama 1983; Itoh et al. 1984d; Itoh et al. 1984a; Itoh et al. 1984b) solved this unsatisfactory situation by calculating the bremsstrahlung neutrino energy loss rate using the Weinberg-Salam theory and taking into account the ionic correlation effects both in the liquid phase and crystalline phase accurately. In Figure 5 I show an example of the results of the calculation. The top curve corresponds to the neutrino energy loss rate that does not take into account the ionic correlation effects and coincides with the result of Dicus, Kolb, Schramm, and Tubbs(1976). The lower curves correspond to the results of the calculation which takes into account the ionic correlation effects accurately. It is readily seen that the ionic correlation effects reduce the neutrino energy loss rate by a factor $\sim$ 3 near the crystallization temperature in the case of $^{56}Fe$ . Therefore the ionic correlation effects are important factors that decide the quantitative feature of the bremsstrahlung neutrino energy loss rate.

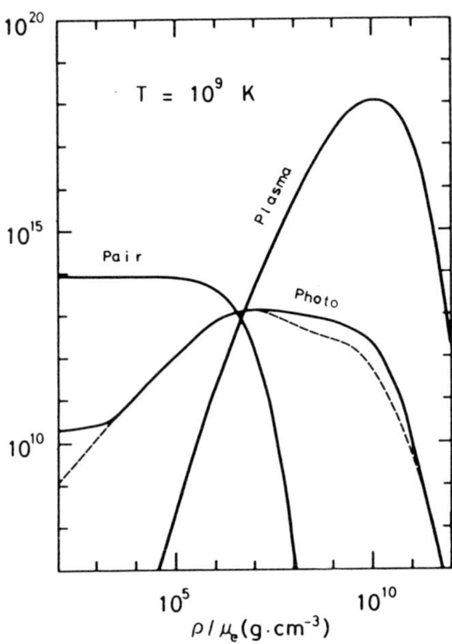


FIG.4. Neutrino energy loss rates $Q_{pair}^+$, $Q_{photo}^+$, and $Q_{plasma}^{BPS}$ in erg $s^{-1}$ $cm^{-3}$ as functions of density. (For the definitions of $Q_{pair}^+$, $Q_{photo}^+$, and $Q_{plasma}^{BPS}$ see Munakata, Kohyama, and Itoh(1985).)

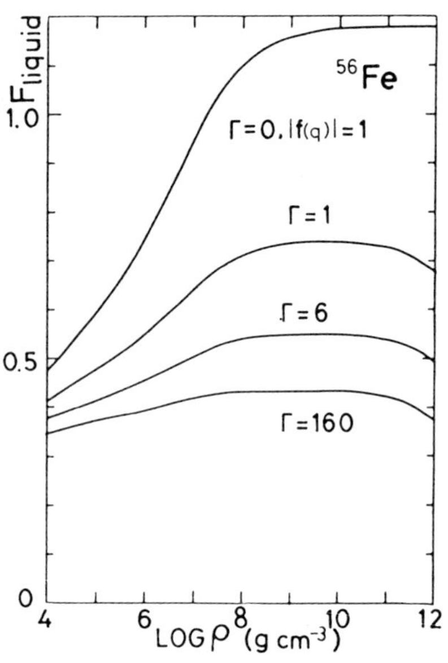


FIG.5. Bremsstrahlung neutrino energy loss rate for $^{56}Fe$ matter in the liquid metal phase. $F_{liquid}$ is proportional to the neutrino energy loss rate.

## 5. CONCLUDING REMARKS

As reviewed in this paper remarkable progress has been made in recent years on theoretical studies of neutron star cooling. In order to have more crucial test of the neutron star cooling theory we certainly need more accurate observations of the neutron star surface temperature, especially the observations of the X-ray spectrum coming from the neutron star surface.

To conclude this review paper, I wish to quote a poem of Fujiwara-no-Teika (1162-1241). In his famous diary Meigetsu-ki he cites an old Japanese record of the Crab supernova(1054). It is also interesting to note that Meigetsu-ki was the very record that opened up the modern research on the historical records of supernovae. In this way Fujiwara-no-Teika contributed greatly to the modern study of neutron stars. His poem is:

I wait and wait
For a lady I love to see
My heart getting scorched
Like salt on the seashore burnt

My conclusion is:

I wait and wait
For a neutron star I love to see
My heart getting scorched
Like salt on the seashore burnt

## REFERENCES

Beaudet,G., Petrosian,V., and Salpeter,E.E. 1967, Astrophys.J., 150, p.979.
Dicus,D.A. 1972, Phys.Rev., D6, p.941.
Dicus.D.A., Kolb,E.W., Schramm.D.N., and Tubbs,D.L. 1976, Astrophys.J., 210, p.481.
Festa,G.G., and Ruderman,M.A. 1969, Phys.Rev., 180, p.1227.
Feynman,R.P., and Gell-Mann,M. 1958, Phys.Rev., 109, p.193.
Flowers,E., and Itoh,N. 1976, Astrophys.J., 206, p.218.
Flowers,E., and Itoh,N. 1979, Astrophys.J., 230, p.847.
Harnden,F.R., Grant,P.D., Seward,F.D., and Kahn,S.M. 1985, Astrophys.J., 299, p.828.
Harnden,F.R., and Seward,F.D. 1984, Astrophys.J., 283, p.279.
Helfand,D.J. 1981, in IAU Symposium 95, Pulsars,ed. R.Wielebinski and W.Sieber (Dordrecht: Reidel), p.343.
Itoh,N., and Kohyama,Y. 1983, Astrophys.J., 275. p.858.
Itoh,N., and Kohyama,Y., Matsumoto,N., and Seki,M. 1984a, Astrophys.J., 280, p.787.
Itoh,N., Kohyama,Y., Mastumoto,N., and Seki.M. 1984b, Astrophys.J., 285, p.304.
Itoh,N., Kohyama,Y., Matsumoto,N., and Seki,M. 1984c, Astrophys.J., 285, p.758.
Itoh,N., Mastumoto,N., Seki,M., and Kohyama,Y. 1984d, Astrophys.J., 279, p.413.
Itoh,N., Mitake,S., Iyetomi,H., and Ichimaru,S. 1983, Astrophys.J., 273, p.774.
Iwamoto,N. 1984, Phys.Rev.Lett., 53, p.1198.
Kohyama.Y., Nakagawa,M., and Itoh,N. 1986, in preparation.
Mitake,S., Ichimaru,S., and Itoh,N. 1984, Astrophys.J., 277, p.375.
Munakata,H., Kohyama,Y., and Itoh,N. 1985, Astrophys.J., 296, p.197.
Munakata,H., Kohyama,Y., and Itoh,N. 1986, Astrophys.J., 304, p.580.
Nomoto,K., and Tsuruta,S. 1986, Astrophys.J., 305.
Pye,J.P., Pounds,K.A., Rolf,D.P., Seward,F.D., Smith,A., and Willingale,R: 1981, M.N.R.A.S., 194, p.569.
Raffelt,G.G. 1986, Phys.Lett., 166B, p.402.
Salam,A. 1968, in Elementary Particle Physics, ed. N.Svartholm (Stockholm: Almqvist and Wiksells), p.367.

Seward,F.D. 1987, talk given at the IAU Symposium No.125 "The Origin and Evolution of Neutron Stars" (Nanjing, China).
Slattery,W.L., Doolen,G.D., and DeWitt,H.E. 1982, Phys.Rev., A26, p.2255.
Tsuruta,S. 1985, Comments on Ap.
Tuohy,I., and Garmire,G. 1980, Astrophys.J.(Letters), 239, L107.
Yakovlev,D.G., and Urpin,V.A. 1980, Soviet Astr., 24, p.303.
Weinberg,S. 1967, Phys.Rev.Lett., 19, p.1264.

## DISCUSSION

**T. Lu:** Your calculation seems to be sensitive to the number of neutrino flavors. Could the present cooling data be used to set a meaningful upper limit on the number of neutrino flavors?

**N. Itoh:** No. The X-ray observational data is not accurate enough to decide the number of neutrino flavors.

**A. Burrows:** It should be pointed out that it is the modified URCA process that dominates the neutrino cooling of a standard neutron star, not the pair, plasmon, bremsstrahlung, or photo-neutrino processes.

**N. Itoh:** Of course the modified URCA process is very important. But plasmon and bremsstrahlung neutrinos are also important for neutron stars which have extended envelopes.

**A. Burrows:** To truly identify surface emission from a neutron star, obtaining X-ray spectra will be crucial. Can you please summarize the status of neutron star emissivity calculations?

**N. Itoh:** The importance of the emissivity of the neutron star surface has been pointed out by Itoh and Brinkmann. But it is a very difficult problem involving the physics of matter under strong magnetic fields and also the atmosphere.

**R. Narayan:** With respect to A. Burrows' question concerning the spectrum of X-rays emitted by a hot neutron star, I should mention that there is a poster that I have put up on this subject, describing work by Romani, Blandford and Hernquist. (The authors were unfortunately unable to come to this meeting.) Could you comment on how the conductivity results would be modified if there were magnetic-field-induced anisotropy in the neutron star interior?

**N. Itoh:** According to the work of Hernquist and his collaborators, the effects of the magnetic field on thermal conduction due to degenerate electrons would not be very great.

# NEUTRON STARS FORMED FROM SUPERNOVA EXPLOSION AND QUARK MATTER

Y.C.Li, X.J.Kong, C.W.Wei, Y.Z.Ge
Physics Department, Hebei Teachers' University
Shijiazhuang, Hebei
China

We study the possibility of the existence of quark matter during the early stage of hot neutron stars. According to Walecka ( 1978 ) and Shuryak ( 1980 ), we calculate the EOS of neutron matter and quark matter at different temperatures, T, in which we take the coupling constant $\alpha_s$ to be 0.5 ( Rafelski, 1982 ) and the bag constant $B^{\frac{1}{4}}$ to be 145, 170 and 190 Mev ( Chin, 1978 ) due to a slight influence of $\alpha_s$ and a great influence of B on our results. Supposing that neutron matter-quark matter phase transition is the first order phase transition ( Baym and Chin, 1976 ), we obtain the phase transition pressures and densities at T = $3\times10^{10}$ and $10^{12}$ K, respectively, from the relation between the pressure and the chemical potential. Then we try to determine the existence of quark matter thru the comparison of these densities with those of stable hot neutron stars.

Our conclusions are as follows: a) There may exist a core of quark matter during the early stage of hot neutron stars as long as its temperature is higher than $10^{10}$ K. The higher the T is, the larger the quark core is; b) B is the important parameter affecting the size of quark core. At a given T, the quark core will become larger in the case of smaller B. When T = $10^{12}$ K and $B^{\frac{1}{4}}$ = 145 Mev, the whole star will almost be composed only of quark matter; c) We choose 1.4 $M_\odot$ as the mass of neutron stars at T = $10^{10}$ K and suppose that there is no mass ejection during the cooling from $10^{12}$ K to $10^{10}$ K. We then obtained the initial mass of neutron stars at T = $10^{12}$ K is equal to 1.68 $M_\odot$ for pure neutron matter and 1.51 $M_\odot$ for pure quark matter ( $B^{\frac{1}{4}}$ = 145 Mev ) according to the conservation of total baryon number. The total energy released by a neutron star during such a cooling process is about $5\times10^{53}$ erg and $2\times10^{53}$ erg, respectively. Besides, we also estimate that the neutron matter-quark matter phase transition will offer $10^{53}$ erg for SN explosion

References

Baym,G., and Chin,S.A.: 1976, Phys.Lett., 62B, 241
Chin,S.A.: 1978, Phys.Lett., 78B, 552
Rafelski,J.: 1982, Phys.Rep., 89, 239
Shuryak,V.: 1980, Phys.Rep., 61, 71
Walecka,J.D.: 1978, Phys.Lett., 78B, 552



*D. J. Helfand and J.-H. Huang (eds.), The Origin and Evolution of Neutron Stars, 447.*

# THE PROBLEM OF SOLIDIFICATION IN NEUTRON STARS

W.H. Huang, S.H. Gao
Department of Physics
Beijing Normal University
Beijing, China

The possibility existing crystal lattice in the interior of neutron stars is an interesting topic, Vela pulsar's glitches phenomenon is explained by means of this structure. There have been many articles in this respect in recent more than ten years. We are doing some work on the same topic.

The pure neutron matter in density region of about $0.4 - 3.6\ fm^{-3}$ at zero temperature is examined. We assume that it is possible to exist diamond lattice with " parallel-spin " configuration in neutron stars.

The neutron-neutron ( N-N ) interaction is chosen to be the central part of the Reid soft-core potential, the interaction of all singlet states are taken to be $V(^1D_2)$, all triplet states are taken to be $V_c(^3P_2 - {}^3F_2)$. In addition, HJ and BKR potentials are examined.

Let the single-particle wave functions are Gaussians centered round any lattice site. The uncorrelated two-particle wave function is an antisymmetric wave function constructed from two single-particle wave functions. The N-N interaction exerts an influence on two-particle wave function. The influence is expressed by two-body correlation function.

The enrgies of the solid enutron matter are obtained by using Pandharipande's lowest-order constrained variation method. There isn't any indication of solidification in neutron stars for this three potentials.

We calculate the loss of energy because of $V(^1D_2)$ instead of $V(^1S_0)$ at $\rho = 1.6\ fm^{-3}$ and correct the energy at this density. The resulting energy is more unfavorable to appearing solidification. This shows that the choice of the interaction has a substantial effect on the appearing of solidification in neutron stars.

Finally, we discuss the effect of crystal lattice on solidification in neutron stars. We compare our results obtained from the Reid potential with the energy of b.c.c. and f.c.c. lattices obtained by Mittet ( 1983 ) using the same method and potential. In order to illustrate the problem further, we compare with their packing fractions either. We find that the crystal lattice is marginally favorable to solidification when the packing fraction is large.

Reference

Mittet,R., et al.:1983, Nucl.Phys., A411,417



*D. J. Helfand and J.-H. Huang (eds.), The Origin and Evolution of Neutron Stars, 448.*

# INTERIOR STRUCTURES FOR TWO TYPES OF PULSARS

J.-H. Huang
Astrophys. Inst., Nanjing Univ., Nanjing, China
Z.-G. Deng
Phys. Dept., Graduate School, Academia Sinica, Beijing, China
X.-Y. Xia
Phys. Dept., Tianjin Normal Univ., Tianjin, China

The Monte-Carlo simulation of the distribution of $\lg(\dot{p}/p)$ vs. $\lg \frac{p\dot{p}}{1+5p^3}$ for type I and that of $\lg(p^2\dot{p})$ vs. $\lg(p\dot{p})$ for type II indicate that EOS for type I is stiff and that for type II is soft.

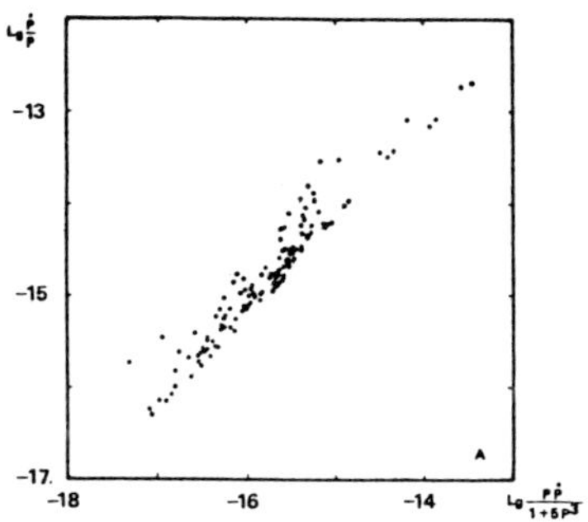


Fig.1a. The observational distribution of $\lg(\dot{p}/p)$ vs. $\lg[p\dot{p}/(1+5p^3)]$ for type I pulsars.

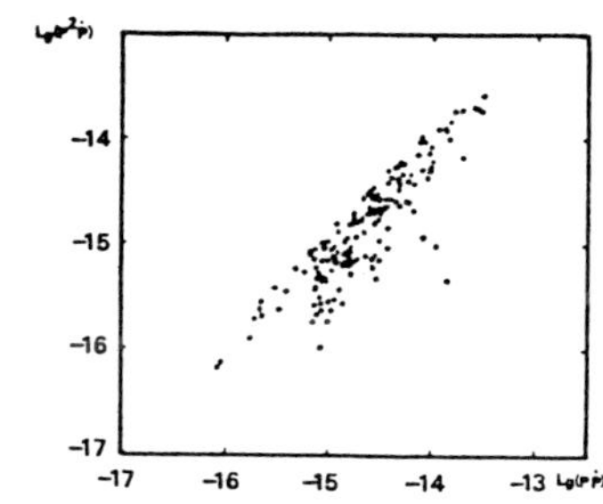


Fig.2a. The observational distribution of $\lg(p^2\dot{p})$ vs. $\lg(p\dot{p})$ for type II pulsars.

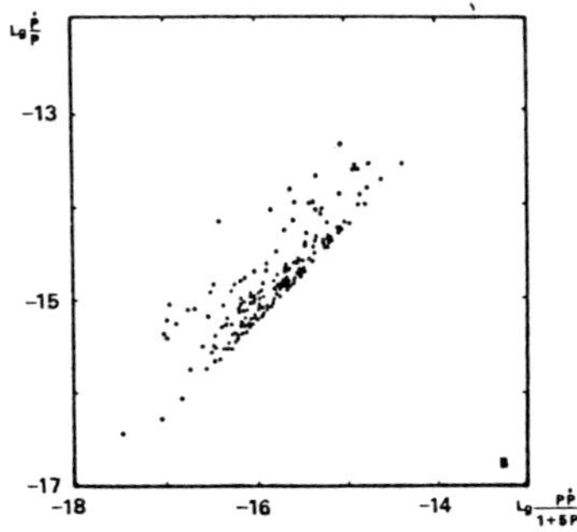


Fig.1b. The M.-C. simulation distribution to fit the observational one showing in Fig.1a.

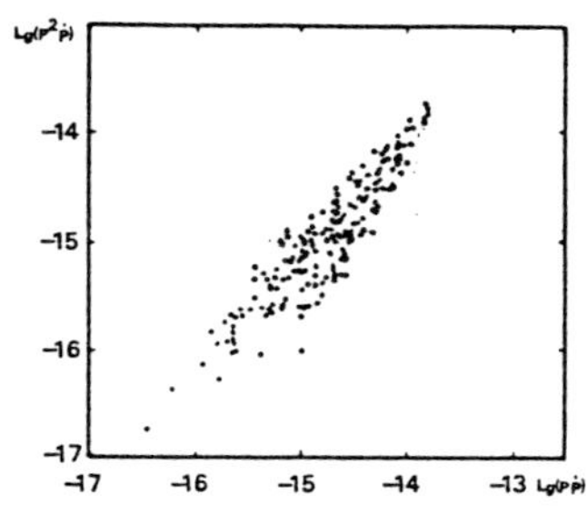


Fig.2b. The M.-C. simulation distribution to match the one showing in Fig.2a.



*D. J. Helfand and J.-H. Huang (eds.), The Origin and Evolution of Neutron Stars, 449.*

# Modes of Energy Loss from Isolated Magnetized Neutron Star

S. Shibata*
Astronomical Institute
Tohoku University
Sendai 980
Japan

Pulsar may be regarded as a discharge tube by electron-positron pair creation. On this viewpoint we carry out two numerical calculations. The obtained magnetic field is consistent with the flow. We find that pulsars emit their rotational energy through three modes simultaneously. The three modes are (1)relativistic acceleration and following gamma-ray emission in the closed current circuit in the magnetosphere, (2)wind of the electron-positron pair plasma, and (3)dipole radiation.

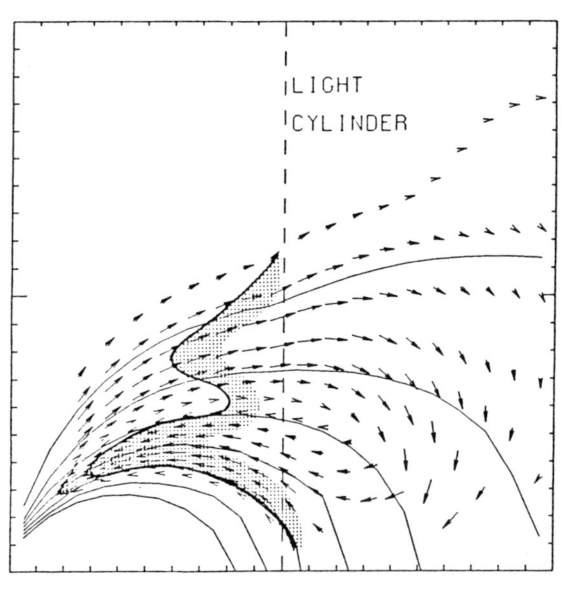


Fig. 1. A numerical model with the circular flow (the closed current circuit). The mechanism of acceleration is that given by Mestel in IAU Symposium No. 95 (1982). We confirm the strong relativistic acceleration causing the trans-field flow in the transition layer (shaded region). Note that the magnetic field, which is consistent with the flow, is closed. Our model suggests the pair creation in the region of the trans-field flow.

* present address: Gunma Technical College,
Maebashi Gunma 371 Japan.



*D. J. Helfand and J.-H. Huang (eds.), The Origin and Evolution of Neutron Stars, 450.*

# EFFECTS OF MAGNETIZATION ON STRUCTURE PARAMETERS OF NEUTRON STARS

Q.M. Wang, S.H. Gao
Department of Physics
Beijing Normal University
Beijing, China

We took the magnetic susceptibility $\chi$ as a criterion in this study and supposed that the system of neutron matter has a ferromagnetic transition as $1/\chi$  0. The magnetic susceptibility of pure neutron matter at zero temperature was calculated by means of Owen's lowest order constraint variation method. The following results were obtained: if the interaction between neutron and neutron was the Reid soft-core potential, then no transition to ferromagnetic state was found to occur; if the interaction was HJ,IY potentials, the neutron matter would undergo a transition. These results indicate that the existence of ferromagnetic state depends on the form of potentials. If the interaction between particles is attractive, it will not profit the existence of ferromagnetic state, while the interaction between neutrons has a repulsive core, it will profit the existence of ferromagnetic state. We also calculated the equation of state and structure parameters of neutron stars. It showed that the energy of ferromagnetic state is lower than that of nonferromagnetic one and the ferromagnetic state is more stable. In other words, the ferromagnetic state may exist in neutron stars. We can readily find that ferromagnetic state has some influence on structure parameters of neutron stars, the magnitude of this effect depends on the form of potentials and the values of these structure parameters with ferromagnetism are within the allowed range. In this paper, the possibility existing the ferromagnetic state has been dsicussed. By a rough estimate, the magnetic field strength coming from the complete ferromagnetic state is about $10^{15}$ Gauss at $\rho \sim 10^{15}$ g/cm$^3$. We assume that this is a possible origin of the strong magnetic fields in neutron stars. If there exists a ferromagnetic state in neutron stars, it will have a substantial influence on the gravitaional collapse theory, neutron superfluid and proton superconductivity.



*D. J. Helfand and J.-H. Huang (eds.), The Origin and Evolution of Neutron Stars, 451.*

# SYNCHRO-CURVATURE RADIATION AND MAGNETIC PAIR PRODUCTION OF RELATIVISTIC ELECTRONS IN STRONG CURVED MAGNETIC FIELDS

S.-Y. An
Phys. Dept., Beijing Teachers College
Beijing
China

Most of common curved magnetic fields with axial symmetry in astrophysics can be expressed in a local cylindrical coordinate system $(R, \varphi, Z)$ as

$$\vec{H} = \vec{H}(0, H_\varphi, 0), H_\varphi/H_o = (R/R_o)^N, \quad (1)$$

where $H_o$, $R_o$ and N are real constants. It has been known that the non-zero curvature of the magnetic field lines can impose an additional pitch angle on the electrons moving in the field (1).

Based on the discussions about the critical pitch angle $\psi_{cr} \simeq \beta c/R_{mc}\omega_H$, where $\omega_H$ is the relativistic cyclotron frequency and $R_{mc}$ the curvature radius of field lines, a criterion of the feature of synchro-curvature radiation from a relativistic electron with pitch angle $\psi$, moving in the field (1), is determined. The results show that when $\psi = \psi_{cr}$, which is generally associated with the primary electrons flying along magnetic field lines out of the polar cap of pulsars, both the curvature radiation mechanism and the synchrotron radiation mechanism would roughly make the same contribution to the total power emitted by the electrons. If $0 < \psi \ll \psi_{cr}$, compared with synchrotron radiation mechanism, the contribution of curvature radiation mechanism to the total radiation power will become much more significant, but if $\psi \gg \psi_{cr}$ the contribution of synchrotron radiation mechanism will be dominant. The magnetic bremsstrahlung from the resulted secondary pairs of cascade, moving in the region far from the stellar surface of pulsars, would mainly belong to the latter.

Under some extreme conditions like that of fast pulsars, the additional pitch angle effect would also affect the estimate of magnetic pair production, magnetic cascade shower and, to a certain extent, the estimate of $\gamma$-ray intensity from fast pulsars, provided that $\psi_{cr}$ is large enough.



*D. J. Helfand and J.-H. Huang (eds.), The Origin and Evolution of Neutron Stars, 452.*

# INVERSE COMPTON SCATTERING IN STRONG MAGNETIC FIELDS: APPLIED TO THE RADIATION MECHANISM OF PSR 0531+21

H. Chen, X.J. Wu, G.J. Qiao
Division of Astrophysics, Geophysics Department
Peking University
Peking, China

In strong magnetic field near pulsar's surface, the quantum effect for electrons is quite complicated. The classical approximation may lose resonance feature, which gives much smaller cross-section.

In this paper, we performed numerical integrations to get the total cross-section and the power spectrum of single electron. Thus considering the resonance, the inverse Compton scattering could be an efficient mechanism in strong magnetic fields. We have carefully calculated the power spectrum of single electron travelling through the isotropic thermal fields.

Taking the electron energy distribution N ( $\gamma$ ) as power law, the same method was used to calculate various non-thermal radaition based on the assumption by Herold that the initial and final states of electrons are in the lowest Lanlau level, which is a good approximation for low temperature ( $< 10^6$ K ). The index of power law spectrum is given by fitting the calculated results to the spectrum of PSR 0531+21. The spark discharge electron number shows us that for hard X-ray ( 5 - 500 kev ), the radiating energy comes from the Compton radiation mechanism with the electron Lorentz factor of 3 - 50, which is quite reasonable and much smaller than that by other models. Some observations indicate that this mechanism may be important in some radaition bands, even in radio band.

Other interesting features are as follows:

a) From 5 to 500 kev, the electron energy strongly affects the radiation power. The higher the electron energy, the more the radiation emitted.

b) Scattered photon directions are quite close to the incoming electron velocity directions, this is so called Doppler beaming effect. From this, pulsars' narrow pulse width could be explained.



*D. J. Helfand and J.-H. Huang (eds.), The Origin and Evolution of Neutron Stars, 453.*

# THE NUTATIONS OF NEUTRON STARS AND CORE-CRUST COUPLING

C. R. Gwinn
Harvard-Smithsonian Center for Astrophysics

Neutron stars, like the earth, are rotating fluid-filled ellipsoids. Poincaré (Bull. Astron. **27**, 321, 1910), Hough (Phil. Trans. R. Soc. **A186**, 469, 1895) and others have discussed the nutations of such objects through a simple model, which treats the crust as rigid and the core as an ideal fluid of uniform density and vorticity. The core and crust are coupled by inertial coupling: the forces which constrain the fluid to its cavity within the crust can produce a net torque, since the cavity is ellipsoidal. Additional torques, and the effects of the elasticity in the crust and density stratification in the core, may be accomodated in such models as well (Sasao *et al.*, Proc. IAU Symposium 78, p. 165, 1980, and references therein).

Such a system has two nutational normal modes, corresponding to the earth's Chandler Wobble (CW) and free core nutation (FCN). The frequencies of the modes in an inertial reference frame are:

$$n_{CW} = \Omega\left(1 - \frac{A}{A_s}e\right),$$

and

$$n_{FCN} = -\Omega\frac{A}{A_s}\left[e_f - i\frac{\kappa}{A_f}\right].$$

Here, $\Omega$ is the star's rotation rate. The equatorial moment of inertia of the entire star is $A$, and the polar moment of inertia is $C$. The dynamical ellipticity of the star is $e = \frac{C-A}{A}$. Analogous quantities $A_f$, $e_f$ refer to the core, and $A_s$ to the crust. Possible perturbations to the inertial coupling are parametrized by $\kappa$.

These modes produce very different motions (Toomre, Geophys. J. R. Astron. Soc. **38**, 335, 1974). In particular, only the FCN involves large relative motion between core and crust, and only it is sensitive to core-crust coupling. Other than inertial coupling, core-crust coupling via the core plasma may be important (Sauls *et al.*, Phys. Rev. D **25**, 967, 1982; Alpar *et al.*, Ap. J. **282**, 533, 1984), particularly in determining the $Q$ of the FCN. The superfluidity of the core fluid in neutron stars should not affect the validity of the model. Shaham (Ap. J. **214**, 251, 1977) has pointed out that pinning of superfluid vortices to crust nuclei may increase the crust's effective moment of inertia; core-crust coupling by vortex pinning is orders of magnitude less than the inertial coupling torque. Observations of the frequency and $Q$ of free nutations of neutron stars, and identification of the mode responsible, should usefully constrain neutron star models.



*D. J. Helfand and J.-H. Huang (eds.), The Origin and Evolution of Neutron Stars, 454.*

# THE SPIN-TORSION COUPLING PRECESSION OF SPIN AND ITS EFFECTS ON SINGLE PULSES OF PULSARS

Y. Zhang
Department of Physics and Chemistry
Nanjing Institute of Technology
China

According to the Gravitational Gauge Field Theories ( GGFT ), the spin $\vec{\mu}$ of elementary particles experiences a precession: $d\vec{\mu}/dt = -3\,\vec{\Omega}\times\vec{\mu}$ in a space-time with torsion tensor $\vec{\Omega}$. This allegation distinguishes the GGFT from the General Relativity. Choosing neutron star as a candidate to judge the theories, we assume an orderly spin alignement model footing on QCD considerations. From detailed calculations we obtained an internal solution for torsion $\vec{\Omega}$ in neutron stars : $|\vec{\Omega}| \sim (\frac{GM}{c^2R})^2\,\omega$, here M is the mass, R the radius and $\omega$ the angular velocity of neutron stars. Every neutron-like particles with spin in neutron stars will experience a spin-torsion coupling precession, which is shown to be a complex function of time variable and spatial coordinates inside the star. It is also shown that in a neutron star every spin magnetic moment has two kinds of precession: one is due to the rotation of the star, another arises from the spin-tortion coupling. These two precessions superpose each other, and give the mode of magnetic dipole radiation fields outside the star. Integrating the low-frequency radiation fields generated by every spin magnetic moment, we obtain the whole electromagnetic radiation field. Our radiation energy current looks like the Pacini-Gold simple rotating magnetic dipole model, except an additional factor $(0.354+1.02\cos 3|\vec{\Omega}|t)$ which is just the need to explain the data of pulses for some pulsars. The illustrations of current-time show a pretty good agreemnet with the observations of single pulses in some pulsars. The main features which could be interpreted with our model are: the periodical forward drifting and restoring of phase of single pulses, the existence of two subpulses ( two peaks ), and some irregular profiles of single pulses, all of which might be the way to tell the GGFT from the General Relativity.



*D. J. Helfand and J.-H. Huang (eds.), The Origin and Evolution of Neutron Stars, 455.*

# NEUTRON STAR COOLING AND THE VELA PULSAR

Ken'ichi Nomoto[1] and Sachiko Tsuruta[2]
[1]Physics Department, Brookhaven National Laboratory; on leave from the Department of Earth Science and Astronomy, University of Tokyo
[2]Department of Physics, Montana State University

We have calculated cooling models of young neutron stars.[3] The theoretical cooling curves for several models are compared with the Einstein X-ray observations of young supernova remnants (Figure 1).

Our most interesting new finding may be that for the Vela pulsar the observed temperature upper limit is below the *standard* cooling curve. This may raise some interesting possibilities for the Vela pulsar, i.e., greatly enhanced cooling through the presence of *exotic* particles such as charged pion condensates and quarks. This may be possible if the Vela pulsar is slightly more massive than a neutron star in, e.g., RCW103. Note that the observed temperature upper limits for point sources RCW103, 3C 58, and the Crab are consistent with the *standard* cooling.

This outcome is interesting in view of the recent report[4] that the temperature of the internal crustal layers of the Crab and Vela pulsar independently estimated by the vortex creep theory is $3.8 \times 10^8$ K and $1.5 \times 10^7$ K, respectively. The corresponding temperatures of these pulsars obtained from the *standard* cooling model are $\sim (3 - 6) \times 10^8$ K and $(2 - 4) \times 10^8$ K, respectively. This leads to a potentially important implication that for the Crab the *standard* cooling model (i.e., no *exotic* particles included) is consistent with the vortex creep theory but it would not be so for the Vela pulsar.

[3]Nomoto K. and Tsuruta S. 1986, *Ap. J. (Lett)*, *305*, L19; 1987, *Ap. J.*, Jan. 15.
[4]Alpar M.A., Nandkumar R., and Pines D. 1985, *Ap. J.*, *288*, 191.

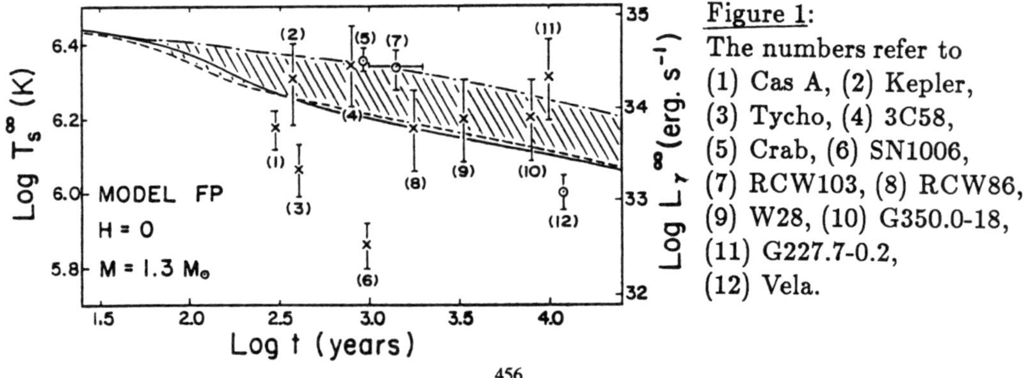


Figure 1: The numbers refer to (1) Cas A, (2) Kepler, (3) Tycho, (4) 3C58, (5) Crab, (6) SN1006, (7) RCW103, (8) RCW86, (9) W28, (10) G350.0-18, (11) G227.7-0.2, (12) Vela.



*D. J. Helfand and J.-H. Huang (eds.), The Origin and Evolution of Neutron Stars, 456.*

# EINSTEIN OBSERVATORY LIMITS ON NEUTRON STAR SURFACE TEMPERATURES

F.R. Harnden, Jr.
Harvard-Smithsonian Center for Astrophysics
60 Garden Street
Cambridge, MA 02138
U.S.A.

ABSTRACT. For years the theoretical models of neutron star formation and evolution had remained largely unconstrained by observation. Following the Einstein X-ray Observatory surveys of supernova remnants and pulsars, however, strict temperature limits were placed on many putative neutron stars. The Einstein search for additional objects in the class of supernova remnants with embedded pulsars has increased the number of such objects by two. For the four objects in this class, the surface temperature limits (see Table 1) provide meaningful logically sound constraints on the neutron star models. For the future, however, still better X-ray observations are needed, both to increase the number of objects available for study and to refine the spatial and spectral capabilities of the X-ray measurements.

TABLE 1
Young Pulsars in Supernova Remnants

| Object | Period (s) | $\dot{P}$ (s/s) | E ($10^{38}$ ergs/s) | age ($10^3$ yr) | $T^a$ ($10^6$ K) |
|---|---|---|---|---|---|
| 0531+21 (Crab) | .033 | 423 | 4.7 | 1.24 | < 2.5 |
| 0540-69 | .050 | 479 | 1.5 | 1.67 | < 2.5 |
| 1509-58 | .150 | 1540 | 0.18 | 1.55 | < 2.5 |
| 0833-45 (Vela) | .089 | 125 | 0.071 | 11.3 | ~ 1.0 |

Note a: Surface temperature determination assumed neutron star radius of 10 km, and blackbody emission for observed counts.

Supported by NASA contract NAS8-30751.



*D. J. Helfand and J.-H. Huang (eds.), The Origin and Evolution of Neutron Stars, 457.*

# THERMAL RADIATION FROM A RADIO PULSAR: PSR1055-52

W. Brinkmann, H. Ögelman
Max-Planck-Institut für Physik und Astrophysik
Institut für Extraterrestrische Physik
8046 Garching
FRG

ABSTRACT. The radiopulsar PSR1055-52 has been observed now by the Exosat observatory for about $10^5$ sec, thus increasing the previously published data base (Brinkmann et al. 1985) by a factor of two. The pulsar was seen in the low energy telescope (0.04 - 2 keV) with the thin Lexan and Al/Par filters, but not in the Boron filter or with the ME experiment.

The data can be fitted with a black body spectrum. Requiring that the emission originates from a region of the size of a neutron star, i.e. $R \sim 10^6$ cm, we obtain temperatures $0.04 \leq kT \leq 0.09$ keV and a value for the interstellar absoption of $0.5 \leq N_H \leq 2.5\times10^{20}$ cm$^{-2}$. A power law fit, as proposed by Cheng and Helfand (1983) results in unacceptable high $\chi^2$-values, unless the power law index $\alpha$ is rather high, $\alpha \geq 3.5$.

The inferred black body surface temperatures of $5.8\times10^5 \leq T_s \leq 9.3\times10^5$ K are compatible with the initial cooling of this $\sim 6\times10^5$ year old neutron star (Tsuruta 1985), with the reheating model of Alpar et al. (1984) - or with a combination of both.

If the interpretation as thermal radiation is correct, this would be the first case of a temperature determination of a neutron star surface by broad band X-ray filter spectroscopy.

## REFERENCES

Brinkmann W., Ögelman H., Aschenbach B. 1985, Sp. Sc. Rev. **40**, p. 527
Cheng A. F., Helfand D. J. 1983, Ap. J. **271**, p. 271
Tsuruta S. 1985, MPA preprint **183**
Alpar A. M., Anderson P W., Pines D., Shaham J. 1984, Ap. J. **276**, p. 325



*D. J. Helfand and J.-H. Huang (eds.), The Origin and Evolution of Neutron Stars, 458.*

# MODEL ATMOSPHERES FOR COOLING NEUTRON STARS

Roger W. Romani and Roger D. Blandford
California Institute of Technology
Pasadena, California 91125

and Lars Hernquist
University of California, Berkeley
Berkeley, California 94720

**Abstract.** The failure of *Einstein* X-ray observations to detect central neutron stars in most young supernova remnants (Helfand and Becker 1984) has provided interesting constraints on cooling theories (*cf.* review by Tsuruta 1985). The comparison of the measured fluxes with the predicted effective temperatures is sensitive to the nature of the emitted spectrum, commonly assumed to be blackbody. The presence of a substantial absorbing atmosphere can, however, produce significant departures. We have calculated model atmospheres for unmagnetized neutron stars with effective temperatures $10^5\mathrm{K} \leq T_{eff} \leq 10^{6.5}\mathrm{K}$ using Los Alamos opacities and equations of state (Romani 1986). We consider a range of surface compositions, since the accretion of $\sim 10^{-19} M_{\odot}$ will cover the surface to the X-ray photosphere and subsequent settling in the strong gravitational field can severely deplete the heavy species. In a low Z atmosphere (*eg.* He) the measured X-ray flux will substantially exceed the blackbody value--the *Einstein* limits on $T_{eff}$ are correspondingly lowered (*eg.* by ~1.6 for SN1006 with a helium surface). For high Z atmospheres, the flux is close to the black body value, but prominent absorption edges are present. Recent calculations of the electron heat transport in magnetized neutron star envelopes (Hernquist 1984, 1985) have shown that, contrary to earlier estimates, magnetic fields will have a small effect on the heat flux ($\lesssim 3$ for parallel field geometries and ~1 for tangled fields). Extension of the atmosphere computations to the magnetic case is important for comparison with X-ray observations of known pulsars.

## References

Helfand, D. J. and Becker, R. 1984. *Nature*, **307**, 215.
Hernquist, L. 1984. *Ap. J. Suppl.*, **56**, 325.
Hernquist, L. 1985. *Mon. Not. R. Ast. Soc.*, **213**, 313.
Romani, R. W. 1986. *Ap. J.*, submitted.
Tsuruta, S. 1985. Max-Planck Institut Für Astrophysik preprint.



*D. J. Helfand and J.-H. Huang (eds.), The Origin and Evolution of Neutron Stars, 459.*

# THE NEUTRON STARS WITH MAGNETIC CHARGES

Y.-J. Wang
Astrophys. Dept., Changsha Railway Inst., Changsha, China
Q.-H. Peng
Astrophys. Inst., Nanjing Univ., Nanjing, China

From KNK metric, we have shown that there may exist some kind of celestial bodies which can not collapse to black holes provided that the proportion of the content of the magnetic monopoles to the nuclei $\xi = N_m/N_B$ exceeds a critical value $\xi_c(1-a^2/R_g^2)$ ,in which $\xi_c = \sqrt{G}\, m_B/g_m \sim 4.4 \times 10^{21}$ , $a = J/Mc$, $R_g = GM/c^2$.

The main results are given below:

a. If the content of magnetic monopoles in celestial bodies is so high that $\chi \equiv \xi/\xi_c > 1$ which is possible for supermassive celestial bodies, no horizon of black holes will appear for the heavily collapsing bodies. Around the celestial body, the force from KNK metric becomes much strong. In the region $R \leqq r \leqq a$, any particle will be repulsed by the central body, which means that no collapse will happen. Because of the RC effect of the magnetic monoples, there may not exist a central singularity. This kind of celestial bodies will be the neutron stars rather than black holes.

b. Because of the coupling between the angular momentum of the central body and that of particles, some sort of latitudinal forces will be produced. Under the combining effect of $F_\vartheta$ and $F_\varphi$ , most of particles on the surface of celestial body including magnetic monoples will move to the polar region near $r \approx \tilde{R}_g(1+a^2/\tilde{R}_g^2)$, forming the bright spots due to the RC effect. The appearance of the celestial body will be a ellipsoid revolving around the polar axis. In view of the same effect, the strong rediation pressure will push matter outward from the two polar regions, forming two jets from opposite diractions.



*D. J. Helfand and J.-H. Huang (eds.), The Origin and Evolution of Neutron Stars, 460.*

# THE NONLINEAR DISPERSION RELATION AND THE RELATIONSHIP OF THE FORMING SOLITON AREA TO THE EVOLUTION OF PULSARS

Mao Xinjie and Tong Yi
Astronomy Department
Beijing Normal University
Beijing, China

Based on Zakharov's equations, Karpman et al. claimed that there were solitons in the magnetosphere. According to the model proposed by Goldreich and Julin, there is a strong induced electric field in the magnetosphere. It seems that we should include the nonlinear effects of the electric field on the polarization of hydrogen atoms if there really are some hydrogen atoms spreading in the magnetosphere of the pulsar. Assuming the magnetosphere is symetric, therefore the electric polarization of hydrogen atoms is of the form $P=\chi^{(1)}E+\chi^{(3)}E^3$. We treat $\chi^{(1)}$ and $\chi^{(3)}$ as scalars because most hydrogen atoms in the universe are in the ground state and $\chi^{(3)}$ is much smaller than $\chi^{(1)}$.

The effect of a strong magnetic field on an electron in orbital motion may be ignored if $B \leqslant 10^8$ Gauss for the case of some pulsars. When the external E-field is comparable with the interior of a hydrogen atom, the induced nonlinear term of P is significant. In laser experiments carried out in labs physicists have observed as high as three-order nonlinear effects when the intensity of coherent lights reaches $10^3$ statvolt /cm. So we may set the interior field intensity of a hydrogen atom as an upper and $10^3$ statvolt/cm as a lower limit, and including the limit of B-field, delimit the region where nonlinear effects dominate.

Although the strong magnetic field makes a plasma anisotropic and the electric conductivity is not a scalar, either, we can expect that $\sigma_\perp$ is much larger than the Hall conductivity so that we obtain a one-dimentional equation derived from the Maxwell equations and get the nonlinear dispersion relation as $\omega=\omega(k, E^2)$. Finally we obtain the nonlinear Schroedinger equation. The localized steady solution of this equation for the case $\kappa>0$ is solitons. Therefore the solitons still exist even though the conditions for establishing the Zakharov's equations are not satisfied.

It is plausible for a pulsar with smaller polar magnetic intensity and a longer rotating period to generate solitons and emit radiation near to the star according to the curvature radiation of solitons, but for an opposite case, near the light cylinder. The area emitting energy in the magnetosphere may be associated with the age and further the evolution of pulsars.



*D. J. Helfand and J.-H. Huang (eds.), The Origin and Evolution of Neutron Stars, 461.*

# VI. OTHER MANIFESTATIONS OF NEUTRON STARS AND THEIR PLACE IN NEUTRON STAR EVOLUTION

*CHAIR: S. Colgate*

# Neutron Stars and Gamma-Rays

Giovanni F. Bignami
Istituto di Fisica Cosmica del C.N.R.
Via Bassini, 15
20133 Milano, ITALY

## 1. Introduction.

Neutron stars are difficult to observe. This has always been known, since the time of their conception back in 1932, just after the laboratory discovery of the neutron, and also since the time of the first pulsar observation in 1968. And yet they must be abundant in our Galaxy, if they are a frequent endpoint of stellar evolution: may be as many as $10^9$, if they have been appearing at the rate of $\sim 10^{-1}$/year over a Hubble time. This, for any reasonable disc distribution, implies that the nearest to us is only $\sim$10 pc away, and, what's more, that there could be $\lesssim$ 1,000 within 100 pc from us. There also seems to be a reasonable consensus (see Manchester, 1987; and Naranyan 1987) that the total number of pulsars in the Galaxy, i.e. of those neutron stars that are currently in the active pulsar phase, is 1-3 x $10^5$, and that of these some 5-7 x $10^4$ should be observable with an "infinite sensitivity" telescope. The current grand total of radio pulsars detected in our Galaxy is 435 (Manchester, 1987), and to these observations we owe much of what we know about neutron stars. Gamma rays, however, are also coming up strong as an observational channel for neutron stars, and one that is unique in its width: it covers nearly ten decades of photon energy (from $10^6$ to $10^{16}$ eV), and as such it also probably covers a rich variety of physical processes happening on and near the neutron stars. The total number of objects observed in gamma-rays, and surely or most probably associated with neutron stars is similar to that of the observed radio pulsars, featuring:

- 400 gamma-ray "busters", probably associated with a local population of binary neutron stars undergoing weak accretion from underluminous companions.
- two fast radiopulsar (in Crab and Vela), several ($\sim$5 ) X-ray binaries, including the "honorary" binary Cyg X-3 and the "exotic" one SS433.
- an unkown fraction of the 20 Cos-B Unidentified Gamma Objects in the galactic plane, including Geminga.

While an extensive presentation of both the gamma-ray burst phenomenology and of Cyg X-3 are given elsewhere in this volume (see



D. J. Helfand and J.-H. Huang (eds.), The Origin and Evolution of Neutron Stars, 465–475.

Hurley, 1987 and Barnard, 1987, respectively), in what follows we shall briefly review the highlights of the gamma-ray manifestation of the Crab and Vela pulsars, present a summary of the evidence of UHE/VHE ($10^{11}$-$10^{16}$eV) emission from X-ray binaries, give the latest Cos-B Collaboration results on high-energy ($\gtrsim$100 Mev) galactic sources, and finally present an update on the work in progress toward the identification of Geminga, our best case for nailing an UGO. Much more detailed work on the energetic emission on neutron stars, both from an observational and interpretative viewpoint, can be found in the recent review by Ruderman (1986).

## 2. Highlights of recent gamma-ray observations of the Crab and Vela pulsars.

The gamma-ray emission from the two fast radiopulsars in Crab and Vela (PSR 0531+21 and PSR 0833-45) has been investigated for about one and half decade now, with a wealth of data summarized e.g. in Bignami and Hermsen (1983), and more recently, in Ruderman (1986). Briefly, the data now consist of:

- strong, pulsed emission in the $10^{6}$- $10^{9}$ eV region, studied in detail in balloon and satellite missions.

This gamma-ray emission is important for the overall energy budget of the neutron stars, being, in the case of the Crab, $10^{-4}$of the total e.m. pulsar energy loss, and as such only slightly inferior to the energy channeled in pulsed X-rays. In fact, up to the upper limit of the Cos-B energy range, and starting from the hard X-ray region, PSR 0531+21 shows a remarkably constant power-law spectrum with $dN/dE \sim E^{-2}$, i.e. with a costant energy content per decade. In this context, the only new information is that on detailed spectrum from the Crab pulsar in the final Cos-B data base.
Fig 1, taken from Clear et al (1986), shows the pulsed gamma-ray emission of PSR 0531+21 in several energy intervals over the Cos-B range and for the totality of the data of the $\lesssim$ 7-year mission. Possible variability of emission has also been observed by Cos-B, albeit at the $\sim$99% confidence level (Wills et al, 1982).

- VHE ( $10^{11}$ eV) gamma-ray have been observed from the Crab (e.g. Dowthwaite et al, 1984), and, with a smaller degree of confidence also from the Vela pulsar .

Here the overall energy involved is somewhat smaller, $10^{-2}$ of that at e.g., $10^{8}$eV, and the overall spectrum does indeed show a steepening.

- Finally, there have been recent reports of UHE ($\gtrsim 10^{15}$ eV) gamma-rays observed from the Crab Nebula region, with no evidence as yet of pulsed emission.

If confirmed (see e.g. Dzikowsky et al.1981, Kirov et al. 1985), this all-important observation would imply the presence at or around the pulsar of particles with energies in excess of $10^{15}$-$10^{16}$ eV.

Obviously, young, rapidly spinning neutron stars like Crab and Vela can act as potent accelerators, in the sense that for them the $\Omega^2$B term can sustain the strong potential drops $\Delta$V's eventually responsible for the particle acceleration. However, the VHE/UHE observations have

Fig.1

Phase histogram of the total gamma-ray emission of PSR 0531+21 in six energy intervals.
Interval boundaries are given.

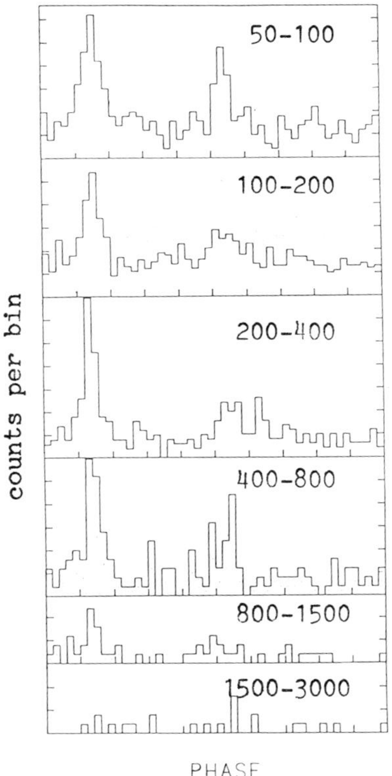


implications on the acceleration sites and / or mechanisms: photons of $\sim 10^{15}$ eV cannot escape from the intense magnetic field regions around the neutron stars, and need to be originated at distances comparable to that of the light cylinder radius ($\sim c\Omega^{-1}$) in order not to be absorbed by the pair production in the strong B .This becomes even more stringent if the $10^{15}$ eV emission form the Crab is shown to be pulsed.

## 3. Gamma-Rays from X-ray Binaries

One of the greatest results of ground-based gamma-ray astronomy in recent years has been the detection of very energetic emission from a number of binary systems containing a neutron star undergoing strong accretion from the non-collapsed companion. As such, these systems have all been well studied through X-ray astronomy, and it's mostly through their X-ray signatures that the gamma-ray identifications have been possible. Table 1 of Rudermann (1986)gives a summary of the results reported so far for the VHE and UHE observations; while it is very difficult to assess accurately the confidence level of the various results, based generally on limited statistics, it is hard to see how they could all be wrong, especially for the cases were several independent confirmations now exist. Critical rewiews of the recent results are available, but especially for Cyg X-3, the 'honorary' X-ray binary, the VHE/UHE evidence is now overwhelming, and confirms the brillant observations of the Crimean group (e.g. Stepanian et al.1977), who have been claiming observations of this source for well over a decade. The energetic emission of these systems, in general terms, must be linked to the presence of a rotating neutron star, but by no means to it alone. Indeed, neutron stars in these X-ray binaries are relatively old and are all slow rotators, so that their $\Omega^2 B$ is certainly not capable of producing the necessary $\Delta V$ for acceleration. For the specific case of Cyg X-3 many models exist and are discussed by Barnard (1986). In general, two basic approaches have been so far proposed for the creation of UHE/VHE gamma-ray in neutron-stars X-ray binaries. They have in common, as mentioned for the case of radiopulsars, the need for producing the energetic gamma-ray far enough from the intense neutron stars magnetic field; this can be accomplished either by creating the accelerating conditions in regions where $B \lesssim 10^3$ G (the limiting field for $\gtrsim 10^{15}$ eV gamma's), or by extracting energy from high-B regions in the form of nuclear beams, made to interact subsequently to yield electromagnetic by-products where they have a chance to escape. One way of producing the required $\Delta V$ for particle (proton) acceleration is to have charged matter moving at high speed in strong magnetic field: this (Chanmungam and Brecher, 1985) is accomplished near the pulsar magnetosphere at the inner edge the accretion disc, if this is available in the system. Note that the B at the magnetospheric radius is in general much higher than the value ($10^3$ G) required for escape - so also in this case energy trasport in the form of proton beam up to the production region is substantially required. The other proposed accelerator mechanism (e.g. Vestrand and Eichler, 1982) is in the accretion shock generated by infalling matter near (e.g. $\lesssim$ 1 stellar radius away) the neutron star surface. One needs to put the accelerator near the neutron star surface if the $10^{15}$-$10^{16}$ eV luminosity is to be directly accretion powered because it represents a big ( 10%) fraction of the overall accretion luminosity $G\dot{M}MR^{-1}$ . Some difficulty arises then with electron synchrotron losses in the standard high magnetic fields expected ($\gtrsim 10^8$ G), so that a lower surface magnetic field neutron star may be helpful in this picture.

## 4. Cos-B Sources vs. Unidentified Gamma Objects.

The picture of the $\gtrsim$100 mev gamma-ray sources in the Galaxy, as proposed by the Cos-B mission (Swanenburg et al, 18981), and further discussed by Bignami and Hermsem (1983), has recently undergone some serious changes (see, e.g. Pollock et al, 1985a). This is due to a re-analysis of the complete Cos-B data set in the light of new radio-astronomical information bearing on the gaseous content of the galactic disc. In particular, the recent mapping in the CO mm. line of low-latitude regions has permitted to evaluate the content of H2, an abundant but elusive state of galactic hydrogen, and one with a quite different distribution than that of the 21 cm. line data tracking the HI. The latter represented, until recently, the only known ISM on which to let the galactic CR interact to produce the flux of diffuse galactic radiation. It was against such diffuse radiation, taken as a background, that the Cos-B sources were found as significant excesses, unresolved by the instrument's PSF. Clearly, the detection of more gas, and all at very low galactic latitude, has changed the situation, and several excesses in the 2CG catalogue can now be well explained by interaction with the ISM of CR taken to have a local spectrum. Accordingly, these 2CG objects are to be dropped from a list of sources proper, but, on the other hand, also a few other true sources now appear which had not been seen before. Table 1 of Pollock et al.(1985b) summarizes the situation, showing which 2CG objects have fallen below the required significance and which new ones have popped up; however, the work has only been done so far for those parts of the galactic disc covered by the CO surveys. More work is in progress, leading eventually to a complete new list of true galactic sources; meanwhile, it would be premature to speculate on what their final number will be and on how different will this be from the current $\sim$ 20 or so.

The general characteristics of the bona fide sources are:

- $L_{\gamma} \simeq 10^{36}$ erg/sec
- $\langle E_{\gamma} \rangle \sim 250$ Mev
- in some cases, at least, time variability

and it is therefore logical to associate them with compact objects, e.g. with some manifestation of neutron stars. They remain, however, UGO's. They are not associated with those more obvious neutron star symptoms observable at other wavelengths: they are not bright X-ray binaries (or bright X-ray sources at all), nor radio pulsars. And yet, of course, such symptoms are available only for a tiny fraction ($\sim 10^{-6}$) of all the neutron stars in our Galaxy, so that one should really not be too surprised if, with gamma-ray astronomy, we may have stumbled onto a new manifestation or a different physical scenario. It is nevertheless the observer's duty to search for accurate identifications so that, in parallel with radio searches, it was also resorted to X-ray searches (see, e.g. Caraveo, 1983). As a result of the exploration with the Einstein Observatory of several Cos-B error boxes, many new X-ray sources were discovered, and a subsequent program of optical identification and understanding was initiated. Many of the new sources have now been identified and are accounted for in terms of known objects, and as such with no special reason for being responsible for gamma-ray emission. A few, however, are still unidentified, and as such remain potential candidates. Of these, unfortunately so far only one case could be studied in detail, that of Geminga. For the others, still pending , the high resolution instruments on Eistein or EXOSAT did not make it to supply us with good enough positioning to start optical identification work.

## 5. Work in Progress on Geminga.

As a result of the X-ray exploration of the 2CG195+4 (Geminga), Bignami Caraveo and Lamb (1983) reported the discovery of 4 new IPC sources very near to or within the half square degree Cos-B error box. Of these, two were immediately identified with a field star and an external radio galaxy, while a third, 1E0630+18, has been only recently identified through an EXOSAT CMA 10" positioning (Bignami et al, 1985.). The fourth, 1E0630+178, and by far the brightst source ($\sim 2 \times 10^{-12}$ ergs/cm$^2$ sec in the IPC range), remains the toughest candidate for optical identification, but is the best candidate for the gamma-ray source and is, in any case, most likely associated with a compact object, probably a neutron star. This is the first case of an object, discovered through its gamma-ray emission, being tracked down to x-ray and optical wavelengths; not altogether unexpectedly, it's turning out to be quite different from other known manifestations of neutron stars. In order to better understand this, a short summary of very recent work follows, in fact anticipating forthcoming complete presentations of results.

After the surprising result of the periodicity evidence in the Einstein 1979, 1981 and EXOSAT 1983 X-ray data of 1E0630+178 (Bignami, Caraveo ,Paul, 1984), a new long (64,000 secs) EXOSAT observation was performed in 1985, yielding about 1200 source photons. Their arrival times were periodicity-analyzed in the interval from P= 60.20 sec to P= 60.35 sec., i.e. in that interval obtained from the extrapolation to the measurement date, of the previous three periodicity measurements, implying a high $\dot{P} \sim 4 \times 10^{-9}$ sec/sec. A very significant positive result at 60.28 secs was found in this small interval, covered in 30 independent steps of $5 \times 10^{-3}$ secs each . A posteriori, a wide scan was performed (from 10 to 110 secs in period), to check on the presence of other, possibly spurious, signals. No other signal of comparable significance was found, except for the harmonics at exactly 1/2 and 1/4 of the 60.28 secs signal. Fig 2 shows the "secular" X-ray period change plot. As mentioned, a more detailed presentation of these data is currently being prepared, including a study of the possible time-variability of the periodicity effect within the 64,000 secs observation.

As to the problem of the optical identification of 1E0630+178, a first proposal had been made by Caraveo et al (1984) and Sol et al (1985) for an $m_V \sim 21$ object, located at the edge of the HRI error box, at 4.2" from its centre. This object was then subject to spectral and time variability studies (e.g. Halpern et al, 1985 Kulkarni and Djorgovski, 1986), which, although difficult because of its faintness, did not reveal any peculiarity, all data being consistent with a G type field star. This fact, coupled with the marginal positional coincidence, cast some doubt on the proposed identification; however, only recently the final word has been said by IR data on this object, kindly taken for us by M.Lebofski at the MMT. The observed K and J magnitues allow finally to join with a smooth Plankian curve all the optical-IR data; the resulting best fit yields a temperature of 5500$^\circ$K (consistent with the scant continuum spectral observations) but requires an IS absorption Av $\geq$ 1 mag. This is clearly incompatible with the X-ray deduced NH ($< 10^{20}$ cm$^{-2}$), and it thus represents a strong, and probably final, argument against this object being associated with the X-ray source. The optical observations have yielded at least one and possibly two other objects in the HRI error circle (of $\sim$ 50 sq. arc secs), but of such faint magnitudes ( 24.5 and 25.5), that their probability of chance presence is very high (see e.g. Djorgovski and Kulkarni,1986 and Bignami, 1986). Early reports of possible proper motion of either of the

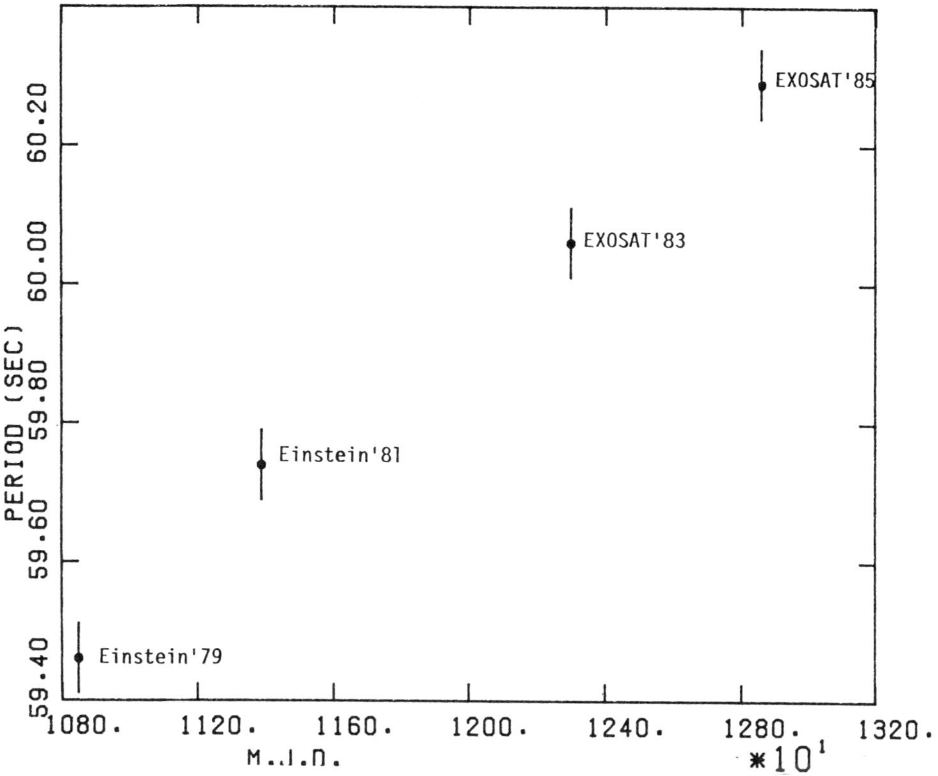

EXOSAT'85
EXOSAT'83
Einstein'81
Einstein'79
PERIOD (SEC)
59.40
59.60
59.80
60.00
60.20
1080.
1120.
1160.
1200.
1240.
1280.
1320.
M.J.D.
*10[1]

Fig 2
Time history of the 59/60 sec periodicity.

EXO 85
HRI 81
EXO 83

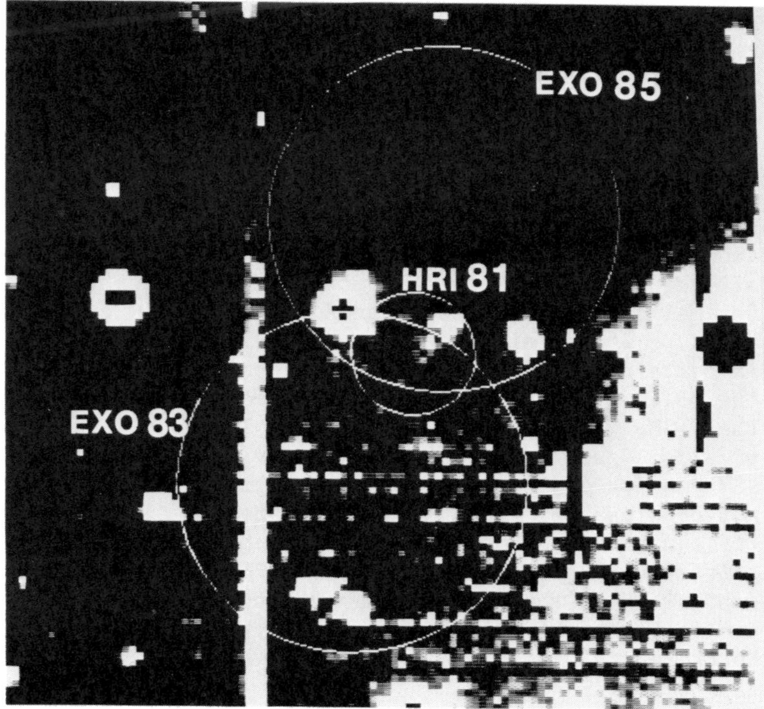

Fig 3
Stack of 12 CCD plates of 15 m each taken at the CFHT in January '84 (N at bottom E at left).

two faint objects still await confirmation, but may have to wait until the Space Telescope becomes operational. Fig. 3 is a stack of many CCD exposure taken at the 3.6 CFHT telescope. The HRI error box has the 'old' mv 21 candidate at the SE edge, and the other faint objects inside. The EXOSAT error boxes of 1983 and 1985 are also shown.

What do the data gathered so far on Geminga/1E0630+178 and its possible optical counterpart tell us ?

- The 50 Mev-few Gev gamma-ray flux is $\sim 2\times10^{-9}$ erg/cm$^{2}$sec, or 1,000 higher than the ~ Kev X-ray flux
- In turn, Fx/Fv $>$ 1,000 or 2,000, or 3,000, depending whether the counterpart is either or none of the two faint objects in the HRI box.
- The IPC spectrum is soft, well consistent, e.g. with a BB spectrum with T~ $9\times10^{5}$ °K; the observed NH, certainly less than $10^{20}$ cm$^{-2}$, and consistent with zero, places the source nearer than 100-200 pc. Support to this comes from remembering the NH=$3\times10^{21}$ cm$^{-2}$ to the Crab, located at 2 Kpc in a similar l, b position, and from the Nh $2\times10^{20}$ cm$^{-2}$ to the galactic pole, i.e. through ~ 200 pc of decreasing density disc.
- The source periodicity (and its $\dot{P}$), now apparently confirmed for the X-ray data, but no longer seen (possibly owing to instrumental limitations) in the gamma-ray data (see Buccheri et al, 1985).
- Deep radio searches have not revealed the existance of a pulsar or of any radio source compatible with HRI position.

As already pointed out by several authors, this sum of evidence represents a unique set of properties, not found so far in astronomy, even independently of the gamma-ray emission. It is tempting to try to interprete this set of properties with a model involving a neutron star - even if a model based on, e.g., a black hole does exist (Bisnovathi - Kogan, 1985). The simplest would be, for the X-ray and optical data at least, that of thermal emission from an isolated neutron star (presently non active as radio pulsar) at a distance of less then several tens of parsecs, within which hundreds of such objects should exist. At 40 pc, for example, the observed X-ray flux is explained with a surface of $\lesssim 10^{10}$ cm$^{2}$ at the observed temperature of $9\times10^{5}$°K, possibly the heated polar cap of a neutron star . The optical emission around mv 25 would then be accounted for by thermal emission of the rest of the neutronstar surface at the same distance and at a much lower temperature ($\gtrsim$50,000 °K). Several questions arise: can such a thermal gradient exist? According, e.g., to Greenstein and Hartke (1983) it could be not impossible, but it is certainly difficult to be more precise. More importantly, what heats the polar cap? In an isolated neutron star, it could be accretion from the ISM, which, in a scenario à la Bondi, could provide Lx $10^{29}$-$10^{30}$ ergs/sec (just the right amount) for space velocities of 100 Km/sec and ISM density of 1 cm$^{-3}$, i.e. for parameter values slightly favourable but not unthinkable. However, a fundamental difficulty with ISM accretion of a magnetized neutron star is posed by the so called "propeller effect" (Illarionov and Sunyaev, 1975). The case for IS accretion being impossible on the neutron star if it retains a magnetic field is in fact explicitly treated in that seminal paper, where it is shown that, owing to the weak IS mass transfer, the neutron star would have to slow down to periods so long as could only be achieved in more than a Hubble time. Alternatively, it can also be shown that, even for P$\sim$100 secs, the gravitational capture radius (equivalent to the magnetospheric radius if the B field is confined by ISM pressure) is much greater than the corotation radius, and thus no accretion can take place. If the ISM accretion, invoked to heat the polar cap, is not viable, the simple model of thermal emissions from a slowly rotating neutron star is also untenable (which is just as well, because it would have been very difficult to explain at least $10^{33}$ ergs/sec of gamma-rays from it). Some form of non-thermal emission can certainly be postulated (e.g. Katz, 1985), but it would seem more natural to turn one's attention to binary systems, obviously with an underluminous companion to a collapsed object. The case for two neutron stars has been considered by Nulsen and Fabian (1984), and that of a black hole/white dwarf system by Bisnovathi-Kogan (1984), (based on the "old" optical identification). Certainly, a binary model would also have the advantage of greater latitude for explaining the gamma-ray emission and the (period)/(period derivative) combination, unique for this object. Another qualitative consideration against Geminga-like objects being too common (as, e.g., old isolated neutron stars) is that if their gamma-ray luminosity is at least $10^{33}$ ergs/sec (that for a Geminga distance $\sim$ 100 pc), their current total number cannot exceed $10^5$-$10^6$ (but could also be much less), or their summed gamma-ray emission would exceed that of the total Galaxy of $< 10^{39}$ ergs/sec. Such maximum number $\sim 10^{-3}$ of the total galactic neutron stars could indeed point to a relatively rare neutron star binary combination, exceedingly luminous in gamma-rays, of which current gamma-ray astronomy has detected the nearest specimen; note that Geminga would not have been seen by SAS-2 or Cos-B if it were only a factor of $\sim$ 3 further away, owing to the current sensitivity limitations.This helps in explaining its apparent uniqueness. On the other hand, as mentioned above, sources with similar gamma/X/optical combinations could be buried in the existing data, which are very limited in sensitivity and angular resolution.

It is perhaps appropriate to close by remarking that the type of astrophysical scenario described above is strikingly reminiscent of those imagined for gamma-ray bursters; even if Geminga has never been seen to burst in gamma-rays, it is true that a similar "steady" high-energy gamma-ray luminosity would still be undetectable for the vast majority of the gamma-ray burst population. No optical flashes are visible at the Geminga position in the Harvard plate collection (that was checked ,Bignami et al.1983) , and no convincing steady-state X-ray source has been associated with bursters, and so the similarity remains a speculation - still it provides a clear indication for future observational work.

It is a pleasure to thank J. Arons for enlightening discussions and D. Helfand for stimulating scientific conversations some of which were held in China's remote and beautiful corners.

## REFERENCES

Barbard, J. this volume
Bignami, G.F. and Hermsen W.: Annual Rev. Astr. Astrophys. 21,67 (1983)
Bignami, G.F., Caraveo,P.A., Lamb,R.C. : Ap.J. 272,L9 (1983)
Bignami, G.F., Caraveo, P.A., Paul,J.A.: Nature 310,464 (1984)
Bignami, G.F. et al. IAUC N.4081 (1985)
Bignami, G.F.:Proceedings of the NATO ASI held in Cargese Corsica Sept.85 ed. F. Pacini(1986)
Bisnovati Kogan, G.S. : Nature 315,555 (1985)
Buccheri,R. et al (Caravane Collaboration) 19th ICRC 1,169(1985)
Caraveo, P.A. : Space Science Rev.36,207 (1983)
Caraveo, P.A. et al.: Ap.J. 276,L45 (1984)
Chanmungam, G. Brecher K. : Nature 313,767 (1985)
Clear J. et al. (Caravane Collaboration) submitted to Astr. Ap. (1986)
Dowthwaite, J.C. et al. 1983 :Astr. Ap. 126,1 (1983)
Djorgovsky, S. and Kulkarni, S.R. :A.A. 91,90 (1986)
Dzikowsky ,T. et al. :17th ICRC 1,8 (1981)
Greenstein, G. and Hartke, G.J.: Ap.J. 271,283 (1983)
Halpern, J. et al. : Ap.J. 296,190 (1985)
Hurley ,K. : this volume
Katz, J.I. : Astrophys. Lett. 24,183 (1985)
Kirov, I.N. et al. : 19th ICRC 1,135 (1985)
Kulkarni, S.R. and Djorgovski,S. :A.A. 91,98 (1986)
Illarionov,A.F. and Sunyaev, R.A.: Astr. Ap.39,185 (1975)
Manchester, R. : this volume
Narayan, R. : this volume
Nulsen, P.E. and Fabian, A.C.: Nature, 312,48(1984)
Pollock, A.M. et al.(Caravane Collaboration):Astr. Ap.146,352(1985a)
Pollock, A.M. et al.(Caravane Collaboration):19th ICRC,1,338(1985b)
Rudermann, M.:Proceedings of the NATO ASI held in Cargese Corsica Sept.85 ed.F.Pacini (1986)
Sol, H. et al.: Astr. Ap. 144,109(1985)
Stepanian, A.A. et al.:15th ICRC 1,135(1977)
Swanenburg, B.N. et al. (Caravane Collaboration):Ap.J. 243,L69(1981)
Vestrand, W.T. and Eichler, D.:Ap.J.,261,251 (1982)
Wills, R.D. et al.(Caravane Collaboration): Nature 296,723(1982)

## DISCUSSION

**J. Grindlay:** What constraints, if any, can be placed on variability of the 100 MeV emission from Geminga. Do the Cos-B vs. SAS-2 data suggest long-term variability? Is there any indication of changes in spectra in the Cos-B data?

**G. Bignami:** No evidence of long term flux and/or spectral variability from Geminga is visible in the Cos-B data base. However, the rather large (say 20%) error margin within which data from the 7 year mission can be compared does not allow for a very strict statement on this.

**G. Frontera:** What do you think of the proposal by Chinese astronomers to identify Geminga with a SNR, with explosion recorded in ancient records?

**G. Bignami:** As you can clearly see from the enclosed figure, the record only speaks of a guest star as big as an orange, i.e., not of the more classical "melon" size associable with SN explosion and moreover shining for only a short time. As to the positional coincidence, the recorded error box is even bigger than that of Cos-B. Finally, let me underline that this proposal was first made by Vladimirskii as mentioned in Bignami et al. 1984.

魏太延三年正月壬午，有星
晡前昼见东北，在井左右，
色黄，大如橘

魏收 572年 魏书105卷

元嘉十四年正月，有星晡前
昼见东北维，在井左右，
黄赤色，大如橘

**J. Dolan:** With regard to the optical counterpart of 1E0630+178, what type of UV spectrum would the various models you discussed predict?

**G. Bignami:** The (otherwise untenable) single neutron star model would predict a strong UV flux relative to the optical, if we are looking at an object with surface temperature $\gtrsim 10^4$ K. In any binary model it is not completely clear where the optical light would be coming from, but my guess is that it would also have a strong UV component. We have looked at Geminga with IUE, of course, without success, but we have great hopes in the ST.

# A REVIEW OF GAMMA-RAY BURST OBSERVATIONS

W. Doyle Evans and John G. Laros
Los Alamos National Laboratory
Earth and Space Sciences Division
P.O. Box 1663
Los Alamos, NM 87545

ABSTRACT. Gamma-ray bursts are generally believed to originate in the vicinity of neutron stars, but the phenomenology is still not understood. In this paper we review the known characteristics of gamma bursts and give new observational results on temporal and spectral properties. We suggest that a class of repeating bursters exists that are spectrally harder than x-ray bursters but significantly softer than "classical" gamma bursts. The March 5, 1979, burst may be the prototype of this class of bursters.

## 1. INTRODUCTION

Excellent reviews of the current observational situation in transient gamma-ray astronomy have been provided recently in three AIP Conference Proceedings Volumes.[1-3] These volumes also contain many details on the current research in this field. This paper will be a summary of these results in terms of temporal properties, spectral properties, and source locations. In addition, significant new results in each of the three areas will be discussed in more detail. To efficiently use the time available for the discussion of gamma bursts, we will limit our remarks to the observational data. Kevin Hurley, in the paper that follows, will then describe the inferences concerning neutron stars that can be drawn from the data.

## 2. TEMPORAL PROPERTIES

Gamma-ray bursts (GRBs) are characterized by short duration and extreme variability. Most bursts have a total duration of 1-10 s. Generally the time profiles consist of one or more intense peaks of varying widths separated by intervals of weak emission. Figure 1 is of historical interest and shows the type of time profiles first obtained by the Vela satellites. The instrumental characteristics prevented the acquisition of data during the initial rise and provided high-resolution data only during very early times. Figure 2 shows data from the Third International Sun-Earth Explorer (ISEE-3) instrument that provides a sampling period with pretrigger



D. J. Helfand and J.-H. Huang (eds.), The Origin and Evolution of Neutron Stars, 477–487.

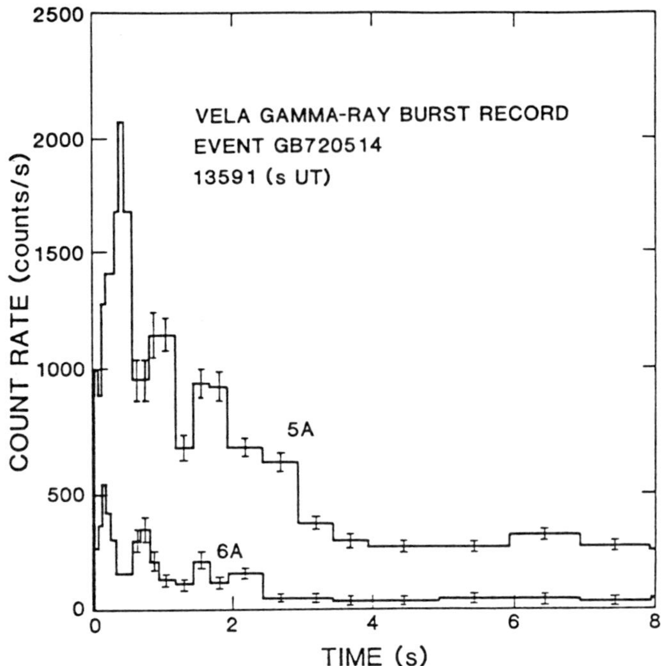


Figure 1. A gamma-ray burst time history recorded by detectors on the Vela 5A and Vela 6A satellites.

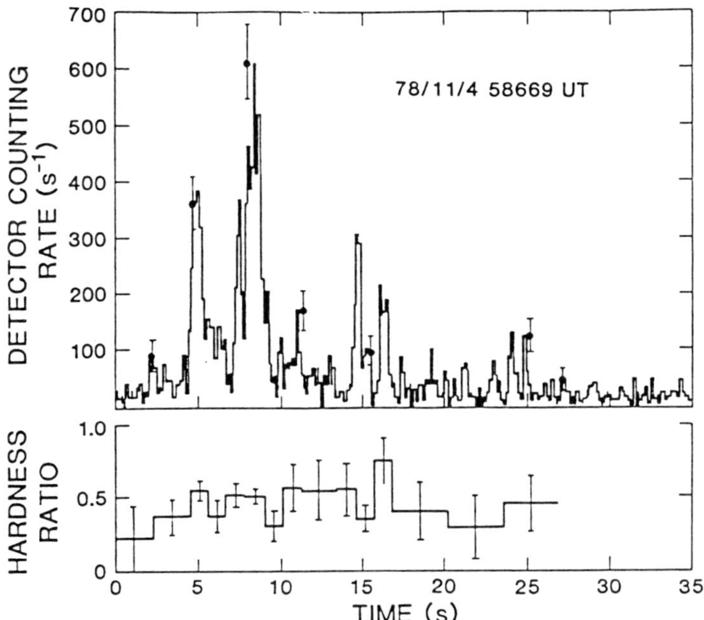


Figure 2. A gamma-ray burst time history recorded by the ISEE-3 detector.

information and uniform time resolution throughout. The presence of well-defined peaks throughout the event is clearly shown.

There are notable exceptions to the typical 1-10-s duration, both shorter and longer. In particular several short events appear to consist of a single peak lasting for a few tenths of a second. A notable example of this type was the "March 5 Event," GB790305. Until recently it was the only short burst that was intense enough to allow detection of fine structure within the burst profile. Such fine structure was of low amplitude for this event--the burst had a very fast rise with a fairly smooth but irregular decay, followed by the well-known periodic afterglow. Recently Laros et al.[4] have published data from another very intense short burst that occurred on 1984 December 15. Unlike GB790305, GB841215 consisted of at least seven distinct, well-separated peaks with weak, nonperiodic emission following the main burst. This suggests that there could be at least two types of short bursts. It also suggests that we have not yet been able to observe the true characteristic time scale of the variability because of instrumental limitations. Since GB790305 and GB841215 were differentiated by their spectra as well as by their fine structure, we will return to this topic in the following section.

In another recent observation, Itoh et al.[5] have presented results obtained on 1984 March 4, from the Pioneer Venus Orbiter (PVO) instrument. As shown in Figure 3, this burst consisted of two

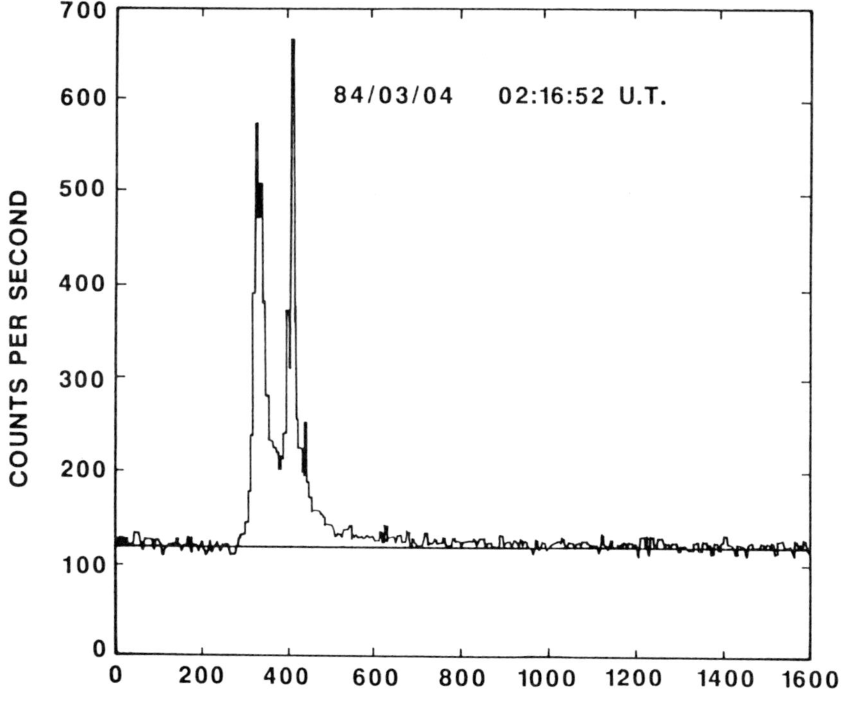


Figure 3. A gamma-ray burst recorded by the Pioneer Venus Orbiter. The two intense peaks are followed by more than 1000 s of spectrally hard emission.

distinct peaks followed by over 1,000 s of weaker, but spectrally very hard, emission. Since GRB cooling times are believed to be extremely short, this observation implies that an energy source is active for at least 1000 s. Burster models are severely constrained by this finding.

Searches for periodicities in gamma bursts have been largely unsuccessful; an obvious exception is the extended emission from GB790305. Also, Wood et al.[6] presented evidence for possible periodic structure in a burst that occurred on 1977 October 29. The 1979 January 13 event has been suggested by Barat et al.[7] as having a possible periodicity. U. Desai at Goddard has proposed that several additional bursts show evidence for periodicities or quasi periodicities, but the evidence is not overwhelming. However, in a collection of several hundred gamma bursts, it might be expected that, simply due to chance, several will have peaks that are approximately evenly separated.

## 3. SPECTRAL PROPERTIES

As with other properties of gamma bursts, continued observation has revealed a rich diversity in the spectral shape of the emission and the spectral variability during the burst. Notwithstanding the wide differences between bursts, however, it is still true that, as a group, the spectra are distinctly different from any other astrophysical radiation source. Figure 4 from Epstein (private

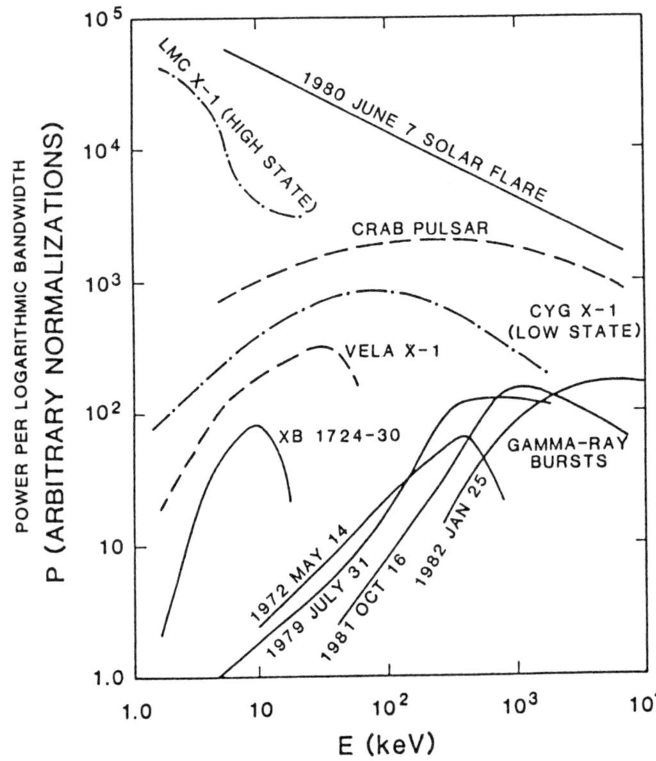


Figure 4. Typical spectra of various high-energy astrophysical sources. Gamma-ray bursts form a distinct class.

communication) illustrates this difference clearly. Examples of the spectral shape of the emission from several types of transients are plotted in units of logarithmic intensity. Vertical normalization is arbitrary. The several gamma-burst spectra shown in the lower right are grossly different from x-ray bursts, x-ray transients, and solar flares. Most of the power in the bursts came from above 1 MeV. Matz[8] has recently published excellent results from the Solar Maximum Mission showing a number of bursts with emission extending to several MeV without any apparent turnover in the continuum spectra. Figure 5 shows one such spectrum, best fit by a power law. Again, instrumentation limits our knowledge at the highest energies.

In the x-ray region of the spectrum the observations are limited, but sufficient data exist to support some important conclusions.[9,10] Very little of the power is emitted below 30 keV--approximately 1%. The soft emission has a very different time profile than does the harder portion of the spectrum; i.e., the overall spectral shape varies widely during a burst, even though the shape above 100 keV remains relatively unchanged. Some investigators have suggested that the observations are most easily understood as two distinct components--soft and hard--that are not strongly coupled at the source.

Golenetskii et al.[11] have suggested a correlation between intensity and spectral shape and have published several examples illustrating this correlation. Other investigators have not been able to confirm this relationship and have instead suggested that the spectra are hardest on the leading edges of the individual peaks.[12,13] GB840304 has been analyzed and the interesting result is shown in Figure 6. These data are for the two peaks shown in Figure 3. Each point is

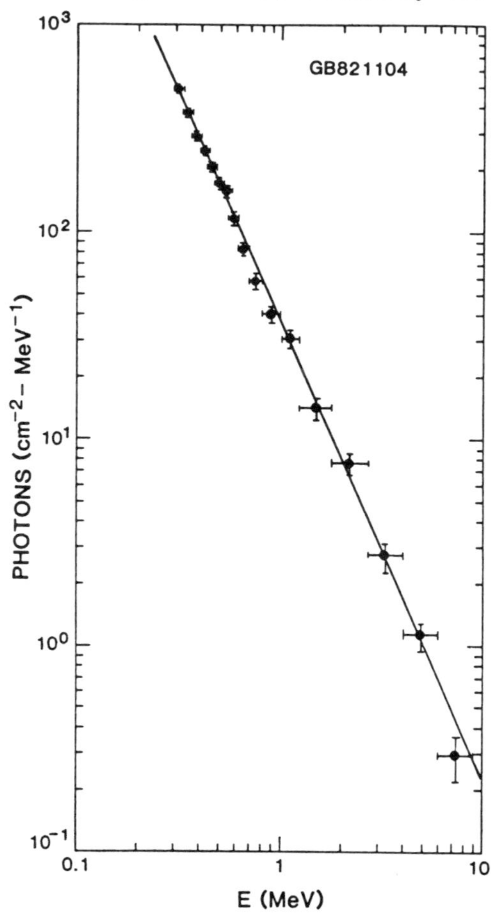


Figure 5. Gamma-ray burst spectrum obtained from the Solar Maximum Mission (Matz, ref. 8) showing emission extending to 10 MeV.

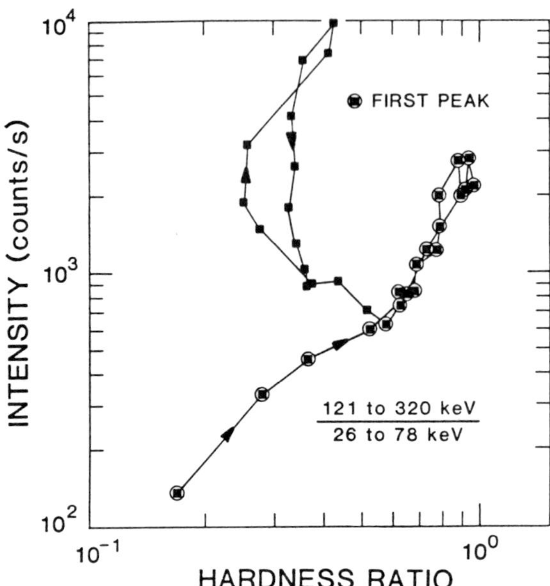


Figure 6. Hardness ratio vs. intensity for the two peaks of the GB840304 (See Figure 3). During the first peak (circled squares) there is a strong correlation. This correlation completely disappears during the second peak (solid squares).

derived from 4.5 s of data, and the arrows show the direction of time. During the early portion of this peak, there is a strong positive correlation between intensity and spectral hardness. However, this relationship breaks down completely after approximately 90 s, and the intensity becomes roughly spectrum independent. Apparently there are bursts, or at least portions of bursts, for which the intensity-spectrum correlation holds, but as is usual for gamma bursts, there is no consistent behavior. Certainly, it has been shown that GRB spectra can vary as rapidly as their overall time histories and that future spectral analyses must deal with this fact.

The presence (or absence) of spectral features has been the subject of considerable debate in the past. Several spectra published by Mazets et al.,[14] show significant departures from a smooth continuum in the energy range between 400 and 500 keV. The presence of gravitationally shifted annihilation radiation could produce the observed spectral features, and this was interpreted as strong evidence that the emission originated at the surfaces of neutron stars. Also, apparent absorption features below 100 keV were explained as synchrotron absorption in magnetic fields of several x $10^{12}$ G--again consistent with magnetized neutron stars. However, this interpretation was questioned by Fenimore et al.,[15] who argued that an incorrect assumption concerning the continuum shape coupled with instrumental effects could produce the observed results, and by Epstein (private communication), who pointed out that $>10^{12}$-G magnetic fields would significantly deplete the continuum emission at high energies. Alternatively, the features could be caused by the superposition of two or more different emission components.[10] Confirmation of the features by other observational groups has been inconsistent. Certainly at this time even the reality of the spectral features is not completely established.

Spectral measurements have helped provide evidence that a few of the catalogued GRBs might belong to a separate class of gamma-ray transients. It has already been noted that GB790305 had, in addition to its other remarkable properties, an unusually soft spectrum for a GRB. Subsequently, it was discovered that GB790305 and another soft burster, GB790324, were repetitive.[16] (Normal, hard-spectrum GRBs have never been observed to repeat.) Based on its soft spectrum and short duration, GB790107 was pointed out as a possibly similar object.[17,18] Very recently we discovered perhaps more than 50 repetitions of GB790107 that occurred in late 1983. Thus, it appears likely that a new class of short, rapidly repeating, soft gamma-ray transients exists.

## 4. SOURCE LOCATIONS

Although the directions to ~100 GRBs have been determined,[16,19] their locations in space (i.e., their distances) remain unknown. Since the directions do not seem to favor either the Galactic center or the Galactic plane, the sources must be either very nearby or very distant, compared with the diameter of the Galaxy. Attempts to resolve this ambiguity through analysis of the logN-logS or logN-logP data have been inconclusive. Jennings[20] has devoted considerable effort to this study and concludes that an extended Galactic halo best explains the observations. Conversely, Higdon and Lingenfelter[21] claim that the departure of logN-logS from the 3/2 power-law slope is totally the result of selection effects produced by instrumental design and variations. The most recent results from Klebesadel (private communication) using PVO data show no significant departure of the logN-logP curve from the 3/2 slope (Figure 7), except that due to the instrumental threshold. These results, all from the same instrument and all based on peak intensity rather than total fluence, are compatible with either distance scale. Apparently an increase in instrumentation sensitivity of at least an order of magnitude will be needed for logN-logS or logN-logP curves to be meaningful distance indicators.

GRB observers have attempted to follow the example of x-ray astronomy, where association of x-ray sources with optically identifiable astronomical systems has produced tremendous progress. Interplanetary networks, in a commendable example of international cooperation, have provided several very precise (~arcmin) error boxes through time-of-arrival analysis techniques.[19] With only a few notable exceptions, these error boxes have contained no interesting

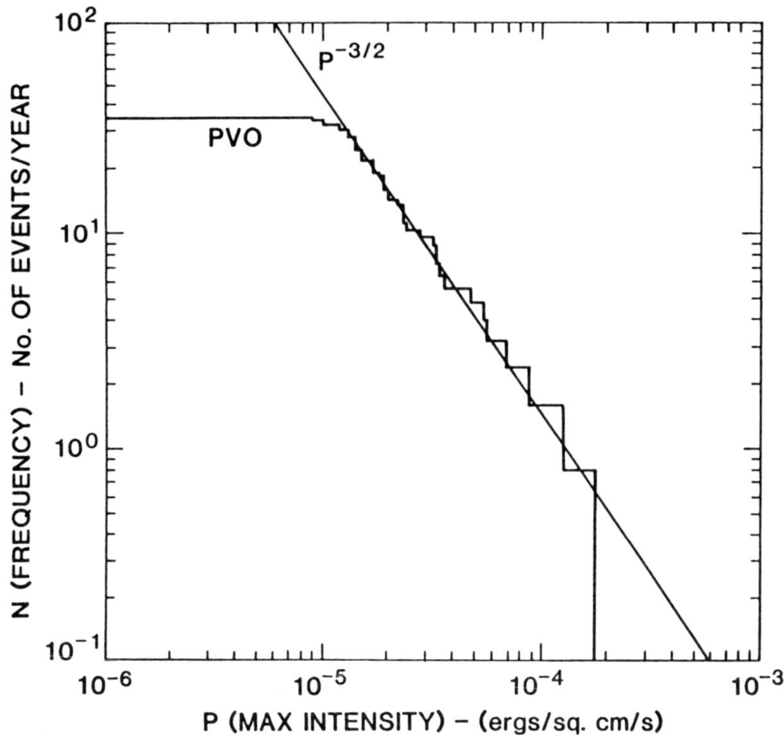


Figure 7. LogN vs. LogS data for all Pioneer Venus Orbiter data. The 35 events are consistent with a -3/2 law.

objects or no observable objects at all. The first exception was the March 5 event, whose direction was found to be coincident with the LMC supernova remnant N49. More recently, Schaefer[22] has discovered probable optical flashes on archival photographs of three GRB error boxes. However, the LMC association has been questioned on energetics grounds, and the optical flashes have not led to the firm identification of any quiescent counterpart. In any case, we have learned that gamma-ray bursters in quiescence are extremely faint objects.

## 5. CONCLUSIONS

New temporal observations have shown that even the shortest class of gamma burst, previously observed only as single peaks, can be composed of many faster, distinct peaks that are observable when there is sufficient count rate. Even better observations may be required to determine the characteristic time scale of the variability. Also, the extended, hard emission from GB840304 seems to require an energy source that remains available for over 1000 s, thus adding a major constraint to virtually all gamma-burst models.

New time-resolved spectral measurements indicate that interpretations of spectral data have been seriously limited by the fact that the burst spectra evolve on time scales at least as short as the fastest accumulation times of the present spectroscopic instruments. Time-averaged spectra may contain artifacts produced by this variability and must therefore be treated with caution. Even so, the high-energy spectral data from the Solar Maximum Mission are important in that they constrain important parameters of the emitting region, particularly the magnetic field. Correlations between intensity and spectral hardness are clearly present during some portions of some bursts, but they are not always present and their origins are unknown. Spectral measurements have also led to the identification of a possible new class of short, rapidly repeating, soft gamma-ray transient.

The new GRB location information probably has been less illuminating. Attempts to use the ~100 approximately known directions, together with logN-logS or logN-logP curves to establish a distance scale, have been inconclusive. Furthermore, the several available precise directions have not led to the identification of a gamma-ray burst counterpart.

## References


1. R. E. Lingenfelter, H. S. Hudson, and D. M. Worrall, eds., AIP Conf. Proc. No. 77, La Jolla (1982).
2. S. E. Woosley, ed., AIP Conf. Proc. No. 115, Santa Cruz (1984).
3. E. P. Liang and V. Petrosian, eds., AIP Conf. Proc. No. 141, Stanford (1986).
4. J. G. Laros et al., Nature 318, 448 (1985).
5. M. Itoh et al., Bull. Am. Astron. Soc. 17, 850 (1985).
6. K. Wood et al., Astrophys. J. 247, 632 (1981).
7. C. Barat et al., Astrophys. J. (Lett.) 286, L5 (1984).
8. S. M. Matz, 'On the Spectra of Gamma-Ray Bursts at High Energies,' Ph.D. dissertation, U. of New Hampshire (1986).
9. J. G. Laros et al., Astrophys. J. 286, 681 (1984).
10. J. G. Laros and J. Nishimura, AIP Conf. Proc. No. 141, 79 (1986).
11. S. Golenetskii et al., Nature 307, 41 (1984).
12. J. G. Laros et al., Astrophys. J. 290, 728 (1985).
13. J. P. Norris et al., Bull. Am. Astron. Soc. 16, 447 (1984).
14. E. P. Mazets et al., Nature 290, 378 (1981).
15. E. E. Fenimore et al., AIP Conf. Proc. No. 77, 201 (1982).
16. E. P. Mazets et al., Astrophys. Space Sci. 80, 3 (1981).
17. E. P. Mazets et al., Astrophys. Space Sci. 84, 173 (1982).
18. J. G. Laros et al., Nature 322, 152 (1986).
19. J. -L. Atteia et al., Astrophys. J. (Supp.), in press (1986).
20. M. Jennings, AIP Conf. Proc. No. 115, 412 (1984).
21. J. Higdon and R. Lingenfelter, AIP Conf. Proc. No. 115, 568 (1984).
22. B. E. Schaefer, AIP Conf. Proc. No. 115, 406 (1984).

## DISCUSSION

**E. Liang:** Have you looked at the logN-logS distributions of the $VS^2B$'s as a class? Are all of them much weaker than the March 5th event.

**D. Evans:** Not yet. But they are weaker than March 5, 1979.

**S. Woosley:** With the discovery of $VS^2Bs$ the void in energy space you referred to in the beginning of your talk separating X-ray bursts and γ-ray bursts appears to be filled in, suggesting if not a continuity of mechanism, at least that nature has been resourceful in providing transients over a broad, continuous range of energies, i.e., a large fraction of so called "γ-ray bursts" emit most of their energy at "hard X-ray" wavelengths.

**D. Evans:** The $VS^2Bs$ certainly fall between the X-ray bursters and γ-ray bursters, but I don't believe that they "fill in" to produce a continuum. Rather, they form a new "clump" that has little overlap with either X-ray bursters or γ-ray bursters.

**S. Woosley:** Besides its brilliance, the rapid rise time (~0.2 ms) of the March 5 event is a hitherto unique characterisic of this particular gamma-ray burst. Does the class of $VS^2B$ display this same rapid rise time?

**D. Evans:** Rise time data are not yet available from Signe and Prognoz and aren't determined by either the PVO or ISEE observations. This may well be the critical test of whether or not 790305 should be included as a $VS^2B$.

**S. Colgate:** Is there any information concerning the intensity versus spectra on the $VS^2Bs$?

**D. Evans:** Not at this time. Neither PVO nor ISEE-C provide time-resolved data for these events. I believe that only Signe experiments or the Prognoz experiments might have these results later.

**C. Alcock:** Just how singular is 790305? If it had the same light curve, but was much fainter, would it be distinguishable from the rest of the $VS^2Bs$?

**D. Evans:** The spectrum may be a bit harder than the average of soft bursts, but otherwise it would be very similar. 790305 had a very fast rise time (< 1/4 millisecond) and we don't yet know that parameter for the soft bursts. The extended emission from 790305 would not be observable if the overall strength were reduced by a factor of 100 to make it comparable to the soft bursts.

**D. Helfand:** If March 5 is the prototype of the $VS^2Bs$ then the obvious conclusion which I cannot resist stating is that this famous event had absolutely nothing to do with the LMC.

**D. Evans:** That is correct.

**J. Grindlay:** Your assertion that "traditional" gamma burst sources (with hard spectra, etc.) do not recur is in conflict with Schaeffer's optical results. How do you resolve this?

**D. Evans:** I actually made a rather different statement - i.e., that no recurrence from a "traditional" burst has been observed. It's possible to deduce from this that the recurrence time is probably greater than ~10 yrs. I believe that Brad Schaeffer concludes that he needs sources to repeat optically in 3-5 years to explain the optical flashes. I don't rememeber the uncertainty in this result. Maybe 10 yrs isn't inconsistent with his calculation. It's also possible tht optical flashes originate on $\gamma$-burst sources but are not always (or ever?) accompanied by $\gamma$-ray emission.

# SOME CONSTRAINTS ON NEUTRON STAR PROPERTIES FROM GAMMA RAY BURSTER OBSERVATIONS

K. Hurley
Centre d'Etude Spatiale des Rayonnements (CNRS-UPS)
B.P. 4346
31029 Toulouse Cedex, France

ABSTRACT. The results of recent soft X-ray and optical searches for quiescent gamma ray burster counterparts are used to constrain the properties of the neutron stars responsible for bursters. Ages are restricted to the range $2 \times 10^5$ y and above based on temperature upper limits and theoretical cooling curves, or $10^7$ y and above if bursters have evolved from pulsars. Velocities are greater than 20 km/s if the neutron stars are unmagnetized. Practically no main sequence star could have escaped detection in the optical/IR searches, so if the neutron stars are in binary systems, the companion is most likely a degenerate, low mass, low temperature object.

## 1. INTRODUCTION

It has become increasingly clear recently that the more than 400 observations of gamma ray bursts, carried out since their discovery in 1967, represent a significant fraction of the total number of galactic neutron stars observable to us. Other examples are the 437 pulsars now known (R. Manchester, this meeting), the ~150 galactic supernova remnants detected by their radio emission, not all of which show direct evidence for a neutron star (Helfand, 1985), the high mass X-ray binaries, which include about two dozen pulsating sources and as many X-ray burst sources (D. Lamb, 1985; F. Lamb, 1985; Trümper, 1985), the 93 binary X-ray sources and 22 X-ray supernova remnants in the HEAO A-1 catalog (Wood et al., 1984), and 2-4 high energy gamma ray sources in the COS-B data (Bignami, 1985). Qualitatively, it is still too early to determine exactly what the study of gamma ray bursts will teach us about the galactic neutron star population. One thing, however, is certain, namely that gamma bursters involve a quite different set of neutron stars from those observable in the radio, X-ray, or high energy gamma ray ranges. It is presently unknown whether they are related to those neutron stars, e.g. as earlier or later evolutionary phases, or not.

In this paper, some of the recent observations of bursters, particularly (but not exclusively) quiescent counterpart searches in



*D. J. Helfand and J.-H. Huang (eds.), The Origin and Evolution of Neutron Stars, 489–500.*

the optical, infrared, and soft X-ray ranges, will be used to establish constraints on some of the properties of the neutron stars assumed to be related to gamma bursters: ages, velocities, distances, and possible membership in binary systems. Little of the observational data on bursters will actually be reviewed here (see, e.g., Hurley, 1985); likewise, nothing will be said about the possible mechanisms for energy generation and radiation which are, to a certain extent, independent of the topics considered here. The reader is referred instead to a comprehensive review edited by Liang and Petrosian (1986). Finally, another, new topic which will not be reviewed here for lack of space is the question of what can be learned about neutron star equations of state from burster observations: see Liang, 1985.

## 2. NEUTRON STAR AGES

Observations of burst time histories, and soft X-ray observations of bursters in the quiescent state, may be used to obtain lower limits to the ages of the neutron stars responsible for gamma bursts. The results of Einstein observations of 5 gamma-ray burst sources (Pizzichini et al., 1986) and EXOSAT observations of 2 sources (Atteia et al., 1986a) have recently appeared (Figure 1). With one possible exception, no point sources associated with bursters were detected. As explained in Pizzichini et al. (1986), these observations may be used to derive relatively model-independent upper limits to the temperatures of the neutron star surfaces. These limits may in turn be used to infer the minimum age of the neutron star if a cooling model is assumed. Using the Glen and Sutherland (1980) cooling curves for a $1.25M_{\odot}$, 16 km radius neutron star for the cases B=0 Gauss and $B=5\times10^{12}$ Gauss gives the minimum ages in Figure 2. The temperature upper limit is a function of the (unknown) distance to the neutron star, and hence so is the minimum age. But in all cases the limits are reasonable ones, in the sense that both younger (e.g. SNR-associated) and older (e.g. pulsar) neutron stars are known.

The relation between age and time histories relies on the observations of periodicities. Two gamma ray burst time histories have been demonstrated convincingly to display periodicities (Barat et al., 1979; Wood et al., 1981), while a third has only marginally convincing evidence (Barat et al., 1984), and several more examples (as yet unpublished) exist in the Franco-Soviet SIGNE experiment data base. In all cases the periods fall in the range 4-10 s. The most natural explanation for these periodicities is the rotation of a magnetized neutron star. Some insight into the evolution of such an object can be obtained from the study of pulsars. Based on a study of 256 pulsars, Lyne, Manchester, and Taylor (1985) have shown that for a neutron star with a mean initial field of $0.75\times10^{12}$ Gauss, the e-folding decay time for the field is $9.1\times10^{6}$ years, and the average terminal period of the pulsar (i.e., the period for zero field) is 0.4 s. Using the fits to their data, a pulsar whose terminal period is 4 s must have started out with a minimum field of about $7.5\times10^{12}$ Gauss, and is about $10^{7}$ years old. Mechanisms for pulsar spindown other than those considered in the study of Lyne et al. (1985) have also been proposed (e.g., Peng, Huang, and Huang, 1982) which may be efficient for long period pulsars. Thus if gamma bursters have evolved from lone pulsars, as some models predict (e.g. Shklovskii and Mitrofanov, 1985), their minimum ages should be around $10^7$ years. Is it normal

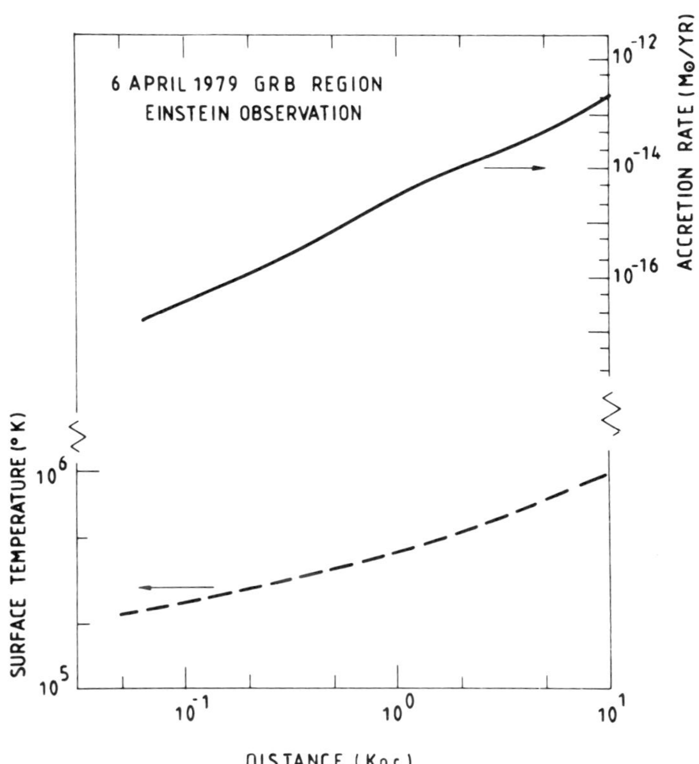


Figure 1. Typical result of an Einstein observation of a gamma ray burst source in the quiescent state (after Pizzichini et al., 1986). Dashed curve, left hand scale: upper limit to the blackbody temperature of the surface of the neutron star, as a function of the (unknown) distance. Solid curve, right hand scale: upper limit to the polar cap accretion rate as a function of distance. Note that the rates are at least 3 orders of magnitude below those proposed for X-ray bursters.

that such an object should not be detectable as an Einstein or EXOSAT soft X-ray source, either due to the residual blackbody radiation from its entire surface, or due to polar cap heating from accretion? The answer appears to be "yes" in both cases. The Glen and Sutherland (1980) cooling model predicts surface temperatures well below $10^5$ °K, which are undetectable. Illarionov and Sunyaev (1975) have considered the fate of old pulsars, and have shown that the "propeller" mechanism prevents accretion onto a spinning, magnetized neutron star until the period is sufficiently long. Spindown by the propeller mechanism itself is not efficient enough to slow the star down on timescales shorter than the age of the galaxy.

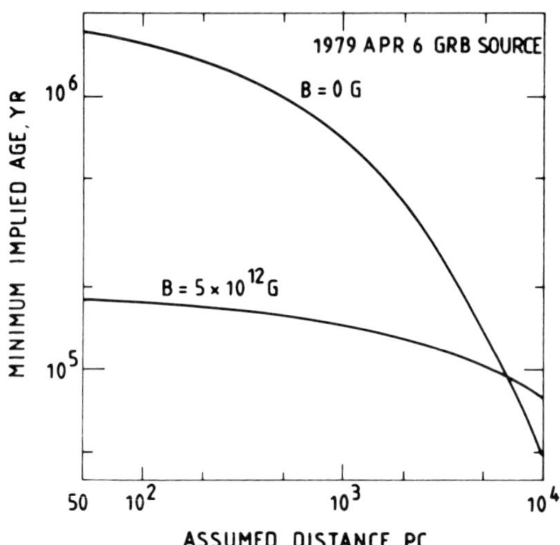


Figure 2. Minimum age of the 1979 Apr 6 burster, as a function of distance. The Einstein upper limits of Figure 1 have been used, as well as the cooling model of Glen and Sutherland (1980) for two values of the magnetic field.

The above discussion indicates that no fundamental inconsistency arises in the interpretation of soft X-ray and time history data if gamma bursters are assumed to be lone neutron stars, either magnetized or unmagnetized.

## 3. VELOCITIES

The soft X-ray observations of bursters in the quiescent state may also be used to estimate minimum neutron star velocities under some conditions. Here it is assumed that accretion is taking place over the entire surface of the neutron star, which results in heating and radiation in the soft X-ray region (Pizzichini et al., 1986). Higdon and Lingenfelter (1984) and Hameury (1984) have discussed the conditions under which accretion may take place onto the surface of a lone unmagnetized neutron star of mass M traveling through the interstellar medium at a velocity w and Hameury (1984) has shown that the rate is

$$dM/dt=5.8\times10^{-15}(10\ \mathrm{km\ s^{-1}}/w)^3(M/M_\odot)^2 n\ M_\odot/y$$

where n is the number density of the medium, and where w is never less than the sound speed (10 km/s). The result is shown in Figure 3, which gives the minimum allowed velocities for two densities n. As before, the accretion rate is a function of the (unknown) distance to the neutron star, and therefore so is the minimum allowed velocity. Here, as in Figure 2, the constraints are reasonable ones, since the minimum velocities are less than typical pulsar velocities.

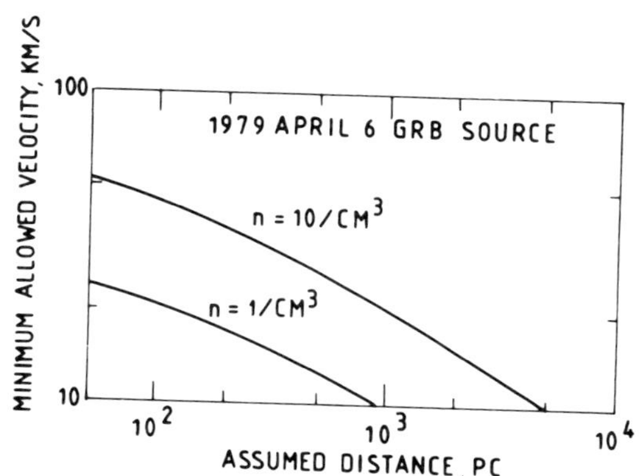


Figure 3. Lower limits to the velocities of lone, unmagnetized neutron stars in the interstellar medium (density n) as a function of distance. These curves are based on the Einstein upper limits.

4. DISTANCE AND MEMBERSHIP IN BINARY SYSTEMS

Some interesting constraints on burster distances and possible binary companions may be obtained from the results of deep optical searches for quiescent counterparts. Taking $m_V$=24 for the brightest possible counterpart, as obtained for two bursters (Chevalier et al., 1981; Pedersen et al., 1983; Schaefer, Seitzer, and Bradt, 1983; Motch et al., 1985), the absolute visual magnitude of a main sequence counterpart must be

$$M_V=29-5\log(d)-A$$

where d is the distance in pc and A the absorption in magnitudes. An unabsorbed main sequence star closer than $10^4$ pc must thus have $M_V>9$. This excludes all classes except M (see, e.g., Allen 1976). The allowed distances of some M stars are shown in Figure 4.

A degenerate star is also possible as a binary companion. Again, using the statistics from Allen (1976), the minimum possible distances for two types of white dwarfs are shown in Figure 4. Finally, one can ask whether the deep optical searches could have detected a lone neutron star. Integrating the redshifted blackbody spectrum for a neutron star with $1.3M_\odot$, a true radius of 16 km, and a true temperature of $10^5$ °K to obtain $M_V$ gives the result shown in Figure 4: such an object can be detected out to about 27 pc. According to this calculation, higher temperature neutron stars could of course be detected optically out to larger distances (e.g. 94 pc for $10^6$ °K), but here the soft X-ray constraints (Section 2) are more severe.

The extremely constraining value for $M_V$ might prompt one to speculate that the optical counterparts are either cool or heavily absorbed, and that infrared observations might yield a detection or at least a more severe constraint. Apparao and Allen (1982) have reported infrared scans of two burster regions, which did not result in any likely detectection of a counterpart. One of their distance lower limits is shown as a dashed line in Figure 4. Schaefer et al.

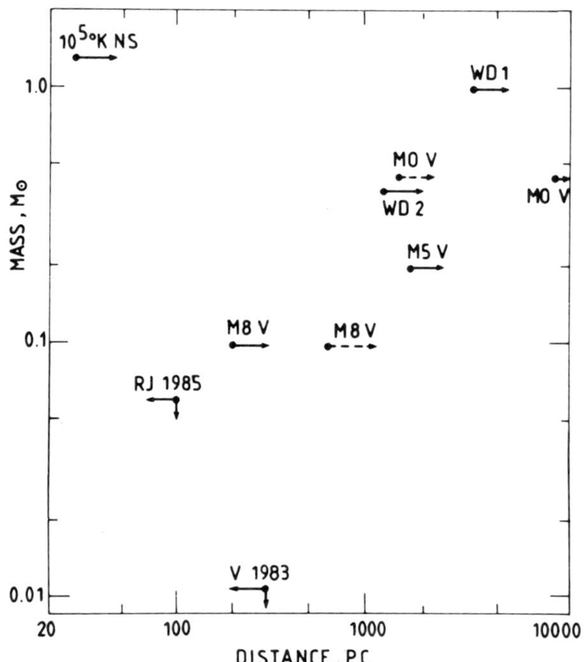


Figure 4. Where the bursters may lie, and what their masses are. The solid lines labelled M8 v, M5 v, M0 v, WD 1 and WD 2 are based on stellar statistics as tabulated by Allen (1976), and assume a maximum $m_V$=24 for burster counterparts. The dashed line labelled M0 v is from the infrared observations of Apparao and Allen (1982), and that for M8 v from Schaefer et al. (1986). The limit for the $10^5$ °K neutron star is calculated as described in the text. The region labelled V 1983 is from the model of Ventura et al. (1983), and that labelled RJ 1985 is from the model of Rappaport and Joss (1985).

(1986) have recently reported their results of infrared observations of 7 burster regions, also with no probable detection of a counterpart.

Several authors have considered the question of whether the

presence of nearby bursters violates the Oort limit (e.g., Barat et al. 1979; Schaefer and Ricker, 1983; Epstein, 1985). Using Bahcall's (1984) estimate of 0.09 $M_\odot/pc^3$ for the invisible matter in the solar neighborhood, all the observed bursters could be placed in a sphere of radius

$$R=(2.65\ n\ t_{rec}\ M)^{1/3}\ pc$$

where n is the number of distinct (i.e. not coming from the same regions of the sky) bursts observed per year, $t_{rec}$ is their recurrence period in years, and M is the mass of a burster in $M_\odot$. n may be estimated from Atteia et al. (1986b) to be 59 bursts/year, and to compare this limit to that set by the lone $10^5$ °K neutron star in Figure 4, M may be taken to be $1.3M_\odot$. However, for the purposes of this calculation, $t_{rec}$ is not significantly constrained by observations. Schaefer and Cline (1985) and Atteia et al. (1986b) conclude that the best lower limit obtainable from the observations is of the order of 10 years. This leads to R=13 pc above, which is certainly not in conflict with the point in Figure 4. Note, though, that a quantity of greater interest here is just the distance to the nearest burster, and that this depends only on the value of the Oort limit and the assumed mass of the neutron star:

$$d_{min}=(M/0.09)^{1/3}\ pc=2.3\ pc$$

for the case considered here. Thus this can never give a more constraining lower limit to the distance than the optical observations, as interpreted in Figure 4.

Before membership in a binary system can be assumed, a consistency check should be carried out: should a magnetized neutron star in a binary system, rotating with a period of 4-10 s not be visible as a soft X-ray source? Neutron star spindown and accretion in a binary system have been studied by a number of authors (e.g., Illarionov and Sunyaev, 1975; Davies and Pringle, 1981). These models aim to explain the evolution of bright binary X-ray sources, which is obviously not the case of gamma bursters. However, as the critical parameters are the spin periods, which evolve through the periods of interest here to hundreds of seconds, the strength of the wind from the companion, of order $10^{-11}$ $M_\odot/y$, and the magnetic moment of the neutron star, of order $10^{30}$ G $cm^3$, some features of the models may be applicable. Of particular interest is the prediction that accretion cannot take place onto the compact object, even in the presence of a strong wind, until it has been spun down to rather long periods. This would therefore appear to be consistent with the properties of gamma bursters. On the other hand, these systems evolve into X-ray sources while the massive companion is still on the main sequence, which casts some doubt on the applicability to bursters.

As constraining as the optical limits are, it is still possible to imagine binary systems in which bursts may be produced. Ventura et al. (1983) and Rappaport and Joss (1985) have studied the evolution of very low mass binary systems for the specific purpose of explaining bursters. The system of Ventura et al. (1983), after passing through a phase as a bright X-ray source, ends up after $10^{10}$ y as a detached binary system whose X-ray and optical luminosities are undetectable at 300 pc. In the model of Rappaport and Joss (1985), after about 5 x $10^9$ y the system is undetectable optically, although in principle it is undergoing mass transfer at a rate which would make it a bright X-ray source unless the transfer can be cut off somehow. But in any case, from the above discussion, it is clear that membership in a binary system, albeit a peculiar one, need not contradict the observational data.

## 5. MAGNETIC FIELDS

It is clear from the above discussion that the magnetic field strength is a crucial parameter in the interpretation of the observations. Two lines of observational evidence support the idea that a field should indeed exist. The first comes from the observations of periodicities. As pointed out in Section 2, however, relatively few events display them; even though this may just be an observational selection effect, its magnitude is difficult to estimate. In addition, the observation of periodicities does not provide an accurate estimate of the field strength. The second line of evidence is the observation of spectral features at energies around 50 keV (Mazets et al., 1981a), which have been interpreted as cyclotron absorption in a strong (several times $10^{12}$ G) magnetic field. Several questions concerning these observations must be considered. The first is, "Are the observations reliable?". Although one observation of such a spectral feature has been made, non-simultaneously, by an independent experiment (Hueter, 1984), simultaneous observations in another case failed to confirm a feature (Fenimore, Klebesadel, and Laros, 1983; Dennis et al, 1982). It has been demonstrated that in some cases such features could be artifacts (Fenimore, Klebesadel, and Laros, 1983). It appears, however, quite unlikely that all features can be explained in this way, and therefore that the reality of them should be accepted. The second is "Can they be interpreted as being due to effects other than cyclotron absorption?". Here too, the answer is yes. Two component spectra have been proposed (Fenimore et al., 1982; Lasota and Belli, 1983) which could mimic the observed features, and it has been pointed out that they could also be caused by photoelectric absorption of heavy atoms (Trümper, 1982; Bussard and Lamb, 1982). But in almost all of these cases, strong fields are still required. A final question concerns the observational statistics. Some 20 gamma bursts out of 143 (Mazets et al. 1981a) have been observed to display spectral features around 50 keV. What does this imply for the population of gamma bursters as a whole? A number of observational selection effects will tend to reduce the number of features detected. One might be the burst strength. Mazets et al. (1981a) report the observation of features only in bursts with fluences greater than 3 x $10^{-6}$ erg/cm$^2$, of which there are about 100 in the KONUS catalog (Mazets et al., 1981b). Thus _if_ the detection were related to the fluence, the true ratio might be nearer to 20/100. Another is the energy at which the features appear in the spectra. The KONUS detectors had a lower energy threshold of 30 keV in general (but 17 keV in some cases), and many features appear in the second energy channel. Features with energies below about 27 keV (the lowest energy reported by Mazets et al., 1981a), if they existed, would go undetected. Finally, the geometry of the emission region itself may play a role. The detection of 20 features out of 143 does not seem unreasonable when compared to the statistics of pulsating X-ray sources, which are clearly associated with rotating magnetized neutron stars (e.g., Trümper, 1986). Here two sources out of about two dozen have been observed to display cyclotron features, Her X-1 at 40 keV and possibly 80 keV, and 4U0115+63 at 11.5 and 23 keV. It therefore does not seem unreasonable, based on the observations of bursts, to assume that all of them are produced on or near high magnetic field objects.

## 6. CONCLUSION

The deep optical and soft X-ray searches for burster counterparts reported over the last year have begun to impose significant constraints on the parameter space which GRB systems may occupy. There is still no general concensus on the question of whether bursts are generated in binary systems or by lone neutron stars, or even both, but it is now clear that any binary companion must be quite cool, have very low mass, be very distant, or a suitable combination of the above.

No mention has been made of the Mar 5 source, which remains as enigmatic as before. Recent optical studies by H. Pedersen, C. Motch and colleagues, as yet unpublished, have set limits of about $m_V > 17.7$ for an optical counterpart. At the distance of the LMC, this would imply $M_V > -1$, which would allow many stellar types as counterparts. However, the well known 8 s period of the source would still appear to conflict with the age of the N49 supernova remnant, $10^4$ y. On the other hand, if the source is nearby, the Einstein observations make it difficult to understand the repeating bursts from the source in terms of an accretion-driven process (Pizzichini et al., 1986).

Further understanding of bursters may have to await the arrival of new instrumentation, particularly the Space Telescope and AXAF.

## 7. REFERENCES

Allen, C. 1976, Astrophysical Quantities, Athlone Press, London
Apparao, K., and Allen, D. 1982, Astron. Astrophys. Lett. 107, L5
Atteia, J.-L., Boer, M., Hurley, K., Niel, M., Chevalier, C., Ilovaisky, S., Motch, C., Taylor, B., Sims. M., Pizzichini, G., Mason, K., Branduardi-Raymont, G., Cordova, F., Evans, W., Klebesadel, R., Laros, J., and Martin, C. 1986a, XXVI COSPAR Topical Meeting E.4 on Gamma Ray Bursts, and New Results on Gamma Radiation from the Galaxy, Toulouse, France, to be published in Adv. Space Res.
Atteia, J.-L., Barat, C., Hurley, K., Niel, M., Vedrenne, G., Evans,

W., Fenimore, E., Klebesadel, R., Laros, J., Cline, T., Desai, U., Teegarden, B., Estulin, I., Zenchenko, V., Kuznetsov, A., and Kurt, V. 1986b, Ap. J. (submitted)
Bahcall, J. 1984, Ap. J. **276**, 169
Barat, C., Chambon, G., Hurley, K., Niel, M., Vedrenne, G., Estuline, I., Kurt, V., and Zenchenko, V. 1979, Astron. Astrophys. Lett. **79**, L24
Barat, C., Hurley, K., Niel, M., Vedrenne, G., Cline, T., Desai, U., Schaefer, B., Teegarden, B., Evans, W., Fenimore, E., Klebesadel, R., Laros, J., Estulin, I., Zenchenko, V., Kuznetsov, A., Kurt, V., Ilovaisky, S., and Motch, C. 1984, Ap. J. Lett. **286**, L5
Bignami, G. 1985, Proc. NATO Advanced Study Institute on High Energy Phenomena Around Collapsed Stars, Cargese, Corsica, D. Reidel Pub. Co., Dordrecht, Holland
Bussard, R. and Lamb, F. 1982, in Gamma Ray Transients and Related Astrophysical Phenomena, Eds. R. Lingenfelter, H. Hudson, and D. Worrall, AIP Conference Proc. No. 77, AIP Press, New York, p. 189
Chevalier, C., Ilovaisky, S., Motch, C., Barat, C., Hurley, K., Niel, M., Vedrenne, G., Laros, J., Evans, D., Fenimore, E., Klebesadel, R., Estulin, I., and Zenchenko, V. 1981, Astron. Astrophys. Lett. **100**, L1
Davies, R., and Pringle, J. 1981, Mon. Not. R. astr. Soc. **196**, 209
Dennis, B., Frost, K., Kiplinger, A., Orwig, L., Desai, U., and Cline, T. 1982, in Gamma Ray Transients and Related Astrophysical Phenomena, Eds. R. Lingenfelter, H. Hudson, and D. Worrall, AIP Conference Proceedings No. 77, AIP Press, New York, p. 153
Epstein, R. 1985, Ap. J. **297**, 555
Fenimore, E., Klebesadel, R., and Laros, J. 1983, Adv. Space Res. **3(4)**, 207
Fenimore, E., Klebesadel, R., Laros, J., Stockdale, R., and Kane, S. 1982, Nature **297**, 665
Glen, G. and Sutherland, P. 1980, Ap. J. **239**, 671
Hameury, J.-M. 1984, Thesis, Universite de Paris VII
Helfand, D. 1985, Proc. NATO Advanced Study Institute on High Energy Phenomena Around Collapsed Stars, Cargese, Corsica, D. Reidel Pub. Co., Dordrecht, Holland
Higdon, J. and Lingenfelter, R. 1984, in High Energy Transients in Astrophysics, Ed. S. Woosley, AIP Conf. Proc. No. 115, AIP Press, N.Y., p. 568
Hueter, G. 1984, in High Energy Transients in Astrophysics, Ed. S. Woosley, AIP Conference Proceedings No. 115, AIP Press, N.Y., p. 373
Hurley, K. 1985, Proc. NATO Advanced Study Institute on High Energy Phenomena Around Collapsed Stars, Cargese, Corsica, D. Reidel Pub. Co., Dordrecht, Holland
Illarionov, A. and Sunyaev, R. 1975, Astron. Astrophys. **39**, 185
Lamb, D. 1985, Proc. NATO Advanced Study Institute on High Energy Phenomena Around Collapsed Stars, Cargese, Corsica, D. Reidel Pub. Co., Dordrecht, Holland
Lamb, F. 1985, ibid
Lasota, J. and Belli, B. 1983, Nature **304**, 139
Liang, E. 1985, UCRL Preprint 93454, Lawrence Livermore Nat'l. Lab.
Liang, E. and Petrosian, V. 1986, Gamma Ray Bursts, AIP Conf. Proc. No. 141, AIP Press, New York

Lyne, A., Manchester, R., and Taylor, J. 1985, Mon. Not. R. astr. Soc. **213**, 613
Mazets, E., Golenetskii, S., Aptekar, R., Gur'yan, Yu., and Il'inskii, V. 1981a, Nature **290**, 378
Mazets, E., Golenetskii, S., Il'inskii, V., Panov, V., Aptekar, R., Gur'yan, Yu., Proskura, M., Sokolov, I., Sokolova, Z., Kharitonova, T., Dyatchkov, A., and Khavenson, N. 1981b, Astrophys. and Space Sci. **80**, 3
Motch, C., Pedersen, H., Ilovaisky, S., Chevalier, C., Hurley, K., and Pizzichini, G. 1985, Astron. Astrophys. **145**, 201
Pedersen, H., Motch, C., Tarenghi, M., Danziger, J., Pizzichini, G., and Lewin, W. 1983, Ap. J. Lett. **270**, L43
Peng, Q., Huang, K., and Huang, J. 1982, Astron. Astrophys. **107**, 258
Pizzichini, G., Gottardi, M., Atteia, J.-L., Barat, C., Hurley, K., Niel, M., Vedrenne, G., Laros, J., Evans, W., Fenimore, E., Klebesadel, R., Cline, T., Desai, U., Kurt, V., Kuznetsov, A., and Zenchencko, V. 1986, Ap. J. **301**, 641
Rappaport, S. and Joss, P. 1985, Nature **314**, 242
Schaefer, B., and Ricker, G. 1983, Nature **302**, 43
Schaefer, B., Seitzer, P., and Bradt, H. 1983, Ap. J. Lett. **270**, L49
Schaefer, B., and Cline, T. 1985, Ap. J. **289**, 490
Schaefer, B., Cline, T., Desai, U., Teegarden, B., Atteia, J.-L., Barat, C., Hurley, K., Niel, M., Evans, W., Fenimore, E., Klebesadel, R., Laros, J., Estulin, I., and Kuznetsov, A. 1986, Ap. J. (submitted)
Shklovskii, I. and Mitrofanov, I. 1985, Mon. Not. R. astr. Soc. **212**, 545
Trümper, J. 1982, in Gamma Ray Transients and Related Astrophysical Phenomena, Eds. R. Lingenfelter, H. Hudson, and D. Worrall, AIP Conference Proceedings No. 77, AIP Press, New York, p. 179
Trümper, J. 1985, Proc. NATO Advanced Study Institute on High Energy Phenomena Around Collapsed Stars, Cargese, Corsica, D. Reidel Pub. Co., Dordrecht, Holland
Trumper, J. 1986, in Cosmic Radiation in Contemporary Astrophysics, Ed. M. Shapiro, D. Reidel Publishing Co., Dordrecht, Holland, p. 217
Ventura, J., Bonazzola, S., Hameury, J., and Heyvaerts, J. 1983, Nature 301, 491
Wood, K., Byram, E., Chubb, T., Friedman, H., Meekins, J., Share, G., and Yentis, D. 1981, Ap. J. **247**, 642
Wood, K., Meekins, J., Yentis, D., Smathers, H., McNutt, D., Bleach, R., Byram, E., Chubb, T., Friedman, H., and Meidav, M. 1984, Ap. J. Supp. **56**, 507

## DISCUSSION

**C. Alcock:** I would like to play devil's advocate for the fun of it. This comment applies to the $VS^2Bs$ directly, and less directly to all GRB's. The 1979 March 5 burst is the brightest burst, so maybe it's the nearest. This is further suggested by it having the only optical identification. Why then are we assuming a galactic population?

**K. Hurley:** As long as we are only talking about soft bursts, I think that there is no hard observational evidence which constrains them to be local, at least not yet. But the gamma burster population as a whole is rather more constrained: we believe that the 400 keV features, for example, are not related to cosmological redshifts, but rather to gravitational redshifts at the neutron star surface. Very high energy emission (> 1 MeV) is also a reality, and is difficult to explain in distant sources. But these arguments don't apply to the soft bursts, where no emission is observed above a few hundred keV, or in the case of 5 Mar, ~1 MeV.

**S. Woosley:** Would you comment on the status of the various experiments that are presently searching for optical transients accompanying γ-ray bursts?

**K. Hurley:** George Ricker informs me that part of his Explosive Transient Camera has come on line at Kitt Peak, and has looked for optical transients from the Aries flasher. He will be presenting a review of his results at the COSPAR meeting in Toulouse in July. Holger Pedersen's telescope array is now starting to operate at ESO (La Silla) and he will also talk about his experiment at COSPAR.

# THEORY OF GAMMA RAY BURSTERS

G.S.Bisnovatyi-Kogan
Space Research Institute
Acad. Sci. USSR, Moscow,
Profsoyuznaja 84/32,
USSR

ABSTRACT. Gamma ray bursters are interpreted as nuclear explosions under the surface of neutron stars. The explosions occur after transportation of the matter with nonequilibrium composition during starquakes in outer layers where the matter becomes unstable and explodes as a result of a developing chain reaction. In the nonequilibrium matter nuclear charge $Z$ decreases with increasing density, contrary to the equilibrium composition. Formation of a shock wave and total energy output are calculated. Parameters of the mighty burst of 5 March 1979 are estimated. Formation of observed lines in different partions of the $\gamma$-range and a possible nature of accompanying optical bursts are discussed with the proposed model as a basis.

## 1. INTRODUCTION

Soon after the discovery of gamma ray bursters [1] about ten very different theories have been proposed for their explanation [3]. The discovery of a pulsar stage in the gamma ray burst of 5 March 1979 [3] was a stromg support to the theory related to neutron stars [4,5]. Intensive investigations of gamma-ray bursters in late 70s-early 80s gave vast information about complex temporal and spectral behaviour of gamma-ray bursts, their statistical properties connected with the log N - log S diagram, exact localization on the sky, etc. [6-9]. Despite numerous attempts of optical identification no burster have ever been associated with any stationary optical source [7,8]. Neither radio nor X-ray sources have been found in the error boxes of several well-localized sources *). In this indicates that the neutron stars, responsible for gamma-ray

*) The projection [7] of the 5 March 1979 burster on the edge of SNR N 49 in LMC seems accidental.



D. J. Helfand and J.-H. Huang (eds.), The Origin and Evolution of Neutron Stars, 501–519.

bursts are single and very old with moderate ($B \lesssim 10^{10}$ Gs) magnetic fields. The model of bursts from such objects has been developed in [4,5, 10-14]. It is necessary in this model that a neutron star have a nonequilibrium layer of superheavy elements, stable at high densities and unstable at low ones. The transfer of the nonequilibrium matter to low densities must occur during starquakes which are often observed in young neutron stars (Crab and Vela pulsars) and may be very rare in old ones. This may explain almost total absence of recurrence in gamma bursters.

It is impossible now to explain with only one theoretical model all striking features of gamma-ray bursters. We try here to fit only the most important features: the energy supply, short rising time, spectra including formation of $\gamma$-lines, emission in hard and soft regions. We give quantitative estimations of this characteristics and several observational predictions. The construction of the model seems to be a hard job, but it could have been enormously harder if the connection with the neutron stars had not been firmly established.

## 2. FORMATION OF A NONEQUILIBRIUM LAYER AND ITS STRUCTURE

The dense matter may have an Equilibrium Composition only at high temperatures when all possible channels of reactions are open. The channels become shut during cooling so that strictly speaking, the equilibrium in the cold matter (minimum energy state) is impossible. The cold matter composition is always nonequilibrium and depends on the way it reaches the state with given $\rho$ and (low) T.

Two extreme approaches have been considered: 1) contraction of the cold matter [10,15-18] which occurs during the accretion on a neutron star; 2) cooling of the matter at a given density, which occurs after the neutron star is formed.

The contraction of cold matter continues with A=const and decreasing Z up to the neutron drip line where the stripping energy of the neutron is $Q_n = 0$.

Further the contraction leads to cold evaporation of neutrons and decreasing both A and Z along the neutron drip line. Using nuclear parameters from [19] and beginning the compression from iron $Fe^{56}$ we obtain the nucleus

(A,Z) = (56,16) at $\rho = 4{,}2 \cdot 10^{11}$ g/cm$^3$ on the neutron drip line. Calculations [18] based on the nuclear matter theory from [20] give nuclei along the neutron drip line

from (A,Z) = (56,16) at $\rho = 5{,}01\ 10^{11}$g/cm$^3$ up to

$(A,Z) = (32,9)$ at $\rho = 1,41\cdot 10^{12}$ g/cm$^3$. As estimated in [18], further contration leads to fusion of the nuclei in picnonuclesr reactions.

Analytical estimations of parameters along the neutron drip line were made in [15,16]. The stripping energy of the proton $Q_p$:

$$Q_p(A,Z) = (m_{A,Z-1} - m_{A,Z})C^2 + Q_n(A,Z-1) - (m_n - m_p)C^2 \quad (1)$$

is approximated along the neutron drip line by relation

$$Q_p = 33 - Z/7 \text{ Mev}, \quad A = 4Z, \quad Z \geqslant 6 . \quad (2)$$

Using consition of beta-equilibrium

$$(m_{A,Z-1} - m_{A,Z}) c^2 = \mathcal{E}_{fe} + m_e c^2 \quad (3)$$

and expression for Fermi-energy $\mathcal{E}_{fe}$, we obtain from (1)-(3) the relation between $Z$ and $\rho$ at $Q_n = 0$:

$$\left[1 + \left(\frac{Z\rho}{10^6 A_o}\right)^{2/3}\right]^{1/2} = 1.96\left(33 - \frac{Z}{7}\right) + 2.54, \quad A = 4Z \quad (4)$$

The compositions along the neutron drip line from (4) are: $(A,Z) = (56,14)$ at $\rho = 10^{12}$ g/cm$^3$ up to $(A,Z) = (24,6)$ at $\rho = 2.6\cdot 10^{12}$ g/cm$^3$.

When cooling proceeds at a constant density much higher $A$ and $Z$ values may be reached [15,16]. They depend on the free-neutron concentration which may be found from the calculations of the kinetic of cooling. Such calculations have not been made, so we suppose that the mass concentration of neutrons is $X_n = 1/2$. The final composition along the neutron drip line, using (1)-(3), is [15,16]:

$$Z = 7\left\{33 - 0.511\left[\left(\frac{\rho}{\mu_z 10^6}\right)^{2/3} + 1\right]^{1/2} + 1.293\right\},$$

$$\mu_z = A/Z(1-X_n);\ A = 4Z \tag{5}$$

The curve $Z(\rho)$ for $X_n = 1/2$ is presented in Fig.1. The value $Q_{p,max}$ is reached for $C_{24}$ when $\rho = \rho_2 = 2.24\cdot 10^{12}$ g/cm$^3$. At $\rho = \rho_1 = 1.2\ 10^{11}$ g/cm$^3$, when $Z = 150$ we have $Z^2/A = Z/4 = 37.5$ with the fission time $\tau_f = 3\cdot 10^7$ years. At densities $\rho < \rho_1$ and $\rho > \rho_2$ new seed nuclei are formed and matter becomes almost equilibrium. How the nonequilibrium composition forms during cooling is schematically in shown the Fig. 2 from [16].

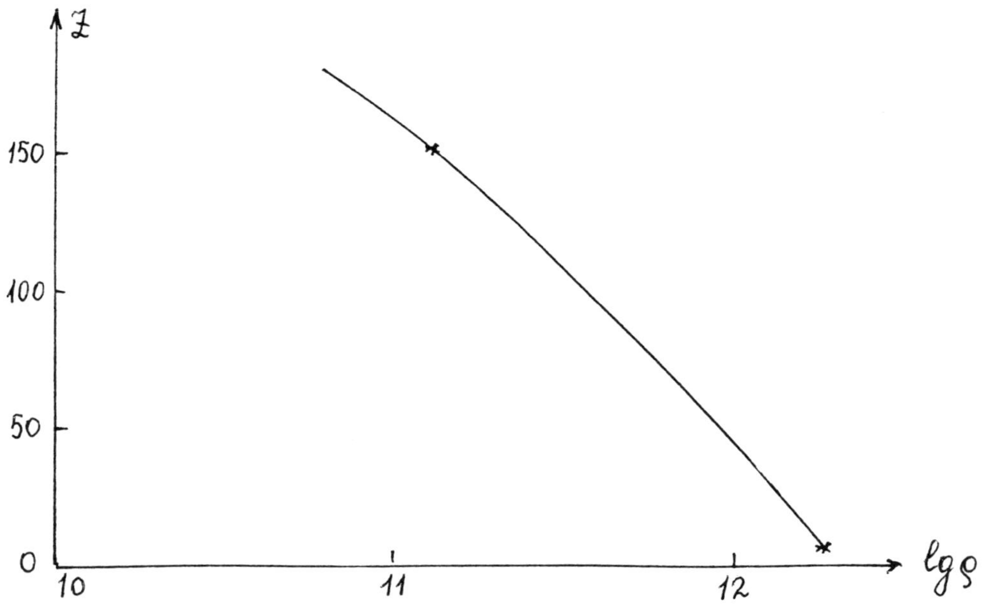


Fig. 1. Composition of neutron star envelope after cooling for $X_n$=½ , nonequilibrium layer is situated at densities in between the asterisks

Note that the nonequilibrium pressure is ~ 1.5 times higher than the equilibrium one at the same density. The total mass and energy stored in this layer are [16]:

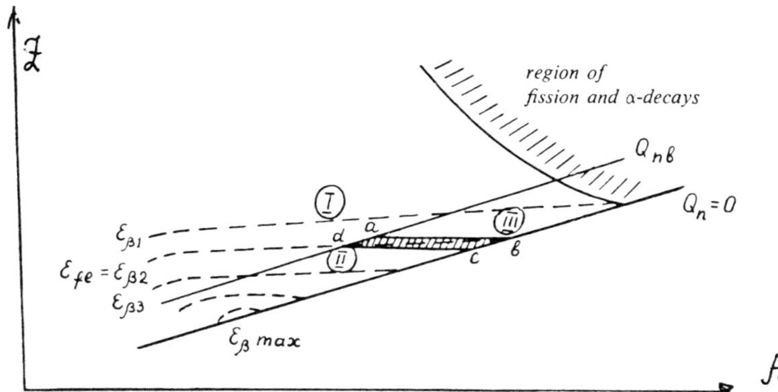


Fig. 2. Formation of non-equilibrium composition; photostripping of neutrons is impossible in region I, where $Q_n > Q_{nb}$ ; in region II - $Q_n < Q_{nb}$, $\varepsilon_{fe} < \varepsilon_{\beta}$, and in region III - $Q_n < Q_{nb}$, $\varepsilon_{fe} > \varepsilon_{\beta}$. On the dash lines $\varepsilon_{\beta} = Q_p - Q_n = \text{const}$, $\varepsilon_{\beta 1} < \varepsilon_{\beta 2} < \varepsilon_{\beta \max}$. Nuclei are situated within the shaded region abcd at given values of $Q_{nb}(T)$ and $\varepsilon_{fe}(\rho)$

$$m_{tot} \simeq 10^{29} \text{ g}, \quad E_{tot} \simeq 3 \cdot 10^{47} \text{ ergs} \quad \text{for } \eta = 3 \cdot 10^{-3} c^2. \tag{6}$$

## 3. NUCLEAR EXPLOSION IN THE OUTER LAYERS OF THE NEUTRON STAR ENVELOPE

A very sharp increase of luminosity observed in the 5 March 1979 burst ($\tau_r < 2.5 \cdot 10^{-4}$ s [21]) is incompatible with

a thermonuclear explosion of the accreted material, where the rise times 0.2-1 s and the spectrum explain well the properties of much softer X-ray bursters identified with neutron stars in low-mass binary systems [22]. The nuclear fission beneath the surface of the neutron star leads to an explosion due to the chain reaction to formmation of the shoch wave and very sharp increase of luminosity in good accordance with observations.

a) Development of a chain reaction [13]

Let us suppose that during a starquake the matter from the layer at $\rho_1 < \rho < \rho_2$ is transported to a lower-density region $\rho \ll \rho_1$ with the depth $d \ll 10^4$ cm *). We now assume for simplicity, that initially the matter consists of the kind of nuclei (A,Z) which are stable against fission and $\alpha$-decay, but not against $\beta^-$-decay at $\rho \ll 10^{10}$ g/cm$^3$. The interaction of these nuclei with rapid neutrons lead to the formation of slow neutrons, so these nuclei act as a moderator material. This property of the nuclei (A,Z) follows from the fact that the nonequilibrium layer at $\rho_1 < \rho < \rho_2$ is stable even in the presence of chance rapid neutrons. The nucleus (A,Z) takes part in two reactions:

$$\begin{aligned} &(A,Z) \longrightarrow (A,Z+1) + e^- + \tilde{\nu}, \\ &(A,Z) + n_r \longrightarrow (A,Z)^* + n_s , \end{aligned} \tag{7}$$

where $n_r$ and $n_s$ are rapid and slow neutrons. Nuclei (A,Z+1) are stable to the spontaneous fission and $\alpha$-decay, but unstable to $\beta^-$-decay and the fission induced by rapid neutrons. So we have

$$\begin{aligned} &(A, Z+1) \longrightarrow (A,Z+2) + e^- + \tilde{\nu}, \\ &(A,Z+1) + n_r \longrightarrow (A_1 Z_1) + (A_2 Z_2) + (\alpha+1) n_r . \end{aligned} \tag{8}$$

The nucleus (A,Z+2) is unstable to $\beta^-$-decay and spontaneous fission and yields $\zeta$ rapid neutrons in one fission:

$$\begin{aligned} &(A,Z+2) \longrightarrow (A,Z+3) + e^- + \tilde{\nu}, \\ &(A,Z+2) \longrightarrow (A_3,Z_3) + (A_4,Z_4) + \zeta n_r . \end{aligned} \tag{9}$$

The picture of the development of explosion is as follows. When bulk of the matter consisting of nuclei (A,Z) comes to a region with low density $\rho \ll 10^{10}$ g/cm$^3$, the $\beta^-$-decay begins, its probability being $W_\beta^Z$. When a sufficient amount of nuclei (A,Z+1) is stored, the absorption of rapid neutrons by the moderator, characterized by the average cross-section multiplied by velocity $\langle \sigma V \rangle_Z$, becomes less intense. The production of rapid neutrons in

*) This transportation does not need energy supply because the entropy of a neutron star is close to constant (zero) which mean that there exists neutral equilibrium to convective (exchange) motion.

the induced fission (8) of (A,Z+1) characterized by $\langle \sigma v \rangle_{z+1}$ and spontaneous fission (9) of (A,Z+2) which follows immediatly the $\beta^-$-decay of (A,Z+1) with probability $W^{z+1}$, prevails as compared with their absorption, a chain reaction develops and a nuclear exposion occurs. For approximate values [13] ($V_n \simeq 3 \cdot 10^9$ cm s$^{-1}$ for $E_{n_r} \simeq$ 10 MeV):

$$W_\beta^z \simeq 10^3 \ s^{-1}, \quad W_\beta^{z+1} \simeq 10^2 \ s^{-1},$$

$$\langle \sigma v \rangle_z \simeq 10^{-15} \ cm^3 \ s^{-1}, \quad \langle \sigma v \rangle_{z+1} \simeq \frac{1}{3} 10^{-17} cm^3 \ s^{-1}$$

$$\alpha = \gamma \simeq 3 \qquad (10)$$

the chain reaction begins when

$$\frac{X_{Z+1}}{X_Z} \simeq \frac{\langle \sigma v \rangle_Z}{\alpha \langle \sigma v \rangle_{Z+1}} \simeq 100 \qquad (11)$$

and concentration of rapid neutrons $X_r$ is

$$X_r = \frac{\gamma \cdot W_\beta^{Z+1}}{(\rho / m_p) \alpha \langle \sigma v \rangle_{Z+1}} \simeq 5 \cdot 10^{-15} \quad \text{for } \rho = 10^6 \ g \cdot cm^{-3} \qquad (12)$$

Chain reaction begins

$$\frac{\ln 100}{W_\beta^Z} \simeq 5 \cdot 10^{-3} \ s \qquad (13)$$

after the starquake and develops practically instantly, so that

$$\tau_{ch} \ll \tau_h \simeq \frac{d}{V_s} \ \frac{10^4}{10^{10}} = 10^{-6} \ s \qquad (14)$$

The average energy output (essiciency) in the fission of superheavy nuclei may be estimated [15] as $\eta =$ $= 3.10^{-3} \ c^2$ ergs/g . Instant heating leads to the formation of a shock wave which may initiate gamma ray burst when it reaches the surface.

b) Formation and output of the shock wave

Hydrodynamical calculations of the shock wave propagation in the neutron star envelope have been done in [14] in a spherically symmetric approximation. The system of equations in Lagrangian coordinates is:

$$\frac{\partial r}{\partial t} = u, \quad \frac{\partial r}{\partial m} = \frac{1}{4\pi \rho r^2} ,$$

$$\frac{\partial u}{\partial t} = -4\pi r^2 \frac{\partial p}{\partial m} - \frac{Gm}{r^2} , \tag{15}$$

$$\frac{\partial \mathcal{E}}{\partial t} = -4\pi p \frac{\partial (ur^2)}{\partial m} + q(m,t) - f_\nu(\rho,\tau) .$$

An approximate equation of state in the analytical form [23] has been used:

$$P = 3.09 \cdot 10^{12} \rho^{5/3} \frac{1+1.59\cdot 10^{-3}\rho^{1/3}}{(1+3.18\cdot 10^{-3}\rho^{1/3})^2} + \rho \mathcal{R} T +$$

$$+ \frac{aT^4}{3} \ \mathrm{dyn/cm^2} , \tag{16}$$

$$\mathcal{E} = 4.635\cdot 10^{12}\rho^{2/3}(1+3.18\cdot 10^{-3}\rho^{1/3})^{-1} + \frac{3}{2}\mathcal{R} T +$$

$$+ \frac{aT^4}{3} \ \mathrm{ergs/g} ,$$

which is valid for $\rho \lesssim 3\cdot 10^9$ g/cm$^3$, $\mathcal{R} = k_B/\mu m_p$ - gas constant, $\mu = 1$, a - energy density constant. The neutrino emission rate $f_\nu(\rho,T)$ takes into account URCA processes and interpolates numerical calculations from [24].

Only the region with $\rho < 3\cdot 10^9$ g/cm$^3$ has been considered in calculations. The star is supposed to be initially cold and in static equilibrium:

$$u = 0, \quad \frac{\partial u}{\partial t} = 0, \quad T = 0 \quad \text{at} \quad t = 0 . \tag{17}$$

The explosion was characterized by the energy $E_{tot}$, which was produced very rapidly $t_b \ll t$ in a spherical layer whose upper edge is at a depth $h_o$ under the surface. The mass of this layer $m_b$ and the depth of its bottom $h_1$ are determined uniquely by the efficiency $\eta$ : $m_b = E_{tot}/\eta$. So, in this layer

$$q(m,t) = \frac{\eta}{t_b} \; , \quad 0 \leq t \leq t_b \; ,$$
$$= 0 \qquad\qquad t > t_b \; , \tag{18}$$

with $t_b \simeq 10^{-8}$ s . At $t = t_b$ the temperature distribution is determined by using (16):

$$\frac{3}{2} \mathcal{R} \, T_b + a \, T_b^4 / \rho = \eta \; . \tag{19}$$

Typical distributions of parameters at $t=0$ and $t=t_b$ are given in Fig. 3. The results of numerical calculations

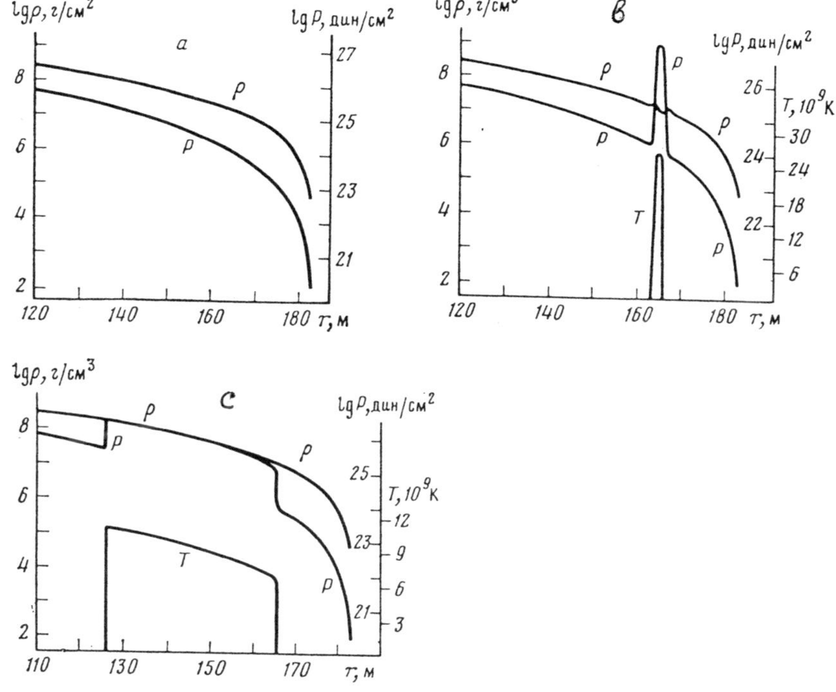


Fig. 3. Density, pressure and temperature distributions in the envelope of the neutron star with $M=1M_\odot$ , $R=10$ km . The distance $r_3$ is measured from the level $\rho = 3\cdot10^9$ g/cm$^3$. a) $t=0$ with $T=0$, b) $t=t_b$, variant 1 from table 1, c) $t=t_b$, variant 6 from table 1.

are presented in Table 1 and Fig. 4. The surface temperature at the moment when the shock wave reaches the surface with optical depth $\tau = 1$ was estimated in [14], using

Table 1. Explosion parameters in the envelope of the neutron star with $M = 1M_\odot$, $R = 10$ km

| N | $\eta/c^2$ | $E_{tot}$ ergs | $h_o-h$ meters | $m_b/M_\odot$ | $M_{hyd}/M_\odot$ (% $m_b$) | $\frac{u_{edge}}{u_p}$ |
|---|---|---|---|---|---|---|
| 1 | 0.3 | $10^{43}$ | 18–20 | $1.86\cdot10^{-11}$ | $8.7\cdot10^{-13}$ (4.7) | > 1 |
| 2 | 0.03 | $10^{43}$ | 18–30 | $1.86\cdot10^{-10}$ | $3.0\cdot10^{-14}$ (0.016) | > 1 |
| 3 | 0.008 | $10^{43}$ | 18–40 | $6.94\cdot10^{-10}$ | ($\lesssim 10^{-3}$) | 0.64 |
| 4 | 0.008 | $10^{43}$ | 40–50 | $6.94\cdot10^{-10}$ | ($\lesssim 10^{-3}$) | 0.86 |
| 5 | 0.003 | $10^{42}$ | 10–30 | $1.86\cdot10^{-10}$ | ($\lesssim 10^{-5}$) | 0.34 |
| 6 | 0.003 | $10^{43}$ | 18–58 | $1.86\cdot10^{-9}$ | ($\lesssim 10^{-5}$) | 0.34 |
| 7 | 0.003 | $10^{43}$ | 68–76 | $1.86\cdot10^{-9}$ | ($\lesssim 10^{-5}$) | 0.61 |
| 8 | 0.003 | $10^{44}$ | 20–174 | $1.86\cdot10^{-8}$ | ($\lesssim 10^{-5}$) | 0.36 |

selfsimilar solution by the method, proposed in [5]

$$p \simeq k_2^2 \rho\, h^{-0.6}\ ,\ \rho \simeq k_1 h^{1.5}\ ,\ u = 0{,}7\ k_2 h^{-0.3}\ . \qquad (20)$$

Constants $k_2$ and $k_1$ are estimated from numerical calculations. Applying (20) at $\tau = \int_0^h \varkappa \rho dh = 1$, where numerical calculations failed, we obtain the surface temperature $T_s = 150$ KeV for models in Table 1 with $\eta = 0.003$. This surface temperature leads to supercritical luminosity and strong mass loss. Supposing that the momentum of radiation is transformed into the momentum of matter running away with velocity close to parabolic one $u_p = (2GM/R)^{1/2} \simeq 0.54$ c, we obtain for the total mass loss $m_{out}$:

$$m_{out} \simeq \frac{\alpha E_{tot}}{0.54c^2} \simeq 2\cdot10^{22}\alpha\,(E_{tot}/10^{43}) \quad g \qquad (21)$$

where $\alpha = 0.5$–$1$ is part of the energy emitted in the su-

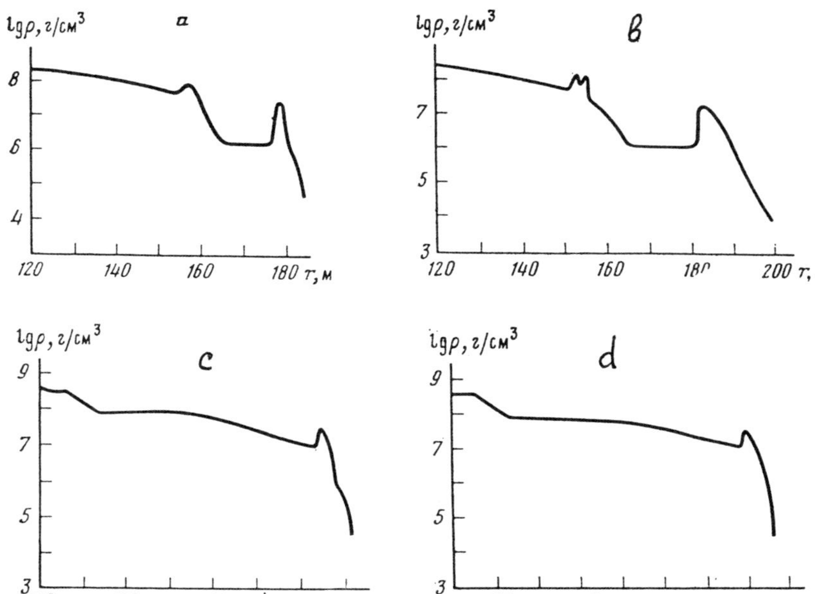


Fig. 4. Density distribution at different moments of the development of the explosion in the star with $M=1M_{\odot}$, R=10 km. Variant 1 from table 1: a) $t=20t_b$ - before the output of the shock wave to the surface; b) $t=30t_b$ - after the output. Variant 6 from table 1: c) $t = 130t_b$ - before the output; d) $t=160t_b$ after the output

percritical regime. For $\eta = 0.003\ c^2$ the part of the running-away matter is equal to $\sim 0.007\alpha$ of the exploded one. Note that in real events the diameter of the bulk of exploding matter is equal about its depth h, so the numbers for energy in Table 1 should be multiplied by

$$\frac{h^2}{4\pi R^2} \simeq 10^9 / (4\pi \cdot 10^{12}) \simeq 10^{-4} \qquad (22)$$

for $h = 3 \cdot 10^4$ cm and R = 10 km. The amount of mass ejected by the supercritical wind from (21) is much larger than a hydrodynamical outburst $m_{hyd}$ by the shoch wave in Table 1.

## 4. MODEL OF THE GAMMA RAY BURST 5 MARCH 1979

The observed flux $Q_{\gamma} = 1.3 \cdot 10^{-3}$ ergs/cm$^2$ from this source leads to the following total energy outputs for different assumed distances $d_{LMC}$ and $d_{Gal}$ :

$$q_{tot} = 4.5 \cdot 10^{44} \text{ ergs for } d_{LMC} = 55 \text{ kpc}$$
$$q_{tot} = 1.5 \cdot 10^{39} \text{ ergs for } d_{Gal} = 100 \text{ pc} \quad . \tag{23}$$

The observed luminosity on the pulsar stage is

$$L_{puls,LMC} = 3.6 \cdot 10^{42} \text{ ergs/s} \gg L_{edd} \ ,$$
$$L_{puls,Gal} = 1.2 \cdot 10^{37} \text{ ergs/s} < L_{edd} \ . \tag{24}$$

The rotational energy of the neutron star with the observed period $p = 8.1$ s is

$$E_{rot} \lesssim \frac{1}{2} I_{max} \Omega^2 \simeq 5 \cdot 10^{44} \text{ ergs} \simeq q_{tot\ LMC} \ . \tag{25}$$

Having in mind that not more than $0.02\ E_{rot}$ (energy of the crust) may be released during a starquake we may exclude the possibility of rotational energy supply in the case of LMC. The explosion of the nonequilibrium matter of such a scale seems to be also improbable. The observed behavior of the burst source on the pulsar stage is rather regular [3, 21] and cannot be related with strongly supercritical regime, where violent outbursts of matter and strong irregularities of luminosity must be expected. So we adopt the galactical origin of this burst .

The existance of spectral features in this burst indicates that optical depth $\tau \lesssim 1$ during their formation. We suppose [4,10] that the spectrum of a $\gamma$-ray burst is a mixture of the emission of a hot spot on the surface of the neutron star and of an expanding cloud which becomes transparent to $\gamma$-radiation at $t \gtrsim t_e$. Spectral features are forming at the $t = t_o > t_e$. For the cloud expanding with velocity $V_c \simeq 0.5\ c$ and supposing $\tau_c < 1$ at $t = t_o$ we estimate cloud parameters as:

$$R_c \simeq V_c t_o \simeq 1.5 \cdot 10^9 \left( \frac{t_o}{0.1c} \right) \ ,$$

$$\rho_c \simeq \frac{\mu_e m_p}{R_c \sigma_e} \tau_c \simeq \frac{10^{-8}}{t_{0.1}} \tau_c \quad \text{g/cm}^3 \ , \tag{25}$$

$$M_c = \frac{4\pi}{3} \rho_c R_c^3 = 1.4 \cdot 10^{20}\ t_{0.1}^2\ \tau_c\ \ g\ ,$$

$$t_{0.1} = t_o/0.1\ \ \sec$$

Here we used $\sigma_e = 0.4\,\sigma_T \simeq 2.6 \cdot 10^{-25}\ cm^2$ for $E_\gamma \simeq$ 0.6 MeV and $\mu_e = 2.3$ . If we suppose that nonthermal emission is generated in the expanding cloud with efficiency $\frac{1}{4}\eta = \frac{1}{4}\,3 \cdot 10^{-3}$ , then we obtain:

$$E_{nonth} = \frac{3 \cdot 10^{-3}}{4} M_c\ c^2 \simeq 10^{38} (t_{o.1})^2\, \tau_c\ \ ergs\ . \quad (26)$$

The total energy of the burst is about two orders of magnitude larger and is equal to

$$E_{tot} \simeq 10^{40} t_{0.1}^2\, \tau_c\ \ ergs \quad (27)$$

For the integral flux on the Earth $F_c \simeq 1.3 \cdot 10^{-3} ergs/cm^2$, we estimate the distance as

$$d \simeq (E_{tot}/4\pi\, F_c)^{1/2} \simeq 8 \cdot 10^{20}\ t_{0.1}\, \tau_c^{1/2}{}_{cm} \simeq 260\ t_{0.1}\, \tau_c^{1/2}{}_{pc} \quad (28)$$

On the pulsar stage the spectum is well approximated by the radiation of optically thin plasma with $T = 4 \cdot 10^8$K [3], so that the emission region should be represented by a corona on the surface of the neutron star. The existence of a corona may be connected with the mechanical flux from a convective layer in the stellar envelope developed after the explosion, and the magnetic field may be important for mechanical energy transformation into heat, like in the heating of the solar corona.

Let us estimate coronal parameters [11,12] . The density distribution in the isothermal corona is represented by a barometric formula

$$\rho = \rho_o\ e^{-(\mu m_p/kT)(GM/R_{ns}^2)x} = \rho_o\ e^{-(X/113)}\ , \quad (29)$$

The flux of free-free emission from the unit surface is

equal to

$$F_{cor} = \int_0^{\infty} 1.4 \cdot 10^{-27} Z^2 \sqrt{T_c}\, n_e n_i\, dx \simeq 8 \cdot 10^{-22} Z\, n_{eo}^2 \quad \text{ergs/cm}^2\,\text{s} \quad . \qquad (30)$$
$$n_e = Z n_i$$

The value of $F_{cor}$ must be less than a critical Eddington flux $F_{Edd}$, so

$$F_{cor} = 8 \cdot 10^{-22} Z n_{eo}^2 = \mathcal{E} F_{edd} = \mathcal{E}\, 4 \cdot 10^{24} \frac{\mu_e m_p}{\sigma_T} , \qquad (31)$$

$$\mathcal{E} < 1, \quad n_{eo} = 2.8 \cdot 10^{22} \sqrt{\mathcal{E}} , \quad \mu_e = 2.3 \quad .$$

We take $\mathcal{E} \leq 0.1$ so that when compton corrections to the spectrum are neglible [11]. On the pulsar stage the flux on the Earth $l_e$ is equal to $l_e \simeq 10^{-5}$ ergs/cm$^2$s . If about half the neutron star surface is heated, then for $\mathcal{E} = 0.07$, $\mu_e = 2.3$ we have from (31):

$$l_e = \frac{1}{2} \left( \frac{R_{ns}}{d} \right)^2 F_{cor} , \quad d = \left( \frac{F_{cor}}{2 l_e} \right)^{1/2} R_{ns} =$$

$$= 2.6 \cdot 10^{20} \text{ cm} = 85 \text{ pc} \quad . \qquad (32)$$

Comparing (32) with (28) and (27) we have

$$t_{0.1} \tau_c^{1/2} \simeq 1/3 , \quad E_{tot} \simeq 10^{39} \text{ ergs} ,$$

$$\tau_c < 1 \implies t_o > 0.033 , \qquad (33)$$

$$t_o < 0.2 \text{ s} \implies \tau_c > 0.02 ,$$

The value of $E_{tot}$ is much less than a total energy reserve of nonequilibrium layer in (6). The qualitative picture of the development of this burst is shown in Fig. 5.

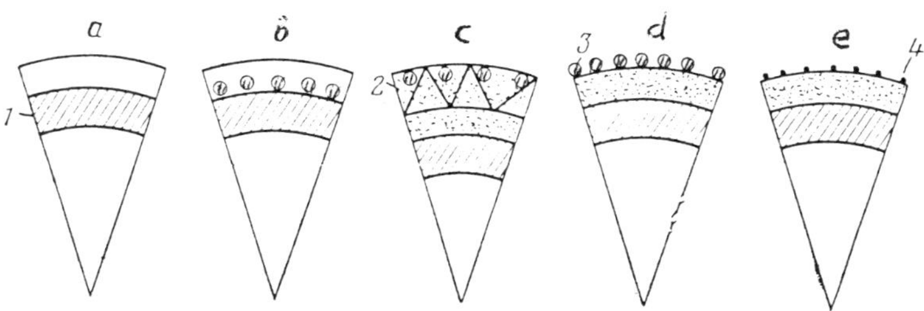


Fig. 5. Schematic picture of gamma-ray burst formation 5 March 1979; a - envelope in static equilibrium, 1 - nonequilibrium layer; b - after starquake, c - after nuclear explosion, 2 - shock wave, d - expansion of the cloud (3) and explosions there, e - pulsar stage, 4 - hot corona

## 5. SPECTRAL FEATURES IN THE EMISSION OF GAMMA-RAY BURSTS

Several kinds of features have been observed in the spectra of gamma-ray bursts. The first are emission features in the hard region which are broad $\Delta E \simeq E$ and have maxima at $E \simeq 400$-$500$ KeV, close to the $e^+e^-$ - annihilation line [25] It was supposed in [25], that these features are emitted in optically thin not regions, separated from the regions which emit most of their energy in softer photons. In our model this region may be an expanding cloud or more probably a corona on the initial stage of the burst, when its temperature may be sufficiently high for intense thermal birth of $e^+e^-$ - pairs.

The second type of features has been observed in a softer region $E_\gamma = 30$-$60$ KeV mostly in absorption [26] and has been interpreted there as cyclotron absorption. It seems improbable for an old neutron star to have very high magnetic field $3 \cdot 10^{12} - 10^{13}$ Gs, nesessary for cyclotron spectral features to form in this region. Besides, such a large field prevents output of hard photons with $E_\gamma \gtrsim 2$ MeV observed in the spectra of bursts. In our model these features may be naturally interpreted as absorption lines of very heavy elements with $Z \simeq 55$ in an expanding cloud. These elements are produced in fission of superheavy nuclei. For example, $Ba_{56}^{40}$, $I_{53}$ with $A = 131$-$135$ and $Ce_{58}$ with $A = 143, 144$ are the main elements produced in fission of $U^{235}$ [27]. The ionization potential of the last K-electron $I_Z \simeq 13.6\ Z^2$ eV $\simeq$ 42.6 eV for $Ba_{56}$. In the cloud expanding with $V = 0.4$ c the blue shifted absorption line may be located close to the observed one at 60 KeV. Temporal behaviour of this absorption feature in some bursts [25] which gradually dissapeared is well explained by the absorption in the expanding cloud, which should disappear when the density in the cloud becomes small. If iron $Fe^{56}$ is the most abumdant element in the cloud, then we may expect in some cases strong absorption in the softer region $E \gtrsim 13$ KeV during the cloud expansion.

In the hard region $E \gtrsim 1$ MeV spectral features may be expected which are associated with the emission of excited nuclei. Most probable is $D^2$ line with $E_\gamma = 2.2$ MeV. Such a line may be formed in our model, because free neutrons are formed during the fission of superheavy nuclei [10]. Indications to the presence of this line have been found in the spectra of the 7 March 1979 burst in some spectra obtained over short periods $\Delta t \lesssim 1$ sec in the Soviet-French signe experiment on Venera 11 and 12 probes.

## 6. OPTICAL EMISSION ACCOMPANYING GAMMA BURST

Intense investigations of the error-box of the 5 March 1979 gamma-ray burst lead to a discovery of optical flashes which may be connected with gamma ray bursts [28]. The shape of one of these flashes on 8 Feb. 1984 is close to that of the main burst (Fig. 6). Several weaker recurrent gamma events have been observed in the same direction [29] The optical flash may be associated with one of the recurrent rather feeble gamma-ray bursts.

The shape of this optical flash is well interpreted by the optical emission of the expanding cloud. The emission prior to the maximum (Fig. 6) may be connected with a Reyleigh-Jeans region of thermal emission of the cloud which is opaque in the optical region. The optical emission after the maximum may be interpreted as free-free radiation of optically thin iron plasma cloud, irradiated by the gamma-ray emission of the neutron star. Estimations made together with A.F.Illarionov, show that optical emission of the transparant cloud may be about $5 \cdot 10^{-4}$ times the gamma luminosity of the neutron star.

## REFERENCES

1. Klebesadel R., Strong I., Olson R., Astrophys J. Lett., 1973, 182, L85.

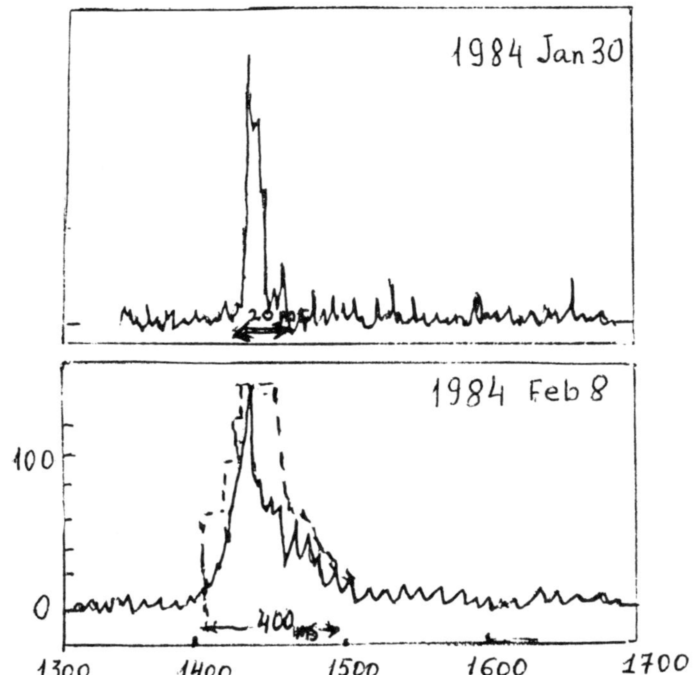


Fig. 6. Optical flash 8 Feb. 1984 and gamma-ray burst 5 March 1979 coming from the same region of the sky. The times of beginning of both flashes are brought into coincide as well as their maxima due to choosen luminocity scales


2. Prilutskii O.F., Rozental I.L., Usov V.V., Uspechi Fiz. Nauk, 1975, 116, 517.
3. Mazets E.P., Golenetskii S.V., Ilyinskii K.N., Panov V.N., Aptekar R.L., Gurjan Yu.A., Sokolova Z.Ya., Charitonova T.V., Pisma Astron. Zh., 1979, 5, 307.
4. Bisnovatyi-Kogan G.S., Gamma-Ray bursts from neutron stars, Report on COSPAR/IAU Symposium on "Fast Transients in X and gamma-ray astronomy", May 29-31, 1975, Varna, Bulgaria; Space Res. Inst. preprint D-203,1975.
5. Bisnovatyi-Kogan G.S., Imshennik V.S., Nadyozhin D.K., Chechetkin V.M. Astrophys. Space Sci., 1975, 35, 23,
6. Mazets E.P. et al. Catalog of cosmic gamma ray bursts from the Konus experiment data. Astrophys. Space Sci., 1981, vol. 80, N 1.
7. Vedrenne G. Cosmic $\gamma$-ray bursts, Phil. Trans. R. Soc. Lond. 1981, A301, 645.
8. Hurley K., Adv. Space Res., 1983, 3, 163.
9. Rozental I.L., Usov V.V., Estulin I.V., Uspechi Fiz. Nauk, 1983, 140, 97.
10. Bisnovatyi-Kogan G.S.,Chechetkin V.M. Sov. Phys. Usp., 1979, 22, 89.

11. Bisnovatyi-Kogan G.S., Chechetkin V.M. Astron. Zh. USSR, 1981, 58, 561.
12. Bisnovatyi-Kogan G.S., Chechetkin V.M. Adv. Space Res. 1981, 1, 153.
13. Bisnovatyi-Kogan G.S., Chechetkin V.M. Astrophys. Space Sci., 1983, 89, 447.
14. Bisnovatyi-Kogan G.S., Blinnikov S.I., Zacharov A.F. Astron. Zh. USSR, 1984, 61, 104.
15. Bisnovatyi-Kogan G.S., Chechetkin V.M. ZhTF, Pisma, 1973, 17, 622.
16. Bisnovatyi-Kogan G.S. , Chechetkin V.M. Astrophys. Space Sci., 1974, 26, 25.
17. Vartanyan Yu.L., Ovakimova N.K. Soobsch . Burakan. Obs. 1976, N 49, p. 87.
18. Sato K. Prog. Theor. Phys. 1979, 62, 957.
19. Grootez H.V., Hilf E.R., Takahashi K., Af. data and nucl. data tables, 1976, 17, 418-427, 476-608.
20. Baym G., Bethe H., Pethick Gh. Nucl. Phys., 1971, A175, 225.
21. Cline T. et al. Astrophys. J. Let. 1980, 237, L1.
22. Joss P., Li F. Astrophys. J., 1980, 238, 287.
23. Bisnovatyi-Kogan G.S., Popov Yu.P., Samochin A.A. Astrophys. Space Sci., 1976, 41, 287.
24. Ivanova L.N., Imshennik V.S., Nadyozhin D.K. Nauch. Inf. Astron. Sov. Acad. Sci. USSR, 1969, 13, 3.
25. Golenetskii S.V. et al. Ioffe Phys.-Tech. Inst. preprint N 959, 1985.
26. Mazets E.P. et al. Ioffe Phys.-Tech. Inst., preprint N 687, 1980.
27. Shirokov Yu.M., Yudin N.P. Nuclear physics, 1980, Nauka, Moscow.
28. Pedersen H. et al. Nature, 1984, 312, 46.

## DISCUSSION

**E. Liang:** You interpret the "cyclotron feature" as due to resonance lines of heavy elements I-Pb. Does this not require an unreasonably high concentration of heavy elements?

**G. Bisnovatyi-Kogan:** We need about 1% of heavy element to Fe ratio. This is much higher than solar abundance but may be obtainable in the expanding cloud after the fission chain reaction.

**K. Hurley:** How are the recurrent bursts from the 5 March source generated in this model?

**G. Bisnovatyi-Kogan:** The recurrent bursts have been observed only from 5 March 1979 gamma-ray burst position and they are at least $10^3$ times weaker than it. Maybe, they are connected with residual activity of the neutron star after the strong starquake.

**S. Woosley:** The weak interaction time scale does not set a limit to the thermonuclear model for γ-ray bursts because helium may burn on a very rapid timescale at high density and temperature. Fryxell and Woosely have shown that a detonation wave will propagate around a neutron star in about a millisecond.

**G. Bisnovatyi-Kogan:** The helium flashes could explain X-ray bursters and it seems improbable that such different phenomena are produced by the same burning. You need a larger magnetic field for your model, which is prohibited by the observed hard γ-ray photons ($E_\gamma$ > 2-5 MeV) which cannot escape from the neutron star with large field.

**S. Woosley:** What is the recurrence time scale for your model? What sets that time scale?

**G. Bisnovatyi-Kogan:** The recurrence time is expected to be between approximately 100-1000 years and is determined by the time between subsequent starquakes in old neutron stars.

**D. Arnett:** In core collapse we consider physical conditions very similar to these you quote: We estimate the relaxation time (to remove nonequilibrium) to be $2\times10^{-5}$ sec. There seems to be a serious disagreement.

**G. Binovatyi-Kogan:** We consider the stages after formation of the neutron star and subsequent cooling. The nonequilibrium layer is formed after the temperature decreases and become less than $\sim2\times10^9$ K, the time for the reactions between charged particles becomes very large, and the number of seed nuclei remains constant. At densities between $10^{11}$ - $10^{12}$ g/cm$^3$ the larger nonequilibrium exists for a very long time.

**A. Burrows:** Do I understand you to mean that many radio pulsars and, in particular, the Crab and Vela should be γ-ray bursters?

**G. Bisnovatyi-Kogan:** Crab and Vela must give the γ-ray bursts, but they are very feeble because of their large distances.

**A. Burrows:** How can matter deep in the crust penetrate all the way to the surface? What is the dynamics?

**G. Bisnovatyi-Kogan:** The matter of an old neutron star is almost at constant entropy (very small) and so is close to neutral stability relative to convection. The elastic energy released in the starquakes may be enough for convective motion which transports matter from the nonequilibrium layer to smaller densities where it becomes unstable.

**S. Kulkarni:** I would like you to clarify some points. If I understand you correctly, we ought to see a γ- and optical-burst from any neutron star when it glitches. So are you suggesting that is worthwhile comparing γ-ray burst catalogs and pulsar catalogs? In addition is it worthwhile cross-checking archival optical plates with radio pulsar positions?

**G.S. Bisnovatyi-Kogan:** That is right. But radio pulsars are too far away, so there is little chance of finding something.

# CYGNUS X-3 AND OTHER ULTRA-HIGH-ENERGY $\gamma$-RAY SOURCES

John J. Barnard
Code 665
NASA-Goddard Space Flight Center
Greenbelt, MD 20771

ABSTRACT. Recently, several binary X-ray sources have been found to be sources of ultra high energy $\gamma$-ray emission. Air shower observations indicate photon energies $\gtrsim 10^{15}$ eV. We review the current status of observations from the source Cygnus X-3, and compare this data with that from the sources Hercules X-1, Vela X-1, and LMC X-4. Current theoretical models for the production of $\gamma$-rays and the acceleration of high energy particles are discussed and the consequences for the evolution of such systems are examined.

## I. INTRODUCTION

When a pencil is dropped onto a tabletop from a height of about one centimeter the energy liberated in the form of sound waves and heat is about $10^{16}$ eV. Individual photons from the binary X-Ray source Cygnus X-3 have been observed (e.g. Samorski and Stamm, 1983; Lloyd Evans et al 1983) to be carrying such macroscopic quantities of energy at a rate such that an area the size of a football field (at the distance of the earth) is bombarded by such photons about 10 times each year. This enormous energy per photon as well as the large implied particle luminosity at the source places constraints on mechanisms for accelerating the particles required to produce such photons. Further, the presence of a high energy particle accelerator in a binary star system has implications on the evolution of the star system itself. The purpose of this review is to examine some of the proposed methods for accelerating particles to ultra-high-energies (UHE; photon energy $E \gtrsim 10^{15}$ eV) and to look at some of the effects that such particles have on binary star systems, such as Cyg X-3.

We begin in section II, with a short synopsis of the observations of Cygnus X-3 from radio through UHE $\gamma$-rays, and a look at the other systems where UHE $\gamma$-rays have been reported. In section III, we examine the stellar-beam-dump model of $\gamma$-ray production (e.g. Vestrand and Eichler, 1979, 1982, Berezinsky, 1980) wherein protons that have been accelerated by a compact star to energies of about $10^{17}$ eV bombard a normal stellar companion. Neutral pions are produced which



D. J. Helfand and J.-H. Huang (eds.), The Origin and Evolution of Neutron Stars, 521–533.

decay into γ-rays observable when the line of sight to the compact star grazes the limb of the companion star. In section IV we examine three proposed methods of accelerating particles: a pulsar (Eichler and Vestrand, 1984), the accretion disk dynamo (Chanmugam and Brecher 1985) and shock acceleration (e.g. Eichler and Vestrand 1985, Kazanas and Ellison 1986). In section V consequences of having a particle accelerator in the stellar system are considered (e.g. Berezinsky 1980, Stecker et al 1985, Gaisser et al 1986) focusing on the formation of a stellar wind, and the internal heating of the star from neutrino absorption.

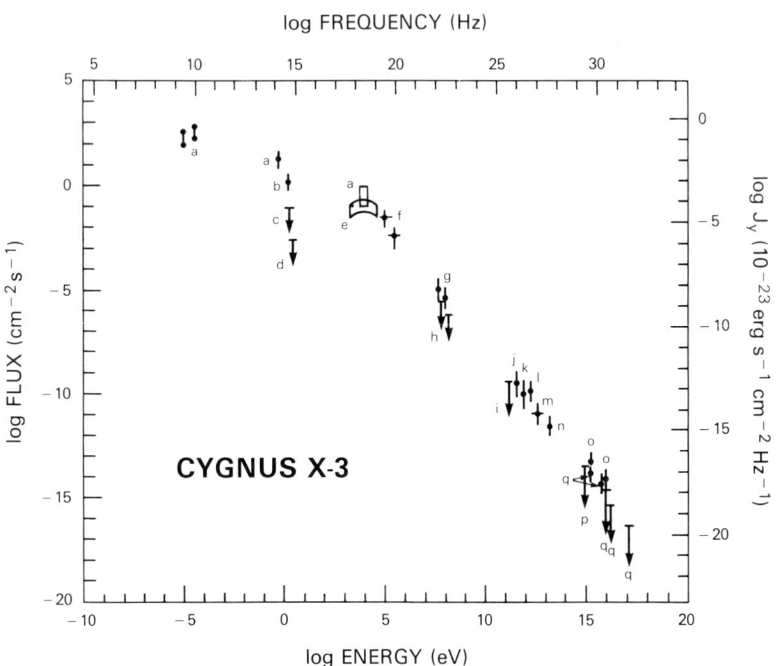


Figure 1. Spectrum of Cygnus X-3. References: a)Mason et al 1976 b) Weekes et al 1982 c)Weekes et al 1981 d)Westphal et al 1972 e) Serlemitsos et al 1975 f) Meegan et al 1979 g) Lamb et al 1977 h) Hermsen 1983 i) Weekes et al 1977 j) Lamb et al 1981 k)Dowthwaite et al 1983 l) Neshpor et al 1979 m)Stepanian et al 1982 n) Morello et al 1983 o)Samorski et al 1983 p) Hayashida et al 1981 q) Lloyd-Evans et al 1983.

## II. OBSERVATIONS

Cygnus X-3 has been observed from the radio to the UHE γ-ray regime. Figure 1 is a composite spectrum of some of the data obtained during the 16 years of observations in the various wavelength bands.

If we assume that all the reported detections are real, then from X-rays (at ~ $10^4$ eV) to UHE γ-rays (at ~ $10^{16}$ eV) the spectrum is a power law. The energy exponent in the differential energy flux (and also the integral number flux) is -1.1 (i.e. EdN/dE ~ $E^{-1.1}$, where dN/dE is the photon number flux per unit energy interval at photon energy E). Note that some of the upper limits contradict the detected fluxes. At ~ $10^8$ eV the SAS-II detections (Lamb et al 1977) (at 4.5σ) contradict the COS-B upper limits (Hermsen 1983). And observations at ~ $10^{11}$ eV by Weekes et al (1977) seem to contradict the other positive detections in this energy range (e.g. Vladimirsky et al 1975, Neshpor et al 1979, Danaher et al 1981, Lamb et al 1981), while Hayashidi et al (1981) report upper limits in the PeV ($10^{15}$ eV) energy range inconsistent with the data of Samorski and Stamm (1983). Time variability of the source may be at work (indeed Bhat et al 1985 report evidence for a secular decrease in flux in the PeV range on a decade long time scale.) Variability is certainly present at lower frequencies, as in the radio (e.g. Gregory et al 1972), IR (Becklin et al 1973), and X-Ray (e.g. Elsner et al 1980). In a more skeptical interpretation, Chardin and Gerbier (1985) question the consistency of the observations and suggest that if the γ-ray signals were actually just statistical fluctuations above the background a power law spectrum would be produced with the same index as is observed.

The upper limits in the optical reflect the ~ 19 magnitudes of extinction (Weekes and Geary 1982) in the direction of Cyg X-3 due to galactic dust while the upper limits at energies greater than ~ $10^{16}$ eV are interpreted as evidence for a real cutoff in the spectrum (Lloyd-Evans et al 1983).

Periodicity on a 4.8 hour timescale has been observed over 16 decades in energy (from IR through UHE γ-ray). In the X-ray regime (~ 10 keV) the light curve is nearly sinusoidal (e.g. Bonnet-Bidaud and van der Klis 1981). At IR energies the depth of the modulation is less (Mason et al 1976) as it is in the high energy X-ray regime (Molnar 1985). In the Very High Energy range (VHE: E >~ $10^{11}$ eV) the light curve appears to have a narrow peak at a phase of ~.6 (relative to phase 0 at X-ray minimum), with a width in phase of about .02 to .05. At UHE the phase also varies with observing group and with time, but phases of about .6 and .2 have been observed with some consistency (Lambert et al 1985). In many of the γ-ray detections a stastically significant signal is not observed until it is folded onto the 4.8 hour period. The resulting signals are typically 2 to 4 standard deviations above the background (cf. Chardin and Gerbier 1985).

Periodicity or variability on other timescales have been reported at a variety of periods. Potentially one of the most significant is the report of a 12.6 ms periodicity in the VHE γ-ray data (Chadwick et al 1985b) In addition this group tentatively reported that their data is consistent with a period derivative of 4.5 x $10^{-15}$ $ss^{-1}$ (Turver et al 1986). These observations have yet to be confirmed by independent groups however.

Periodicity in the radio on a 4.95 hour timescale has been reported (Molnar et al 1985). Molnar et al find that flux density variations (flaring) at wavelengths between 1 and 6 cm exhibits a periodicity with a period that is definitely different than the 4.8 hour X-ray period.

Other periodicities or variability time scales which have been reported are: 19 days (Chadwick et al 1985a) and 34.1 days (Molteni et al 1980). Variability at shorter timescales are indicated in the power spectrum of the X-ray data which is proportional to $\nu^{-1.8}$ (where $\nu$ is frequency) for frequencies between $10^{-4}$ and $10^{-3}$ Hz (Willingale et al 1985). Quasi-periodic-oscilations have been observed (also in the X-Ray regime) with periods in the 50 to 1500 s range (van der Klis and Jansen 1985).

In addition to the electromagnetic signal from Cygnus X-3, two underground experiments (SOUDAN and NUSEX) which detect TeV muons have reported possible time modulated excess muon fluxes from the direction of Cygnus X-3 (Marshak et al 1985, Battistoni et al 1985). Other underground experiments have also searched for muon fluxes and have failed to detect them (cf. Oyama et al 1986, Chudakov 1986). The experiments vary in depth of rock (from about $10^6$ to 5 x $10^6$ g/cm$^2$) and corresponding muon threshold energies (from ~.2 TeV to ~3 TeV). The upper limits of some of the negative experiments contradict the positive results. In addition, in both the SOUDAN and NUSEX experiment the angular windows that were used in the direction of Cygnus X-3 were much larger than the angular resolution of the experiment. This was done in order to maximize the detected signal, but the theoretical justification is not apparent (although cf. Ramana Murthy 1986 and references therein).

The question of what primary particle produces the muons adds more uncertainty to the whole problem. As pointed out by Marshak et al (1985) if the primaries were neutrons they would need energies greater than $10^{18}$ eV to make it to earth without decaying. If they were produced by photons they would need energies >> TeV energies in order to produce a ~0.6 TeV muon. In both cases surface detectors would have detected neutron or photon fluxes several orders of magnitude larger than what is observed. Finally if they were neutrinos, they would not have observed an intensity dependence on the grammage of overlying rock (as was observed) as the earth itself would be optically thin to the neutrinos. In light of these large uncertainties in the muon detections we focus our subsequent discussion strictly on the implications of the UHE and lower energy photons.

In addition to Cygnus X-3, 3 other binary X-ray systems have reported detections of UHE $\gamma$-rays: Vela X-1 (Protheroe et al 1984), LMC X-4 (Protheroe and Clay 1985) and Her X-1 (Balstruaitis et al 1985). Table 1 (from Gaisser et al 1986) summarizes some of the properties of these sources. It is apparent that except for the property of being an X-ray binary, they are not a particularly homogeneous set of objects. If the VHE $\gamma$-ray observations of Cyg X-3 indicate the underlying rotation period of the neutron star, the rotation periods vary over 4 decades, the orbital periods and companion masses vary over nearly 2 decades and the inferred UHE cosmic ray luminosity varies over nearly 4 decades. The inferred cosmic ray luminosity is given by:

$$L_{CR} = 4\pi d^2 F_\gamma (\Delta\Omega/4\pi) (.02/D_\gamma) (.1/\varepsilon) \quad (1)$$

Here d is the distance to the binary system, $F_\gamma$ is the time averaged UHE γ-ray flux, ΔΩ is the solid angle subtended by the cosmic ray proton accelerator (assumed equal to 4π), $D_\gamma$ is the orbital phase interval over which UHE γ-rays are observed, and ε is the assumed efficiency for converting cosmic ray particle energy into γ-ray energy. Equation (1) is based on the "stellar-beam-dump" model, to be discussed in the next section.

Table 1: Parameters of Binary X-Ray/UHE γ-Ray Systems[1]

| System | Rotation Period (s) | Binary Period (days) | Companion Mass (M/M⊙) | a/R† | Distance (kpc) | $L_{CR}$ (erg/s) |
|---|---|---|---|---|---|---|
| Cygnus X-3 | .0126(?) | 0.19 | <4 | >1 | >8 | ~ $10^{39}$ |
| Vela X-1 | 283. | 8.965 | ~23 | 12 | 1.4 | ~ $10^{37}$ |
| LMC X-4 | 13.5 | 1.408 | ~19 | 3.5 | 55 | ~ $10^{41}$ |
| Her X-1 | 1.24 | 1.7 | 2.4 | 6 | 4 | ~ $10^{38}$ |

[1]From Gaisser et al 1986; †a = orbital semi-major axis; R = companion star radius

## III. THE STELLAR BEAM DUMP MODEL

The most developed model to account for the UHE γ-rays from Cyg X-3 is the "stellar beam dump model" (cf. Vestrand and Eichler 1979, 1982; Berezinsky 1980). In this model protons (or in some theories neutrons) accelerated in the vicinity of a black hole or a neutron star strike the companion star producing neutral and charged pions. Neutral pions decay into γ-rays which if directed into the star will be absorbed; if produced near the limb of the star and the grammage traversed is sufficiently low they may escape from the system. Thus γ-rays will be observed as the star goes into and out of eclipse. The width of the γ-ray pulse is determined by the density distribution of the atmosphere and wind of the companion, which may be severely altered by the cosmic-ray bombardment (cf. section V). Hillas (1984) calculates the spectrum of a monoenergetic ($10^{17}$ eV) proton beam bombarding an atmosphere (with a magnetic field of ~ $10^3$ G) and finds a γ-ray spectrum which is consistent with the observed spectrum.

This model requires that the Cygnus X-3 system be an eclipsing X-ray binary. The X-ray light curve however does not show a sharp eclipse but rather is nearly sinusoidal in character. This has led some authors to conclude that matter in the system, in the form of a stellar wind (Pringle 1974, Davidsen and Ostriker 1974) or a shell around the system (i.e. the "cocoon" of Milgrom, 1976), scatters the X-rays and thus smooths the sharp eclipse. See Hertz et al (1978) and Ghosh et al (1981) for a detailed comparison of the two models.

However, in an alternative explanation an accretion disk corona is the source of X-rays which are eclipsed by bulges in the accretion disk associated with the accretion stream (White and Holt, 1982). Molnar (1985) suggests that the frequency dependence of the depth of modulation of the light curves argues for the accretion disk corona model and against the wind or cocoon models. In such a model the bulges associated with the eclipses may provide the grammage for γ-ray production although the details of such a model have not been explored.

In the stellar beam dump model γ-rays are produced in the interaction of UHE particles with target matter. One attractive feature of such models is that the acceleration region which may involve relatively high magnetic fields is distinct from the γ-ray production region which is required to have low magnetic fields ($<\sim 10^3$ G for $E \sim 10^{15}$ eV) in order to avoid magnetic pair creation (cf. Stephens and Verma, 1984).

One problem with the model is that a γ-ray phase of 0.6 is not consistent with the beam dump model which predicts an eclipse from phase 0.75 to 0.25 if $a/R = 1$ and a narrower range about phase 0 for larger, expected values of $a/R$. Attempts (e.g. Hillas 1985) to account for this phase assuming a wake through a stellar wind from the companion have been proposed which try to preserve the basic feature of the model.

## IV. ACCELERATION MODELS

Models for producing UHE particles in the vicinity of a compact star can be divided into three types: pulsar acceleration (Eichler and Vestrand, 1984), shock acceleration (see e.g. Eichler and Vestrand 1985, Kazanas and Ellison 1986) and the accretion disk dynamo (Chanmugam and Brecher 1985).

A pulsar has long been suspected of being a member of the Cyg X-3 system (e.g. Pringle 1974). As the photon energy has gone up, the idea of having a fast young pulsar as the accelerator has become more appealing (e.g. Lamb et al 1977, Eichler and Vestrand, 1984). Goldreich and Julian (1969) pointed out that the maximum potential drop $\Delta\phi$ available along the open field lines above a pulsar yields a maximum energy of:

$$e\Delta\phi \sim eB\left(\frac{R\Omega}{c}\right)^2 R \sim 10^{17}\left(\frac{B}{10^{12}G}\right)\left(\frac{10ms}{P}\right)^2 \text{ eV} \qquad (2)$$

Here $B$ is the surface neutron star field, $P \equiv 2\pi/\Omega$ is the pulsar rotation period and $R \simeq 10^6$ cm is the neutron star radius. However, much of pulsar theory indicates that pair creation may limit the potential drops to a small fraction of this potential (e.g. Ruderman and Sutherland 1975, Arons and Scharlemann, 1979).

The maximum luminosity $L$ is that given by spindown of a magnetized rotating dipole (Gunn and Ostriker 1969) and is approximately given by: $L \sim 10^{39}\ (B/10^{12})^2\ (10ms/P)^4$ erg s$^{-1}$. The lifetime $t$ for this loss mechanism is: $t \sim P/\dot{P} \simeq 1000\ (P/10ms)^2(10^{12}G/B)^2$ years. Thus, a pulsar with rotation period of about 10ms, and surface magnetic field strength of about $10^{12}$ G has a total luminosity, maximum electrostatic potential, and lifetime consistent with what is needed for Cygnus X-3. If observations by Chadwick et al (1985) of P = 12ms and the report by Turver et al (1986) of $\dot{P} \sim 4.5 \times 10^{-14} s^{-1}$ (corresponding to an approximate surface field of $8 \times 10^{11}$ G) are confirmed, the weight of evidence for the pulsar model will be compelling. Indeed, the fact that the Crab pulsar is a TeV source (Grindlay et al 1976) (although not a strong PeV source, cf. Boone et al 1984) provides evidence for particle energies up to TeV energies for isolated neutron stars. On the other hand, in the 3 other UHE sources the maximum potential drops across the polar cap are not large enough to account for the PeV emission.

In the pulsar model particles are accelerated electrostatically, and the energy source is the stored rotational energy of the neutron star. In the accretion disk dynamo model (Lovelace, 1978, applied to Cygnus X-3 by Chanmugam and Brecher, 1985) particles are again accelerated electrostatically but with energy derived from accretion of matter down to the Alfven radius (where the magnetic and kinetic energy densities are equal).

The basic idea is that the matter that is moving azimuthally in the accretion disk around the neutron star encounters a vertical component to the magnetic field. In the frame comoving with the plasma the conductivity is high so that the comoving electric field is approximately zero. In the pulsar frame this requires that $\underline{E} = -(\underline{v} \times \underline{B})/c$ yielding a radial electric field in the accretion disk. Thus the maximum energy available is roughly $eEr_A$:

$$e\Delta\phi \sim \frac{e}{c}\left(\frac{GM}{r_A}\right)^{\frac{1}{2}} r_A B_A \sim 10^{16}\left(\frac{M}{M_\odot}\right)^{\frac{1}{2}}\left(\frac{B_A}{10^8 G}\right)\left(\frac{r_A}{10^6 cm}\right)^{\frac{1}{2}} \text{ eV} \quad (3)$$

Here $B_A \equiv$ the magnetic field at $r_A$, $r_A \equiv$ the Alfven radius $\simeq 10^6$ cm $(B/10^8 G)^{4/7} (R/10^6 \text{ cm})^{10/7} (L/10^{38} \text{erg s}^{-1})^{-2/7}$ (see e.g. Shapiro and Teukolsky 1983 and references therein), M is the neutron star mass. In this picture the inner edge of the accretion disk has a magnetic field strength determined by the dipolar magnetic field of the neutron star. The maximum luminosity in cosmic ray particles is the accretion luminosity $L \sim GM\dot{M}/r_A$ which must be <~ the Eddington luminosity $\simeq 1.3 \times 10^{38} (M/M_\odot)$ erg $s^{-1}$. The basic picture then is that particles will be accelerated along field lines which do not penetrate the disk (i.e. the open field lines), creating a "spray" of high energy particles.

The lifetime for accretion to occur is long: t $\simeq$ $\simeq 10^8 (R/r_A) (10^{38} \text{ erg s}^{-1}/L)(Mc/M_\odot)$ years, where $M_c$ is the companion mass and $\dot{M}$ is the mass loss rate from the companion star (assumed equal to the mass accretion rate onto the neutron star). Note that in this model the highest potential drop occurs when $r_A$ is small (i.e. ~ R). Here the Keplerian velocity is largest resulting in a large electric field. But a small $r_A$ requires a weak magnetic field, some 4 orders of magnitude weaker than in the pulsar case.

The main advantage of appealing to this model is that Vela X-1, LMC X-4, and Her X-1 are slow rotators yet apparently generate UHE particles, and so accretion power is a common link. Disadvantages of the model include the fact that the luminosity is limited to ~ $10^{38}$ erg $s^{-1}$ for a neutron star, apparently in conflict with Cyg X-3 and LMC X-4. Thus, it requires an efficiency better than unity in those cases and close to unity for the other two. Further, the plasma distribution will respond to electric fields generated in and above the disk, perhaps reducing the potential and changing the plasma configuration. A more self-consistent treatment is needed to confirm this idea.

Also utilizing accretion power for the energy source is the shock acceleration model (e.g. Vestrand and Eichler 1982, Eichler and Vestrand, 1985, Kazanas and Ellison, 1986). In this model, a shock is expected to form at the Alfven radius. Particles are caught in a squeeze between magnetic irregularities in the fast moving inflowing upstream matter and the slowly moving irregularities in the post-shocked material. In this Fermi acceleration mechanism (e.g. Bell 1978, Blandford and Ostriker 1978), the average number of scatterings before a particle increases its energy by of order its own energy is ~ $1/\beta^2$, where $\beta$ is the difference between the pre- and post- shocked fluid velocity (in units of c). $\beta$ is assumed to be of order $(GM/r_A c^2)^{\frac{1}{2}}$. The maximum energy achievable is found by balancing the characteristic acceleration time with the energy loss time, which can be from radiative losses, or as in this example, escape from the region with characteristic dimension $r_A$:

$$\left(\frac{r_A}{\lambda}\right)^2\left(\frac{\lambda}{c}\right) = \frac{1}{\beta^2}\left(\frac{\lambda}{c}\right)$$

$$\rightarrow \quad E_{max} \sim 10^{16} \left(\frac{r}{R}\right)^{-\frac{1}{4}}\left(\frac{M}{M_\odot}\right)^{\frac{1}{4}}\left(\frac{L}{10^{38}\ \mathrm{erg/s}}\right)^{\frac{1}{2}}\ \mathrm{eV} \qquad (4)$$

Here $\lambda$ is the mean free path between scatterings off magnetic irregularities assumed to be equal to the particle gyro-radius. The energy is again maximized when the shock radius occurs near the stellar surface (although equation (4) indicates the dependence on radius is relatively weak). As in the accretion disk dynamo this requires the magnetic moment to be small ($B_S$ ~ $10^8$ G) for this to occur. And since this scenario is also accretion powered, the lifetime is the same (t ~ $10^8$ years).

The virtues and problems of this model are similar to those of the accretion disk dynamo in the sense that slow rotators are not a problem, but that Eddington limited accretion onto a neutron star provides a maximum L ~ $10^{38}$ erg/s assuming 100% efficiency, whereas the actual efficiency is not known. Another advantage of shock acceleration is that it has been observed in solar system contexts, whereas accretion disk dynamos are still purely theoretical constructs.

## IV. CONSEQUENCES OF PARTICLE ACCELERATORS IN BINARY STAR SYSTEMS

In addition to being a passive target for the production of $\gamma$-rays, the companion star itself will be altered by the impingent particle energy implied by equation (1). Formation of an expanded atmosphere or stellar wind (cf. Basko and Sunyaev 1973, Berezinsky 1980, Stecker et al 1985, Gaisser et al 1986) can occur due to the atmospheric heating by the bombarding cosmic rays. Heating of the stellar interior from neutrinos produced in the atmosphere (cf. Stecker et al 1985, Gaisser et al 1986, Harding et al 1987) can alter the stellar structure, possibly even disrupting the system.

A $10^{17}$ eV photon is thermalized after traversing a grammage X ~300 g $cm^{-2}$. This corresponds to a linear depth of several tens of thousands of kilometers beneath the surface of a solar mass main sequence star. If the radiation pressure gradient produced by the thermalized particle energy exceeds gravity (i.e. locally super-Eddington heating) a wind will be formed with terminal velocity approximately given by:

$$v \sim \left(\frac{L_B \kappa}{2\pi f R c}\right)^{\frac{1}{2}} \sim 10^8 \left(\frac{L_B}{10^{39}\ \mathrm{erg/s}}\right)^{\frac{1}{2}} \left(\frac{R_\odot}{R}\right)^{\frac{1}{2}} f^{-\frac{1}{2}}\ \mathrm{cm/s} \qquad (5)$$

Here $L_B$ is the cosmic ray power incident upon the star, $\kappa$ is the radiative opacity ($\cong$ .4 $cm^2/g$ for electron scattering) and f is the fraction of the surface area of the star illuminated by cosmic rays. Since the optical depth $\tau \sim \kappa X >> 1$ the radiation does not escape and the beam energy will largely be converted into kinetic energy of mass motion:

$$\dot{M} \sim \frac{2L_B}{v^2} \simeq \frac{4\pi f R c}{\kappa} \simeq 10^{-3}\ f \left(\frac{R}{R_\odot}\right)\ M_\odot yr^{-1}. \qquad (6)$$

Note that this is independent of the power being dissipated into the star, as long as the stellar heating rate is super-Eddington (Stecker et al 1985). The orbital period of Cygnus X-3 is increasing (e.g. Bonnet-Bidaud and van der Klis 1981) with the derivative satisfying $\dot{P}/P \simeq 10^{-6}\ yr^{-1}$. If the orbital angular momentum of the wind of the companion star is lost then $\dot{P}/P \simeq \frac{1}{2}(\dot{M}/M_T)$ (Davidsen and Ostriker 1974), where $M_T$ is the total mass of the system. In order that $\dot{M}$ of equation (6) not produce a $\dot{P}$ larger than is observed, either a small fraction of the star is illuminated by cosmic rays (f ~ $10^{-3}$) or the incident cosmic ray flux is sub-Eddington.

When a high energy proton collides with a target nucleus, in addition to the neutral pions that are produced (which decay into $\gamma$-rays) charged pions are also produced. The charged pions will do one of two things: If the decay time is less than a mean free collision time they will decay into a neutrino and a muon, the latter further decaying into two neutrinos and an electron or positron. If the decay time is greater than a mean free collision time the charged pions will interact and cascade suppressing the production of the high energy neutrinos that are produced when the pions decay. This results in a neutrino spectrum in which only neutrinos with energy less than $E_{\nu c} \simeq 1.3\ (10^{-6}\ g\ cm^{-3}/\rho)$ TeV will be produced. Here $\rho$ is the density in the atmosphere where the high energy neutrinos are produced (at X >~ 30 g $cm^{-2}$). The density at X = 30 g $cm^{-2}$ is ~ $10^{-6}\ (M_c/M_\odot)^{-0.9}$ g $cm^{-3}$ (Gaisser et al 1986). Thus $E_{\nu c} \sim .7\ (M_c/M_\odot)^{0.9}$ TeV which is a much smaller energy than the $10^5$ TeV assumed for the incident cosmic ray particles. For neutrino energies $E_\nu$ <~ 10 TeV the neutrino cross section $\sigma_\nu$ is linear with $E_\nu$: $\sigma_\nu \simeq 5 \times 10^{-36}$ ($E_\nu$/1 TeV) cm². The optical depth through the star $\tau(E_{\nu c}) \sim \sigma_\nu nR \sim .4(M_c/M_\odot)^{0.3}$, implying a relatively large fraction of the neutrinos will be absorbed nearly uniformly throughout the stellar interior. For a wind atmosphere, in which the density falls off as $r^{-2}$ (where r is the radial distance) the density at X = 30 g cm$^{-2}$ is roughly (h/R) times the density in an undistrubed star, where h/R (~ $10^{-3}$) is the ratio of undisturbed scale height to stellar radius. Thus, for a wind atmosphere the high energy neutrino production is not as efficiently suppressed and the optical depth to these higher energy neutrinos is even greater. Gaisser et al find that for main sequence stars ~5 to 20% of the incident cosmic ray energy flux can be carried into the stellar interior by neutrinos and be absorbed. This can provide more energy to the stellar interior than is produced by the star's self generated nuclear burning. The timescale for the star to absorb a gravitational binding energy $t_B$ is given by:

$$t_B \sim \frac{GM_c^2}{\varepsilon RL_B} \sim 6 \times 10^3 \left(\frac{M_c}{M_\odot}\right)^2 \left(\frac{.2}{\varepsilon}\right) \left(\frac{R_\odot}{R}\right) \left(\frac{10^{38} \mathrm{erg/s}}{L_B}\right) \mathrm{yrs.} \qquad (7)$$

Here $\varepsilon$ is the fraction of $L_B$ that is converted into neutrinos and absorbed in the star. For Cygnus X-3 this timescale (~ $10^4$ years) is much shorter than the observed mass loss time scale ($10^6$ years) and the timescales associated with acceleration mechanisms that use accretion as an energy source (~ $10^8$ years) but is comparable with the proposed pulsar lifetime (~ $10^3$ years). Since the radiative diffusion time is longer (~ $10^6$ years) the star will expand and can even be disrupted on this short timescale.

## V. CONCLUSION

It is difficult to draw a firm conclusion from the story that we have just presented. The various UHE data, after years of observations, still do not present a unique picture of the luminosity and phase of emission. The most often cited phase does not fit neatly into $\gamma$-ray production in a stellar companion. The proposed acceleration schemes, although encouraging, are still plausibility arguments rather than full fledged dynamic models. And the inferred mass loss rate and lifetime against neutrino absorption places limits on the sustained cosmic ray luminosity that can exist in a binary system. Clearly, the story of Cygnus X-3 and its bretheren is still unfolding.

Acknowledgements: It is with pleasure that I thank A. K. Harding, F. W. Stecker, and T. K. Gaisser for the enjoyable collaborations we have had on Cygnus X-3. I would also like to thank AKH and D. Kazanas for reading this manuscript and making helpful suggestions. This work is supported by a National Research Council Resident Research Associateship.

REFERENCES


Arons, J. and Scharlemann, E.T.: 1979, Astrophys. J. 231, p. 854.
Balstrusaitis, R.M., et al: 1985, Astrophys. J. Lett. 293, p. L69.
Basko, M.M., and Sunyaev, R.A.: 1973, Astrophys. and Space Science, 23, p. 117.
Battistoni, G., et al: 1985, Phys. Lett. 155B, p. 465.
Becklin, E. et al: 1973, Astrophys. J. Lett. 192, p. L119.
Bell, A.R.: 1978, Mon. Not. Roy. Astr. Soc., 182, p. 147.
Berezinsky, V.S.: 1980, Proc. 1979 DUMAND Summer Workshop at Khabarousk and Lake Baikal, held Aug.22-31, 1979, ed. J.G. Learned, U. Hawaii Press, p 245.
Bhat, C.L., et al: 1985, Proc. 19th Intl. Cosmic Ray Conf. 1, p. 83.
Blandford, R.D. and Ostriker, J.P.: 1978 Astrophys. J. Lett. 221, p. L29.
Bonnet-Bidaud, J.M. and van der Klis, M.: 1981, Astron. Astrophys., 101, p. 299.
Boone, J. et al: 1984, Astrophys. J., 285, p. 264.
Chadwick, P.M. et al: 1985a, Proc. 19th Intl. Cosmic Ray Conf. 1, p. 79.
Chadwick, P.M. et al: 1985b, Nature, 318, p. 642.
Chanmugam, G. and Brecher, K.: 1985, Nature, 313, p. 767.
Chardin, G. and Gerbier, G.: 1985, preprint, Commissariat A L'Energie Atomique, Centre d'Etudes Nucleaires de Saclay, December, 1985.
Chudakov, A.E., 1985, Proc. 19th Intl. Cosmic Ray Conf, 9, p. 441.
Danaher, S. et al: 1981, Nature, 289, p. 568.
Davidsen, A. and Ostriker, J.P.: 1974, Astrophys. J., 189, p. 331.
Dowthwaite, J.C. et al: 1983 Astron. Astrophys. 126, p. 1.
Eichler, D. and Vestrand, W.T.: 1984, Nature, 307, p. 613.
Eichler, D. and Vestrand, W.T.: 1985, Nature, 318, p. 345.
Elsner, R. et al: 1980, Astrophys. J., 239, p.335.
Gaisser, T.K., Stecker, F.W., Harding, A.K., and Barnard, J.J.: 1986, Astrophys. J., 309, (in press).
Ghosh, P., Elsner, R.F., Weisskopf, and Sutherland, P.G.: 1981, Astrophys. J., 251, p. 230.
Goldreich, P. and Julian, W.H.: 1969, Astrophys. J. 157, p. 869.
Gregory, J.E. et al: 1972, Nature Physical Science, 239, p. 114.
Grindlay,J.E., Helmken, H.F., and Weekes, T.C.: 1976, Astrophys. J, 209, p. 292.
Gunn, J.E. and Ostriker, J.P.: 1969, Nature, 221, p. 454.
Harding, A.K., Barnard, J.J., Stecker, F.W., and Gaisser, T.K.: 1987,in IAU Symposium #125: Origin and Evolution of Neutron Stars, ed. D. Helfand (Dordrecht: Reidel) (This volume).
Hayashida, N. et al: 1981, Conf. Papers of 17th Int. Cosmic Ray Conf., Paris, 9, (Paris: Commissariat a l'Energie Atomique and IUPAP) p. 9.
Hermsen, W. 1983 Spa. Sci. Rev., 36, p. 61.
Hertz, P., Joss, P. and Rappaport, S.: 1978, Astrophys. J., 224, p. 614.
Hillas, A.M.: 1984, Nature, 312, p. 50.

Hillas, A.M.: 1985, Proc. 19th Intl. Cosmic Ray Conf., 2, p. 296
Kazanas, D. and Ellison, D.C.: 1986, Nature, 319, p. 380.
Lamb, R.C. et al: 1977, Astrophys. J. Lett., 212, p. L63.
Lamb, R.C., Godfrey, C.P., Wheaton, W.A., Tumer, T.: Nature, 296, p. 543.
Lambert, A. et al: 1985, Proc. 19th Intl. Cosmic Ray Conf., La Jolla, 1, p. 71.
Lloyd Evans, J. et al: 1983, Nature, 305, p. 784.
Lovelace, R.V.E.: 1976, Nature, 262, p. 649.
Marshak, M.L. et al: 1985, Phys. Rev. Lett., 54, p. 2079.
Mason, K.O. et al: 1976, Astrophys. J. 207, p. 78.
Meegan, C.A., Fishman, G.J., and Haymes, R.C.: 1979, Astrophys. J. Lett., 234, p. L123.
Milgrom, M.: 1976, Astron. Astrophys., 51, p. 215.
Molnar, L.A.: 1985, Ph.D. Dissertation, Harvard University.
Molnar, L.A., Reid, M.J., and Grindlay, J.E.: 1985, in Radio Stars (ed. R.M. Hjellming) (Dordrecht:Reidel).
Molteni, D., Rapisarda, M., Robba, R., and Scarsi, L.: 1980, Astron. Astrophys., 87, p. 88.
Morello, C. et al: 1983, Conf. Papers of 18th Int. Cosmic Ray Conf, Bangalore, 1, (TIFR: Bombay) p. 127.
Neshpor, Yu.I. et al: 1979, Ap. Sp. Sci., 61, p. 349.
Oyama, Y. et al: 1986, Phys. Rev. Lett., 56, p. 991.
Pringle, J.: 1974, Nature, 247, p. 21.
Protheroe, R.J. and Clay, R.W.: 1985, Nature, 315, p. 205.
Protheroe, R.J., Clay, R.W. and Gerhardy, P.R. 1984, Astrophys. J. Lett. 280, p. L47.
Ramana Murthy, P.V.: 1986, Physics Letters B, 173, p. 107.
Ruderman, M.A. and Sutherland, P.G.: 1975, Astrophys. J., 196, p. 51.
Samorski, M. and Stamm, W.: 1983, Astrophys. J. Lett., 248, p. L17.
Serlemitsos, S.P. et al: 1975, Astrophys. J. Lett., 201, L9.
Shapiro, S.L. and Teukolsky, S.A.: 1983, Black Holes, White Dwarfs, and Neutron Stars (New York: Wiley).
Stecker, F.W., Harding, A.K. and Barnard, J.J.: 1985, Nature, 316, p. 418.
Stepanian, A.A. et al: 1982, in Proceedings of the Int. Workshop on VHE Gamma Ray Astronomy, Ootacamund, India (eds. Ramana Murthy and Weekes) (TIFR and Smithsonian Inst.), p. 43.
Stephens, S.A. and Verma, R.P.: 1984, Nature, 308, p. 828.
Turver K.E. et al: 1986, Talk at a Cygnus X-3 Workshop given in Washington, D.C. in April, 1986.
van der Klis, M. and Jansen, F.A.: 1985, Nature, 313, p. 768.
Vestrand, W.T. and Eichler, D., 1979, AIP Conf. Proc. No. 56 Particle Acceleration Mechanisms in Astrophysics, La Jolla, held Jan. 3-5, 1979, ed. Arons, Max, and McKee (New York: AIP) p. 285.
Vestrand, W.T. and Eichler, D., 1982, Astrophys. J., 261, p. 251.
Vladimirsky, B.M., Neshpor, Yu.I., Stepanian, A.A., and Fomin, V.P.: 1975, Conf. Papers 14th Int. Cosmic Ray Conf., Munich, 1, p. 118.
Weekes, T.C. and Helmken, H.F.: 1977, in Recent Advances in Gamma Ray Astronomy, ESA SP-124, p. 39.
Weekes, T.C. et al: 1981, Pub. Ast. Soc. Pac., 93, p. 474.

Weekes, T.C. and Geary, J.C.: 1982, Pub. Ast. Soc. Pac., 94, p. 708.
Westphal, J.A. et al: 1972, Nature of Physical Science, 239, p. 134.
White, N.E. and Holt, S.S.: 1982, Astrophys. J., 257, p. 318.
Willingale, R., King, A.R., and Pounds, K.A.: 1985, Mon. Not. Roy. Ast. Soc., 215, p. 295.

DISCUSSION

Colgate: If the beam is locally super-Eddington at the companion star, then mass loss would be large. This was discussed at Moriond. Have you come to a conclusion about this mass loss?

Barnard: Equation (6) indicates that if a large fraction of the star is illuminated by a super-Eddington cosmic ray beam the mass loss rate ($\sim 10^{-3}$ M$_\odot$ yr$^{-1}$) would indeed contradict the observed $\dot{P}/P \sim 10^{-6}$ yr$^{-1}$. However, even if $L_{CR} \sim 10^{39}$ erg/s a .5M$_\odot$ main sequence star will subtend a solid angle of $\sim .02 \times 4\pi$, implying an absorbed luminosity of only $2 \times 10^{37}$ erg/s, which is sub-Eddington. Thus, we may expect the mass loss rate to be substantially less than is given by equation (6) because the assumption of super-Eddington heating may not occur at the companion star even if the implied cosmic ray luminosity is super-Eddington at the compact star.

# A NEW ASPECT OF GALACTIC RIDGE X-RAY EMISSION
## --SNRs IN A TENUOUS MEDIUM ? --

Katsuji Koyama
Institute of Space and Astronautical Science
6-1, Komaba 4-chome,
Meguro-ku,
Tokyo, 153
Japan

ABSTRACT. Thin thermal natures of a plasma temperature of several keV are observed in the diffuse X-ray spectra from the galactic ridge region by the Tenma satellite. Within the constraints imposed by the observational results, possible origins of the ridge X-ray emission are discussed. We show that unidentified young supernova remnants in a tenuous medium could be a significant contributor to the ridge emission.

## 1. INTRODUCTION

The first systematic survey of the diffuse galactic ridge X-ray emission above 2 keV was carried out by the HEAO-A satellite ( Iwan et al. 1982, Worrall et al. 1982). Most of the X-ray emission was found to come from point sources. However, an excess emission, which is not resolved into point sources was also observed along the galactic plane. The origin of this diffuse X-ray emission is still unknown.

Recently Tenma and EXOSAT satellites confirmed this finding with better energy resolution (Tenma) and better spatial resolution (EXOSAT) (Koyama 1984, Warwick et al. 1985 , Koyama et al. 1986a). The diffuse emission is concentrated near the inner galactic disk known as the galactic ridge. In this paper, we report on the observational results by the Tenma gas scintillation proportional counters (GSPC) and try to give a new interpretation on the origin of the ridge emission.

## 2. X-RAY SPECTRUM FROM THE GALCTIC RIDGE

The observed positions were selected so as to exclude cataloged X-ray sources and radio supernova remnants. Figure 1 shows the observed positions in the galactic coordinate. Here, large circles show the full field of view of the detector(GSPC); small open circles indicate cataloged X-ray sources ( Forman et al. 1978, Wood et al. 1984) with intensity greater than 2 m Crab ; and small closed circles indicate radio supernova remnants ( Clark and Caswell 1976).



*D. J. Helfand and J.-H. Huang (eds.), The Origin and Evolution of Neutron Stars, 535–543.*

Fig.1

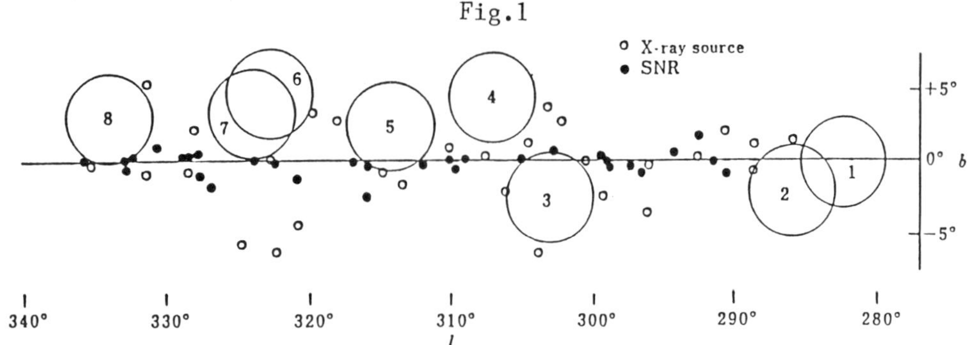


We can observed an intense emission line at about 6.7 keV from every position. This emission line is, without doubt, K-emission line from helium-like iron. Therefore this observation strongly supports the interpretation of of a thin thermal origin for the ridge emission. Furthermore, the continuum spectrum can be fit by a thermal bremsstrahlung model of several keV as is given in the solid curve in figure 2. The best-fit temperatures are also given in the figure.

Fig.2

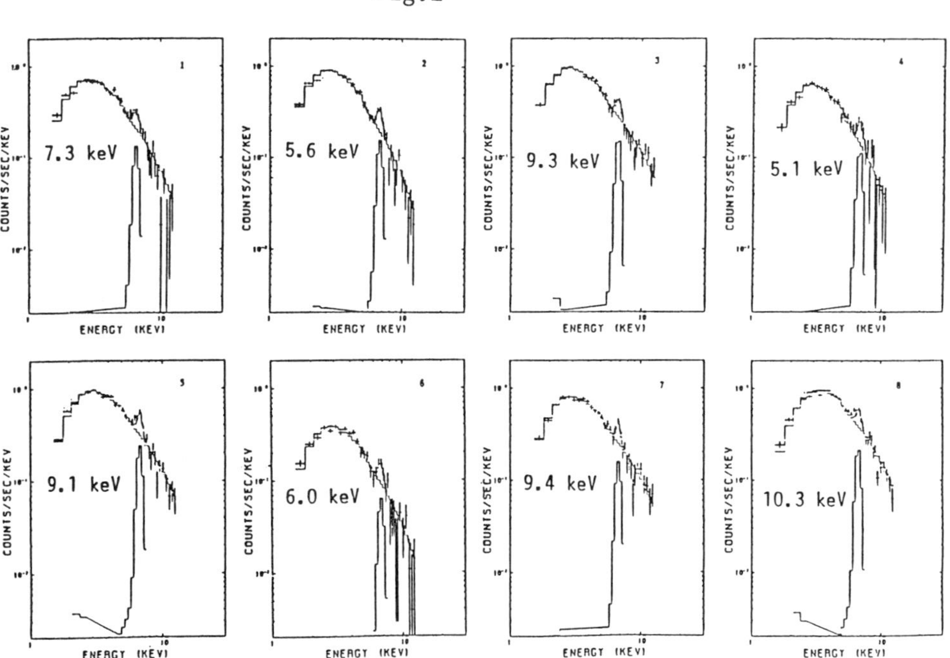


From this, we are able to draw the following conclusion:

(1) The ridge emission mainly comes from thin hot plasma of several keV.

## 3. SIZE OF THE EMISSION REGION AND TOTAL LUMINOSITY

In figure 3, the intensity of iron line and continuum emission are plotted as a function of angular distance from the plane. Solid lines are the estimated curve assuming disk-like emission of 10 kpc radius with scale height of 100 and 300 pc. The lower graph shows the longitudial distribution of iron and that of continuum intensity. Again the solid curves assume the same disk-like emission as that assumed in the upper graph. From this figure, we can conclude that the emission comes from the inner galactic disk at a radius of about 10 kpc and a scale height of about 100 pc. Thus, we can estimate a total luminosity of about $10^{38}$ erg/sec. Therefore we are able to draw a second conclusion:

(2) The ridge emission is concentrated near the inner disk and has a total luminosity of about $10^{38}$ erg/sec.

This conclusion is consistent with the results of both HEAO-A and EXOSAT.

Fig.3

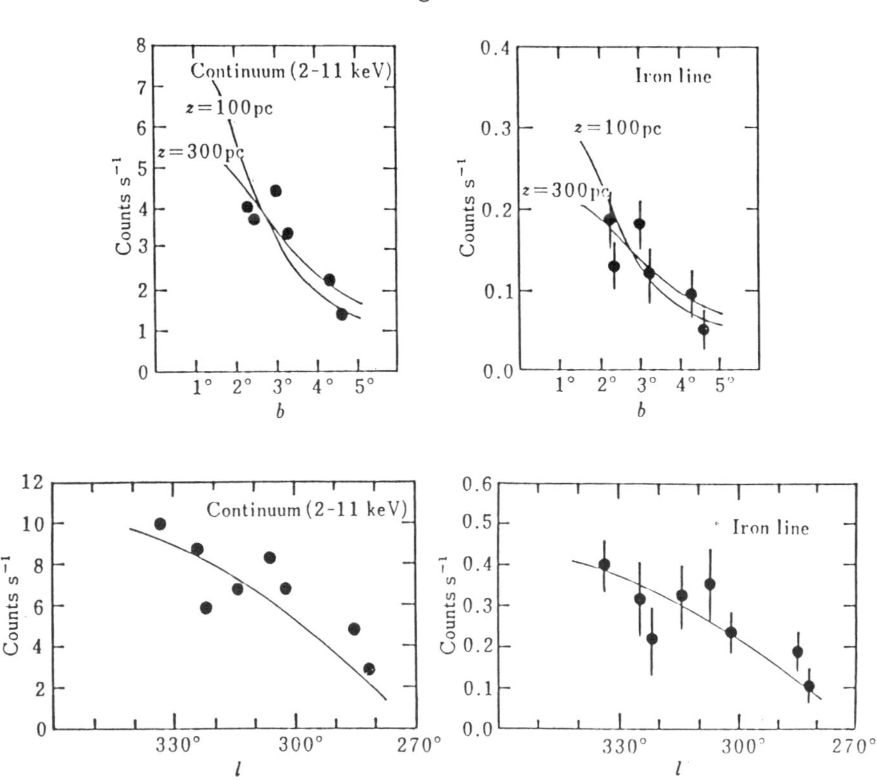


The third conclusion that we can draw based on the results shown in the figure 2 is ;

(3) The observed plasma temperature differs from place to place.

## 4. ORIGIN OF THE RIDGE EMISSION

The ridge emission is probably an integrated effect of unresolved discrete sources. Given the emission volume and total luminosity of the ridge emission, we are able to relate the luminosity of discrete source to the space density. Figure 4 shows this relation. The candidate source should lie on the solid line. As an example, we show the known class of X-ray sources; neutron star binaries (NS) , RS CVn satrs (RS CVn) and cataclysmic variables (CV).

Since the observed temperature differs from place to place, the space density of the candidate source should not be very large. By a simple calculation, we can conclude that the space density should be lower than the dotted line in the figure. Therefore the luminosity of an individual source should be higher than $10^{33}$ erg/sec. On the other hand, the luminosity of individual discrete source is required to be less than $10^{36}$ erg/sec, given the fact that they can not be resolved into point sources.

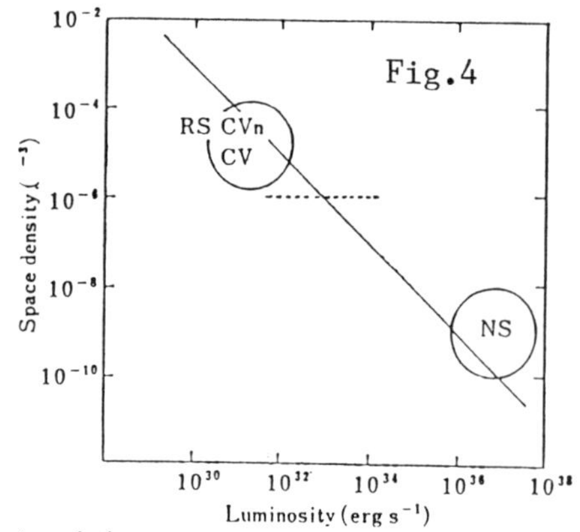


Thus the proper candidate is charactarized by ;

(1) A thin thermal spectrum of several keV with an intense iron line at about 6.7 keV.
(2) An individual luminosity in the range $10^{33}$ - $10^{36}$ erg/sec
(3) A total luminosity of $10^{38}$ erg/sec.
(4) Being spatially concentrated near the inner galactic disk with scale height of about 100pc.

These conditions are depicted graphically in figure 5.

Many authors have suggested M-dwarfs as candidate sources. However this suggestion is inconsistent with several of our findings. For example, X-ray spectrum of M-dwarfs is often very soft, their luminosity is about $10^{28}$ erg/sec and the scale height is very large. Therefore it is inconsistent with conditions (1) ,(2) and (4).

RS-CVn and Cataclysmic Variable has also been suggested as candidate sources . In this case, the continuum spectrum is marginally consistent with that of the ridge emission. The individual luminosity of these objects is typically $10^{32}$ erg/sec and their total luminosity may exceed $10^{37}$ erg/sec. Therefore these X-ray sources could be marginal candidates. The key observation, however, is that of the iron line feature of these objects. A systematic spectral observation of this class of objects would be useful in settling this point.

A third possible candidate is the newly discovered thin thermal source ' star forming region'( Agrawal et al. 1986 , Koyama 1986c). The spectrum , individual luminosity and scale height of this objects is consistent with conditions (1) , (2) and (4). However the total luminosity may be less than $10^{37}$ erg/sec.

Fig.5

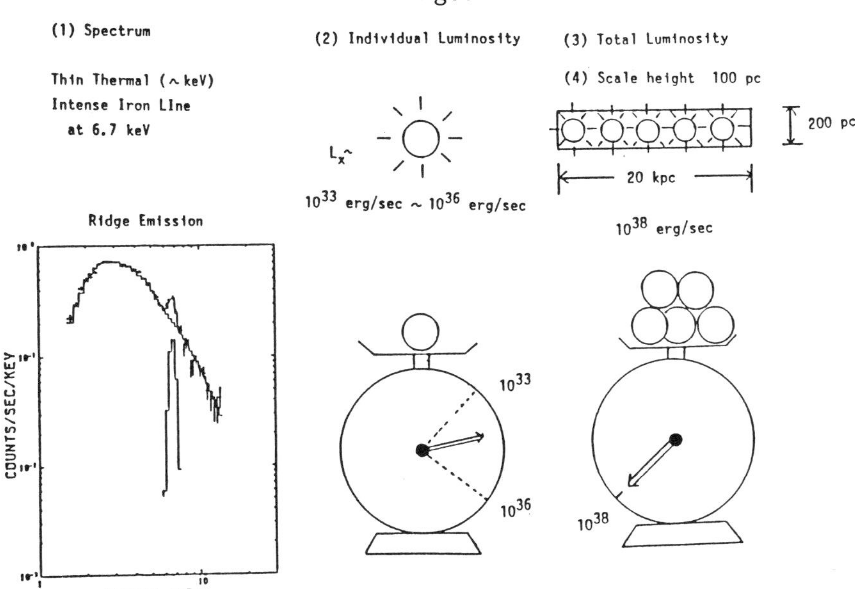


## 5. CONTRIBUTION OF SUPERNOVA REMNANTS

The spectrum of young SNR is similar to the ridge emission. Therefore, uncataloged SNRs may also be possible candidate sources of the ridge emission (Koyama et al 1986b).

Figure 6 shows the number distribution of radio SNR as a function of surface brightness in the ridge region and other regions ( following Milne 1979). From this figure, we note that there could be lots of radio SNRs on the ridge below the current detection limits about $10^{-20}$ ($Wm^{-1}Hz^{-1}sr^{-1}$).

We will estimate the X-ray luminosity from these uncataloged radio SNRs. We derived the surface brightness of radio SNRs semi-empirically as a function of density (n/cc) of the interstellar medium and the diameter of the SNR (D pc) (Tomisaka et al. 1980).

Fig.6

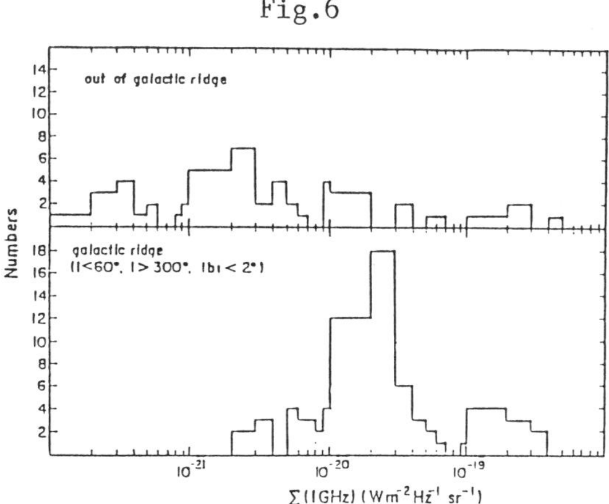

$$\Sigma(1\text{GHz}) = 2.88 \times 10^{-14}\ D^{-3.8}\ n^2 \qquad (\text{Wm}^{-2}\text{Hz}^{-1}\text{sr}^{-1})$$

Assuming a simple Sedov model (1959) with an initial explosion energy of $10^{51}$ erg, the diameter of the SNR is given as a function of density and age (year);

$$D = 0.62 \times n^{-0.2}\ E_{51}^{0.2}\ \tau^{0.4} \qquad (\text{pc})$$

Then the condition that these radio SNRs escape from the current survey is:

$$n\ \tau_3^{-0.55} < 0.11 \times E_{51}^{0.28} \qquad (1)$$

This relation is given in figure 7 on the density (n/cc) and age (τ year) plane.

The X-ray luminosity and temperature of the SNR are given as functions of density and age using the Sedov model.

$$Lx = 1.6 \times 10^{34}\ n^{1.2}\ E_{51}^{0.8}\ \tau^{0.6} \quad (\text{erg/sec}),$$

$$T = 2.0 \times 10^{11}\ n^{-0.4}\ E_{51}^{0.4}\ \tau^{-1.2} \quad (\text{K}).$$

The observations require that the luminosity be in the range $10^{33}$ - $10^{36}$ ergs/ sec, and that the temperature be higher than 1 keV. We therfore arrive at the following conditions;

$$n\ \tau_3^{3} < 57 \times E_{51} \qquad (2),$$

$$10^{-3}\ E_{51}^{-0.8} < n^{1.2}\ \tau_3^{0.6} < E_{51}^{0.8} \qquad (3).$$

Fig.7

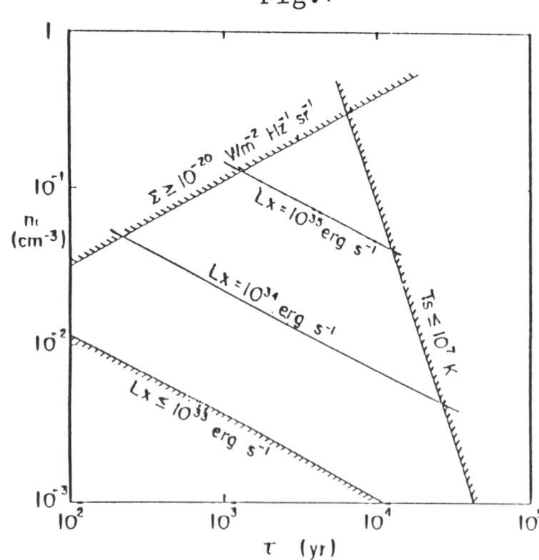


The conditions (1), (2) and (3) are given in the n - τ plane (figure 7), where hatched sides are forbidden region. Based on an inspection of this figure, we can conclude that SNRs in a tenuous medium less than 0.1 /cc and age of 1000-10000 year could be candidate sources of the ridge emission. The total luminosity can be given as a function of $t_{10}$, where $t_{10}$ is the interval of supernova explosion in units of 10 years.

$$Lx \sim 10^{38}/t_{10} \quad \text{erg/sec} \qquad (n = 0.1)$$

$$Lx \sim 10^{37}/t10 \quad \text{erg/sec} \qquad (n = 0.01)$$

In order to estimate the luminosity more accurately, we carried out a numerical calculation of a supernova explosion in a tenuous medium of 0.1 and 0.01 /cc. The results are given in figure 8 with solid lines ( X-ray luminosity from the ejecta) and dashed lines ( blast wave), where E- band and F-band indicate the energy range 1.5–8 keV and 8–25 keV, respectively. The total luminosity above 1.5 keV is then given as:

$$Lx \sim 1.5 \times 10^{38}/t_{10} \quad \text{erg/sec} \quad (n = 0.1)$$

$$Lx \sim 0.5 \times 10^{38}/t_{10} \quad \text{erg/sec} \quad (n = 0.01)$$

Fig.8

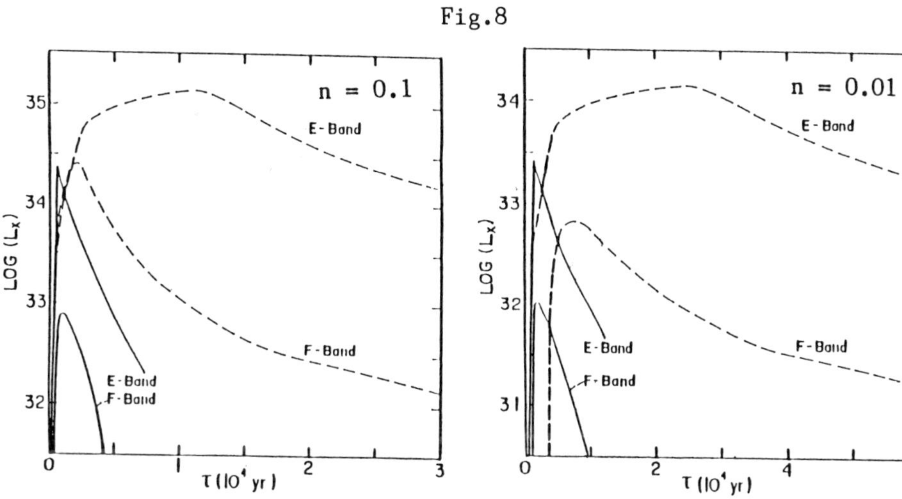


Therefore, the current rate of one explosion per 30–50 years already contributes significantly to the ridge emission. If the rate is as high as one ten years, then we can explain the total ridge emission.

In any case, we can say that there are many uncataloged SNRs on the galactic ridge.

The surface brightness of the X-ray SNRs in a tenuous medium is less than $10^{-13}$ erg/s/cm$^2$/arcmin.$^2$. This value may be at the detection limit of the Einstein medium survey. Furthermore, on the galactic ridge, the heavy

Fig.9

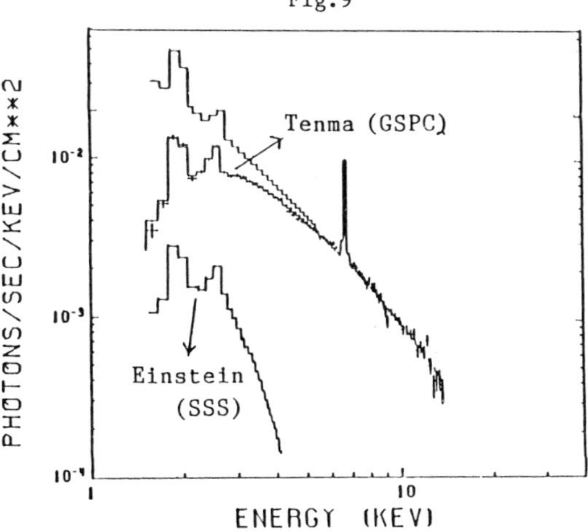

interstellar absorption on the galactic ridge makes it difficult to detect young X-ray SNR, because the Einstein Observatory has high sensitivity only in the low energy band.

We show the efficiency of both the Tenma (GSPC) and the Einstein Observatory (SSS) assuming an $N_H$ value of $3 \times 10^{22}$ H/cm$^2$ using the observed ridge spectrum. Figure 9 shows that the current detection limit of X-ray SNRs by the Einstein Observatory is still limited. If we were able to combine Einstein's spacial resolution and Tenma's effective area above several keV, then the resulting sensitivity would be about 10-100 times larger than that achieved in the current survey of the Einstein Observatory. We would then able to locate many SNRs on the galactic ridge.

The autor thanks J.R.Najita for careful reading of the manuscript.

## References

Agrawal, P.C., Koyama,K., Matsuoka,M., and Tanaka,Y. 1986, Publ. Astron. Soc. Japan, 38,in press.
Clark,D.H., and Caswell,J.L. 1976, Monthly Notices Roy. Astron. Soc., 174,267.
Forman, W., Jones,C., Cominsky,L., Julien,P., Murray,S., Peters,G., Tananbaum,H., and Giacconi,R. 1978, Astrophys.J. Suppl., 38, 357
Iwan,D., Marshall,F.E., Boldt,E.A., Mushotzky,R.F.,Shaher,R.A., and Stottlemyer,A. 1982, Astrophys. J., 260, 111.
Koyama,K. 1984, in X-Ray Astronomy 1094, ed. M.Oda and R.Giacconi (Institute of Space and Astronautical Sience ,Tokyo), p.325.
Koyama,K., Makishima,k., Tanaka,Y., and Tsunemi,H. 1986a, Publ. Astron. Soc. Japan, 38,121
Koyama,K., Ikeuchi, S., and Tomisaka,K. 1986b, Publ. Astron. Soc. Japan, 38, in press.
Koyama,K. 1986c, submitted to Publ. Astron. Soc. Japan
Milne, D.K. 1979, Australian J., 235, 4.
Sedov,L.I. 1959, Similarity and Dimensional Methods in Mechanics (Academic Press,New York), p 210.
Tomisaka,K, Habe,A. and Ikeuchi,S. 1980, Prog. Theor.Phys., 64, 1587.
Warwick,R.S., Turner,M.J.L., Watson,M.G., and Willingale,R. 1985, Nature,317,218.
Wood,K.S., Meekins,J.F., Yentis,D.J., Smathers,H.W., McNutt,D.P., Bleach,R.D., Byram,E.T., Chubb,T.A., Friedman,H., and Meidav,M. 1984, Astrophys.J. Suppl., 56,507.
Worrall,D.M., Marshall,F.E., Boldt,E.A., and Swank,J.H. 1982, Astrophys.J., 255,111.

## DISCUSSION

**D. Helfand:** In the Magellanic Clouds, we see a total population of over 40 SNRs and the sample is complete to $10^{35}$ ergs $s^{-1}$, but there are none with a temperature as high as that of your diffuse emission. I believe you should consider the alternative of Be star binaries which have a scale height, luminosity range, galactic population and harder spectrum consistent with your requirements.

# EXOSAT News on Geminga

Patrizia A. Caraveo and Giovanni F. Bignami
Istituto di Fisica Cosmica del C.N.R.
Via Bassini, 15
20133 Milano ITALY

1E0630+178 was the target of a 64,000 sec EXOSAT observation on March 17-18, 1985. The 1200 photons recorded were folded around the value predicted by previous Einstein '79, '81 and EXOSAT '83 data. The region between 60.20 and 60.35 sec. was searched with independent steps of 0.005 sec, computed according to the relation :
step = (period)$^2$/observations length/number of bins in the light curve. A reduced $\chi^2$ of 3.85 (9 d.o.f.) was found for 60.285 sec. The data were then divided into five segments and the same search was performed yielding for the third segment a red. $\chi^2$ of 4.33. Moreover, the a posteriori periodogram shows also the presence of harmonics at 30.14 and 15.07 (see figure) wich appear to be even more significant ( red. $\chi^2$ of 5.06 and 4.37,respectively) than the effect at 60.28.

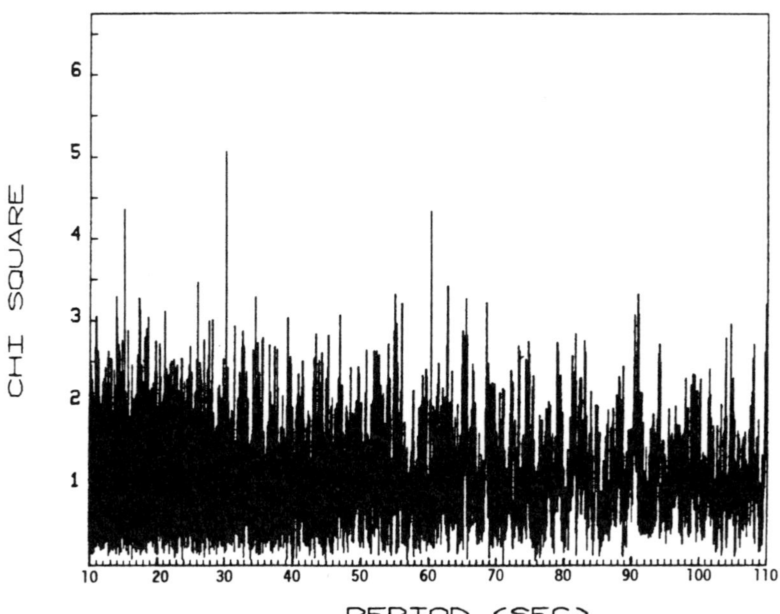




*D. J. Helfand and J.-H. Huang (eds.), The Origin and Evolution of Neutron Stars, 545.*

# THE ORIGIN OF THE ENHANCED DISSIPATION "$\alpha$" IN ACCRETION DISCS AND ITS RELATION TO GAMMA BURSTS

Stirling A. Colgate
Los Alamos National Laboratory
Los Alamos, New Mexico 87545 U.S.A.

ABSTRACT. The source of the enhanced dissipation or "$\alpha$" viscosity of Keplerian accretion discs is central in all putative mechanisms for large energy release by matter accreting onto condensed objects and for the basic mechanism of star formation. The circumstances of gamma burst formation on neutron stars suggest convective buoyancy as a necessary condition for the large $\alpha$. This is because the total mass $\cong 10^{19}$ g necessary to supply the energy of a gamma burst derived from infall is a natural limit for the mass stored in a disc without $\alpha$ viscosity. This suggests that buoyancy driven convective turbulence is the source of the enhanced transport in disc evolution models (Shakura and Sunyaev 1973). In support of this conjecture we find that the maximum possible energy released by ideal friction operating on the velocity shear of a disc is twice that required to destabilize the angular momentum distribution of such a disc. The heat energy available from an $\alpha$ viscosity is twice that necessary to create $\alpha$ in the first place. Hence, a nonlinear instability--nonlinear to create convective turbulence and nonlinear to create shear viscosity heating--is sufficient to drive $\alpha$. One characteristic that would prevent the formation of such an instability is degeneracy of the disc matter as it accumulates near a neutron star (Paczynski and Jaroszynski 1978). Degeneracy inhibits strong convection because a given energy release within degenerate matter results in a large temperature, and hence large energy transport without convection. Convection occurs in an accretion disc whenever the energy which is dissipated in the disc requires a superadiabatic temperature gradient for its radiative or conductive transport to the surface. Some gamma burst mechanisms require exactly such a mechanism as a degenerate disc close to the neutron star in the correct mass ($\cong 10^{19}$ g), at the correct radius several times the neutron star radius, to supply gravitational energy for a gamma burst. The degenerate disc accumulates mass stably until the density is great enough that degenerate fluid viscosity evolves the disc into contact with the neutron star. The large energy released by velocity shear at contact heats the disc causing rapid evolution and a gamma burst.



*D. J. Helfand and J.-H. Huang (eds.), The Origin and Evolution of Neutron Stars, 546.*

# INVERSE COMPTON MODEL OF GAMMA RAY BURST SPECTRA*

W. M. Howard and E. P. Liang
Physics Department, LLNL
University of California
P.O. Box 808
Livermore, CA 94550 U.S.A.

ABSTRACT. We study gamma ray spectra produced by the inverse Compton upscattering of soft photons by relativistic electrons with a one dimensional momentum distribution, which is relevant to gamma ray burst if the source magnetic field is strong enough so that the synchrotron cooling time of transverse energy becomes much shorter than isotropization time via couloub or Compton collisions. We find that for high electron longitudinal temperatures the output power is strongly beamed in the momentum direction and the spectrum softens rapidly with increasing view angle from the momentum direction.

Even if the gamma rays of gamma bursters originate from regions of magnetic fields as low as $10^{10}$ - $10^{11}$ G, the synchrotron cooling time of transverse momentum ($\sim 10^{-14}$ s$(10^{11}$ G/B$)^2$) may still be much shorter than the isotropization time via coulomb or Compton collisions. In that case the electron momentum distribution will be essentially 1-D along the field lines. The hardest photons will be produced by inverse Compton upscattering of synchrotron and other soft photons by these 1-D relativistic electrons. We have computed the inverse Compton spectra with Monte Carlo techniques for Maxwellian electron distributions, varying the scattering depth, longitudinal temperature and input soft photon spectrum. The major findings include:
a) upscattered power is strongly beamed along momentum direction;
b) beam narrows with increasing electron temperature, with half angle less than 2° for temperatures above 1 MeV;
c) for a given temperature, beam narrowest near unit scattering depth;
d) spectrum softens rapidly for increasing view angle away from momentum direction. For further details refer to Ref. 1.

*Work performed under the auspices of the U.S. DOE by the LLNL under contract number W-7405-ENG-48.

Reference

1. W. M. Howard, E. P. Liang and E. Canfield, Astrophys. J. submitted (1986).



*D. J. Helfand and J.-H. Huang (eds.), The Origin and Evolution of Neutron Stars, 547.*

# A MODEL FOR THE 1979 MARCH 5 GAMMA-RAY TRANSIENT

C. Alcock, E. Farhi, and A. Olinto
Department of Physics
Massachusetts Institute of Technology
Cambridge, MA 02139

We have constructed a model for the 1979 March 5 $\gamma$-ray burst (Cline et al. 1980; Evans et al. 1980). This event is characterized by a rapid rise ($\leq$ 250 $\mu$s), an intense flash of duration ~0.15s, followed by a decaying, lower intensity phase which lasted a few minutes, and was modulated with a period of 8s. The timing position error box includes a supernova remnant in the LMC, which suggests a distance of 55 kpc. At this distance the energy in the event is $\sim 10^{44}$ erg and the peak luminosity $\sim 10^{7}$ greater than the Eddington limit for a 1.4 $M_{\odot}$ object.

We propose that a $10^{-6}$ $M_{\odot}$ lump of "strange matter" fell into a "strange star" (strange matter is described in this volume and in Alcock, Farhi, and Olinto 1986). The lump was not tidally disrupted because of its high density, and the duration of the impact was ~1 $\mu$s. Most of the kinetic energy of impact goes into the excitation of normal modes of the star, but $\sim 10^{45}$ erg is dumped into a small hot spot. This hot spot radiates its internal energy in ~0.1s, and is responsible for the high intensity phase of the event. The radiation is produced at an exposed quark matter surface which, since it is held together by the strong force, is not subject to the Eddington limit.

The impacting lump punched a hole through the thin crust on the star. The hot spot radiates through this hole. Another consequence of the local heating is the photo disintegration of nuclei in the crust. The neutrons liberated in this way diffuse out of the crust and react with the strange matter, releasing ~20 MeV per neutron in the form of heat. The timescale for the diffusion is a few minutes, and this process is responsible for the low intensity phase of the event. The 8s modulation is caused by rotation.

Cline, T. L., et al. 1980, Ap. J. (Letters), **237**, L1.

Evans, W. D., et al. 1980, Ap. J. (Letters), **237**, L7.

Alcock, C., Farhi, E., and Olinto, A. 1986, to appear in Ap. J.



*D. J. Helfand and J.-H. Huang (eds.), The Origin and Evolution of Neutron Stars, 548.*

# NEW DISTANCE LIMIT FOR THE 3-5-79 SOURCE*

E. P. Liang
Physics Department, LLNL
University of California
P.O. Box 808
Livermore, CA 94550 U.S.A.

ABSTRACT. We provide new distance limits for the March 5, 1979 gamma ray transient source based on the synchrotron interpretation and the lack of low frequency self-absorption cutoff. This leads to upper limits of 0.6-11 kpc for emission areas ranging from 1 squared km polar caps to a 10 km radius stellar surface.

The burst spectrum of the March 5, 1979 event shows no self-absorption cutoff down to 30 keV. This provides an upper limit to the synchrotron emission column density as a function of electron distribution (temperature or injection energies) and magnetic field. Combining this with the observed flux at earth, spectral shape and the requirement that magnetic field energy density exceeds radiation energy density, we obtain a conservative upper limit to the synchrotron luminosity distance modulo the emission area for steady thermal and cooling electron distributions. It ranges from 0.6 kpc for a 1 squared km polar cap to 11 kpc for a 10 km radius surface.

If in addition we interpret the hard tail above 300 keV (minus the 400 keV annihilation line) as due to thermal self-Compton of the synchrotron photons, we obtain a distance estimate (not upper limit) of 0.1 - 2 kpc for the above emission area ranges. The above distance limit can be pushed to 55 kpc only if the emission surface area is increased to $10^{14}$ sq. cm. These results strongly suggest, but do not prove, that the source is Galactic and the association with N49 of LMC is chance coincidence. Future observations (e.g. optical) should be able to settle this issue definitively. For further details see Ref. 1.

*Work performed under the auspices of the U.S. DOE by the LLNL under contract number W-7405-ENG-48.

Reference

1. E. P. Liang, Astrophys. J. Lett., **308** L00 (1986).



*D. J. Helfand and J.-H. Huang (eds.), The Origin and Evolution of Neutron Stars, 549.*

# WHAT TYPE OF BINARY SYSTEM IS CYGNUS X-3 ?

J.M. Bonnet-Bidaud
Service d'Astrophysique
DPhG/Ap, CEN Saclay
91191 Gif/s/Yvette, France

ABSTRACT. Results of the final analysis of COS-B X-ray observations of Cygnus X-3 (more than 220 days of observations from 1975 to 1982) are presented. Variations of the 4.8h period have been investigated and a new ephemeris including previous and later (Exosat) results, is established. Both long and short-term changes in the period are apparent. The long term variation is found to be consistent with orbital changes expected from mass transfer _and_ mass loss in a standard low-mass binary system with an accreting neutron star. Short term period fluctuations (of the order of 4 $10^{-3}$ d.) are definitively present and not fully explained.
A critical discussion of the recent claimed high energy gamma ($E > 10^{12}$eV) detections of the source is also presented. It is argued that, if confirmed, these results are inconsistent with the typical X-ray picture of Cygnus X-3. In particular, the eclipse duration required in the Vestrand and Eichler (1982) imply a totally unphysical companion for a 4.8h orbit.



*D. J. Helfand and J.-H. Huang (eds.), The Origin and Evolution of Neutron Stars, 550.*

# ON THE PHASES OF PERIOD VARIATION OF SS 433

W.Q. Luo
Department of Physics, Sichuan Teachers' University
Chendu, Sichuan
China

In a previous paper ( Luo, 1986 ), we have improved the method by Katz ( 1982 ) and Fan et al ( 1983 ). We have also found some important features in a ring model with precession and nodding motion, i.e., all the higher periodical components contain the ones with n = 1,2,3. So we can call $T_1$,$T_2$, and $T_3$ fundamental periods.

Recently, Fan and Li ( 1983 ) calculated the short period variations in SS 433 using a ring model with precession and nodding motion, then compared it with the observed results given by Mammano ( 1982 ).

In this paper, we dealt with the same topic using the method by Luo ( 1986 ), some observed values given by Newsom and Collins ( 1982 ), and Mammano ( 1982 ).

It can be seen that the theoretical results are coincident with the observed ones well, therefore, it seems that the ring model with precession and nodding motion is reasonable,

References
Fan,L.Z., and Li,S.X.: 1983, Scientia Sinica, No.5, 447
Katz, J.I.: 1982, Astrophys.J., 260, 780
Luo,W.Q.: 1986, Kue Xue Tongbao, 31,523
Newsom,G.H., and Collins,G.W.: 1982, preprint
Mammano,A.: 1982, preprint



*D. J. Helfand and J.-H. Huang (eds.), The Origin and Evolution of Neutron Stars, 551.*

# PRODUCTION AND INTERACTION OF HIGH ENERGY NEUTRINOS IN CLOSE X-RAY BINARIES

A. K. Harding, J. J. Barnard, F. W. Stecker
Code 665
NASA/Goddard Space Flight Center
Greenbelt, MD 20771
USA

T. K. Gaisser
Bartol Research Foundation
University of Delaware
Newark, DE 19716
USA

ABSTRACT. Reports of air showers with $E > 10^{15}$ eV from Cygnus X-3, LMC X-4, Vela X-1 and Hercules X-1 have been interpreted as requiring production of neutral secondaries by cosmic rays accelerated by the compact partner in these systems. If neutral pions are the source of photons that produce the observed air showers, then charged pions must also be produced, and they will give rise to neutrinos. We consider limits that may be placed on binary systems like Cygnus X-3 in which a neutron star is a strong source of ultra-high energy (UHE) particles that produce photons, neutrinos and other secondary particles in the companion star through nuclear interactions. The highest energy neutrinos (> 1 TeV), which have the largest interaction cross sections, are absorbed deep in the companion. From a detailed numerical calculation of the hadronic cascade induced in the atmosphere of the companion star, we estimate the neutrino production spectrum from an isotropic flux of monoenergetic $10^{17}$ eV protons and we estimate the resulting neutrino absorption in the stellar core. In the case of Cyg X-3 and LMC X-4, the cosmic-ray luminosities required to produce the observed gamma rays would result in energy deposition from neutrino absorption exceeding the intrinsic stellar luminosity of the companion. Over a timescale of $10^4$-$10^5$ yr, the star would absorb its own binding energy and be disrupted. On shorter timescales, the energy deposition will cause significant expansion of the star, perhaps leading to quenching of high-energy signals from the source. From these results, we conclude that systems requiring intense UHE proton fluxes are either very young or the companion star is not the site of observed gamma-ray production. Alternatively, if the gamma-ray source is highly variable, the proton flux requirements would be lower, providing some relaxation of the above constraints.
[See Gaisser et al. 1986, Ap. J. (Oct. 15), in press].



*D. J. Helfand and J.-H. Huang (eds.), The Origin and Evolution of Neutron Stars, 552.*

# AN EXPERIMENT FOR OBSERVING VHE GAMMA RAY SOURCES

Y.L. Jang, C.X.He, C.X.Xu, Y.K.Yuan, Y.G.Li
D.B.Chen, R.T.Zheng, A.X.Huo, M.H.Ye
( Institute of High Energy Physics, Academia Sinica,
P.O. Box 918, Beijing, China )
S.Z.Tang, L.J.Zhang, S.X.Meng, H.L.Shi, S.Y.Jiang, Q.B.Li
( Beijing Observatory, Academia Sinica, Beijing, China )
J.Yang, H.Q.Zhang
( Purple Mountain Observatory, Academia Sinica,Nanjing,China )

ABSTRACT. An apparatus design is described. It is for detecting VHE gamma ray point sources by means of the atmospheric Cerenkov technique. Obviously, the improvement of flux sensitivity and discrimination between gamma ray and isotropic proton showers is still a key problem. Of course, it is necessary to set up more observatories and to track an object continuously with several facilities. With this in mind, we decided to develop an experiment for observing VHE gamma ray sources in China. As a first step, we will set up an apparatus which consists of three 1.5 m diameter searchlight morrors at Xinglong station of Beijing Observatory,Xinglong county, Hebei province ( 40°. 4N,117°.5E, altitude 940 m ). The observation will start in 1988.Then, the second apparatus will be set up at Delingha station of Purple Mountain Observatory, Delingha county, Qinghai porvince ( 37°.22N, altitude 3204 m ). Both the sites are far from air and light pollution , and have suitable meteological condition for Cerenkov light detection as well as quite convenient facilities for transportation. Some probable technical improvements are also discussed in this paper.



*D. J. Helfand and J.-H. Huang (eds.), The Origin and Evolution of Neutron Stars, 553.*

# HIGH ENERGY COSMIC RAYS FROM YOUNG NEUTRON STARS

Shigeki Miyaji*,**
Space Science Laboratory, NASA/Marshall Space Flight Center
Huntsville, Alabama 35812, USA
*On leave from Department of Natural History,
Chiba University, Chiba, 260, Japan
**NAS/NRC Resident Research Associate

Cosmic ray spectrum has an intensity enhancement at energy range $10^{14-16}$ eV/nuc. Recently Takahasi et al. (1986) called an attention to chemical composition there. Although the data still contain large uncertainties, they argued an overabundance of calcium at high energies (Ca/Fe $\geqslant$ 2 above $10^{14}$ eV/nucleus) and some enhancements of medium heavy nuclei (C $\sim$ Ar) instead of no anomalous p, He, and Fe abundances.

There are only two possibilities to produce such type of overabundance, i.e., nova and type II supernova explosions. In the case of nova explosion, however, it is hard to accelerate its ejecta up to such a high energy, and we can decline this possibility. We can expect, on the other hand, overabundance of calcium on both two proposed schemes of type II supernovae which produce neutron stars, i.e., strong and weak bouncing shock schemes (Takahashi et al. 1986). From a strong shock scheme, prompt mass ejection takes place and neutron rich isotopes (ex. $^{48}$Ca) are ejected. From a weak shock scheme, the stars do not explode promptly but later by neutrino heating. The resulting Ca/Fe ratio could be $\sim$ 1. The enhancements of other middle heavy nuclei (Si and S) are also expected.

Takahaski et al. (1986) calculated expected cosmic ray energy spectrum from type II supernova remnants including the effects of photodisintegration and leakage from our galaxy and showed good agreement with the observed data.

Clearly further observations of heavy nuclei at energies above $10^{12}$ eV/amu are needed. Isotope identification would be especially important as one would expect an overabundance of both $^{40}$Ca and $^{48}$Ca. X-ray line spectrum observation of Crab-like supernova remnants will also give the composition in the vicinity of neutron stars. These observations would be a conclusive signature of nucleosynthesis in type II supernovae and indeed could serve as a probe of the physical conditions in the NSE core, i.e., formation of neutron stars.

## REFERENCE

Takahaski, Y., Miyaji, S., Parnell, T. A., Weisskopf, M. C., Hayashi, T., and Nomoto, K., 1986 Nature (in press).



*D. J. Helfand and J.-H. Huang (eds.), The Origin and Evolution of Neutron Stars, 554.*

# COSMIC RAY PARTICLE ACCELERATION IN PULSAR MAGNETOSPHERES

K.O. Thielheim
Institut für Reine und Angewandte Kernphysik
Abt. Mathematische Physik
University of Kiel
Olshausenstr. 40, 2300 Kiel
West Germany

A plausible approach to the theory of pulsars as cosmic ray particle accelerators is to integrate numerically the Lorentz-Dirac-equation, using the vacuum field of a rotating orthogonal magnetic dipole as a model field configuration. Typical parameter values are: angular velocity $\omega = 20\pi$/sec and magnetic dipole moment $\mu = 10^{30}$G $cm^3$ (K.O. Thielheim, Proc. ESO-CERN Conf. 1986).

Protons starting sufficiently near to the magnetic dipole are found to be focussed to one of the polar regions. Protons starting sufficiently far from the magnetic dipole finally evade to very large distances. Within a certain range of initial radial distance the ultimate fate of the proton depends on its initial angular position. This phenomenon can be exploited to define the "critical surface for protons", dividing the space around the magnetic dipole into two regimes: an interior one, from which protons are drawn towards the pulsar and an exterior one, from which protons are expelled to the interstellar space. Under given parameter values the latter is found at about 2 light radii distance from the dipole.

The ability of the field configuration to accelerate protons up to very high energies is found to break down beyond a certain range of radial distance from the magnetic dipole. Outside this "acceleration boundary for protons" the pulsar looses its ability to act as a cosmic ray proton accelerator. Under given parameter values the latter is at about 10.000 light radii distance from the dipole.

Our results therefore suggest that there is a certain region in space surrounding a pulsar from where charged particles can be accelerated to become primary cosmic ray particles. This region is limited for small values of radial distance by the "critical surface", and for larger values of radial distance by the "acceleration boundary".

According to the model suggested here, neutral atoms of the interstellar gas can invade the region between the "critical surface" and the "acceleration boundary" and be ionized there. They then become subject to the influence of the electromagnetic field accelerating them to become primary cosmic ray particles, the mass spectrum of which largely reflecting the chemical composition of the interstellar gas. Alternatively ions can be drawn from certain regions of the pulsar surface.



*D. J. Helfand and J.-H. Huang (eds.), The Origin and Evolution of Neutron Stars, 555.*

# OBSERVATIONS OF NEUTRON STARS PLANNED BY THE HIGH SPEED PHOTOMETER TEAM USING SPACE TELESCOPE

Joseph F. Dolan
NASA Goddard Space Flight Center
Laboratory for Astronomy & Solar Physics
Greenbelt, MD 20771
U. S. A.

The unique capabilities of the High Speed Photometer (HSP) on the Hubble Space Telescope (10 microsecond time resolution; 25 filters covering wavelengths from 1220 to 9000 angstoms; apertures as small as 0.4 arcsec in all filters; no scintillation noise induced by the atmosphere) will be used by the HSP Team to investigate several problems related to the origin and evolution of neutron stars. Among the observational programs planned by the team are:

- a search for pulsations in the ultraviolet from binary systems containing X-ray pulsars. The UV pulsations should arise from reprocessed radiation in the X-ray heated hemisphere of the companion star. Comparisons of the UV period to the X-ray period can yield the mass ratio of the binary.

- monitoring the photometric and polarimetric light curves of X-ray binary systems in the UV. Any variability detected can be related to the orbital parameters and accretion mechanism of the neutron stars.

- a search for the remnant star in supernova remnants. The results will place new constraints upon the mechanisms by which neutron stars can originate.

- optical and UV observations of optically detected radio pulsars (the Crab, Vela, and LMC pulsars) and the attempted detection of other radio pulsars (the millisecond pulsars, two binary pulsars, and the pulsar with the largest measured value of dp/dt).

- monitoring the LMC gamma-ray burst source in N49 in the UV at the zero phase of its periodic optical bursts. The results may confirm that this unusual gamma-ray burst source is a neutron star.

The HSP Investigation Definition Team members are: Principal Investigator Robert C. Bless (University of Wisconsin), Joseph F. Dolan, James L. Elliot (MIT), Edward L. Robinson (University of Texas), G. Wayne van Citters (National Science Foundation) and Richard L. White (Space Telescope Science Institute).



*D. J. Helfand and J.-H. Huang (eds.), The Origin and Evolution of Neutron Stars, 556.*

# *EPILOGUE*

# WHERE NEUTRON STARS COME FROM, HOW NEUTRON STARS EVOLVE, AND WHERE NEUTRON STARS GO

L. Woltjer
European Southern Observatory
Garching b. München

At the end of this symposium, I shall briefly review our current knowledge on neutron stars.

## Where have neutron stars been observed ?

430 Radio pulsars with periods between 1.5 ms and 4.3 s and magnetic fields $B_o \sim 10^{12 \pm 1}$ Gauss. Short periods are undoubtedly underrepresented in the sample because of more difficult discovery.

7 Radio pulsar binaries with $B_o$ in the range $10^9 - 10^{12}$ G. Their luminosity is lower by a factor of 10 than for the singles, and essentially all are in the northern hemisphere. Hence, they are underrepresented in the observed samples by a factor of 15 - 20 in comparison with the singles. Additional selection effects may have to be considered. Binaries may therefore be somewhat less common than single pulsars but not by a large factor.

50 X-ray binaries, some with pulse periods ~100 ms - 1000 s. Quasi Periodic Oscillations (QPO) have been reported in several cases; if the beat models apply, rotation periods as low as 10 ms are involved. Cyclotron line observations in a couple of sources indicate fields up to $10^{12}$ G; if magnetospheric models apply to the QPO, more typical fields might be around $10^9$ G.

15 Pulsar driven Nebulae, with in 3 cases a radio pulsar at the center. This might be taken as evidence of beaming factors around 0.2, but the systematics of these objects are still too poorly understood for firm conclusions.

~100 Gamma-ray bursters, the precise number depending on the repeat frequency, and the gamma-ray source Geminga; the association with neutron stars results mainly from the lack of plausible alternatives. Evidence has been presented for a periodicity of 8 s in the 1979 gamma-ray burst (from the source projected on N 49 in the Large Magellanic Cloud) and of 60 s in Geminga. The



D. J. Helfand and J.-H. Huang (eds.), The Origin and Evolution of Neutron Stars, 559–562.

log N - log S relation for gamma-ray bursters may indicate distances of the order of 100 pc. If the 300 bursts observed in a decade were all from different sources, the phenomenon must be frequent in old neutron stars since their number within 100 pc from the sun is probably in the range $10^3 - 10^4$.

## Where do neutron stars come from ?

Undoubtedly many and perhaps all neutron stars originate in supernova (SN) events. Various types of SN are now recognized: I, $I_{pec}$ (somewhat subluminous), $II_{fast}$, $II_{slow}$, $II_{pec}$, V (Zwicky's very slow type); SN I are the only ones to also occur in elliptical galaxies. A few dozen SN have been observed optically in adequate detail and only a few from the radio to the X-ray parts of the spectrum. Advances in stellar evolution calculations combined with observation indicate which stars are responsible for SN, but a detailed understanding of types still eludes us.

Stars with a main-sequence mass $M_{ms}$ below 6 $M_{\odot}$ may become white dwarfs; this follows from the presence of white dwarfs in clusters with a main-sequence turn off mass of this magnitude. Of course, it does not follow that all stars with $M_{ms} < 6\ M_{\odot}$ become white dwarfs.

The situation for stars with $M_{ms}$ between 6 and 8 $M_{\odot}$ is unclear, but calculations indicate that in the range 8 - 10 $M_{\odot}$ neutron stars are likely to be produced. Arguments have been presented that the Crab Nebula resulted from such a star; problems remain with the CNO abundances, unless these are concentrated in dust grains for which the IRAS data have given some evidence.

Stars with $M_{ms}$ between 10 and 25 $M_{\odot}$ may account for the typical SN II, although the precise "bounce" mechanism is still in some doubt; neutrinos probably play an important role. Nucleosynthesis calculations account in a quantitatively satisfactory way for some fifty solar system isotope abundances. Neutron stars probably are commonly formed.

For $M_{ms} > 25\ M_{\odot}$ the situation is unclear. A black hole collapse is a definite possibility, since with increasing mass it becomes progressively more difficult to have a sufficiently strong bounce for the resulting shock to reach the surface. The number of such stars is too small in any case to affect neutron star statistics significantly.

Evolved stars in binaries are also likely important in neutron star formation. In particular, accretion induced collapse and explosion of white dwarfs may perhaps account for SN I; whether a neutron star remains after such events is uncertain.

While some work has been done on the effect of rotation on the collapse, much remains to be done in this area.

How can properties of neutron stars be understood ?

We shall now see how well the observed properties of neutron stars can be accounted for on the basis of the simple evolutionary picture.

Rotation periods. Much angular momentum has to be lost during stellar evolution; convection and/or magnetic fields are effective in transporting it outwards. There is no particular problem therefore in accounting for the observed periods, even if the initial periods were rather long (0.5 s) as has been suggested. If magnetic fields play the dominant role, weak field pulsars would be expected to rotate fastest. While there seem to be observational indications for this, the effects of field decay and recycling of old pulsars complicate the interpretation. The whole question of angular momentum in the later evolutionary phases needs further study.

Magnetic fields. Fields of the order of $10^{12}$ G are not at all unexpected. Internal fields built up by convection in the core on the main sequence or later would, with subsequent flux conservation, have such values.

Masses. Predicted masses from evolutionary calculations appear not unreasonable although perhaps on the low side by 10 - 20 per cent. The observed masses for the two neutron stars in the binary pulsar (1.45 and 1.38 $M_{\odot}$) are by far the most accurately determined masses outside the solar system.

Space velocities. Binary effects certainly account for some of the observed velocities (which range up to 400 km/sec), but it is doubtful that they account for all. Modest asymmetry in the supernova explosion may well be of much importance. Such asymmetry might result from an asymmetrical magnetic field distribution. Strong asymmetries between the two magnetic poles are frequently observed in magnetic A stars, and the resulting pressure distribution could start the explosion on one side. An observed tendency of low velocities to coincide with low magnetic fields as well as the lower field values in binaries are consistent with such a picture, but again field decay complicates the interpretation. Finally, the observed very asymmetrical location of the X-ray Crab Nebula and of the related optical activity indicates that asymmetrical particle acceleration also may play a role.

Formation rates. Data on supernovae, supernova remnants and pulsars in our galaxy are all consistent with a formation rate of one per 50 years. However, the uncertainties remain very large. The supernova rate is uncertain by a factor of about three because of the small sample and because of the possible importance of subluminous SN (like 3C 58 and Cas A). With a distance scale D and lifetime t, the supernova remnant formation rate in a disk galaxy is proportional to $D^{-2}\ t^{-1}$ or $D^{-3}$ if the expansion velocities are known from X-ray temperatures. An uncertainty of a factor of three can therefore not be excluded. Finally, the pulsar formation rate varies as $D^{-2}\ t^{-1}\ \psi^{-1}$ with $\psi$ the opening angle of the beam. With D, determined from dispersion measures, requiring a model of the electron density distribution in the galaxy and t depending on a model for the field evolution, and with $\psi$ and its evolutionary variation being still very much in doubt, an uncertainty in the pulsar formation rate of a factor of five seems plausible. In fact, just a few years ago, formation rates as high as one per 7 years in our galaxy were sometimes quoted. Obviously with such large uncertainties, the statistics do not allow us to answer questions like whether all supernovae result in pulsars or whether all pulsars come from supernovae.

## What is the further evolution of neutron stars ?

An isolated neutron star with a magnetic field will generally spin down by angular momentum loss from its surface. However, "glitches" may occur, related to the coupling mechanism between the crust and the interior. While such glitches have been frequently observed in young pulsars, a remarkably large event in an older pulsar was reported here.

Magnetic fields generated by currents in the crust of a neutron star are likely to decay on time scales of a few million years. Some evidence was presented that the decay may stop at about $10^9$ G, possibly because of currents in the superconducting interior. The dipolar or multipolar character of the late field is still a subject of debate and affects estimates of its strength.

Cooling calculations for neutron stars have been developed in much detail. While for most neutron stars the predicted thermal X-ray fluxes are well below observed upper limits, in a few cases there appears to be a discrepancy; this may indicate the action of more exotic cooling mechanisms (pion condensation, etc.).

The long term result of the steady production of neutron stars will be the creation of a galactic halo of high velocity objects. Their total mass is unlikely to be of much dynamical significance ($10^8$ - $10^9$ $M_\odot$). But perhaps the gamma-ray events show that there is still much activity possible in old neutron stars and that their nature is not yet fully understood. In fact, it even has been suggested that not all neutron stars are made of neutrons.

# *SUBJECT INDEX*

**A**

absorption,
coefficient 61
lines 236
abundances, of ejecta 289
acceleration models 207,526
accreting neutron stars 162,363
accreting white dwarfs 262,272
accretion 139,350,492
accretion disk 144,200,204,207,247,546
accretion disk corona 371,526
accretion disk dynamo 526
accretion induced collapse (AIC) 109,174,189,265,281,393,400,409,425,560
ages 54,433,461,489
Alfven radius 220,527
atmospheric Cherenkov technique 553

**B**

beaming,
factor 91,93,121,559
mechanism 425
beat frequency model 325,347,400,559
Be star systems 137,203
binary,
evolution 13,23
frequency 114
radio pulsars 5,67,109,127,282,393,407,408,409,410,559
birth, neutron star 275
birthrate, pulsar 47,75,111,255,561
black holes 173,247,460,560
bremsstrahlung 204
bulge sources 399

**C**

carbon, deflagration 293
cataclysmic variables 195,299,365,538
centrifugally driven winds 215
Chandrasekhar limit 109,142,189,262,272,292,386
characteristic ages 26,68
close encounters 188
C+O, white dwarfs 282
cocoon 526
composite supernova remnants 91,310
Compton,
radiation 251,453
scattering 59,145,161,210,247,248,339,372,453,547,549
cooling, neutron stars 407,420,439,447,456,459,489,562
core-halo structure 288
cosmic rays 62,552,554,555
Crab-like Supernova Remnants 81,91,554
Crab Nebula, 87,100,112,122,123,129,290,560
progenitor 281
pulsar 480
supernova 100,123,305,441,456

crust, neutron star 63,415,454
cyclotron,
 absorption 248
 line 227,559

**D**
death, of pulsars 28,113,375,394
decay, of $^{56}Ni$ 291
deflagration,
 conductive 295
 convective 297
dispersion measure 9,14,39,47,124,383,562
distances, of pulsars 39,383,493
distribution, of magnetic fields 112
drifting subpulses 60
dwarf novae 365

**E**
Eddington,
 accretion 246,375
 limit 141,165,207,548
 luminosity 233,363
electron,
 capture 282
 degeneracy 281
 positron pair 450
equation of state 434,449
equilibrium period 113
evolution,
 single pulsars 51,52,110,425
 binary pulsars 393
exchange collisions 189
explosive nucleosynthesis 293

**F**
Fermi acceleration 528
formation mechanisms, X-ray binaries 188

**G**
galactic bulge sources 55
galactic ridge X-ray emission 530
gamma ray,
 burster 375,465,477,489,501,546,547,559
 burst population 473
 emission 59,62,440,442
 sources 3,14,465,489
 UHE/VHE 466,521,550,553
Geminga 315,465,545,559
general relativity 14,387
glitch 14,63,64,413,448,562
globular clusters 18,151,162,173,187,204,363
gravitational collapse 275,451
gravitational radiation 13,299,388,399
guest star 305

**H**
hollow cone emission 60

**I**
initial mass function 111
interferometry 35
interior,
structure 413,449
temperature 447
interpulse 425
interstellar,
medium 14
scattering 6,14,36,47
scintillation 10,18,24,35
iron core 284

**K**
K-emission line 536
kinematics 23,35,52
kinetic age 27

**L**
lifetimes, X-ray binaries 191
low frequency noise (LFN) 321,336,347,364
luminosity,
function 70,194
pulsar 50,54

**M**
magnetic,
alignment 27
braking 27,49,68,399,409
decay 27,51,112,179,356,377,393,407
dipole moment 42,62,375
field 49,50,57,59,61,80,110,246,248,249,264,375,379,452,
453,459,461,485,496,547,559
inclination 376
monopoles 460
susceptibility 451
magnetosphere 207,354,555
March 5 gamma ray burst 477,497,501,548,549,556
maser 227
mass, neutron star 47,263,279,300,387,561
M-dwarfs 538
millisecond pulsars 5,13,55,67,109,115,183,299,347,375,383,389,393,425
Monte Carlo simulation 43,202,449
M31 141,195

**N**
neutrino 255,273,439,524,529,552,560
neutron-rich isotopes 300
nonequilibrium ionization 128
novae 180,400

nuclear wars 209
null 53
NUSEX 524
nutation 454

**O**

OB,
  associations 47,112
  stars 136
Observatory (radio)
  Arecibo 5,14,390
  Cambridge 6
  Deep Space Net 18
  Green Bank 3,18
  Jodrell Bank 3,18
  MERLIN 30
  Molonglo 6,18
  MOST 129
  OOTY 124
  Parkes 30
  VLA 30,129
  Westerbork 18
Observatory (Satellite)
  Cos-B 465,489,523
  EXOSAT 195,203,322,333,364,420,458,470,491,535,545,550
  Einstein (HEAO-B) 180,195,204,457,459,471,491,541,545,550
  HAKUCHO 200,217,321,364
  HEAO-1 (A) 180,336,489,535
  Hubble Space Telescope 473,556
  KONUS 496
  IRAS 560
  ISEE-3 477
  PVO 479
  SAS-2 315,523
  SIGNE 490
  SMM 481
  TENMA 200,217,238,250,339,535
  VELA 5.6 477
O+Ne+Mg white dwarfs 281
optical 99,489,516,523

**P**

P,$\dot{P}$ diagram 87,388,449
pair plasma 86
pair production 61,452
period distribution 9,51,62
pion condensate 416,440,456,562
pions 529
plerion 74,80,124,130
polar cap, 54,55,59,452
  accretion 207,227,245,246,248
polarization, 53,58,389
  limiting radii 56
Population I 23

precession 358,410,420,455,55†
PSR 0329+54 4,25,57
PSR 0355+54 63,87
PSR 0529-66 3,4
PSR 0531+21 (Crab) 3,4,13,25,56,74,91,100,453,457,465,556
PSR 0540+23 40
PSR 0540-69 (LMC) 3,83,91,102,457
PSR 0655+64 4,20,,74,110,384,394,407
PSR 0820-00 20
PSR 0820+02 4,110,384,394,407
PSR 0823+26 57
PSR 0833-45 (Vela) 3,4,38,57,63,64,82,91,440,448,456,457,556
PSR 0950+08 4
PSR 0906-49 6
PSR 1055-52 87,458
PSR 1133+16 24
PSR 1237+25 57
PSR 1508+54 25
PSR 1509-58 (MSH15-52) 4,83,91,102,457
PSR 1541+09 57
PSR 1541-52 114
PSR 1641-45 4
PSR 1642-03 87
PSR 1735-32 47
PSR 1758-23 10
PSR 1758-24 129
PSR 1800-21 124
PSR 1804-08 114
PSR 1806-21 47
PSR 1809-175 47
PSR 1828-10 47
PSR 1830-08 9,124
PSR 1831-00 5,20,75,110,384,394,409
PSR 1841-04 47
PSR 1842-04 47
PSR 1845-19 4
PSR 1850+00 47
PSR 1855+09 4,14,75,116,384,394
PSR 1859+03 10
PSR 1859+07 47
PSR 1913+16 4,16,74,110,127,279,289,384,393,410
PSR 1916+13 255
PSR 1919+21 60
PSR 1937+21 4,14,26,55,383,393
PSR 1953+29 4,14,75,384,393
PSR 2303+46 4,20,74,110,127,384,394,407
pulsar,
  beaming factor 57,91,93
  beams 56
  emission 53,54
  luminosity 388
  periods 5,13,50,264,394,490,559
  profiles 53,58,389,453,455
  progenitors 67,109,408

surveys 3,47,95,390
timing 35
velocities 23,35,49,67,110,114,178,191,255,265,377,408,410,425, 489,561

**Q**

quark 414,456
quark matter 440,447
quasi-periodic oscillations (QPOs) 178,207,241,321,333,347,363, 375,400,524,559

**R**

radio,
emission 59,99
supernovae 81
rapid burster 322,363
recycled pulsar 396
ring model 551
Roche lobe 136,398,409
RS CVn 538

**S**

scale height 115,121,377
shock acceleration 526
soliton 461
SOUDAN 524
spin down, torque 50,199,221,426,492,562
spin up,
line 117,402
torque 199,216
spiral,
galaxies 131
stellar beam dump model 525
strange matter 417,548
superfluid,
neutron 51,416,451
vortices 454
supernova, 29,121,174,256
explosions 35,79,265,273,398,408,447,554,560
frequency 111
remnants (SNR) 3,23,47,87,91,111,121,125,129,131,151,279,300, 420,456,457,459,489,530,548,556
3C58 93,104,123,308,441,456,561
Cas A 107,262,441,561
Crab 100,123,305,441,456
CTB 80 104,311
CTB 109 103
G20.0+0.2 92
G21.5-0.9 92,106,312
G54.1+0.3 92
G74.9+1.2 92,313
G0.9+0.1 92
G24.7+0.6 92
G27.4+0.0 92

G29.7-0.3 92
G39.7-2.0 92
G68.9+2.8 92
G109.1-1.0 92,141
G130.7+3.1 92
G184.6-5.8 92
G227.7-0.2 441,456
G263.9+2.8 92
G291.0-0.1 92
G315.8-0.0 125
G320.4-1.2 (MSH15-52) 92,102
G321.9-0.3 (CIR X-1) 129
G326.3-1.8 92
G328.4+0.2 92
G332.4-0.4 92,125
G350.0-1.8 441,456
G351.2+0.1 92
Kepler 126,127,441,456
Kes 73 106
Kes 75 106
N103B 130
N157B 130
RCW86 441,456
RCW103 105,311,441,456
SN1006 128,441,456,459
Tycho 126,441,456
Vela 101,441,456
W28 441,456
W50 106
Type I 131,161,363
Type Ia 282
Type Ib 262,282
Type II 131,161,255,289,316,363,425
synchrotron nebula 80,122,559

**T**

temperature 59,441,457,458,459
Terzan-2 322
thermonuclear flash 161,174
timing discontinuity 64
timing noise 389

**U**

underground experiments 524

**V**

vacuum polarization 61
vortex creep 456

**W**

white dwarf 5,109,142,173,256,281,375,393,396,409,493,560
Wolf-Rayet stars 127,294

**X**

X-ray, 13
binary 135,149,249,383,408,489,559
binaries,
low mass 18,68,115,149,173,187,199,272,282,299,321,363,393,495,550
massive 114,135,149
bursters 173,233,250,251,477
emission 99,130,249
pulsars 135,203,221,227,248
sources 113,123,425,535
AM Her 142
A0538-66 141
Cen X-3 199
Cir X-1 129,169,322,363,399
Cyg X-1 247,363,480
Cyg X-2 136,202,321,333,336,371,399
Cyg X-3 465,521,550,552
EXO 2030+375 203
GX1+4 199
GX17+2 322
GX3+1 322
GX5-1 238,321,336,347
GX301-2 199
GX339-4 247
GX349+2 238,322,364
Her X-1 199,201,216,221,358,497,521,552
LMC X-1 480
LMC X-4 521,552
MXB1608-522 250
MXB1636-536 250
Sco X-1 136,238,321,339357,368,399
SS433 465,551
Vela X-1 199,221,480,521,552
V0332+53 140
X Per 136
X1608-52 236
X1636-53 236
X1728-34 236
2S0921-63 399
4U0115+63 497
4U1626-67 136,200
4U1820-30 322
spectra 201,248,457,458
transient 195,203